Theoretical and Experimental Modal Analysis

MECHANICAL ENGINEERING RESEARCH STUDIES

ENGINEERING DYNAMICS SERIES

Series Editor: **Professor J. B. Roberts,** *University of Sussex, England*

4. Parametric Random Vibration
   R. A. Ibrahim

5. Statistical Dynamics of Nonlinear and Time-Varying Systems
   M. F. Dimentberg

8. Vibroacoustical Diagnostics for Machines and Structures
   M. F. Dimentberg, K. V. Frolov *and* **A. I. Menyailov**

9. Theoretical and Experimental Modal Analysis
   Edited by **Nuno M. M. Maia** *and* **Júlio M. M. Silva**

10. Modal Testing: Theory, Practice and Application, SECOND EDITION
    D. J. Ewins

Theoretical and Experimental Modal Analysis

Edited by
Nuno Manuel Mendes Maia
and
Júlio Martins Montalvão e Silva
Instituto Superior Técnico, Portugal

RESEARCH STUDIES PRESS LTD.
Baldock, Hertfordshire, England

RESEARCH STUDIES PRESS LTD.
15/16 Coach House Cloisters, 10 Hitchin Street, Baldock, Hertfordshire, England, SG7 6AE

Copyright © 1997, by Research Studies Press Ltd.

Reprinted July 1998

Marketing:
NORTH AMERICA
Taylor & Francis Inc.
325 Chestnut Street, Philadelphia, PA 19106, USA

EUROPE & REST OF THE WORLD
Research Studies Press Ltd.
15/16 Coach House Cloisters, 10 Hitchin Street, Baldock, Hertfordshire, England, SG7 6AE

Distribution:
NORTH AMERICA
Taylor & Francis Inc.
47 Runway Road, Suite G, Levittown, PA 19057 - 4700, USA

EUROPE & REST OF THE WORLD
John Wiley & Sons Ltd.
Shripney Road, Bognor Regis, West Sussex, England, PO22 9SA

Library of Congress Cataloging-in-Publication Data
Theoretical and experimental modal analysis / edited by Nuno Manuel
Mendes Maia and Júlio Martins Montalvão e Silva.
    p. cm. -- (Mechanical engineering research studies.
Engineering dynamics series ; 9)
    Includes bibliographical references and index.
    ISBN 0-86380-208-7 (Research Studies Press : alk. paper). -- ISBN
0-471-97067-0 (Wiley : alk. paper)
    1. Modal analysis. I. Maia, Nuno Manuel Mendes, 1956- .
II. Montalvão e Silva, J. M. (Júlio Martins), 1945- .
III. Series.
TA654. 15. T48 1997
    620.3--dc21   96-38057
                                                       CIP

British Library Cataloguing in Publication Data
A catalogue record for this book is available from the British Library.

ISBN 0 86380 208 7

Contributing Authors

Nuno Manuel Mendes Maia
Associate Professor
Department of Mechanical Engineering
Instituto Superior Técnico / Technical University of Lisbon, Lisbon, Portugal

Júlio Martins Montalvão e Silva
Professor
Department of Mechanical Engineering
Instituto Superior Técnico / Technical University of Lisbon, Lisbon, Portugal

Jimin He
Senior Lecturer
Department of Mechanical Engineering
Victoria University of Technology, Melbourne, Australia

Nicholas Andrew John Lieven
Lecturer
Department of Aerospace Engineering
University of Bristol, Bristol, United Kingdom

Rong Ming Lin
Lecturer
School of Mechanical and Production Engineering
Nanyang Technological University, Singapore, Republic of Singapore

Graham William Skingle
Research Scientist
Aero-Structures Department
Defence Evaluation and Research Agency, Farnborough, United Kingdom

Wai-Ming To
Assistant Professor
Mechanical Engineering Department
The Hong Kong University of Science & Technology, Kowloon, Hong-Kong

António Paulo Vale Urgueira
Associate Professor
Department of Mechanical Engineering
Faculty of Sciences and Technology / New University of Lisbon, Monte da
Caparica, Portugal

Contents

Preface

Probably, the first reaction to this book will be something like: *"What? Eight authors?!"*. It is, *a priori*, a large number of authors, no doubt. The reason for this came naturally on a calm and sunny afternoon in the summer of 95, somewhere on the south bank of Tagus river, at a typical and very old bar facing Lisbon, after a meeting at the University. It was the answer to an innocent question: *"Don't you think there are very few books on Modal Analysis?"*.

In fact, there we were, three of the eight, supposedly experts in the field, wondering why there were so few books on our subject! The logical answer was *"Why don't we write our own one?"*. As it is not hard to imagine, to write a technical book is something quite difficult to take in hand, although we are sure that for us a non-technical one would even be harder. Thus, the next question was *"How to reach that goal within a reasonable time scale?"* Then we thought it might be a good idea to gather some colleagues, each one with the commitment of writing a single chapter. It would be the best solution for the timing problem. Quickly, we agreed that eight chapters would reasonably cover the main areas of Modal Analysis, implying therefore eight people. Then, the final question was *"Who should we choose?"*. Well, that was not very difficult to answer. The three of us there had been colleagues during our Ph.D. studies and we knew of others whose Ph.D. studies more or less overlapped ours in time and whom we knew very well. All of us specialised in a particular subject within Modal Analysis and so it was possible to have all the chapters covered.

In addition, there was a very interesting particularity in choosing these people: we all had studied under the supervision of Professor D. J. Ewins, the author of the first ever book on Modal Analysis. This fact made the project very appealing, as it would be "The book of Professor Ewins' former students". The truth is that it was the first title we thought the book might have. What had begun by being just a light talk became suddenly quite solid. It seemed an excellent idea, a project with all the ingredients to be successful. When the others knew about it, they all were very receptive and enthusiastic.

Now, about the book itself. Having eight chapters written separately by each author could mean just a collection of chapters. That was not the intention and it surely is not the idea we have of what a textbook should be. Although it is quite a hard task to combine, adjust, correct and edit all the material between the various authors, the final product should look a whole meaningful piece, to be read from the beginning till the end in the smoothest possible way, as if it were a single author who had written it. We truly hope that our readers will find it this way. To reinforce this, we agreed not to individualise the writing of each chapter and so, on purpose, we do not reveal explicitly who wrote what. We are all responsible for the whole book. However, the readers who know us will probably guess the paternity of each chapter. For those, we leave that as an enjoyable entertainment.

It is our belief that the book will reach a wide scope of readers, from students pursuing their Master or Doctorate studies to research scientists working daily with these subjects in their laboratories and research centres. It can also help by assisting teaching staff in related courses at Institutes and Universities. And, of course, it will interest ourselves. We have a lot to learn from it.

The layout of the book is as follows: Chapter 1 provides quite a comprehensive explanation of the fundamentals of vibration theory with a view to Modal Analysis. All the basics concerning single and multi-degree-of-freedom systems are covered. Chapter 2 follows with Signal Processing concepts necessary to understand Modal Testing practice which, in turn, is covered in Chapter 3, where details about how to conduct proper experimental tests are given. The book proceeds with Chapter 4, an extensive and detailed coverage of the most utilised methods for modal parameter identification, in both time and frequency domains. Chapter 5 provides a clear explanation of the methods used for substructural coupling. Chapter 6 addresses the problems of structural modification, both from the direct and inverse points of view. In Chapter 7, the various available techniques for finite element updating are discussed, including the very recent applications of Genetic Algorithms and Neural Networks. Chapter 8 discusses specific Modal Analysis techniques applied to structures with a nonlinear behaviour. An appendix on the popular Singular Value Decomposition is added, as well as a short one on Orthogonal Functions. A list of about 370 references allows for the interested reader to go deeper into each subject. Finally, a comprehensive index helps the reader to quickly locate a particular subject in the text.

A remark about notation: gathering material from so many sources and authors makes the job of using a unified notation a very difficult task. Nevertheless, a great effort has been made to keep it as consistent as possible, as well as the closest we could to the standard Modal Analysis practitioners nowadays tend to use most, i.e., the one proposed by Lieven and Ewins, "Call For Comments: A Proposal For Standard Notation And Terminology In Modal Analysis", *IJAEMA*, Vol. 7, No. 2, 1992, pp. 151-156. As every symbol is explained in the text, we found that it was unnecessary to write dozens of pages with all the notation used. We hope, with respect to this, that the text is clear enough. However, we thought that a list of the abbreviations used would be of considerable usefulness, so we decided to include it at the end of the book.

We all made a great effort to avoid mistakes and errors in the text. Various proof-reading iterations have brought the level of mistakes to a reasonable minimum (we hope!). Anyway, all comments from readers concerning this issue and also giving suggestions for corrections and improvements to the text will be welcome. We shall pay to them all due attention and consideration for future alterations.

Very special thanks go to our families, who gave us all their support, encouragement and time, so that this project could come to a successful accomplishment.

Finally, we would like this book to be considered as a tribute to our friend and former supervisor, Professor David John Ewins, with whom we had the unique privilege to work.

The authors

November 1996

To

David John Ewins

CHAPTER 1

Fundamentals of Modal Analysis

1.1 INTRODUCTION

Vibrations, or dynamic motion, are inherent to life though generally mankind regards them as unpleasant and unwanted phenomena causing such undesirable consequences as discomfort, noise, malfunctioning, wear, fatigue and even destruction. Earthquakes are perhaps the most frightening manifestations of dynamic motion, caused by forces generated in the earth's crust, and their destructive effect upon the environment and man-made structures is well known. Though on a smaller scale, mankind is itself responsible for creating an environment where vibration problems are constantly present.

It is a fact that, in today's world, machines and structures are almost everywhere. Being systems of elastic components, machines and structures respond to external and internal forces with finite deformations and overall motion. These responses are the subject of dynamics and vibration analysis.

In the last few decades, technology developments have created an increasing need for reliable dynamic analysis. The sophistication of modern design methods together with the development of improved materials instilled a trend towards lighter structures. At the same time, there is a constant demand for larger structures, capable of carrying more loads at higher speeds under increasing drive power. The consequences of all these trends are dramatic increases in dynamic problems in terms of vibration, noise and fatigue, at the same time as requirements for improved environmental factors are being defined and enforced.

Therefore, strong and reliable vibration analysis tools are a basic need of modern engineering. Modal analysis is just one of those tools, providing an understanding of structural characteristics, operating conditions and performance criteria that enables designing for optimal dynamic behaviour or solving structural dynamics problems in existing designs. Modal analysis, as an engineering tool, was first applied around 1940, in the search for a better understanding of aircraft dynamic behaviour. In the two following decades, developments were slow. Known as the mechanical impedance era, experimental techniques were based on the use of expensive and cumbersome narrow band analogue spectrum analysers. The modern era of modal analysis can be taken as the last twenty five years, based upon the commercial availability (since the early seventies) of the Fast Fourier Transform (FFT) spectrum analysers, transfer function analysers (TFA) and discrete acquisition and analysis of data, together with the availability of increasingly smaller, less expensive and more powerful digital computers to process the data.

In the early stages, most of the applications of modal analysis were concerned with trouble-shooting, dominated by a physical approach to the understanding of the vibration problem. More recently (mid and late eighties), development and verification of modal models have been of higher concern, involving a more integrated physical and mathematical approach. It was in that decade that structural modification based on modal analysis begun to be developed. The trend in the past few years has been for an increasing dominance of the mathematical approach to the modal analysis problem. For example, updating of analytical and finite elements models, using experimental data, is of current concern. Finally, it can be said that integration of modal analysis in a total engineering approach may be taken as the present goal.

Modal analysis is primarily a tool for deriving reliable models to represent the dynamics of structures. In general, it can be said that the applications of modal analysis today cover a broad range of objectives, namely: identification and evaluation of vibration phenomena; validation, correction and updating of analytical dynamic models; development of experimentally based dynamic models; structural integrity assessment, structural modification and damage detection; model integration with other areas of dynamics such as acoustics, fatigue, etc.; establishment of criteria and specifications for design, test, qualification and certification. In short, modal analysis aims to develop reliable dynamic models that may be used with confidence in further analysis.

Understanding modal analysis implies knowledge of a broad range of physical laws and mathematical concepts. As always, it is not possible to cover all the known possibilities in a single text book. Therefore, and before proceeding to develop the theory, three basic assumptions will be established here: i) the structure is a linear system whose dynamic behaviour can be described by a model represented by a set of second order differential equations; ii) the structure obeys Maxwell's reciprocity theorem; iii) the structure is time invariant. The above assumptions will be taken as given throughout this book except in chapter 8 where the subject of nonlinear modal analysis will be approached.

1.2   BASIC CONCEPTS. SINGLE DEGREE-OF-FREEDOM (SDOF) SYSTEMS

1.2.1   Free vibration

All dynamic properties of mechanical systems are distributed in space. These properties are mass, stiffness and damping, responsible respectively for inertia, elastic and dissipative forces. Modelling a real mechanical system is therefore a very complex or even impossible task if one tries to describe how all the features of the system interact with one another. However, in most cases, satisfactory results may be achieved if the basic properties are considered as separated into simple discrete elements which, properly combined, can represent the dynamic properties of the system to sufficient accuracy.

Let us take the simplest possible discretisation i.e., a system with just a single degree-of-freedom (single DOF or SDOF), whose properties are represented by the elements in figure 1.1 (inertia represented by an infinitely rigid constant mass m, elasticity represented by an ideal massless spring of constant stiffness k, and damping represented by an ideal massless viscous damper with constant damping coefficient c).

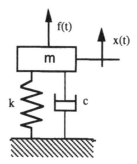

Fig. 1.1 Discretised representation of a single degree-of-freedom (SDOF) system.

The corresponding spatial model is described by the following equation of motion

$$m\,\ddot{x}(t) + c\,\dot{x}(t) + k\,x(t) = f(t) \tag{1.1}$$

where $f(t)$ and $x(t)$ are respectively the time dependent excitation force applied to the system and the corresponding displacement response. Let us also assume initial ($t = 0$) displacement and velocity conditions $x(0)$ and $\dot{x}(0)$.

From differential equation theory, we know that the solution of (1.1) is the sum of the solution of the corresponding homogeneous equation with a particular integral of the nonhomogeneous equation. Setting $f(t) = 0$, the homogeneous form of equation (1.1) is given by

$$m\,\ddot{x}(t) + c\,\dot{x}(t) + k\,x(t) = 0 \tag{1.2}$$

corresponding to the so-called free (there are no external forces applied) vibrating system. Equation (1.2) has the following general solution:

$$x(t) = X\,e^{st} \tag{1.3}$$

where s, known as the Laplace variable as will be seen in Section 1.2.5, is a complex quantity to be determined. Substituting into (1.2)

$$(m s^2 + c s + k) X e^{st} = 0 \qquad (1.4)$$

It can be seen that there is a trivial solution ($x(t) = X e^{st} = 0$) which corresponds to no motion at all, and therefore is of no interest, and a non-trivial solution corresponding to

$$m s^2 + c s + k = 0 \qquad (1.5)$$

Equation (1.5) is the so-called characteristic equation, yielding two roots s_1 and s_2, given by:

$$s_{1,2} = -\frac{c}{2\,m} \pm \sqrt{\left(\frac{c}{2\,m}\right)^2 - \frac{k}{m}} \qquad (1.6)$$

Therefore, the general solution of the homogeneous equation (1.2) is given by

$$x(t) = C_1 e^{s_1 t} + C_2 e^{s_2 t} \qquad (1.7)$$

where C_1 and C_2 are constants determined by the initial conditions imposed on the system at $t = 0$. It is obvious that the two roots s_1 and s_2 may fall in the following cases:

- damping forces are the primary ones governing the motion ($(c/2m)^2 > k/m$) and both roots are real. In this case, the system is said to be overdamped;

- inertia and elastic forces prevail ($(c/2m)^2 < k/m$) and the two roots are complex conjugate. In this case the system is said to be underdamped;

- the square root in (1.6) equals zero ($(c/2m)^2 = k/m$) and there are two equal real roots. In this case the system is said to be critically damped.

The above analysis shows that any vibrating system has an important parameter defined as the critical damping coefficient c_c, which is obtained from $(c/2m)^2 = k/m$ and establishes the border between the underdamped and the overdamped situations:

$$c_c = 2 \sqrt{k\,m} = 2\,m \sqrt{\frac{k}{m}} = 2\,m\,\omega_n \qquad (1.8)$$

where $\omega_n = \sqrt{k/m}$ is the undamped natural frequency. Defining a dimensionless quantity ξ as the damping ratio

$$\xi = \frac{c}{c_c} \tag{1.9}$$

the roots of the characteristic equation may be written as

$$s_{1,2} = -\omega_n \xi \pm \omega_n \sqrt{\xi^2 - 1} \tag{1.10}$$

and therefore

- overdamped system : $\xi > 1$
- critically damped system : $\xi = 1$
- underdamped system : $\xi < 1$

The corresponding time domain solutions of (1.2) may be written as:

overdamped system:

$$x(t) = e^{-\xi \omega_n t} \left(C_1 e^{\omega_n t \sqrt{\xi^2 - 1}} + C_2 e^{-\omega_n t \sqrt{\xi^2 - 1}} \right) \tag{1.11}$$

or, considering the initial conditions:

$$x(t) = e^{-\xi \omega_n t} \left[x(0) \cosh\left(\omega_n t \sqrt{\xi^2 - 1} \right) + \frac{\dot{x}(0) + \xi \omega_n x(0)}{\omega_n \sqrt{\xi^2 - 1}} \sinh\left(\omega_n t \sqrt{\xi^2 - 1} \right) \right] \tag{1.12}$$

and similarly

critically damped system:

$$x(t) = e^{-\omega_n t} \left(C_1 + C_2 t \right) \tag{1.13}$$

or

$$x(t) = e^{-\omega_n t} \left[x(0) \left(1 + \omega_n t \right) + \dot{x}(0) t \right] \tag{1.14}$$

underdamped system:

$$x(t) = e^{-\xi \omega_n t} \left(C_1 e^{i \omega_n t \sqrt{1 - \xi^2}} + C_2 e^{-i \omega_n t \sqrt{1 - \xi^2}} \right) \tag{1.15}$$

where $i = \sqrt{-1}$, or

$$x(t) = e^{-\xi\,\omega_n\,t}\left[x(0)\cos\left(\omega_n\,t\,\sqrt{1-\xi^2}\right)\right.$$

$$\left. + \frac{\dot{x}(0) + \xi\,\omega_n\,x(0)}{\omega_n\,\sqrt{1-\xi^2}}\sin\left(\omega_n\,t\,\sqrt{1-\xi^2}\right)\right] \tag{1.16}$$

Figure 1.2 shows typical time responses, to an initial non-zero displacement $x(0) = 1$ and initial velocity $\dot{x}(0) = 0$, of the referred three different cases. It is clear that the underdamped system is the only one where the free motion is oscillatory whereas in the other cases the system simply returns to the static equilibrium position without oscillating (the motion in the latter cases is known as relaxation).

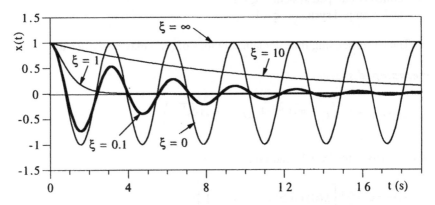

Fig. 1.2 Example of typical time domain free vibration responses to an initial displacement $x(0) = 1$, for different values of the damping ratio.

While the undamped solution ($\xi = 0$) corresponds to a harmonic motion, of frequency ω_n (known as the undamped natural frequency) and with constant amplitude, the oscillating damped solution ($0 < \xi < 1$) is closer to reality, tending exponentially to zero. In the latter case the frequency of oscillation is given by

$$\omega_d = \omega_n\,\sqrt{1-\xi^2} \tag{1.17}$$

which is known as the damped natural frequency.

The decaying characteristic of this behaviour may be used as a means to evaluate the damping ratio associated with a given system. From the plot of a simple free vibration test (figure 1.3) one may take the value of the peak amplitude X_i at a certain instant of time and the value of the peak amplitude X_{i+n} taken after n complete cycles of vibration. From these data and from (1.16) it is possible to derive a quantity known as the logarithmic decrement which is given by

$$\delta_n = \ln\frac{X_i}{X_{i+n}} = \frac{2\,n\,\pi\,\xi}{\sqrt{1-\xi^2}} \tag{1.18}$$

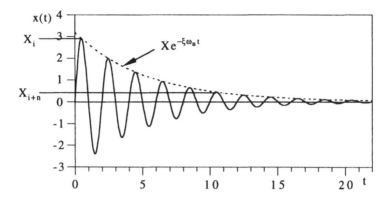

Fig. 1.3 Free decaying damped oscillation (in this example n = 5).

There are many practical situations where the dynamic behaviour of real systems may be represented by a single degree-of-freedom model (like the one represented in figure 1.1). The basic problem then is the derivation of the spatial properties of the model. This is possible, for example, through the realisation of a simple test yielding a plot of a decaying free vibration. From this plot and from (1.18) one extracts the value of ξ. Counting n complete cycles of vibration and reading, from the plot, the corresponding time interval, enables us to derive the value of the damped natural frequency ω_d.

The next step is the calculation of the undamped natural frequency ω_n from (1.17). The identification of the spatial properties m, k and c will then be an easy task provided one knows the value of one of these quantities. In most cases it is easy to know (or measure) m and therefore the problem is solved. The value of k will subsequently be obtained from $\omega_n = \sqrt{k/m}$ and finally, the value of c will be derived from (1.8) and (1.9).

Extracting dynamic characteristics from experimental data is the aim of the so-called identification procedures. This subject will be dealt with, in more depth, in Chapter 4. For the moment it will suffice to emphasise that an experimental plot of the free vibration response of a system is not sufficient for extracting all the dynamic properties (one of them must be known beforehand).

Another interesting and important aspect to emphasise is the fact that, for most real structures, the damping ratio is small (typically below 10%). Thus, if one plots the number n of cycles of free vibration necessary to reduce the corresponding amplitude by a given factor $N = X_i/X_{i+n}$ against the value of ξ, as shown in figure 1.4, it is possible to conclude that the free vibration (also known as transient) component of the complete response solution of equation (1.1) tends to disappear very rapidly.

The consequences of the above reasoning are that, for most of the situations encountered in practice, it is only of interest to consider the particular integral of the forced response solution i.e., the initial transient component may be ignored without loss of accuracy.

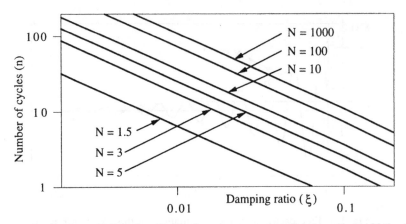

Fig. 1.4 Number n of cycles of vibration necessary to reduce by a factor N the free vibration amplitude of a SDOF system, plotted against the damping ratio.

1.2.2 Forced vibration

The forced vibration problem is described by (1.1) with $f(t) \neq 0$. We already know the solution of the corresponding homogeneous equation and therefore, to obtain the complete solution, one must derive the particular solution of (1.1). Assuming that the forcing function is of the form,

$$f(t) = F e^{i\omega t} \tag{1.19}$$

where F and ω are two constants (the harmonic excitation force amplitude and frequency respectively) and $i = \sqrt{-1}$, the particular solution is given by

$$x(t) = \overline{X} e^{i\omega t} \tag{1.20}$$

where $\overline{X}$ is a complex amplitude also known as phasor i.e., it allows inclusion of a phase angle of the motion response with respect to the forcing function f(t):

$$\overline{X} = X e^{i\theta} \tag{1.21}$$

Substituting (1.20) into (1.1), it follows that

$$\overline{X} = \frac{F}{\left(k - \omega^2 m\right) + i\omega c} \tag{1.22}$$

As any complex number of the form $x + iy$ can be written as $R e^{i\theta}$, with $R = (x^2 + y^2)^{1/2}$ and $\tan\theta = y/x$, (1.22) can be written as:

$$\overline{X} = \frac{F}{\sqrt{(k - \omega^2 m)^2 + (\omega c)^2}} e^{i\theta} \tag{1.23}$$

with

$$\tan \theta = \frac{-\omega c}{k - \omega^2 m} \tag{1.24}$$

The particular solution of (1.1), for the harmonic forcing function defined by (1.19), is therefore given by

$$x(t) = \frac{F}{\sqrt{(k - \omega^2 m)^2 + (\omega c)^2}} e^{i(\omega t + \theta)} \tag{1.25}$$

which is a harmonic function with constant amplitude, as is the exciting force. Furthermore, (1.24) and (1.25) indicate that the response x(t) is delayed with respect to the forcing function f(t), this delay being described, in angular terms, by θ. This solution and the vibration it represents, in this case, are called steady-state solution and steady-state vibration.

The complete solution is given by the sum of (1.25) with the solution (1.15) of the homogeneous equation:

$$x(t) = e^{-\xi \omega_n t} \left(C_1 e^{i \omega_n t \sqrt{1 - \xi^2}} + C_2 e^{-i \omega_n t \sqrt{1 - \xi^2}} \right)$$
$$+ \frac{F}{\sqrt{(k - \omega^2 m)^2 + (\omega c)^2}} e^{i(\omega t + \theta)} \tag{1.26}$$

or

$$x(t) = e^{-\xi \omega_n t} \left(C_1 e^{i \omega_n t \sqrt{1 - \xi^2}} + C_2 e^{-i \omega_n t \sqrt{1 - \xi^2}} \right)$$
$$+ \frac{F}{k} \frac{1}{\sqrt{(1 - \beta^2)^2 + (2 \xi \beta)^2}} e^{i(\omega t + \theta)} \tag{1.27}$$

where $\beta = \omega/\omega_n$ is a dimensionless parameter representing the ratio of the forcing frequency to the undamped natural frequency of the system. The previous equation shows that the motions corresponding to the particular solution of equation (1.1) and to the solution of the homogeneous equation are superposed i.e., they are added algebraically (figure 1.5). As shown in the previous section (see also figure 1.4), it is obvious that the transient solution is important only for a limited initial period of time and therefore, after enough time has elapsed, practically only the steady-state solution remains.

Taking only the steady-state part of the solution of the forced vibration problem, it is common, in basic literature, to consider not the magnitude X of the response but the quantity given by

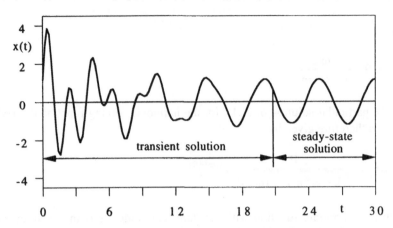

Fig. 1.5 Complete solution of the equation of motion for a SDOF system.

$$\frac{X}{X_s} = Q = \frac{1}{\sqrt{\left(1-\beta^2\right)^2 + \left(2\,\xi\,\beta\right)^2}}$$ (1.28)

where X_s is the ratio F/k, corresponding to the static deformation of the system if loaded by a constant force F. Equation (1.28) has the advantage of being totally dimensionless and therefore its graphical representation is valid for any SDOF system (figure 1.6).

It can be seen that when $\xi = 0$ and $\beta = 1$ (i.e., $\omega = \omega_n$), the denominator of (1.28) is zero, meaning that steady-state vibration has infinite amplitude X[†] no matter how small the exciting force amplitude F is. This particular situation is called resonance. Avoiding resonance is therefore of great importance[‡] and is, in fact, one of the goals in design as it is obvious that such a situation can lead to the ruin of any structure.

Fortunately, in practice, ξ is never zero because there is always some degree of energy dissipation in real systems. This means that any dynamic model should include a damping mechanism and therefore a non-zero value of c. In this case, the amplitude at resonance is not infinite, though for low damping it can have very large values. It can be proven that the maximum value of the amplitude of the steady-state vibration (minimum of the denominator of (1.25) or of (1.28)) occurs for $\omega = \omega_n \sqrt{1-2\xi^2}$. This particular feature is observable in figure 1.6 by the peak amplitudes occurring to the left of the $\beta = 1$ line, the shift being larger for larger damping values. As stated previously, most real structures have low damping and therefore resonance is usually taken as occurring for $\omega = \omega_n$. The error is

[†] Note that the infinite (or very high, in practice) amplitude can be shown to take some time to be reached. It is not an instantaneous result.

[‡] Unless one wishes, specifically, to design some kind of vibrating machine, taking advantage of the vibration phenomena.

below 1% for $\xi = 0.1$ and below 10% for $\xi = 0.5$, thus justifying the assumption. The quantity Q is known as the amplification factor. At resonance, $Q = 1/2\xi$ and can be taken as a 'measure' of the sharpness of resonance. Finally, it is important to note that damping is only effective near resonance. Away from resonance the response is hardly influenced by damping and all the response curves coincide.

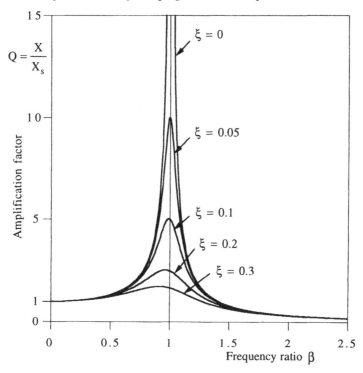

Fig. 1.6 Graphical display of the amplification factor $Q = X/X_S$, plotted against β.

Plotting the phase angle θ against the frequency ratio β, as shown in figure 1.7, one may note that the response has a phase shift from an initial $0°$ value to a final $-180°$ value when it passes through resonance (where $\theta = -90°$). The meaning of this phase shift is that the response is delayed in time relative to the forcing function. In the ideal theoretical case where ξ is zero, the phase shift is instantaneous whereas it becomes increasingly gradual for larger values of ξ.

In the above analysis, the objective was to take a given system and calculate its dynamic response x(t) to a given forcing function f(t). An alternative way of looking at the equations derived for the steady-state case is to consider the dynamic properties of our system which are contained in the mathematical expression relating the output to the input

$$\frac{\overline{X}}{F} = H(\omega) = \frac{1}{(k - \omega^2 m) + i\,\omega\,c} \tag{1.29}$$

The complex function of the frequency denoted by $H(\omega)$ is called the system's Frequency Response Function (FRF).

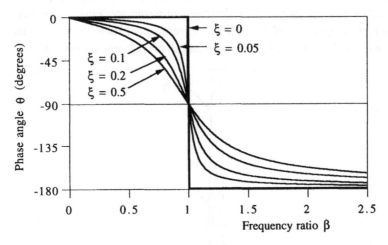

Fig. 1.7 Graphical display of the phase of the response $x(t)$, relative to the forcing function, plotted against the frequency ratio β.

1.2.3 Nonharmonic excitation. Fourier analysis

So far, a particular type of forcing function has been used in order to derive the complete solution of the equation of motion, using what can be regarded as a standard method present in all text books. However, excitations may be of many different types other than harmonic. In fact if one considers real excitation sources such as, for example, earthquakes, wind, sea waves, rough road surfaces, and all types of machinery, it is easy to understand that the forcing functions are of many different types and may only be harmonic in very particular cases.

Dynamic signals may be generally classified as deterministic or random (figure 1.8). The former can be described by an analytical expression of their magnitude, as a function of time, while the latter cannot. Random signals are characterised by analysing their statistical properties and may be classified as stationary (i.e., they have constant statistical properties along their length) and non-stationary. Random vibration analysis has a place of its own and will not be considered in depth in this book. However, random signals may occur in experimental modal analysis and their treatment and application are mentioned in Section 1.2.7 and in Chapters 2 and 3.

Deterministic signals may be periodic or transient. A periodic signal is one that repeats itself after a period T of time, i.e., $f(t) = f(t + T)$. Harmonic signals are therefore particular types of periodic signals. A transient signal is one that occurs only during a short period of time. As will be seen in Section 2.2, if a function is periodic and satisfies certain conditions then it can be represented by a summation of harmonic functions known as the Fourier series. Hence, Fourier analysis of periodic functions yields discrete frequency spectra representing the amplitudes (and phases) of the discrete harmonic components plotted against frequency.

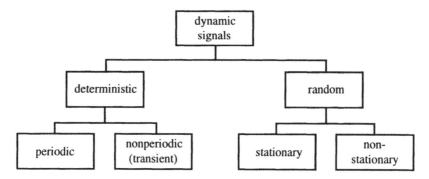

Fig. 1.8 Types of dynamic signals.

Recalling the fact that we are considering linear behaviour, the mathematical statement of the so-called principle of superposition applies. This principle has already been applied in 1.2.2 when the complete response was obtained by summing the solution of the homogeneous equation of motion with the particular forced response solution. Thus, the validity of the application of the principle of superposition means that the response of a linear system to a periodic forcing function f(t) can be obtained by adding the responses to the separate harmonic forcing functions obtained from the decomposition of f(t) into its harmonic components. Each distinct steady-state response is obtained from the application of (1.25). The previous reasoning may appear an impossible task, given the fact that Fourier series have an infinite number of harmonic terms. However, in practice, consideration of just a few (typically less than 10) initial terms of the Fourier series yields results that are sufficiently accurate for most applications.

When the forcing functions are nonperiodic (transient), they cannot be handled directly through the use of Fourier series. However, it is not difficult to accept that a transient signal may be viewed as a periodic signal with period $T = \infty$. By considering the limit which is approached by a Fourier series as the period becomes infinite, it will be found that, under certain conditions, an arbitrary function f(t) can be described by an integral $F(\omega)$ given by (see also Section 2.2):

$$F(\omega) = \mathscr{F}\left[f(t)\right] = \int_{-\infty}^{+\infty} f(t)\, e^{-i\omega t}\, dt \tag{1.30}$$

where $F(\omega)$ is known as the Fourier transform of f(t). Conversely, the time dependent function f(t) can always be obtained from $F(\omega)$ through the inverse Fourier transform

$$f(t) = \mathscr{F}^{-1}\left[F(\omega)\right] = \frac{1}{2\pi} \int_{-\infty}^{+\infty} F(\omega)\, e^{i\omega t}\, d\omega \tag{1.31}$$

Equations (1.30) and (1.31) constitute what is known as a Fourier transform pair. Applying this same reasoning to equation (1.29), f(t) representing an arbitrary nonperiodic excitation, will yield

$$X(\omega) = H(\omega)\, F(\omega) \tag{1.32}$$

i.e., the Fourier transform of the response is simply the product of the complex frequency response function $H(\omega)$ and the Fourier transform of the excitation. The response $x(t)$ is then obtained from $X(\omega)$:

$$x(t) = \mathscr{F}^{-1}\left[X(\omega)\right] = \frac{1}{2\pi} \int_{-\infty}^{+\infty} X(\omega)\, e^{i\omega t}\, d\omega \tag{1.33}$$

Note that the Fourier transform and its inverse are frequently written as follows:

$$X(f) = \int_{-\infty}^{+\infty} x(t)\, e^{-i2\pi f t}\, dt \tag{1.34}$$

$$x(t) = \int_{-\infty}^{+\infty} X(f)\, e^{i2\pi f t}\, df \tag{1.35}$$

where $f = \omega/2\pi$ denotes the frequency expressed in Hertz (or cycles per second).

It is important to note that the frequency spectrum is now a continuous function of ω, in contrast to the frequency spectrum obtained for periodic time functions which consists of discrete components only.

As a consequence, the $F(\omega)$ values represent an amplitude which is continuously distributed along the frequency and therefore represent units of amplitude per unit frequency, i.e., what is known as a spectral density. Finally, as the Fourier integrals are defined in complex form, the spectrum is two-sided, that is, there is a spectrum for negative frequencies which is the mirrored spectrum of that of the positive frequencies.

To obtain $x(t)$ it is then necessary to evaluate the integral in (1.33) which often leads to difficulties from a mathematical point of view. On the other hand, there are a number of situations where the Fourier transform analysis is inadequate, yielding completely meaningless solutions of the integral (as in the case where f(t) is a step function, for example). Attempting to solve these situations, through adequate mathematical manipulations, results in the use of a 'modified' Fourier transform method known as the Laplace transform method (see Section 1.2.5).

In practice, the forcing function may be quite irregular, even if it is periodic, and may be determined only experimentally. Such cases correspond to having a graphical representation of the signal and no analytical expression to describe it. These situations can still be handled by means of adequate discretisation and numerical procedures applied to the signal, as will be explained in Chapter 2.

1.2.4 Time Domain. Impulse Response Function (IRF)

An alternative to Fourier analysis is the use of a time domain approach for estimating a system's response to an arbitrary input. The simplest form of a nonperiodic forcing function is the unit impulse or Dirac δ-function

$$f(t) = \delta(t - \tau) \tag{1.36}$$

which is zero for all values of t except for $t = \tau$, where

$$\lim_{\Delta t \to 0} \int_{\tau}^{\tau + \Delta t} f(t)\, dt = 1 \tag{1.37}$$

This function may be visualised (figure 1.9) as represented by a rectangular area of width Δt and height $1/\Delta t$ taken to the limit as Δt goes to zero.

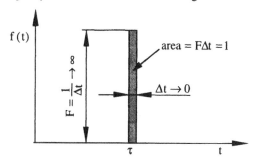

Fig. 1.9 Definition of a unit impulse forcing function.

Considering our SDOF system (figure 1.1) at rest before the unit impulse excitation is applied ($x(t) = \dot{x}(t) = 0$ for $t < \tau$ or at $t = \tau^{-}$), we obtain, from the impulse momentum relation,

$$\lim_{\Delta t \to 0} F\, \Delta t = 1 = m\, \dot{x}\big|_{t=\tau} - m\, \dot{x}\big|_{t=\tau^{-}} = m\, \dot{x}\big|_{t=\tau} \tag{1.38}$$

and therefore, we may conclude that the response to the unit impulse forcing function is nothing but a free vibration with zero initial displacement and initial velocity equal to $1/m$. Thus, from (1.16)

$$x(t) = h(t - \tau) = e^{-\xi \omega_n (t - \tau)}\, \frac{1}{m\, \omega_d} \sin\big[\omega_d\, (t - \tau)\big] \qquad \text{for } t > \tau \tag{1.39}$$

where $h(t - \tau)$ denotes the unit impulse response function (IRF).

The response to an arbitrary input f(t) may now be taken as the superposition (summation) of the responses to a series of impulses which represent the original forcing function (figure 1.10). This is true given the fact that we are considering linear systems and therefore the principle of superposition applies.

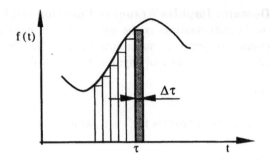

Fig. 1.10 Arbitrary nonperiodic forcing function.

Hence:

$$x(t) \approx \sum_\tau f(\tau)\, h(t-\tau)\, \Delta\tau \qquad \text{for } t > \tau \tag{1.40}$$

Letting $\Delta\tau \to 0$, the summation is replaced by integration (note that $x(t) = 0$ for all t except for $t > \tau$), and therefore

$$x(t) = \int_0^t f(\tau)\, h(t-\tau)\, d\tau \qquad \text{for } t > \tau \tag{1.41}$$

The integral in (1.41) is called the convolution or Duhamel's integral. In cases where the function f(t) does not have a form that permits an explicit integration, it can be evaluated numerically. Substituting (1.39) into (1.41), we obtain

$$x(t) = \frac{1}{m\,\omega_d} \int_0^t f(\tau)\, e^{-\xi\omega_n(t-\tau)} \sin\left[\omega_d\,(t-\tau)\right] d\tau \qquad \text{for } t > \tau \tag{1.42}$$

which represents the response of an underdamped SDOF system to an arbitrary forcing function f(t).

Let us now consider the problem in terms of Fourier analysis, assuming that our forcing function is the Dirac δ-function. Recalling (1.41) and noting that since $h(t-\tau) = 0$ for $t < \tau$, the lower limit of the integral can be extended to minus infinity. Thus,

$$x(t) = \int_{-\infty}^t f(\tau)\, h(t-\tau)\, d\tau \tag{1.43}$$

Next, we may change the variable of integration to ϑ by using the relationship $\tau = t - \vartheta$. As a consequence, $d\tau = -d\vartheta$ and (1.43) becomes

$$x(t) = -\int_{+\infty}^0 f(t-\vartheta)\, h(\vartheta)\, d\vartheta \tag{1.44}$$

which is the same as

$$x(t) = \int_0^{+\infty} f(t - \vartheta)\, h(\vartheta)\, d\vartheta \qquad (1.45)$$

Finally, stating that $h(t - \tau) = 0$ for all $t < \tau$ is equivalent to stating that $h(\vartheta) = 0$ for all $\vartheta < 0$ and therefore the lower limit of the integral in (1.45) can also be extended to minus infinity, i.e.,

$$x(t) = \int_{-\infty}^{+\infty} f(t - \vartheta)\, h(\vartheta)\, d\vartheta \qquad (1.46)$$

What we have in (1.46) is the convolution of the forcing function $f(t)$ with the impulse response function $h(t)$. Thus (1.46) may be rewritten as

$$x(t) = h(t) * f(t) \qquad (1.47)$$

where the symbol $*$ denotes the convolution operation. Taking the Fourier transform of (1.47) leads to

$$\mathscr{F}\left[x(t)\right] = X(\omega) = \mathscr{F}\left[h(t) * f(t)\right] \qquad (1.48)$$

At the right-hand side of (1.48) we have the Fourier transform of a convolution which has the following property:

$$\mathscr{F}\left[h(t) * f(t)\right] = \mathscr{F}\left[h(t)\right] \mathscr{F}\left[f(t)\right] \qquad (1.49)$$

and therefore

$$X(\omega) = H(\omega)\, F(\omega) \qquad (1.50)$$

which is the same as equation (1.32) as should be expected. Only, in this case, $F(\omega)$ is the Fourier transform of the Dirac δ-function for which, by definition

$$\int_{-\infty}^{+\infty} \delta(t - \tau)\, g(t)\, dt = g(\tau) \qquad (1.51)$$

for any function $g(t)$. Therefore, the Fourier transform of the Dirac δ-function is given by

$$F(\omega) = \int_{-\infty}^{+\infty} \delta(t - \tau)\, e^{-i\omega t}\, dt = e^{-i\omega\tau}\Big|_{\tau=0} = 1 \qquad (1.52)$$

Inserting (1.52) into (1.50) and applying (1.33) yields

18

$$x(t) = \mathscr{F}^{-1}\left[H(\omega)\,F(\omega)\right] = \mathscr{F}^{-1}\left[X(\omega)\right] = \frac{1}{2\pi}\int_{-\infty}^{+\infty} H(\omega)\,e^{i\omega t}\,d\omega \qquad (1.53)$$

which, by definition, must be identical to h(t). Thus, it may be concluded that the frequency response function H(ω) and the impulse response function h(t) constitute a Fourier transform pair. This is a very important conclusion as it allows derivation of the FRF for a given system just by taking the Fourier transform of its impulse response function.

1.2.5 The Laplace Domain. Transfer Function

One way of deriving the dynamic response of a system under any type of excitation, including obviously the periodic and harmonic ones, is by means of the Laplace transform method. Basically, the Laplace transform method converts differential equations into algebraic ones which are easier to manipulate. Another great advantage of the method is that it can treat discontinuous functions and automatically takes into account the initial conditions.

The Laplace transform of a function x(t), denoted as X(s), is defined as

$$X(s) = \mathscr{L}\left[x(t)\right] = \int_0^{+\infty} e^{-st}\,x(t)\,dt \qquad (1.54)$$

where s is, in general, a complex quantity known as the Laplace variable. Taking the Laplace transform on each side of (1.1), one obtains

$$\mathscr{L}\left[m\,\ddot{x}(t) + c\,\dot{x}(t) + k\,x(t)\right] = m\left[s^2\,X(s) - s\,x(0) - \dot{x}(0)\right] + c\left[s\,X(s) - x(0)\right] + k\,X(s)$$

$$= (m s^2 + c s + k)\,X(s) - m\,s\,x(0) - m\,\dot{x}(0) - c\,x(0)$$

$$(1.55)$$

and

$$\mathscr{L}\left[f(t)\right] = F(s) \qquad (1.56)$$

or

$$\left(m s^2 + c s + k\right)X(s) = F(s) + m\,\dot{x}(0) + (m s + c)\,x(0) \qquad (1.57)$$

where x(0) and ẋ(0) are the initial displacement and velocity respectively and the right-hand side of equation (1.57) can be regarded as a generalised transformed excitation. If the initial conditions are zero, which is equivalent to ignoring the solution of the homogeneous equation (1.2), the ratio of the transformed response to the transformed excitation can be expressed as

$$H(s) = \frac{X(s)}{F(s)} \qquad (1.58)$$

where

$$H(s) = \frac{1}{m s^2 + c s + k} \tag{1.59}$$

is known as the system transfer function. $H(s)$ is a complex valued function of s and is represented as a surface in the Laplace domain (figures 1.11). The denominator of (1.59) is the characteristic equation (already defined as such in (1.5)) yielding two roots as expressed by (1.6).

For an underdamped system, the roots s_1 and s_2 of the characteristic equation can be written as

$$s_{1,2} = \sigma \pm i \omega_d \tag{1.60}$$

with

$$\sigma = -\xi \omega_n \tag{1.61}$$

and

$$\omega_d = \omega_n \sqrt{1 - \xi^2} \tag{1.62}$$

as already defined in Section 1.2.1 (equation (1.10), for the case where $\xi < 1$). The transfer function can now be rewritten as

$$H(s) = \frac{1}{m (s - s_1)(s - s_2)} \tag{1.63}$$

where $s_1 = \sigma + i \omega_d$ and $s_2 = s_1^* = \sigma - i \omega_d$ (the two roots of the characteristic equation) are the so-called poles of the transfer function which can be viewed looking down on the s-plane as shown in figure 1.12.

Through partial fraction expansion, (1.63) can be expressed as

$$H(s) = \frac{1}{m (s - s_1)(s - s_1^*)} = \frac{A}{(s - s_1)} + \frac{A^*}{(s - s_1^*)} \tag{1.64}$$

where the complex conjugates A and A^* are defined as the residues of the transfer function being, as will be shown later (equation (1.69)), directly related to the amplitude of the impulse response function. The residues can be easily found and are given by

$$A = \frac{1}{i 2 m \omega_d} \tag{1.65}$$

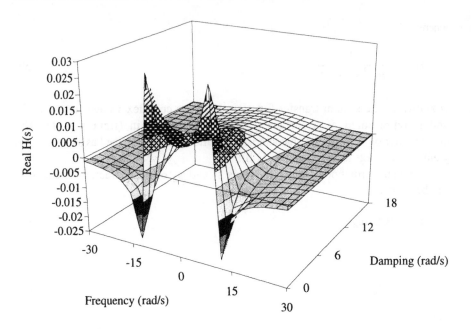

Fig. 1.11 a) Three-dimensional plot of Real H(s) against damping and frequency (m=1 kg, k=100 N/m and c=0.6 Ns/m).

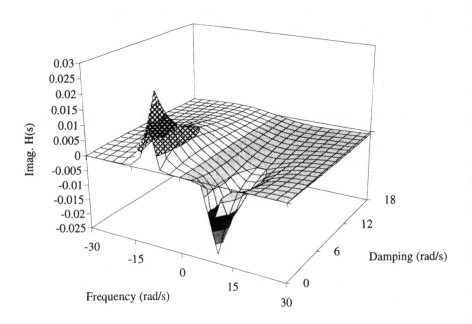

Fig. 1.11 b) Three-dimensional plot of Imag. H(s) against damping and frequency (m=1 kg, k=100 N/m and c=0.6 Ns/m).

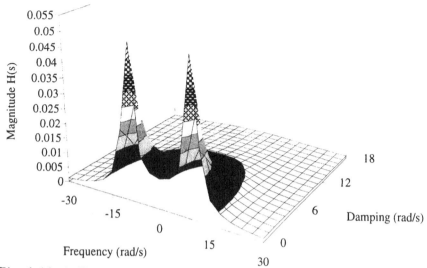

Fig. 1.11 c) Three-dimensional plot of Magnitude of H(s) against damping and frequency (m=1 kg, k=100 N/m and c=0.6 Ns/m).

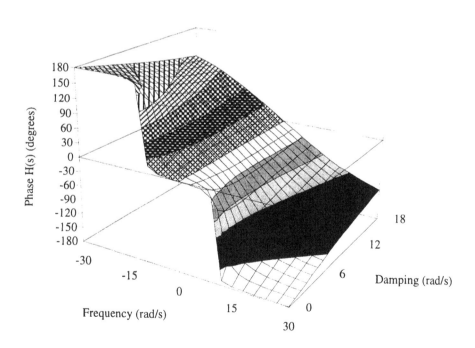

Fig. 1.11 d) Three-dimensional plot of Phase of H(s) against damping and frequency (m=1 kg, k=100 N/m and c=0.6 Ns/m).

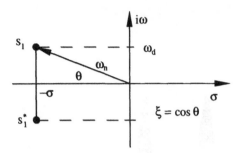

Fig. 1.12 Pole representation on the Laplace plane.

Though for a single degree-of-freedom A is purely imaginary, for multiple degree-of-freedom systems the residues are, in general, complex quantities.

1.2.6 The Frequency Response Function (FRF)

As seen previously, the Laplace domain describes the system under analysis in terms of poles and residues. Evaluating now the transfer function only in the frequency domain, i.e. along the frequency axis, one obtains

$$H(\omega) = H(s)\big|_{s=i\omega} = \left[\frac{A}{(s-s_1)} + \frac{A^*}{(s-s_1^*)}\right]\Bigg|_{s=i\omega} = \frac{A}{i\omega - s_1} + \frac{A^*}{i\omega - s_1^*}$$

$$= \frac{A}{i(\omega - \omega_d) + \xi\omega_n} + \frac{A^*}{i(\omega + \omega_d) + \xi\omega_n} \tag{1.66}$$

Equation (1.66) represents the partial fraction expansion form of the Frequency Response Function (FRF) of a SDOF system and can be seen as describing it in terms of periodicities. The same FRF could have been obtained from (1.59) under a form more commonly presented in the basic literature:

$$H(\omega) = \frac{1}{(k - \omega^2 m) + i\omega c} \tag{1.67}$$

which is nothing but equation (1.29). Thus, the FRF is just a particular case of the transfer function.

Free vibration behaviour may be obtained assuming that the system was excited by an impulse type forcing function at time t=0. The impulse response function of a single degree-of-freedom system can be easily determined from (1.58) and (1.64) assuming zero initial conditions and that $F(s) = 1$ for an impulse forcing function. Thus:

$$X(s) = H(s)\big|_{F(s)=1} = \frac{A}{(s-s_1)} + \frac{A^*}{(s-s_1^*)} \tag{1.68}$$

and therefore

$$x(t) = \mathscr{L}^{-1}[X(s)] = A\,e^{s_1 t} + A^* e^{s_1^* t} = e^{-\xi\omega_n t}\left(A\,e^{i\omega_d t} + A^* e^{-i\omega_d t}\right) \qquad (1.69)$$

which is precisely the same as the result obtained by classical means (see (1.15)). Equation (1.69) represents a decaying oscillation (figure 1.13) of frequency ω_d. Thus, the oscillation frequency corresponds to the imaginary part of the pole, the residue A controls the initial amplitude of the impulse response and the real part of the pole controls the decay rate.

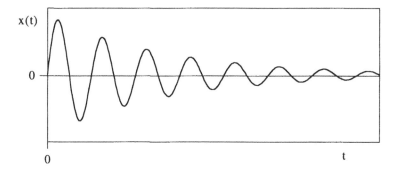

Fig. 1.13 Impulse response function of a single degree-of-freedom system.

1.2.7 Random excitation

Random signals cannot be treated in the same way as the deterministic signals so far discussed. By nature they are nonperiodic and it might be thought they could be analysed assuming a periodicity of infinite period. However this is not possible because they do not obey the Dirichlet condition which states that Fourier transforms can only be used if the following relationship is satisfied:

$$\int_{-\infty}^{+\infty}|x(t)|\,dt < \infty \qquad (1.70)$$

Given their inherent properties, the analysis of random signals entails the use of probabilistic concepts. The way to sidestep some mathematical difficulties is to assume that our random signals are stationary and ergodic, i.e., both averaging across many time history records at a given instant in time and averaging over time using just one time history give the same mean properties (mean, mean square and statistical distributions). This assumption is made in most practical cases and is one that we are going to consider in this book. Also, it is assumed that the definition and meaning of the above mentioned properties are known.

Let us then take a signal $f(t)$ as our random forcing function (as exemplified in figure 1.14) and compute, along the time axis, the average (or 'expected') value of the product $f(t)f(t+\tau)$

24

$$R_{ff}(\tau) = \lim_{T \to \infty} \frac{1}{T} \int_{-T/2}^{+T/2} f(t)\, f(t+\tau)\, dt \qquad (1.71)$$

where f(t) is the magnitude of our function at an instant t of time and $f(t+\tau)$ designates the magnitude of the same function observed after a time delay τ has elapsed.

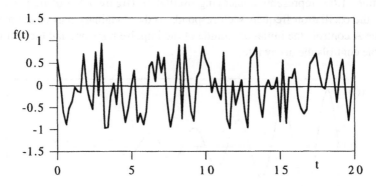

Fig. 1.14 Example of a sample of the time history of a random signal.

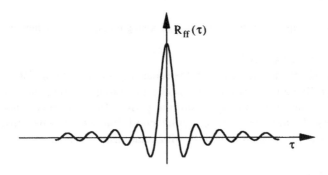

Fig. 1.15 Example of an auto-correlation function.

Equation (1.71) is the definition of the so-called random auto-correlation function. This function takes, generally, the form illustrated in figure 1.15. In physical terms, the auto-correlation function describes how a particular instantaneous amplitude value of our random time signal depends upon previously occurring instantaneous amplitude values. For instance, for an ideal random process (no frequency limits), the auto-correlation function consists of a δ function at $\tau = 0$ as shown in figure 1.16 a). In practice, real signals are all frequency limited and the narrower the frequency limits the wider the spreading of the auto-correlation function (figure 1.16 b) and 1.16 c)).

Thus, the original stationary random time history has been transformed into a new function of time, $R_{ff}(\tau)$, which is even, real valued, goes to zero as τ

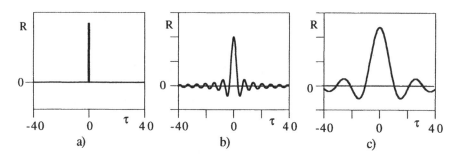

Fig. 1.16 Physical meaning of auto-correlation function: a) ideal random process; b) random process with wide frequency limits (wide band process); c) same as in b) but with narrower frequency limits.

becomes large (both positively and negatively) and obeys the Dirichlet condition (1.70). As a consequence, $R_{ff}(\tau)$ can be Fourier transformed. Applying (1.30) we obtain:

$$S_{ff}(\omega) = \mathscr{F}\left[R_{ff}(\tau)\right] = \int_{-\infty}^{+\infty} R_{ff}(\tau)\, e^{-i\omega\tau}\, d\tau \qquad (1.72)$$

known as the auto-spectral density (ASD) or power spectral density (PSD), which is also a real and even function of frequency. Therefore, we can describe our stationary random process through the following Fourier transform pair (well known as the Weiner-Khintchine relationships):

$$S_{ff}(\omega) = \int_{-\infty}^{+\infty} R_{ff}(\tau)\, e^{-i\omega\tau}\, d\tau$$
$$R_{ff}(\tau) = \frac{1}{2\pi} \int_{-\infty}^{+\infty} S_{ff}(\omega)\, e^{i\omega\tau}\, d\omega \qquad (1.73)$$

The ASD provides a frequency description of the original signal f(t). However, its physical meaning is not exactly the same as the one provided by the Fourier transform of a transient signal (Section 1.2.3).

We have already seen that a periodic signal can be described by a series of harmonic components, hence by a discrete spectrum, and that we may interpret the significance of the Fourier transform of a transient (nonperiodic) signal as a continuous amplitude description along the frequency. In that case we have a continuous spectrum and we can talk about a spectral density as the Fourier transform has units of amplitude per unit frequency. In the case of a random process, the physical meaning of the ASD can be better understood if we set $\tau = 0$ in (1.71) and combine the result with (1.73):

$$R_{ff}(0) = \lim_{T\to\infty} \frac{1}{T} \int_{-T/2}^{+T/2} f^2(t)\, dt = \frac{1}{2\pi} \int_{-\infty}^{+\infty} S_{ff}(\omega)\, d\omega \qquad (1.74)$$

It is clear that the ASD has units of amplitude mean square per unit frequency and, for that reason, it is sometimes referred to as the mean square spectral density. What is now described, continuously along the frequency range, is not the signal amplitude but rather a quantity squared that can be taken as an energy content indicator. This is the reason why this function is also called the power spectral density. In graphical display against the frequency, it can be easily seen that the narrower the auto-correlation function the wider the PSD (wider frequency limits).

The previous concepts can be extended in order to consider simultaneously the random force and the random response functions. Thus we may define

$$R_{fx}(\tau) = \lim_{T \to \infty} \frac{1}{T} \int_{-T/2}^{+T/2} f(t)\, x(t+\tau)\, dt = \frac{1}{2\pi} \int_{-\infty}^{+\infty} S_{fx}(\omega)\, e^{i\omega\tau}\, d\omega$$

$$S_{fx}(\omega) = \int_{-\infty}^{+\infty} R_{fx}(\tau)\, e^{-i\omega\tau}\, d\tau$$

$$(1.75)$$

and, conversely

$$R_{xf}(\tau) = \lim_{T \to \infty} \frac{1}{T} \int_{-T/2}^{+T/2} x(t)\, f(t+\tau)\, dt = \frac{1}{2\pi} \int_{-\infty}^{+\infty} S_{xf}(\omega)\, e^{i\omega\tau}\, d\omega$$

$$S_{xf}(\omega) = \int_{-\infty}^{+\infty} R_{xf}(\tau)\, e^{-i\omega\tau}\, d\tau$$

$$(1.76)$$

as the cross-correlation and the cross-spectral density functions, respectively. It is important to note that the cross-spectral densities are complex frequency spectra, containing real and imaginary parts (or magnitude and phase information) whereas the PSD is a real function containing only magnitude (squared) information. Also, it can be seen that the cross-correlation functions obey the relationship

$$R_{xf}(\tau) = R_{fx}(-\tau) \tag{1.77}$$

and that the cross-spectral density functions are complex conjugates:

$$S_{xf}(\omega) = S_{fx}^{*}(\omega) \tag{1.78}$$

Having established the previous concepts, let us recall that the response of a SDOF system to an arbitrary input may be calculated using the impulse response function (1.41), as defined in Section 1.2.4. From this knowledge we can define $x(t)$ and $x(t+\tau)$ and therefore the response auto-correlation function R_{xx}

$$R_{xx}(\tau) = \lim_{T \to \infty} \frac{1}{T} \int_{-T/2}^{+T/2} x(t)\, x(t+\tau)\, dt \tag{1.79}$$

Taking now the PSD definition, for the response

$$S_{xx}(\omega) = \int_{-\infty}^{+\infty} R_{xx}(\tau)\, e^{-i\omega\tau}\, d\tau \tag{1.80}$$

and substituting (1.79)

$$S_{xx}(\omega) = \int_{-\infty}^{+\infty} \left[\lim_{T\to\infty} \frac{1}{T} \int_{-T/2}^{+T/2} x(t)\, x(t+\tau)\, dt \right] e^{-i\omega\tau}\, d\tau \tag{1.81}$$

Next, let us substitute $x(t)$ in the integral using equation (1.46). For the sake of generality we use now two variables ϑ_1 and ϑ_2 . We obtain,

$$S_{xx}(\omega) = \int_{-\infty}^{+\infty} \left\{ \lim_{T\to\infty} \frac{1}{T} \int_{-T/2}^{+T/2} \left[\int_{-\infty}^{+\infty} f(t-\vartheta_1)\, h(\vartheta_1)\, d\vartheta_1 \right.\right.$$
$$\left.\left. \int_{-\infty}^{+\infty} f(t-\vartheta_2+\tau)\, h(\vartheta_2)\, d\vartheta_2 \right] dt \right\} e^{-i\omega\tau}\, d\tau \tag{1.82}$$

and therefore,

$$S_{xx}(\omega) = \int_{-\infty}^{+\infty} \left\{ \lim_{T\to\infty} \frac{1}{T} \int_{-T/2}^{+T/2} \left[\int_{-\infty}^{+\infty}\int_{-\infty}^{+\infty} f(t-\vartheta_1)\, h(\vartheta_1) \right.\right.$$
$$\left.\left. f(t-\vartheta_2+\tau)\, h(\vartheta_2)\, d\vartheta_1\, d\vartheta_2 \right] dt \right\} e^{-i\omega\tau}\, d\tau \tag{1.83}$$

Changing the order of integration,

$$S_{xx}(\omega) = \int_{-\infty}^{+\infty} \left\{ \int_{-\infty}^{+\infty}\int_{-\infty}^{+\infty} h(\vartheta_1)\, h(\vartheta_2) \right.$$
$$\left. \left[\lim_{T\to\infty} \frac{1}{T} \int_{-T/2}^{+T/2} f(t-\vartheta_1)\, f(t-\vartheta_2+\tau)\, dt \right] d\vartheta_1\, d\vartheta_2 \right\} e^{-i\omega\tau}\, d\tau \tag{1.84}$$

and because the excitation random process is stationary, we have:

$$\lim_{T\to\infty} \frac{1}{T} \int_{-T/2}^{+T/2} f(t-\vartheta_1)\, f(t-\vartheta_2+\tau)\, dt$$
$$= \lim_{T\to\infty} \frac{1}{T} \int_{-T/2}^{+T/2} f(t)\, f(t-\vartheta_2+\tau+\vartheta_1)\, dt = R_{ff}(\tau+\vartheta_1-\vartheta_2) \tag{1.85}$$

and therefore,

$$S_{xx}(\omega) = \int_{-\infty}^{+\infty} \left\{ \int_{-\infty}^{+\infty}\int_{-\infty}^{+\infty} h(\vartheta_1)\, h(\vartheta_2) \right.$$
$$\left. \left[R_{ff}(\tau+\vartheta_1-\vartheta_2) \right] d\vartheta_1\, d\vartheta_2 \right\} e^{-i\omega\tau}\, d\tau \tag{1.86}$$

Taking now into consideration that the auto-correlation function of the excitation process, $R_{ff}(\tau + \vartheta_1 - \vartheta_2)$, can be expressed as the inverse Fourier transform

$$R_{ff}(\tau + \vartheta_1 - \vartheta_2) = \frac{1}{2\pi} \int_{-\infty}^{+\infty} S_{ff}(\omega)\, e^{i\omega(\tau + \vartheta_1 - \vartheta_2)}\, d\omega \tag{1.87}$$

we may insert (1.87) into (1.86) and obtain:

$$S_{xx}(\omega) = \int_{-\infty}^{+\infty} \left\{ \int_{-\infty}^{+\infty}\int_{-\infty}^{+\infty} h(\vartheta_1)\, h(\vartheta_2) \right.$$
$$\left. \left[\frac{1}{2\pi} \int_{-\infty}^{+\infty} S_{ff}(\omega)\, e^{i\omega(\tau + \vartheta_1 - \vartheta_2)}\, d\omega \right] d\vartheta_1\, d\vartheta_2 \right\} e^{-i\omega\tau}\, d\tau \tag{1.88}$$

Recalling that, at the end of Section 1.2.4, we demonstrated that the frequency response function $H(\omega)$ and the impulse response function $h(t)$ constitute a Fourier transform pair, i.e., that

$$H(\omega) = \int_{-\infty}^{+\infty} h(t)\, e^{-i\omega t}\, dt \tag{1.89}$$

equation (1.88) can be further manipulated so that:

$$S_{xx}(\omega) = \int_{-\infty}^{+\infty} \left\{ \frac{1}{2\pi} \int_{-\infty}^{+\infty} S_{ff}(\omega) \left[\int_{-\infty}^{+\infty} h(\vartheta_1)\, e^{i\omega\vartheta_1}\, d\vartheta_1 \right. \right.$$
$$\left. \left. \int_{-\infty}^{+\infty} h(\vartheta_2)\, e^{-i\omega\vartheta_2}\, d\vartheta_2 \right] e^{i\omega\tau}\, d\omega \right\} e^{-i\omega\tau}\, d\tau \tag{1.90}$$
$$= \int_{-\infty}^{+\infty} \left\{ \frac{1}{2\pi} \int_{-\infty}^{+\infty} S_{ff}(\omega)\, H(-\omega)\, H(\omega)\, e^{i\omega\tau}\, d\omega \right\} e^{-i\omega\tau}\, d\tau$$

where

$$H(-\omega) = H^*(\omega) \tag{1.91}$$

is the complex conjugate of the FRF $H(\omega)$. Rewriting (1.90) as

$$S_{xx}(\omega) = \int_{-\infty}^{+\infty} \left\{ \frac{1}{2\pi} \int_{-\infty}^{+\infty} S_{ff}(\omega)\, H^*(\omega)\, H(\omega)\, e^{i\omega\tau}\, d\omega \right\} e^{-i\omega\tau}\, d\tau \tag{1.92}$$

it may be concluded that this equation is nothing other than an inverse Fourier transform followed by a Fourier transform and therefore,

$$S_{xx}(\omega) = \mathscr{F}\left\{\mathscr{F}^{-1}\left[S_{ff}(\omega) H^*(\omega) H(\omega)\right]\right\} \tag{1.93}$$

Hence, equation (1.92) may be finally written as

$$S_{xx}(\omega) = H(\omega) H^*(\omega) S_{ff}(\omega) \tag{1.94}$$

or

$$S_{xx}(\omega) = \left|H(\omega)\right|^2 S_{ff}(\omega) \tag{1.95}$$

Thus, (1.95) establishes a relationship between the auto-spectral density of the force excitation and the auto-spectral density of the response. In short, given a stationary random forcing input, it is possible to calculate the stationary random response characteristics, provided one already knows the system frequency response function $H(\omega)$. However, it is clear that (1.95) cannot be used if the objective is to extract the system's FRF from force and response data. In fact, this equation would only yield the magnitude of $H(\omega)$. To obtain information on the phase of $H(\omega)$ also, we need another equation.

Fortunately, a reasoning process similar to the above one, but based also on the cross-correlation functions, will lead to:

$$S_{fx}(\omega) = H(\omega) S_{ff}(\omega) \tag{1.96}$$

and

$$S_{xx}(\omega) = H(\omega) S_{xf}(\omega) \tag{1.97}$$

Equations (1.96) and (1.97) are alternative adequate relationships that enable the frequency response function of a system to be determined from knowledge of the input force and output response characteristics. Chapter 2 will discuss their practical use in more detail.

1.2.8 Viscous and Hysteretic damping mechanisms

In general, real systems dissipate energy, while vibrating, by several different mechanisms. The dissipative process is therefore the simultaneous result of all those mechanisms, and is difficult to identify and to model accurately. By including a viscous damper in our model we are merely trying to represent the dissipative mechanism through the use of an equivalent linear element.

The viscous damper is the simplest damping element from a theoretical point of view. It is the only strictly linear damper, in the sense that the equations of motion of a system incorporating this damping mechanism may be solved for any type of input. This characteristic makes viscous damping very simple to deal with mathematically and the consequence is its widespread use in most textbooks and papers. In actual fact, it often bears little resemblance to damping mechanisms encountered in practice, its use in steady-state vibration only being justified when it represents an actual source of viscous damping such as an oil-filled dashpot.

By definition, the viscous damper is a device that opposes the relative velocity between its ends with a force which is proportional to that velocity ($f = c\dot{x}$). Considering the system represented in figure 1.17, the corresponding harmonic dynamic load/deflection curve exhibits an elliptic loop denoting the energy dissipation phenomenon. The energy ΔE dissipated per cycle of oscillation is given by the area enclosed in the oscillation loop, i.e.,

$$\Delta E = \int_0^{2\pi/\omega} f(x)\,dx = \pi\,X^2\,c\,\omega \qquad (1.98)$$

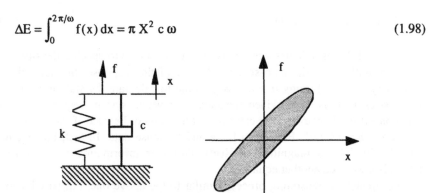

Fig. 1.17 Typical dynamic load/deflection curve for a system with viscous damping.

As shown in (1.98), the energy dissipated per cycle of oscillation is directly proportional to the damping coefficient, the square of the response amplitude and the driving frequency, i.e., it is frequency dependent or, in other terms, the viscous model yields a loop which grows in size with frequency. This frequency dependence differs from that observed in most common materials and real structures whose behaviour is found to be closer to a frequency independent or weakly dependent dissipation mechanism which is roughly proportional to the square of the displacement amplitude.

Though rubber and other viscoelastic materials do exhibit frequency dependence (and temperature dependence also), it is far less pronounced than that associated with a viscous damper and, over a limited range of frequencies, the properties may often be assumed constant.

Thus, what is apparently required, is a different damper model which opposes the relative motion between its ends with a force that is proportional to displacement and not to velocity (though still in phase with velocity). This is equivalent to using a viscous damper but making the viscous damping rate vary inversely with frequency, i.e., $c = d/\omega$. This is known as a hysteretic, solid or structural damper and the parameter d is called the hysteretic damping coefficient. This designation results from the fact that such a mechanism closely describes the load/deflection hysteresis behaviour of most materials.

It must be stressed that the hysteretic damper is only a linear device in the sense that it is described by a linear frequency response relationship. Its use is limited to steady-state vibrations as it presents difficulties to rigorous free vibration or shock response analyses. In such cases there is little option but to return to the viscous damper, however inadequate this may be. The alternative damping model given by the hysteretic damper has the advantage of not only describing more closely the energy dissipation mechanism exhibited by most real structures, but also of providing a much simpler analysis for multiple degree-of-freedom systems as will be shown later on (Section 1.4.2).

In this case, the energy dissipated per cycle of oscillation is given by

$$\Delta E = \int_0^{2\pi/\omega} f(x)\,dx = \pi\,X^2\,d \tag{1.99}$$

and the steady-state equation of motion of a hysteretically damped single degree-of-freedom system, harmonically excited, may be written as

$$(-\omega^2\,m + k + i\,d)\,\overline{X}\,e^{i\omega t} = F\,e^{i\omega t} \tag{1.100}$$

giving

$$H(\omega) = \frac{\overline{X}}{F} = \frac{1}{(k - \omega^2 m) + i\,d} = \frac{1}{k\left(1 - \beta^2 + i\,\eta\right)} = \frac{1}{m\,(\omega_n^2 - \omega^2 + i\,\eta\,\omega_n^2)} \tag{1.101}$$

where

$$\eta = \frac{d}{k} \tag{1.102}$$

is known as the damping loss factor. The similarities between the FRF expressions for the viscous and hysteretic damping cases are clear from (1.67) and (1.101). It is important to note that, for hysteretic damping, the amplification factor Q is given by

$$Q = \frac{X}{X_s} = \frac{1}{\sqrt{\left(1 - \beta^2\right)^2 + \eta^2}} \tag{1.103}$$

and therefore the maximum amplitude X of the steady-state response is always obtained for $\omega = \omega_n$ while for viscous damping the maximum occurs at $\omega = \omega_n\sqrt{1 - 2\xi^2}$. A plot of Q *versus* frequency such as the one in figure 1.6 will not show a shift to the left of the peak value of Q with increasing damping values.

However, and as stated in 1.2.2, for low damping values, a system with viscous damping may be assumed with sufficient accuracy as exhibiting its maximum steady-state response amplitude at $\omega = \omega_n$ and therefore its response amplitude behaviour may be assumed as equivalent to the behaviour of a hysteretically damped system. Thus, taking (1.67) and (1.101) at $\omega = \omega_n$, it may be concluded that the viscous and hysteretic models are approximately equivalent with $\eta = 2\,\xi$.

The hysteretic model also exhibits other small discrepancies when compared with the viscous model. For example, when ω approximates zero, Q approximates a value given by $1/\sqrt{1 + \eta^2}$ which is less than 1. The phase angle, given by

$$\tan \theta = -\frac{\eta}{1 - \beta^2} \qquad (1.104)$$

approximates a non-zero value when β approaches zero. Low values of η will obviously lead to $1/\sqrt{1 + \eta^2} \approx 1$ and $\tan^{-1}(-\eta) \approx 0$ and the equivalence between the damping models holds.

1.3   REPRESENTATION AND PROPERTIES OF AN FRF
1.3.1 Receptance
The frequency response function previously defined and discussed is just one of the possible forms of an FRF and is called Receptance[†], being usually denoted by $\alpha(\omega)$ or $\alpha(i\omega)$. This complex quantity fully describes the relationship between the displacement response and the excitation force applied to a system and thus fully characterises its dynamic properties.

Being a complex function of the frequency, there are three quantities - real part, imaginary part and frequency - to be taken into account whenever plotting FRF data. Thus, a full representation of an FRF in a single plot can only be done using a three-dimensional display as illustrated in figure 1.18.

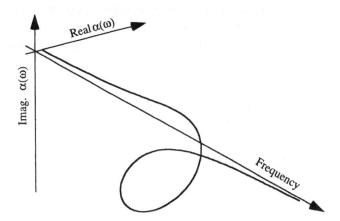

Fig. 1.18 Three-dimensional plot of the receptance of a SDOF system.

---

† Receptance is also called 'Dynamic Compliance' by several authors.

It is obvious that this is not a convenient way of graphically representing the FRF. As an alternative, we may display the FRF data in two separate plots - real and imaginary parts against frequency - as shown in figures 1.19 and 1.20 respectively (note that in the displayed examples, $\omega_n = 10$ rad / s). Each of these plots corresponds to the projection of the curve in figure 1.18 into the Real/Frequency and Imag./Frequency planes, respectively. It is interesting to note that the real part of the receptance $\alpha(\omega)$ crosses the frequency axis at resonance while, in the same frequency region, the imaginary part reaches a minimum[†].

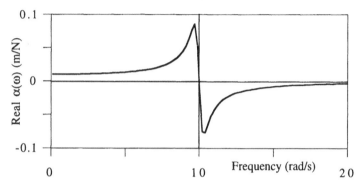

Fig. 1.19 Real part of the receptance plotted against frequency (example with m = 1 kg, k = 100 N/m and c = 0.6 Ns/m).

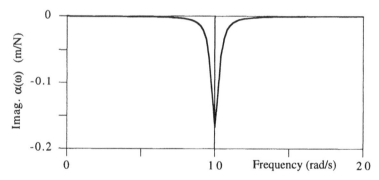

Fig. 1.20 Imaginary part of the receptance plotted against frequency (example with m = 1 kg, k = 100 N/m and c = 0.6 Ns/m).

If one takes the projection of $\alpha(\omega)$ into the Real/Imag. plane, i.e., into the Argand or complex plane, the net result is a circular loop that contains all the information. The inconvenience is that one is normally unable to identify the frequency value corresponding to any point on the curve. Each data point (as shown in figure 1.21, which displays some equally spaced frequency data points), should

---

[†] Note that this is only exactly true for hysteretic damping. In the case of viscous damping, the statement may only be taken as true for low damping values (which is the case of the presented examples).

then be accompanied by a caption indicating the corresponding frequency value. The latter representation is known as a Nyquist plot and it has the particularity of enhancing the resonant region as the circular loop occurs only close to resonance (corresponding to the 180° phase shift of the FRF). This characteristic makes the Nyquist plot very popular in experimental modal analysis.

Actually, the most common representation of a frequency response function is not any of the preceding. In fact, the Bode diagram, plotting the FRF magnitude and phase as functions of the frequency (figures 1.22 and 1.23), allows an easier visual interpretation of the information contained in $\alpha(\omega)$.

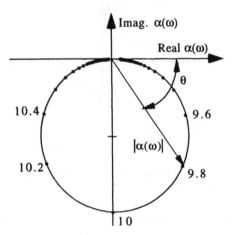

Fig. 1.21 Nyquist plot of receptance (example with $m = 1$ kg, $k = 100$ N/m and $c = 0.6$ Ns/m).

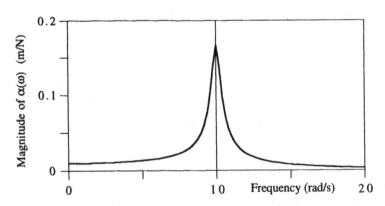

Fig. 1.22 Magnitude of receptance plotted *versus* frequency (example with $m = 1$ kg, $k = 100$ N/m and $c = 0.6$ Ns/m).

Finally, it is also interesting to emphasise that all these plots (except the Nyquist one) can be easily visualised in figures 1.11 a) to 1.11 d) if one considers the line plot resulting from the intersection of the corresponding transfer function

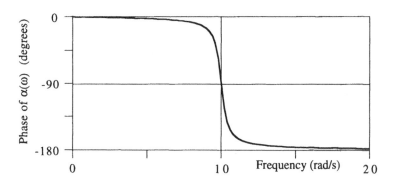

Fig. 1.23 Phase of receptance plotted *versus* frequency (example with m = 1 kg k = 100 N/m and c = 0.6 Ns/m).

surface with the vertical plane containing the frequency axis. The information contained in figures 1.22 and 1.23 is almost exactly the same as in figures 1.6 and 1.7. The only difference is that we are now plotting magnitude of $\alpha(\omega)$, i.e., the ratio X/F, instead of the dimensionless ratio X / X_s. Bode plots are constantly being used in modal analysis. However, the magnitude-frequency graph is not presented as in figure 1.22 but rather using logarithmic scales or, at least, a magnitude (vertical) logarithmic scale. In fact, the dynamic range of the responses may very easily be of the order of 10,000 to 1, thus encompassing a relatively wide range of values. As a consequence, linear scales tend to completely lose detail at lower levels of the response. The use of the logarithmic scale enables plot detail to be maintained at all levels.

If, in addition, one uses logarithmic frequency scales, data displayed as curves on linear scales become asymptotic to straight lines. This behaviour is advantageous as it provides a simple means of checking the validity of a plot and also allows for easily identifying the mass and stiffness characteristics of the system under study.

Let us consider a rigid mass m, free in space, upon which a force f is applied. The corresponding equation of motion is given by Newton's law

$$m \ddot{x} = f \tag{1.105}$$

and therefore, assuming f is harmonic, the receptance of this simple system is given by

$$\alpha(\omega) = -\frac{1}{\omega^2 m} \tag{1.106}$$

Thus, the magnitude of $\alpha(\omega)$ becomes, in logarithmic terms:

$$\log |\alpha(\omega)| = -\log (m) - 2 \log (\omega) \tag{1.107}$$

A log-log plot of the magnitude of $\alpha(\omega)$ against frequency is therefore a straight line with a slope of -2. Conversely, considering a simple isolated massless spring element, the corresponding receptance is given by

$$\alpha(\omega) = \frac{1}{k} \qquad (1.108)$$

and therefore

$$\log|\alpha(\omega)| = -\log(k) \qquad (1.109)$$

which means that a log-log plot of the magnitude of $\alpha(\omega)$ against frequency is also a straight line but now with zero slope. Figure 1.24 presents the log-log magnitude against frequency plot of the receptance of a SDOF system, like the one represented in figure 1.1, taken as having 1 kg mass and 100 N/m spring stiffness. Damping was assumed as being equal to 0.6 Ns/m, i.e., $\xi = 0.03$.

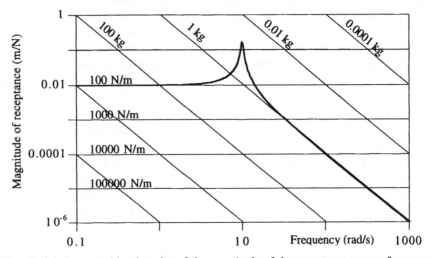

Fig. 1.24 Annotated log-log plot of the magnitude of the receptance *versus* frequency for a system with m = 1 kg, k = 100 N/m and c = 0.6 Ns/m.

The plot is annotated with constant mass and constant stiffness lines. It is obvious that, outside the resonance region ($\omega_n = 10$ rad/s in the given example), the system response is asymptotic to the straight stiffness and mass lines. The meaning of this, for low frequencies ($\omega << \omega_n$), is that the system response is dominated by spring stiffness (at very low frequency values the system approaches a static load/deflection behaviour) whereas for high frequency values ($\omega >> \omega_n$) it is mass inertia that dominates the system response to the forcing excitation.

Hence, if figure 1.24 represented a plot of experimental data, it would be possible to derive the mass and spring spatial characteristics of a SDOF model for the system under analysis. Damping characteristics can be obtained as well, as will be explained further on. Though at a very simple level, this reasoning can be viewed as a first step towards the so-called system identification techniques which aim at deriving the dynamic characteristics from experimental data. Chapter 4 will deal in depth with this subject.

The stiffness and mass lines intersect at a point corresponding to the natural frequency of the system. This is to be expected if we recall that spring force and inertia force cancel when the system is oscillating at its natural frequency ($\omega = \omega_n$), i.e., $k - \omega_n^2 m = 0$. Under these conditions, the only dominant force left in the system to counterbalance the applied force is the damping force. Though not so useful, one could also define damping lines in a log-log plot. The receptance is, in the case where the damping model is viscous, given by

$$\alpha(\omega) = \frac{1}{i \, \omega \, c} \qquad\qquad (1.110)$$

and therefore

$$\log |\alpha(\omega)| = -\log(c) - \log(\omega) \qquad\qquad (1.111)$$

corresponding to a straight line with slope -1. In the case of a hysteretic damping model, the damping line has the same slope as a stiffness line and therefore would have zero slope corresponding to

$$\log |\alpha(\omega)| = -\log(d) = -\log(k) - \log(\eta) \qquad\qquad (1.112)$$

As at resonance the inertia and stiffness forces cancel each other, the damping line passing on the resonant peak of the receptance, in a log-log plot, would yield the damping characteristics of the system. The damping loss factor value is given by the distance between the response peak damping line and the stiffness line corresponding to the low frequency behaviour.

It is important to point out that magnitude log scales are not normally expressed in linear units (as in figure 1.24) but rather in dB (decibel). The dB log scale used in vibration analysis is defined in analogy to the dB log scale used in acoustics.

Taking a given variable such as, say, the receptance amplitude α, its value in dB is defined as

$$\alpha(dB) = 20 \log_{10} \left(\frac{\alpha}{\alpha_{ref}} \right) \qquad\qquad (1.113)$$

where α_{ref} is a reference value that has to be known beforehand. When there is no indication of the reference value to be used it is usual to assume it as unity.

Use of dB scales also leads to referring to the slopes in terms of the variation of the dB value per octave or per decade. An octave corresponds to doubling (or halving) the frequency, whereas a decade corresponds to multiplying (or dividing) the frequency value by 10. For example, a mass line, in a receptance plot, has a negative slope of 12 dB/octave or 40 dB/decade; the same line in a mobility (see next section) plot, has a negative slope of 6 dB/octave or 20 dB/decade.

1.3.2 Alternative forms of the FRF

The dynamic properties of a system may be expressed in terms of any convenient response characteristics, and not necessarily in terms of displacement as we have been doing so far.

Usually vibration is measured in terms of motion and therefore the corresponding FRF may be presented in terms of displacement, velocity or acceleration. This simple motion-force relationship is also found to be described, in some old literature, not by the ratio motion/force but by its inverse i.e., the ratio force/motion. To make matters worse, nomenclature may be found to vary from author to author despite a current effort to standardise. In order to avoid confusion, the terminology used in this book is the following one:

$$\alpha(\omega) = \frac{\text{displacement response}}{\text{force excitation}} = \text{Receptance}$$

$$Y(\omega) = \frac{\text{velocity response}}{\text{force excitation}} = \text{Mobility}$$

$$A(\omega) = \frac{\text{acceleration response}}{\text{force excitation}} = \text{Accelerance}$$

$$\frac{\text{force excitation}}{\text{displacement response}} = \text{Dynamic Stiffness}$$

$$\frac{\text{force excitation}}{\text{velocity response}} = \text{Mechanical Impedance}$$

$$\frac{\text{force excitation}}{\text{acceleration response}} = \text{Apparent Mass}$$

Accelerance is also commonly referred to as Inertance[†]. The use of the inverse relationships may lead to confusion and therefore should be avoided. The reason for this will be better understood when studying multiple degree-of-freedom (MDOF) systems. Finally, it is important to note that the designation Mobility is also

---

[†] International standardisation recommends the use of the term Accelerance. Inertance is said to be unacceptable because it is in conflict with the common definition of acoustic inertance and also contrary to the implication carried by the word 'inertance'. However, the term inertance keeps being widely used by the modal analysis community.

widely accepted as a general designation for any of the motion/force FRF forms. In a similar way, Mechanical Impedance is used for the inverse relationships.

As displacement, velocity and acceleration are mathematically interrelated response quantities, knowledge of an FRF in terms of any one of the motion parameters will allow immediate derivation of the other FRF forms. Considering harmonic vibration and taking, for example, the mobility as our FRF:

$$Y(\omega) = \frac{\dot{x}(t)}{f(t)} = \frac{i\,\omega\,\overline{X}\,e^{i\omega t}}{F\,e^{i\omega t}} = i\,\omega\frac{\overline{X}}{F} = i\,\omega\,\alpha(\omega) \tag{1.114}$$

Therefore:

$$|Y(\omega)| = \omega\,|\alpha(\omega)|$$
$$\arg\left[Y(\omega)\right] = \arg\left[\alpha(\omega)\right] + \frac{\pi}{2} \tag{1.115}$$

Similarly reasoning, for acceleration, yields:

$$A(\omega) = \frac{\ddot{x}(t)}{f(t)} = \frac{-\omega^2\,\overline{X}\,e^{i\omega t}}{F\,e^{i\omega t}} = -\omega^2\,\alpha(\omega) \tag{1.116}$$

leading to

$$|A(\omega)| = \omega^2\,|\alpha(\omega)|$$
$$\arg\left[A(\omega)\right] = \arg\left[\alpha(\omega)\right] + \pi \tag{1.117}$$

and therefore

$$|A(\omega)| = \omega\,|Y(\omega)|$$
$$\arg\left[A(\omega)\right] = \arg\left[Y(\omega)\right] + \frac{\pi}{2} \tag{1.118}$$

It follows that log-log plots of mobility or accelerance will show some differences with respect to the receptance plot, resulting from the fact that mass and stiffness, though still displayed as straight lines, have different slopes (-1 and 1 respectively for mobility plots; 0 and 2 respectively for accelerance plots) as shown in figures 1.25 and 1.26, where the mobility and accelerance magnitudes of the system described by figure 1.24 are plotted against frequency. Note that in both these graphs a dB scale has been used on the magnitude (vertical) axis.

One must not forget that the magnitude plot does not contain all the information. There is still the need to consider the phase or argument of the complex FRF as shown in figure 1.27. Phase plots may use logarithmic scales

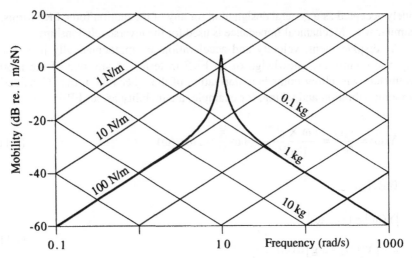

Fig. 1.25 Annotated log-log plot of the magnitude of mobility *versus* frequency for a system with m = 1 kg, k = 100 N/m and c = 0.6 Ns/m.

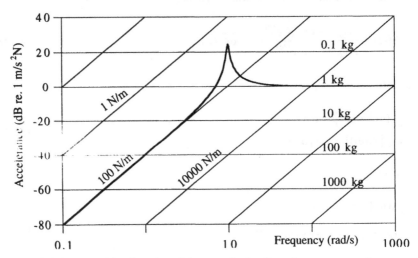

Fig. 1.26 Annotated log-log plot of the magnitude of accelerance *versus* frequency for a system with m = 1 kg, k = 100 N/m and c = 0.6 Ns/m.

only for the frequency axis. All present a phase shift of $180°$ (viscous damping model) near resonance and they differ only on the vertical axis range. Care must be taken not to forget that the hysteretic model receptance phase angle starts at a value which is not exactly zero and therefore the mobility phase angle starts not exactly at $90°$ and the accelerance phase angle starts not exactly at $180°$.

We have shown how FRF log-log Bode plots of the receptance, mobility and accelerance compare and how the mass and stiffness characteristics can be derived from the knowledge that they are plotted as straight lines.

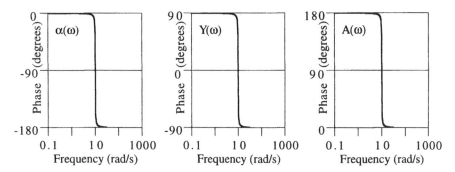

Fig. 1.27 Phase plots of receptance, mobility and accelerance.

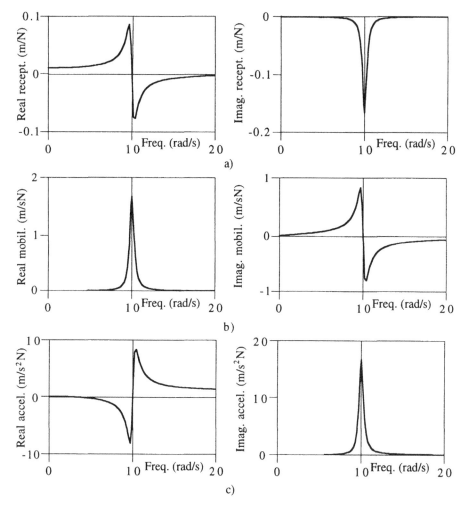

Fig. 1.28 Comparison of the plots of the real and imaginary parts of an FRF for a SDOF system; a) receptance; b) mobility; c) accelerance.

Though not so widely used, it is interesting to compare the companion plots for real and imaginary parts against frequency for the three FRF forms under analysis. Figure 1.28 presents all three forms and it can be seen that they all show that the phase change in the resonance area corresponds to a sign change in one of the parts whereas the other part presents a peak value (either a maximum or a minimum). Unlike the magnitude representation in the Bode plots which is a positive number, the real and imaginary parts can be positive or negative. As a consequence, logarithmic scales cannot be used. Apart from the effect of the small discrepancies already mentioned, viscously and hysteretically damped systems yield similar plots.

Finally, figure 1.29 displays the Nyquist plots corresponding to all three FRF forms describing the viscously damped SDOF system illustrated in figure 1.28. In all cases the plots were modified in order to show discrete data points because, as stated previously, plots of this type cannot be conveniently interpreted if one does not include some means of reading the information on the values of the data points' frequencies. Taking data points equally spaced in frequency, it is clear that only those points that are close to resonant frequency can be distinctly identified. Data points away from resonance are too close together and therefore are not clearly

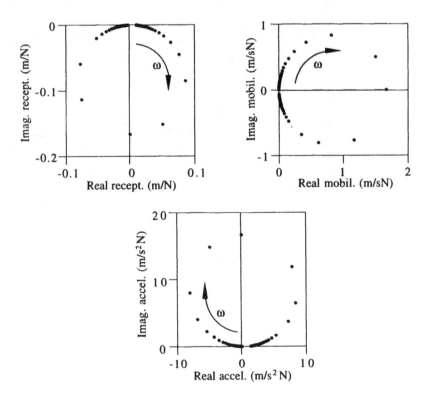

Fig. 1.29 Receptance, mobility and accelerance Nyquist plots of the system described in figure 1.28 (data points are equally spaced in frequency).

identifiable. This particular feature of the Nyquist plot corresponding to the enhancement of the resonant region makes this type of display very convenient in a number of modal testing applications.

However, if one recalls that the circular loop described by the data corresponds to the phase shift suffered by the response relative to the force excitation, and that this phase shift tends to take place within an increasingly narrower frequency range with increasingly lower values of the damping, it is easy to conclude that Nyquist plots cease to be useful when damping is very low. This is shown in figure 1.30 where the receptance of the same SDOF system is plotted for two different damping values. The data points displayed are equally spaced in frequency and the frequency increment is 0.2 rad/s. For the case where $\xi = 0.003$, the data points are all concentrated near the axis origin and there is no visible loop on the display. Nyquist plots are therefore not convenient for use in the analysis of lightly damped systems.

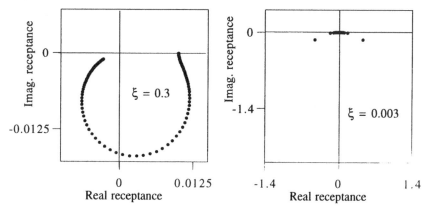

Fig. 1.30 Nyquist plots of the receptance of a given system, for two different values of the damping. Both plots display exactly the same data points, from 0.1 to 20.3 rad/s, with 0.2 rad/s frequency increments.

Another interesting and useful feature of FRF Nyquist plots results from the shape of the path traced out by the data. It is clear from figure 1.29 that, in each plot, the data describes a loop that looks like a circle. This particular behaviour is shown by both the hysteretic and viscous damped models that look very similar at a first glance. In fact, as will be shown later on in this chapter, hysteretically damped systems yield FRF data that plot exactly as a circle when receptance is considered. Mobility and Accelerance data plots are distorted circles. Conversely, for viscously damped systems, it is the mobility which traces out an exact circle while receptance and accelerance trace out as distorted circles. In both cases, the amount of distortion depends heavily on the damping values.

1.3.3 Damping estimates. Special properties of the Nyquist plots
We have seen how annotated FRF log-log plots may provide immediate information on the mass and stiffness characteristics of a SDOF model. However,

for a complete definition of our model it is also necessary to derive the damping value. If we recall the definition of the amplification factor Q (equations (1.28) and (1.103)) and how it relates to the magnitude of the receptance

$$|\alpha(\omega)| = \frac{1}{k} Q \tag{1.119}$$

it is obvious that

$$\log Q = \log |\alpha(\omega)| - \log\left(\frac{1}{k}\right) \tag{1.120}$$

and therefore, at resonance ($\omega = \omega_n$),

$$\log Q_{max} = \log\left(\frac{1}{k\,\eta}\right) - \log\left(\frac{1}{k}\right) = -\log(\eta) \tag{1.121}$$

for a hysteretically damped system. This means that the damping loss factor can be easily derived from the distance between the peak amplitude and the stiffness line in a log-log plot as shown in figure 1.31. The same method may be applied to the viscous damping model for which we can assume

$$\log Q_{max} = \log\left(\frac{1}{2\,k\,\xi}\right) - \log\left(\frac{1}{k}\right) = -\log(2\,\xi) \tag{1.122}$$

provided that the damping value is small. Similarly, one could use mobility or accelerance plots.

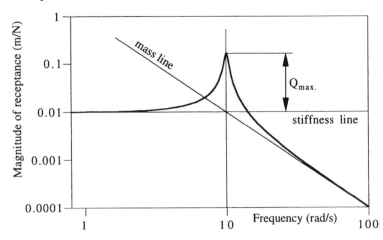

Fig. 1.31 Basic characteristics of a log-log plot of receptance (example with m = 1 kg, k = 100 N/m and c = 0.6 Ns/m).

However, though the location and therefore the value of the resonant frequency can be accurately derived from the receptance plot, the precise value of the peak amplitude is by no means easy to measure. Also, the low frequency region of real receptance data is difficult to define due to measurement errors introduced by background noise and by the electronics of the measuring equipment. This means that the stiffness line may be prone to inaccurate evaluation and therefore Q_{max} will also be affected by this inaccuracy, due to the testing procedures and inherent difficulties.

An alternative means of evaluating the damping value is based on the use of the so-called half-power points. Taking, for example, a SDOF system with hysteretic damping under steady-state harmonic vibration, the energy dissipated per cycle of oscillation (see Section 1.2.8) at resonance is given by

$$\Delta E_{max} = \pi \, X_{max}^2 \, d = \pi \, |\alpha(\omega)|_{max}^2 \, F^2 \, k \, \eta \qquad (1.123)$$

Now, considering measurements on the flanks of the response peak (where damping is still effective) for which the energy dissipated per cycle of vibration is half of ΔE_{max}, we have two such points (figure 1.32). For these data points, denoted by subscripts 1 and 2,

$$\Delta E_{1,2} = \frac{\Delta E_{max}}{2} \qquad (1.124)$$

and therefore

$$|\alpha(\omega)|_{1,2} = \frac{|\alpha(\omega)|_{max}}{\sqrt{2}} \qquad (1.125)$$

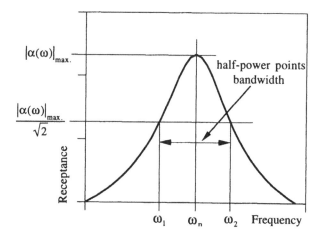

Fig. 1.32 Half-power points definition.

These points are called the half-power points (though, in fact, one should be speaking of energy and not of power). From (1.101) and (1.103) it can be easily found that

$$\eta = \frac{\omega_2^2 - \omega_1^2}{2\,\omega_n^2} \qquad (1.126)$$

Equation (1.126) is exact and allows η to be calculated based only on frequency values. Recalling that, for low damping, $\eta \approx 2\xi$, the previous reasoning is also adequate for the derivation of viscous damping characteristics, though we would be dealing with an equation which is no longer exact. Low damping means also that, if we take (1.126) and rewrite it as

$$\eta = \frac{(\omega_2 - \omega_1)(\omega_2 + \omega_1)}{2\,\omega_n^2} \qquad (1.127)$$

it is easy to conclude that one may accept, as having sufficient accuracy, that

$$\omega_2 + \omega_1 \approx 2\,\omega_n \qquad (1.128)$$

and therefore

$$\eta = \frac{\omega_2 - \omega_1}{\omega_n} \qquad (1.129)$$

Hence, the damping value may be obtained by simply measuring the frequency bandwidth at the half-power points amplitude and dividing it by ω_n. Note that, when using logarithmic amplitude scales, the half-power points amplitudes are exactly 3 dB lower than the peak amplitude.

Despite using frequency values and though we can measure frequency with sufficient accuracy, all the previous methods also rely on the knowledge of the exact peak amplitude which, as stated before, is not so accurately measured. In order to overcome these difficulties it is necessary to recall the receptance Nyquist plot and look at its shape. Let us take equation (1.101) and rewrite it as:

$$\alpha(\omega) = \frac{k - \omega^2\,m}{(k - \omega^2\,m)^2 + d^2} - i\,\frac{d}{(k - \omega^2\,m)^2 + d^2} \qquad (1.130)$$

$$= \mathrm{Re}\left[\alpha(\omega)\right] + i\,\mathrm{Im}\left[\alpha(\omega)\right]$$

Knowing that

$$\left\{\mathrm{Re}\left[\alpha(\omega)\right]^2 + \mathrm{Im}\left[\alpha(\omega)\right]^2\right\} - \left\{\mathrm{Re}\left[\alpha(\omega)\right]^2 + \mathrm{Im}\left[\alpha(\omega)\right]^2\right\} = 0 \qquad (1.131)$$

we obtain

$$\left\{ \operatorname{Re}\left[\alpha(\omega)\right]^2 + \operatorname{Im}\left[\alpha(\omega)\right]^2 \right\} - \frac{1}{\left(k - \omega^2 m\right)^2 + d^2} = 0 \tag{1.132}$$

and therefore

$$\operatorname{Re}\left[\alpha(\omega)\right]^2 + \left\{ \operatorname{Im}\left[\alpha(\omega)\right] + \frac{1}{2\,d} \right\}^2 = \left(\frac{1}{2\,d}\right)^2 \tag{1.133}$$

Equation (1.133) is nothing but the equation of a circle. Therefore the locus of the receptance data on the Argand plane, for a SDOF system with hysteretic damping, is an exact circle with radius $1/2\,d$ and centre on the imaginary axis at $-1/2\,d$ (figure 1.33). Thus, the circle passes through the real and imaginary axis origin and therefore the resonant (natural) frequency corresponds to the lower point where the circle crosses the imaginary axis (at $\operatorname{Im}\left[\alpha(\omega)\right] = -1/d$).

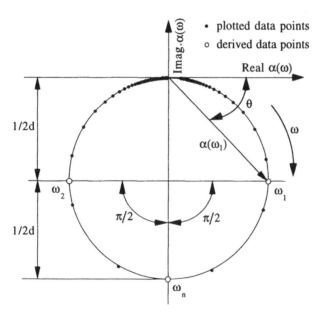

Fig. 1.33 Nyquist plot of the receptance for a SDOF system with hysteretic damping.

This means that, even without having measured it, one may derive both the resonant frequency and the resonant amplitude directly from the Nyquist plot. Due to the geometric properties of this plot, the half-power points (ω_1 and ω_2) are also easy to locate as they correspond to the points where the circle is intersected by its diameter parallel to the real axis.

These particular features of the receptance Nyquist plot, for the case where damping is hysteretic, make it very popular as a simple means of deriving the dynamic properties of a system from experimental data. In Section 4.4.1 this subject will be discussed in greater detail, and will take MDOF systems into account.

Mobility and accelerance data for hysteretically damped systems do not plot as exact circles on the complex plane. The same is true if one considers receptance and accelerance for viscously damped systems. However, the mobility FRF of a SDOF system with viscous damping will plot as an exact circle in the Argand plane. In fact, for such a system, the mobility is described by

$$Y(\omega) = \frac{i\,\omega}{(k - \omega^2\,m) + i\,\omega\,c} \tag{1.134}$$

or

$$Y(\omega) = \frac{\omega^2\,c}{(k - \omega^2\,m)^2 + (\omega\,c)^2} + i\,\frac{\omega\,(k - \omega^2\,m)}{(k - \omega^2\,m)^2 + (\omega\,c)^2}$$

$$= \mathrm{Re}[Y(\omega)] + i\,\mathrm{Im}[Y(\omega)] \tag{1.135}$$

and after some mathematical manipulations similar to the ones performed for hysteretic damping we obtain

$$\left\{\mathrm{Re}[Y(\omega)] - \frac{1}{2\,c}\right\}^2 + \left\{\mathrm{Im}[Y(\omega)]\right\}^2 = \left(\frac{1}{2\,c}\right)^2 \tag{1.136}$$

Thus a plot of the real part against the imaginary part of the mobility $Y(\omega)$ traces out a circle on the Argand plane (figure 1.34). This circle is of radius $1/2\,c$ and its centre is on the real axis at $\mathrm{Re}[Y(\omega)] = 1/2\,c$. The resonant frequency ω_n is derived from the point where the circle crosses the real axis.

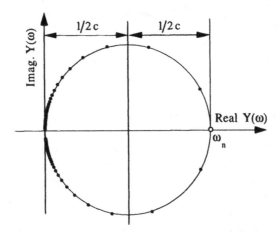

Fig. 1.34 Nyquist plot of the mobility for a SDOF system with viscous damping.

1.4   MULTIPLE DEGREE-OF-FREEDOM (MDOF) SYSTEMS

In the previous sections, the simplest possible discretisation of a system, denoted as a SDOF system, was introduced as a model capable of describing its dynamic behaviour in the simplest possible terms. The advantage of this initial approach is that it makes it a lot easier to understand most of the basic concepts and their physical meaning. However, most real mechanical systems and structures cannot be modelled successfully by assuming a single degree-of-freedom, i.e., a single coordinate to describe their vibratory motion.

Real structures are continuous and nonhomogeneous elastic systems which have an infinite number of degrees of freedom. Therefore, their analysis always entails an approximation which consists of describing their behaviour through the use of a finite number of degrees of freedom, as many as necessary to ensure enough accuracy. Adequate choice of the motion coordinates corresponds therefore to an initial decision that the analyst must take, which is of paramount importance as the success of the subsequent analysis depends on this choice.

Usually, continuous nonhomogeneous structures are described as lumped-mass (i.e., discretised) multiple degree-of-freedom (MDOF) systems. It is worth recalling that the degrees of freedom of a system are the number of independent coordinates necessary to completely describe the motion of that system. For example, let us consider the model in figure 1.35 representing a viscously damped system described by its spatial mass, stiffness and damping properties. A total of N coordinates $x_i(t)$ $(i = 1, 2, ..., N)$ are required to describe the position of the N masses relative to their static equilibrium positions and the system is said to have N degrees of freedom.

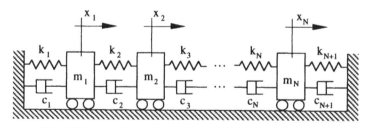

Fig. 1.35   Example of a model with N degrees of freedom.

Assuming that each mass may be forced to move by an external force $f_i(t)$ $(i = 1, 2, ..., N)$ and establishing the equilibrium of the forces acting on them, the motion of the system is found to be governed by the following system of simultaneous equations:

$$m_1\,\ddot{x}_1 + (c_1 + c_2)\,\dot{x}_1 - c_2\,\dot{x}_2 + (k_1 + k_2)\,x_1 - k_2\,x_2 = f_1$$

$$m_2\,\ddot{x}_2 - c_2\,\dot{x}_1 + (c_2 + c_3)\,\dot{x}_2 - c_3\,\dot{x}_3 - k_2\,x_1 + (k_2 + k_3)\,x_2 - k_3\,x_3 = f_2$$

$$\vdots$$

$$m_N\,\ddot{x}_N - c_N\,\dot{x}_{N-1} + (c_N + c_{N+1})\,\dot{x}_N - k_N\,x_{N-1} + (k_N + k_{N+1})\,x_N = f_N$$

$$(1.137)$$

Equations (1.137) consist of N second order differential equations, each of which requires two initial conditions in order to resolve the complete response at the N motion coordinates. It is immediately obvious that neither equation can be solved by itself because they are coupled, i.e., the motion response at a single coordinate depends on the motion at the other coordinates. This dependency is expressed by the fact that each equation includes terms involving more than one coordinate. A convenient method for solving the above system of equations is to use matrices:

$$
\begin{bmatrix}
m_1 & 0 & \cdots & 0 \\
0 & m_2 & \cdots & 0 \\
\vdots & \vdots & \ddots & \vdots \\
0 & 0 & \cdots & m_N
\end{bmatrix}
\begin{Bmatrix}
\ddot{x}_1 \\
\ddot{x}_2 \\
\vdots \\
\ddot{x}_N
\end{Bmatrix}
+
\begin{bmatrix}
c_1+c_2 & -c_2 & \cdots & 0 \\
-c_2 & c_2+c_3 & \cdots & 0 \\
\vdots & \vdots & \ddots & \vdots \\
0 & 0 & \cdots & c_N+c_{N+1}
\end{bmatrix}
\begin{Bmatrix}
\dot{x}_1 \\
\dot{x}_2 \\
\vdots \\
\dot{x}_N
\end{Bmatrix}
$$

$$
+
\begin{bmatrix}
k_1+k_2 & -k_2 & \cdots & 0 \\
-k_2 & k_2+k_3 & \cdots & 0 \\
\vdots & \vdots & \ddots & \vdots \\
0 & 0 & \cdots & k_N+k_{N+1}
\end{bmatrix}
\begin{Bmatrix}
x_1 \\
x_2 \\
\vdots \\
x_N
\end{Bmatrix}
=
\begin{Bmatrix}
f_1 \\
f_2 \\
\vdots \\
f_N
\end{Bmatrix}
$$

(1.138)

which can be condensed in a more compact form, as

$$[M]\{\ddot{x}\}+[C]\{\dot{x}\}+[K]\{x\}=\{f\} \tag{1.139}$$

where [M], [C] and [K] are NxN mass, damping and stiffness symmetric matrices, respectively, describing the spatial properties of the system. The column matrices $\{\ddot{x}\}$, $\{\dot{x}\}$ and $\{x\}$ are Nx1 vectors of time-varying acceleration, velocity and displacement responses, respectively, and $\{f\}$ is an Nx1 vector of the time-varying external excitation forces.

1.4.1 Natural Frequencies and Mode Shapes

We have seen that when an underdamped SDOF system is subjected to an initial perturbation and subsequently left free to move, it oscillates around the static equilibrium position in what can be described as its own natural mode of vibration.

Dynamically, the system was fully characterised through an unique property described by its free vibration natural frequency. We shall examine what happens now that we have a MDOF system.

Undamped MDOF systems

Let us start by assuming that the system is undamped and consider the free vibration solution of (1.139) which becomes:

$$[M]\{\ddot{x}\}+[K]\{x\}=\{0\} \tag{1.140}$$

Given the fact that the N simultaneous equations in (1.140) are homogeneous, it will be seen that if $x_1(t)$, $x_2(t)$, ..., $x_N(t)$ represent a solution $\{x\}$ of the system, then $\gamma x_1(t)$, $\gamma x_2(t)$, ..., $\gamma x_N(t)$, where γ is an arbitrary non-zero constant, also represent a solution. This means that the solution of (1.140) can only be found in terms of relative motions.

It is known that (1.140) has solutions where the time-dependent motions of the system coordinates are synchronous, i.e., they all obey the same time-variation law, and that those solutions are of the form

$$\{x(t)\} = \{\overline{X}\} e^{i\omega t} \tag{1.141}$$

where $\{\overline{X}\}$ is an Nx1 vector of time-independent response amplitudes (note that allowance is made for the amplitudes to be complex). Substituting into (1.140), we obtain

$$\left[[K] - \omega^2 [M]\right]\{\overline{X}\} e^{i\omega t} = \{0\} \tag{1.142}$$

As $e^{i\omega t} \neq 0$ for any instant of time t, then:

$$\left[[K] - \omega^2 [M]\right]\{\overline{X}\} = \{0\} \tag{1.143}$$

What we have in (1.143) is a generalised eigenvalue problem. It is clear now that if a solution $\{\overline{X}\}$ exists then $\gamma\{\overline{X}\}$, where γ is an arbitrary non-zero constant, is also a solution. Pre-multiplying both sides of (1.143) by the inverse of $[[K] - \omega^2 [M]]$, we obtain

$$\left[[K] - \omega^2 [M]\right]^{-1} \left[[K] - \omega^2 [M]\right]\{\overline{X}\} = \left[[K] - \omega^2 [M]\right]^{-1} \{0\} \tag{1.144}$$

i.e., if $[[K] - \omega^2 [M]]^{-1}$ exists,

$$\{\overline{X}\} = \{0\} \tag{1.145}$$

which corresponds to the so-called trivial solution of (1.143) and is of no interest because it implies that there is no motion at all. As a consequence, for (1.143) to have a non-trivial solution, the inverse of $[[K] - \omega^2 [M]]$ must not exist and therefore, for this condition to be satisfied,

$$\det\left[[K] - \omega^2 [M]\right] = 0 \tag{1.146}$$

where 'det' stands for determinant. This is an algebraic equation, known as the characteristic equation for the system, which yields N possible positive real solutions ω_1^2, ω_2^2, ..., ω_N^2 also known as the eigenvalues of (1.143). The values ω_1, ω_2, ..., ω_N are the undamped natural frequencies of our system.

Substituting each natural frequency value in (1.143) and solving each of the resulting sets of equations for $\{\overline{X}\}$, we obtain N possible vector solutions $\{\psi_r\}$ (r = 1, 2, ..., N), known as the mode shapes of the system under analysis, which are the eigenvectors of our problem.

Each $\{\psi_r\}$ contains N elements that are real quantities (positive or negative) and are only known in relative terms. Therefore we know the direction of the vectors but not their absolute magnitude.

In physical terms, we have found that our system can indeed vibrate freely, with synchronous motion, for N particular frequency values ω_r, each of which implies a particular configuration or 'shape' of the free motion, described by $\{\psi_r\}$. Each pair ω_r and $\{\psi_r\}$ is known as a mode of vibration of the system[†]. The subscript r denotes the mode number and varies from 1 to N.

A graphical representation of a model in its static equilibrium position, superimposed on the same model with the coordinates displaced by values proportional to the values of the elements of $\{\psi_r\}$, is often used as it gives a clear view of how the system moves at that particular mode. This representation is very easy to perform given the fact that the elements in $\{\psi_r\}$ are real (positive or negative), the change in their sign indicating a phase shift of 180°, i.e., that the motion is in opposite directions.

In figures 1.36 a) and b) an example is given where the floor of a railway car shell, modelled as a simple structure built up from plates and a beam frame, is displayed showing its two first modes of vibration. It is immediately recognisable that the first mode shape corresponds to a simple bending motion whereas the second mode shape corresponds to torsion.

The complete free vibration solution is very often expressed in two NxN matrices

$$\left[\,\raisebox{0.5ex}{$\diagdown$}\,\omega_r^2\,\raisebox{-0.5ex}{$\diagdown$}\, \right] = \begin{bmatrix} \omega_1^2 & 0 & \cdots & 0 \\ 0 & \omega_2^2 & \cdots & 0 \\ \vdots & \vdots & \ddots & \vdots \\ 0 & 0 & \cdots & \omega_N^2 \end{bmatrix} \tag{1.147}$$

and

$$[\Psi] = \begin{bmatrix} \{\psi_1\} & \{\psi_2\} & \cdots & \{\psi_N\} \end{bmatrix} \tag{1.148}$$

which contain a full description of the dynamic characteristics of the system. Hence, (1.147) and (1.148) constitute what is known as the Modal Model, i.e.,

---

[†] Though normally we just call mode shape to $\{\psi_r\}$.

they describe the system through its modal properties (natural frequencies and mode shapes), as opposed to the Spatial Model where the system was described by its spatial properties ([M], [C] and [K]). [Ψ] is commonly known as the modal matrix.

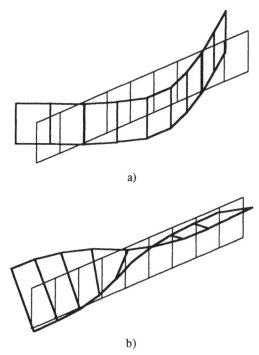

a)

b)

Fig. 1.36 Example of the graphical representation of the two first modes of vibration of a model (extracted from experimental data) for the floor of a railway car shell; a) first bending mode; b) first torsion mode.

The mode shape vectors, being nothing but the eigenvectors satisfying the symmetric eigenvalue problem described by (1.143), possess special and very important properties known as the orthogonality properties. Taking (1.143) and two particular modes r and s we may write

$$\left[[K] - \omega_r^2 [M]\right] \{\psi_r\} = \{0\} \tag{1.149}$$

and

$$\left[[K] - \omega_s^2 [M]\right] \{\psi_s\} = \{0\} \tag{1.150}$$

Pre-multiplying (1.149) by $\{\psi_s\}^T$ it follows that

$$\{\psi_s\}^T \left[[K] - \omega_r^2 [M]\right] \{\psi_r\} = 0 \tag{1.151}$$

On the other hand, if we transpose (1.150) and post-multiply it by $\{\psi_r\}$ we obtain:

$$\{\psi_s\}^T \left[[K]^T - \omega_s^2 [M]^T\right]\{\psi_r\} = 0 \qquad (1.152)$$

which is the same as

$$\{\psi_s\}^T \left[[K] - \omega_s^2 [M]\right]\{\psi_r\} = 0 \qquad (1.153)$$

due to the fact that [M] and [K] are symmetric. Combining (1.151) and (1.153), yields

$$(\omega_r^2 - \omega_s^2)\{\psi_s\}^T [M]\{\psi_r\} = 0 \qquad (1.154)$$

which can only be satisfied for $\omega_r \neq \omega_s$ if

$$\{\psi_s\}^T [M]\{\psi_r\} = 0; \qquad (r \neq s) \qquad (1.155)$$

In addition, from (1.155) and (1.153) it follows that

$$\{\psi_s\}^T [K]\{\psi_r\} = 0; \qquad (r \neq s) \qquad (1.156)$$

Finally, if we take $r = s$ and consider either (1.151) or (1.153), we obtain

$$\{\psi_r\}^T [K]\{\psi_r\} = \omega_r^2 \{\psi_r\}^T [M]\{\psi_r\} \qquad (1.157)$$

or

$$\omega_r^2 = \frac{\{\psi_r\}^T [K]\{\psi_r\}}{\{\psi_r\}^T [M]\{\psi_r\}} = \frac{k_r}{m_r} \qquad (1.158)$$

where k_r and m_r are commonly known as the modal or generalised stiffness and mass, respectively, of mode r.

Thus, considering all the possible combinations of r and s we may state the modal model orthogonality properties as follows:

$$[\Psi]^T [M][\Psi] = \left[\,\ddots\, m_r \,\ddots\,\right]$$
$$[\Psi]^T [K][\Psi] = \left[\,\ddots\, k_r \,\ddots\,\right] \qquad (1.159)$$

The mode shape vectors, due to their orthogonality properties, are linearly independent (no vector in the set can be obtained by a linear combination of the remaining ones) and therefore they form a basis in the N-dimensional space. As a consequence, any other vector in the same space can be expressed as a linear combination of the N linearly independent mode shape vectors. This statement constitutes what is usually known as the expansion theorem and its usefulness will become clear later on when we find the response of MDOF systems to arbitrary forcing conditions.

It has been shown that, in contrast with the natural frequencies which are unique fixed quantities, the mode shapes are known within an indeterminate scaling factor. Thus, k_r and m_r cannot be taken separately as their values are also known within a scaling factor. It is the ratio $k_r/m_r = \omega_r^2$ that has a well defined value. Presentation of the mode shape vectors is therefore always subjected to a previous scaling or normalisation procedure. This normalisation is often based on making the largest element in each vector equal to unity. However, in modal analysis, it is common to scale the mode shape vectors so that

$$[\Phi]^T[M][\Phi] = [I] \tag{1.160}$$

where $[I]$ is the identity matrix (which, by definition, is diagonal) and $[\Phi]$ is the mass-normalised modal matrix built up from mode shape vectors $\{\phi_r\} = \gamma_r\{\psi_r\}$ each of which obey the relationship

$$\begin{aligned}\{\phi_r\}^T[M]\{\phi_r\} &= \{\gamma_r\,\psi_r\}^T[M]\{\gamma_r\,\psi_r\} \\ &= \gamma_r^2\{\psi_r\}^T[M]\{\psi_r\} = 1\end{aligned} \tag{1.161}$$

for each mode r. From (1.161) we obtain

$$\gamma_r = \frac{1}{\sqrt{\{\psi_r\}^T[M]\{\psi_r\}}} = \frac{1}{\sqrt{m_r}} \tag{1.162}$$

Therefore, the mass-normalised modal matrix orthogonality properties may be described by:

$$[\Phi]^T[M][\Phi] = [I]$$
$$[\Phi]^T[K][\Phi] = \left[\,\ddots\,\omega_r^2\,\ddots\,\right] \tag{1.163}$$

These particular properties of the modal matrix may be used to our advantage in order to find the free vibration solution of (1.140). Let us therefore define the coordinate transformation

$$\{x(t)\} = [\Phi]\{q(t)\} \tag{1.164}$$

and substitute into (1.140)

$$[M][\Phi]\{\ddot{q}(t)\} + [K][\Phi]\{q(t)\} = \{0\} \qquad (1.165)$$

Pre-multiplying (1.165) by $[\Phi]^T$,

$$[\Phi]^T[M][\Phi]\{\ddot{q}(t)\} + [\Phi]^T[K][\Phi]\{q(t)\} = \{0\} \qquad (1.166)$$

and taking into consideration (1.163), it follows that our matrix representation of the set of equations of motion becomes

$$\{\ddot{q}(t)\} + \left[`\omega_r^2 \, \right]\{q(t)\} = \{0\} \qquad (1.167)$$

which represents a set of N uncoupled SDOF equations of motion. Note that if we used an arbitrarily scaled modal matrix $[\Psi]$ we would arrive at

$$\left[`m_r \, \right]\{\ddot{q}(t)\} + \left[ `k_r \, \right]\{q(t)\} = \{0\} \qquad (1.168)$$

Thus, through a simple coordinate transformation, our MDOF system has been transformed into N independent SDOF systems (each equation of motion depends solely on a coordinate $q_i(t)$), that can be solved separately using the methods described in Section 1.2.1 (the initial conditions in the new coordinate system $q_i(t)$ are obtained from the application of the coordinate transformation to the initial conditions stated in the original set of coordinates $x_i(t)$).

After solving (1.167) (or (1.168)) for $q_i(t)$, the final free vibration solution, in terms of $x_i(t)$, is easily obtained through the coordinate transformation (1.164). The response coordinates $\{q(t)\}$ are known as the modal or principal coordinates and the mode shape vectors $\{\phi_r\}$ are said to represent the normal modes of the system.

Viscously damped MDOF systems

Recalling the equations of motion of our general MDOF system with viscous damping, as described by (1.139), assuming $\{f\} = \{0\}$ and applying the same techniques as before, based on the modal matrix for the undamped system, we obtain

$$[\Phi]^T[M][\Phi]\{\ddot{q}(t)\} + [\Phi]^T[C][\Phi]\{\dot{q}(t)\} + [\Phi]^T[K][\Phi]\{q(t)\} = \{0\} \qquad (1.169)$$

or

$$\{\ddot{q}(t)\} + [\mathcal{C}]\{\dot{q}(t)\} + \left[`\omega_r^2 \, \right]\{q(t)\} = \{0\} \qquad (1.170)$$

where $[\mathscr{C}]$ is, in general, a non-diagonal NxN matrix. This characteristic, in simple terms, is explainable by the fact that the modal matrix $[\Phi]$ was derived using only mass and stiffness information. It might be said that the mode shape vectors $\{\phi_r\}$ 'knew' nothing about $[C]$ when they were calculated, so there is no reason for them to diagonalise the damping matrix. We are, therefore, confronted with a difficult problem due to the fact that damping is providing additional coupling between the equations of motion which cannot be decoupled by the above modal transformation. Before advancing to see how to deal with this problem, let us analyse the particular situations where damping is said to be proportional.

Proportional damping may be defined as a dissipative situation where the viscous damping matrix $[C]$ is directly proportional to the stiffness matrix, to the mass matrix or to a linear combination of both. Considering the more general case of proportional damping, we may write

$$[C] = \varepsilon[K] + v[M] \tag{1.171}$$

where ε and v are constants. It is immediately obvious that, for this case, the undamped modal matrix orthogonality properties will lead to

$$\{\ddot{q}(t)\} + [\Phi]^T [\varepsilon[K] + v[M]][\Phi]\{\dot{q}(t)\} + \left[{}^\diagdown \omega_r^2 {}_\diagdown \right]\{q(t)\} = \{0\} \tag{1.172}$$

or

$$\{\ddot{q}(t)\} + \left[{}^\diagdown v + \varepsilon\omega_r^2 {}_\diagdown \right]\{\dot{q}(t)\} + \left[{}^\diagdown \omega_r^2 {}_\diagdown \right]\{q(t)\} = \{0\} \tag{1.173}$$

and therefore, taking the SDOF system for analogy, we may write

$$\{\ddot{q}(t)\} + \left[{}^\diagdown 2\xi_r\omega_r {}_\diagdown \right]\{\dot{q}(t)\} + \left[{}^\diagdown \omega_r^2 {}_\diagdown \right]\{q(t)\} = \{0\} \tag{1.174}$$

where

$$\xi_r = \frac{v}{2\omega_r} + \frac{\varepsilon\omega_r}{2} \qquad r = 1, 2, ..., N \tag{1.175}$$

is defined as a modal damping ratio for mode r. We have now in (1.174) a set of N uncoupled damped SDOF equations each of which can be solved using the methods described in Section 1.2.1. Again, the final free vibration response will be derived using the coordinate transformation relationship (1.164) and knowledge of the initial conditions.

In general, damping is not proportional (known as non-proportional damping) and we end up with equation (1.170) and a non-diagonal matrix $[\mathscr{C}]$. In many situations, when damping is small, it is acceptable to neglect the off-diagonal elements in $[\mathscr{C}]$ without a great loss in accuracy and an approximate solution may

58

be reached. However, if damping is large, such an approximation cannot be done. The correct approach is then to consider the homogeneous form of (1.139) and assume a general solution of the form

$$\{x(t)\} = \{\overline{X}\} e^{st} \tag{1.176}$$

Substituting into

$$[M]\{\ddot{x}\} + [C]\{\dot{x}\} + [K]\{x\} = \{0\} \tag{1.177}$$

we obtain:

$$\left[s^2 [M] + s [C] + [K] \right] \{\overline{X}\} = \{0\} \tag{1.178}$$

which constitutes a complex eigenproblem. A more convenient way of solving (1.177) is to define a complex state vector u(t) as

$$u(t) = \begin{Bmatrix} \{x(t)\} \\ \{\dot{x}(t)\} \end{Bmatrix} \tag{1.179}$$

Rewriting (1.177) in terms of this new variable we obtain

$$\begin{bmatrix} [C] & [M] \\ [M] & [0] \end{bmatrix} \{\dot{u}(t)\} + \begin{bmatrix} [K] & [0] \\ [0] & -[M] \end{bmatrix} \{u(t)\} = \{0\} \tag{1.180}$$

or, more simply

$$[A]\{\dot{u}(t)\} + [B]\{u(t)\} = \{0\} \tag{1.181}$$

This formulation is often called the state-space analysis, by contrast with the usual vector-space analysis. [A] and [B] are 2Nx2N real symmetric matrices. In the vector space one looks for a solution in the form (1.176) where $\{\overline{X}\}$ is an Nx1 complex vector representing the amplitude of the response and s is a complex quantity. Hence,

$$\{u(t)\} = \begin{Bmatrix} \{\overline{X}\} \\ s\{\overline{X}\} \end{Bmatrix} e^{st} = \{\overline{U}\} e^{st} \tag{1.182}$$

and

$$\{\dot{u}(t)\} = \begin{Bmatrix} s\{\overline{X}\} \\ s^2\{\overline{X}\} \end{Bmatrix} e^{st} = s\{\overline{U}\} e^{st} \tag{1.183}$$

Substituting (1.182) and (1.183) into (1.181) we obtain, for all time t

$$[s[A]+[B]]\{\overline{U}\}=\{0\} \tag{1.184}$$

representing a generalised eigenproblem whose solution comprises a set of 2N eigenvalues that are real or exist in complex conjugate pairs. For the case in which we are interested, i.e., underdamped systems, the values will always appear in complex conjugate pairs. Denoting the eigenvalues by s_r and s_r^* and the eigenvectors by $\{\psi_r'\}$ and $\{\psi_r'^*\}$, we have

$$\{\psi_r'\}=\left\{\begin{matrix}\{\psi_r\}\\\{\psi_r\}s_r\end{matrix}\right\} \tag{1.185}$$

and

$$\{\psi_r'^*\}=\left\{\begin{matrix}\{\psi_r^*\}\\\{\psi_r^*\}s_r^*\end{matrix}\right\} \tag{1.186}$$

where $\{\psi_r\}$ and $\{\psi_r^*\}$ are the Nx1 complex eigenvectors corresponding to the vector space coordinates $\{x\}$.

As in the case of the undamped or proportionally damped systems, the eigenvectors obey orthogonality properties. Thus, defining now a coordinate transformation

$$\{u(t)\}=[\Psi']\{q(t)\} \tag{1.187}$$

where $[\Psi']$ is a 2Nx2N complex modal matrix (the state-space modal matrix). Substituting into (1.181) and pre-multiplying by $[\Psi']^T$, we obtain

$$[\Psi']^T[A][\Psi']\{\dot{q}(t)\}+[\Psi']^T[B][\Psi']\{q(t)\}=\{0\} \tag{1.188}$$

which becomes

$$\left[\,^\backprime a_r\,\right]\{\dot{q}(t)\}+\left[\,^\backprime b_r\,\right]\{q(t)\}=\{0\} \tag{1.189}$$

Thus, we ended up by having a set of 2N uncoupled equations which is equivalent to having a set of independent 2N SDOF systems. Considering each solution of the form

$$q_r(t)=\overline{Q}_r\,e^{s_r t} \tag{1.190}$$

where $\overline{Q}_r$ depends on the initial conditions, and substituting into (1.187) and (1.189) we have the free vibration response in terms of the state-space coordinates

$$\{u(t)\} = \sum_{r=1}^{2N} \{\psi_r'\} \overline{Q}_r \, e^{s_r t} \tag{1.191}$$

with $s_r = -b_r/a_r$. In (1.191), $\overline{Q}_r$ may be taken as a weighting factor associated with each mode $\{\psi_r'\}$, representing the contribution from each mode to the total response at each coordinate and generally known as the modal participation factor.

We arrived at a solution of the complex eigenproblem where we have 2N eigenvalues occurring in complex conjugate pairs and 2N eigenvectors also occurring in complex conjugate pairs. In other words, we may say that we have a set of N complex eigenvalues s_r and the corresponding N complex eigenvectors $\{\psi_r\}$ plus another set of N complex eigenvalues s_r^* and the corresponding complex eigenvectors $\{\psi_r^*\}$. Each eigenvalue is usually written under the form

$$s_r = -\omega_r \, \xi_r + i \, \omega_r \, \sqrt{1-\xi_r^2} \tag{1.192}$$

taking as analogy the SDOF case. One way to understand the physical meaning of these quantities is to recall that each eigenvalue/eigenvector pair, such as s_r and $\{\psi_r\}$, must satisfy equation (1.178)

$$\left[s_r^2 \, [M] + s_r \, [C] + [K] \right] \{\psi_r\} = \{0\} \tag{1.193}$$

and similarly, for a different pair such as s_p and $\{\psi_p\}$, after transposing the resulting equation

$$\left\{\psi_p\right\}^T \left[s_p^2 \, [M] + s_p \, [C] + [K] \right] = \{0\} \tag{1.194}$$

If we now pre-multiply (1.193) by $\{\psi_p\}^T$ and post-multiply (1.194) by $\{\psi_r\}$, we obtain

$$\left\{\psi_p\right\}^T \left[s_r^2 \, [M] + s_r \, [C] + [K] \right] \{\psi_r\} = 0 \tag{1.195}$$

and

$$\left\{\psi_p\right\}^T \left[s_p^2 \, [M] + s_p \, [C] + [K] \right] \{\psi_r\} = 0 \tag{1.196}$$

Subtracting the resulting equations yields (provided that s_p and s_r are different)

$$(s_r + s_p)\left\{\psi_p\right\}^T [M] \{\psi_r\} + \left\{\psi_p\right\}^T [C] \{\psi_r\} = 0 \tag{1.197}$$

On the other hand, multiplying (1.195) by s_p and (1.196) by s_r and subtracting one from the other leads to

$$s_r \, s_p \left\{ \psi_p \right\}^T [M] \left\{ \psi_r \right\} - \left\{ \psi_p \right\}^T [K] \left\{ \psi_r \right\} = 0 \qquad (1.198)$$

Equations (1.197) and (1.198) constitute a pair of orthogonality conditions that must be satisfied by our MDOF system eigenvectors. Considering (1.192) and assuming that the modes r and p are a complex conjugate pair (i.e., $s_p = s_r^* = -\omega_r \, \xi_r - i \, \omega_r \sqrt{1 - \xi_r^2}$ and $\{\psi_p\} = \{\psi_r^*\}$), and taking into account (1.197) and (1.198), we reach the following expressions:

$$\frac{\left\{ \psi_r^* \right\}^T [C] \left\{ \psi_r \right\}}{\left\{ \psi_r^* \right\}^T [M] \left\{ \psi_r \right\}} = \frac{c_r}{m_r} = 2 \, \omega_r \, \xi_r \qquad (1.199)$$

and

$$\frac{\left\{ \psi_r^* \right\}^T [K] \left\{ \psi_r \right\}}{\left\{ \psi_r^* \right\}^T [M] \left\{ \psi_r \right\}} = \frac{k_r}{m_r} = \omega_r^2 \qquad (1.200)$$

Thus we end up by defining, for our general viscously damped MDOF system, in analogy with the undamped and with the proportionally damped MDOF systems, a modal mass m_r, a modal damping c_r and a modal stiffness k_r. Also, ω_r and ξ_r may be taken as the undamped natural frequency and the damping ratio, respectively, associated with mode r.

Finally, it is important to note that, as in the case of undamped systems, the eigenvectors are not determined in an absolute sense. We have seen that undamped systems exhibit mode shapes with real amplitudes that are only known within a multiplicative constant. Now, in the general case of damped systems, we have complex mode shapes, this meaning that we are dealing both with amplitudes and with phase angles. Thus, the mode shape vectors are not only known within a multiplicative constant, as far as the amplitudes are concerned, but also within a constant angular shift, as far as the phase angles are concerned.

The motion represented by a given mode shape, although synchronous, is not as easy to recognise as in the case of undamped systems. On the other hand, graphical displays such as the ones presented in figure 1.36 are no longer possible, because they are based on the assumption that the motion of each coordinate represented by the eigenvector elements will reach its maximum, either in positive or in negative direction, at exactly the same time. This is only true if the eigenvector elements are real quantities (positive or negative). In the case of damped systems, the existence of phase angles which are not $0°$ or $180°$ means that the coordinates will reach their maximum excursions at different instants of time.

As it is no longer possible to have 'static' graphical displays such as the ones in figure 1.36, the current alternative is to take advantage of computer capabilities, displaying, in sequence, a series of different images of the structure, corresponding to the values of the response coordinates at different instants of time. The result is an 'animated mode shape' display on the screen of the computer, giving the user a very good visual idea of how the structure moves at each of its modes.

Hysteretically damped MDOF systems
Coming back to our MDOF system (figure 1.35) and assuming that the dissipative mechanism is hysteretic, instead of (1.139) we would arrive at

$$[M]\{\ddot{x}(t)\} + i[D]\{x(t)\} + [K]\{x(t)\} = \{f(t)\} \tag{1.201}$$

where [D] is the NxN hysteretic damping matrix. The problem now is the fact that this type of damping has been defined only for the particular case of forced harmonic vibrations and, as stated in 1.2.8, presents some difficulties to rigorous free vibration or shock response analysis. Resolution of (1.201) in the case where $\{f(t)\} = \{0\}$ may be thus considered a questionable decision. However, one must bear in mind that, in most of the problems in modal analysis, we are considering forced vibrations.

Whether or not the free vibration solution is valid is therefore of no immediate concern. What we intend first to look for now is a description of the modes of vibration, i.e., a modal model for our hysteretically damped system. Mathematically there is no difficulty in solving the resulting eigenproblem. Furthermore, we shall see that, when solving for the steady-state forced vibrations, the FRFs may be expressed in terms of the eigenvalues and eigenvectors. Thus, provided we can use the mode description and interpret it in accordance with a physical point of view, despite it being not strictly valid for free vibration, the use of the hysteretic model is justified.

Following a reasoning similar to the viscous damped model, we may start by assuming that damping is proportional, i.e., it can be written as

$$[D] = \varepsilon[K] + v[M] \tag{1.202}$$

where ε and v are constants. Assuming that a solution exists which is of the form

$$\{x(t)\} = \{\overline{X}\} e^{i\lambda t} \tag{1.203}$$

where $\{\overline{X}\}$ is an Nx1 vector of time-independent response amplitudes, and substituting into the homogeneous form of (1.201), we arrive at

$$\left[\left[[K] - \lambda^2[M]\right] + i\left[\varepsilon[K] + v[M]\right]\right]\{\overline{X}\} = \{0\} \tag{1.204}$$

which represents a complex eigenvalue problem leading to a solution in terms of N complex eigenvalues λ_r^2 and N real eigenvectors $\{\psi_r\}$ that are the same as for the undamped case. As the eigenvectors are the same mode shape vectors as for the undamped case, they may be taken as having the same physical meaning. Also, for the case under analysis, it is obvious that their properties of orthogonality will decouple the equations of motion. Thus, we may take λ_r^2 as containing information on the natural frequencies of the system and write

$$\lambda_r^2 = \omega_r^2 (1 + i\, \eta_r) \tag{1.205}$$

where ω_r^2 and η_r are the natural frequency and damping loss factor, respectively, for mode r, defined as being given by

$$\omega_r^2 = \frac{k_r}{m_r} \tag{1.206}$$

and

$$\eta_r = \varepsilon + \frac{v}{\omega_r^2} \tag{1.207}$$

The generalised modal stiffness and modal mass, k_r and m_r respectively, have been already defined in (1.158).

If non-proportional damping is assumed, the corresponding complex eigenproblem yields not only N complex eigenvalues λ_r^2 but also N complex eigenvectors $\{\psi_r\}$. Again, we may choose to write λ_r^2 in the form of equation (1.205) though we have a value for ω_r^2 which may be slightly different from the one in that equation. In fact, what we are doing now is to take ω_r^2 as the real part of λ_r^2 which, due to the orthogonality properties being maintained, is given by the following relationship:

$$\lambda_r^2 = \frac{\{\psi_r\}^T [[K] + i[D]]\{\psi_r\}}{\{\psi_r\}^T [M]\{\psi_r\}} = \frac{k_r}{m_r} \tag{1.208}$$

where k_r and m_r are now complex quantities.

1.4.2 MDOF forced response analysis

We shall now turn our attention to the forced response solution of MDOF systems and of the corresponding set of equations in the cases of both hysteretic and viscous damping.

As in the case of SDOF systems, we are going to neglect the transient part of the complete response and consider solely the steady-state situation. There is an obvious and direct way of deriving the corresponding equations, for the particular case of harmonic excitation. From (1.139), taking the excitation force vector $\{f(t)\} = \{F\}e^{i\omega t}$ and the response vector $\{x(t)\} = \{\overline{X}\}e^{i\omega t}$, we obtain

$$\left[[K] - \omega^2 [M] + i\,\omega\,[C]\right]\{\overline{X}\}\,e^{i\omega t} = \{F\}\,e^{i\omega t} \qquad (1.209)$$

and therefore

$$\{\overline{X}\} = \left[[K] - \omega^2 [M] + i\,\omega\,[C]\right]^{-1}\{F\} = [\alpha(\omega)]\{F\} \qquad (1.210)$$

The same reasoning based on the hysteretic model would lead to

$$\{\overline{X}\} = \left[[K] - \omega^2 [M] + i\,[D]\right]^{-1}\{F\} = [\alpha(\omega)]\{F\} \qquad (1.211)$$

where $[\alpha(\omega)]$ is the NxN system receptance matrix containing all the information on the system dynamic characteristics. Each element α_{jk} of the matrix corresponds to an individual FRF describing the relation between the response at a particular coordinate j and a single force excitation applied at coordinate k. The receptance matrix $[\alpha(\omega)]$ constitutes another form of modelling our system and is known as the Response Model as opposed to the Spatial Model and the Modal Model already mentioned.

Despite their apparent simplicity, equations (1.210) and (1.211) tend to be very inefficient for numerical applications and their usefulness for identification purposes is very limited (see Chapter 4). In fact, although it is possible to calculate the values of $[\alpha(\omega)]$ at any frequency of interest, this operation requires the inversion of an NxN matrix for the chosen frequency value. When dealing with systems with a large number of degrees of freedom (large N), this may be a highly time-consuming operation. The inefficiency of the procedure is enhanced if one is interested only in a limited number of individual receptance elements (individual FRFs). Fortunately, it is possible to derive more useful expressions for $[\alpha(\omega)]$, based on the modal properties, which, as an additional advantage, provide an insight into the form of the various FRF properties. We shall start by considering the hysteretic model which is far easier to manipulate from the mathematical point of view.

Hysteretically damped model

We have seen in the previous section that there is a set of N complex eigenvalues λ_r^2 and associated eigenvectors $\{\psi_r\}$ which satisfy the homogeneous equation

$$\left[[K] + i\,[D] - \lambda_r^2 [M]\right]\{\psi_r\} = \{0\} \qquad (1.212)$$

and that the N eigenvectors form a linearly independent set of vectors in N-space, possessing orthogonality properties. Hence, any vector in N-space, such as $\{\overline{X}\}$, can be expressed as a linear combination of the eigenvectors, i.e.,

$$\left\{\overline{X}\right\} = \sum_{r=1}^{N} \gamma_r \left\{\psi_r\right\} \tag{1.213}$$

Substituting (1.213) into (1.211) and pre-multiplying by $\{\psi_s\}^T$ yields

$$\left\{\psi_s\right\}^T \left[[K]+i[D]\right] \sum_{r=1}^{N} \gamma_r \left\{\psi_r\right\} - \omega^2 \left\{\psi_s\right\}^T [M] \sum_{r=1}^{N} \gamma_r \left\{\psi_r\right\} = \left\{\psi_s\right\}^T \{F\}$$
$$\tag{1.214}$$

Taking now into consideration the orthogonality properties of the eigenvectors, (1.214) becomes

$$\gamma_r \left\{\psi_r\right\}^T \left[[K]+i[D]\right]\left\{\psi_r\right\} - \omega^2 \gamma_r \left\{\psi_r\right\}^T [M]\left\{\psi_r\right\} = \left\{\psi_r\right\}^T \{F\} \tag{1.215}$$

or

$$\gamma_r k_r - \omega^2 \gamma_r m_r = \left\{\psi_r\right\}^T \{F\} \tag{1.216}$$

where k_r and m_r have been defined in (1.208). Thus,

$$\gamma_r = \frac{\left\{\psi_r\right\}^T \{F\}}{k_r - \omega^2 m_r} \tag{1.217}$$

Substituting (1.217) into (1.213) leads to the definition of the steady-state response for hysteretic damping, in terms of the complex modes

$$\left\{x(t)\right\} = \left\{\overline{X}\right\} e^{i\omega t} = \sum_{r=1}^{N} \frac{\left\{\psi_r\right\}^T \{F\}\left\{\psi_r\right\}}{k_r - \omega^2 m_r} e^{i\omega t} \tag{1.218}$$

Recalling (1.205) and (1.208), we have

$$\left\{\overline{X}\right\} = \sum_{r=1}^{N} \frac{\left\{\psi_r\right\}^T \{F\}\left\{\psi_r\right\}}{m_r (\omega_r^2 - \omega^2 + i\,\eta_r\,\omega_r^2)} \tag{1.219}$$

If we are interested in extracting a single receptance element, for example, the response at coordinate j due to a single harmonic force excitation applied at coordinate k, this means that vector $\{F\}$ will have just one non-zero element and therefore we may write

$$\overline{X}_j = \sum_{r=1}^{N} \frac{\psi_{jr}\,F_k\,\psi_{kr}}{m_r (\omega_r^2 - \omega^2 + i\,\eta_r\,\omega_r^2)} \tag{1.220}$$

or

$$\alpha_{jk}(\omega) = \frac{\overline{X}_j}{F_k} = \sum_{r=1}^{N} \frac{\psi_{jr}\,\psi_{kr}}{m_r\,(\omega_r^2 - \omega^2 + i\,\eta_r\,\omega_r^2)} \qquad (1.221)$$

where ψ_{jr} and ψ_{kr} are elements j and k, respectively, of the mode shape vector $\{\psi_r\}$. Thus, we have arrived at a general expression for the elements of the receptance matrix, in terms of the modal properties, that shows some resemblance to the equivalent SDOF equation (1.101). In physical terms, (1.221) may be interpreted as stating that the total response is the result of a summation of contributions from N separated SDOF system responses.

In the general case of non-proportional damping, the numerator of (1.221) is complex whereas in the undamped or proportionally damped cases it is real. Taking into consideration the mass-normalised mode shape vectors

$$\alpha_{jk}(\omega) = \frac{\overline{X}_j}{F_k} = \sum_{r=1}^{N} \frac{\phi_{jr}\,\phi_{kr}}{\omega_r^2 - \omega^2 + i\,\eta_r\,\omega_r^2} \qquad (1.222)$$

or

$$\alpha_{jk}(\omega) = \frac{\overline{X}_j}{F_k} = \sum_{r=1}^{N} \frac{{}_r\overline{A}_{jk}}{\omega_r^2 - \omega^2 + i\,\eta_r\,\omega_r^2} \qquad (1.223)$$

where

$${}_r\overline{A}_{jk} = {}_rA_{jk}\,e^{i\,{}_r\varphi_{jk}} \qquad (1.224)$$

is a complex quantity known as the Modal Constant, for which

$${}_rA_{jk} = \left| \frac{\psi_{jr}\,\psi_{kr}}{m_r} \right| = \left| \phi_{jr}\,\phi_{kr} \right| \qquad (1.225)$$

and

$${}_r\varphi_{jk} = \arg\left(\frac{\psi_{jr}\,\psi_{kr}}{m_r} \right) = \arg\left(\phi_{jr}\,\phi_{kr} \right) \qquad (1.226)$$

are constants for a given r, j and k. Two important conclusions may be extracted from the above derivations. First, it is clear that the receptance matrix is symmetric and therefore

$$\alpha_{jk} = \frac{\overline{X}_j}{F_k} = \alpha_{kj} = \frac{\overline{X}_k}{F_j} \qquad (1.227)$$

this property being known as the principle of reciprocity and second, the modal constants are interrelated, obeying a relationship that is described by the pair of equations

$$_r\overline{A}_{jk} = \phi_{jr}\,\phi_{kr}$$
$$_r\overline{A}_{jj} = \phi_{jr}^2 \quad \text{or} \quad _r\overline{A}_{kk} = \phi_{kr}^2 \tag{1.228}$$

known as the modal constants consistency equations. What (1.227) and (1.228) mean is that if one knows a full line (or column) of the matrix $[\alpha(\omega)]$, then the whole matrix can be evaluated.

Viscously damped model
It has been seen that the viscously damped model described by (1.139), in the general case where damping is non-proportional, is not so easy to deal with. To find the solution of the homogeneous equation we used a formulation known as state-space analysis, converting our NxN complex matrix problem into a 2Nx2N real matrix problem described by (1.181) which, for the forced response analysis, becomes

$$[A]\{\dot{u}(t)\} + [B]\{u(t)\} = \{f'(t)\} \tag{1.229}$$

where

$$\{f'(t)\} = \begin{Bmatrix} \{f(t)\} \\ \{0\} \end{Bmatrix} \tag{1.230}$$

Considering now the coordinate transformation defined by (1.187), equation (1.229) becomes

$$[A][\Psi']\{\dot{q}(t)\} + [B][\Psi']\{q(t)\} = \{f'(t)\} \tag{1.231}$$

Pre-multiplying this equation by $[\Psi']^T$ and considering the orthogonality properties, we obtain

$$\left[\,`a_r`\,\right]\{\dot{q}(t)\} + \left[\,`b_r`\,\right]\{q(t)\} = [\Psi']^T\{f'(t)\} \tag{1.232}$$

which represents a set of 2N uncoupled equations each of which can be written as

$$\dot{q}_r(t) - s_r\,q_r(t) = \frac{1}{a_r}\{\psi_r'\}^T \begin{Bmatrix} \{f(t)\} \\ \{0\} \end{Bmatrix} \tag{1.233}$$

where $s_r = -b_r/a_r$ is usually written under the form (1.192). For a harmonic excitation force vector of the form

$$\{f(t)\} = \{\overline{F}\} e^{i\omega t} \tag{1.234}$$

the response will be of the form

$$\{q(t)\} = \{\overline{Q}\} e^{i\omega t} \tag{1.235}$$

and thus, equation (1.233) becomes

$$(i\,\omega - s_r)\,\overline{Q}_r = \frac{1}{a_r} \{\psi_r'\}^T \begin{Bmatrix} \{\overline{F}\} \\ \{0\} \end{Bmatrix} \tag{1.236}$$

and so,

$$\overline{Q}_r = \left(\frac{1}{i\,\omega - s_r}\right) \frac{1}{a_r} \{\psi_r'\}^T \begin{Bmatrix} \{\overline{F}\} \\ \{0\} \end{Bmatrix} \tag{1.237}$$

Recalling (1.191), into which we may now substitute (1.237), we obtain

$$\{u(t)\} = \sum_{r=1}^{2N} \{\psi_r'\} \left(\frac{1}{i\,\omega - s_r}\right) \frac{1}{a_r} \{\psi_r'\}^T \begin{Bmatrix} \{\overline{F}\} \\ \{0\} \end{Bmatrix} e^{i\omega t} \tag{1.238}$$

or

$$\{u(t)\} = \sum_{r=1}^{2N} \{\phi_r'\} \left(\frac{1}{i\,\omega - s_r}\right) \{\phi_r'\}^T \begin{Bmatrix} \{\overline{F}\} \\ \{0\} \end{Bmatrix} e^{i\omega t} \tag{1.239}$$

where the eigenvectors have been normalised with respect to a_r, i.e.,

$$\{\phi_r'\} = \frac{1}{\sqrt{a_r}} \{\psi_r'\} \tag{1.240}$$

and $\{\phi_r'\}$ is a 2Nx1 eigenvector of the form

$$\{\phi_r'\} = \begin{Bmatrix} \{\phi_r\} \\ \{\phi_r\} s_r \end{Bmatrix} \tag{1.241}$$

Recalling now (1.182) and considering harmonic response (i.e., $s = i\omega$), substituting (1.241) in (1.239) yields

$$\left\{\begin{array}{c}\{\overline{X}\}\\ i\,\omega\,\{\overline{X}\}\end{array}\right\} = \sum_{r=1}^{2N}\left\{\begin{array}{c}\{\phi_r\}\\ \{\phi_r\}\,s_r\end{array}\right\}\left(\frac{1}{i\,\omega - s_r}\right)\left\{\begin{array}{c}\{\phi_r\}\\ \{\phi_r\}\,s_r\end{array}\right\}^T\left\{\begin{array}{c}\{\overline{F}\}\\ \{0\}\end{array}\right\} \tag{1.242}$$

or

$$\left\{\begin{array}{c}\{\overline{X}\}\\ i\,\omega\,\{\overline{X}\}\end{array}\right\} = \sum_{r=1}^{2N}\left\{\begin{array}{c}\{\phi_r\}\\ \{\phi_r\}\,s_r\end{array}\right\}\left(\frac{1}{i\,\omega - s_r}\right)\{\phi_r\}^T\{\overline{F}\} \tag{1.243}$$

Returning to the vector space coordinates and extracting from (1.243) the values of their amplitude responses, it follows that:

$$\{\overline{X}\} = \sum_{r=1}^{2N}\{\phi_r\}\left(\frac{1}{i\,\omega - s_r}\right)\{\phi_r\}^T\{\overline{F}\} \tag{1.244}$$

The modal participation factors for (1.244) correspond now to the product $\{\phi_r\}^T\{\overline{F}\}$. The receptance $\alpha_{jk}(\omega)$, defined as the response displacement at coordinate j due to a force excitation at coordinate k, all other forces being zero, is then given by

$$\alpha_{jk}(\omega) = \frac{\overline{X}_j}{\overline{F}_k} = \sum_{r=1}^{2N}\frac{\phi_{jr}\,\phi_{kr}}{i\,\omega - s_r} \tag{1.245}$$

Because the eigenvalues and eigenvectors appear in complex conjugate pairs, we can write (1.245) under the form

$$\alpha_{jk}(\omega) = \frac{\overline{X}_j}{\overline{F}_k} = \sum_{r=1}^{N}\left(\frac{\phi_{jr}\,\phi_{kr}}{i\,\omega - s_r} + \frac{\phi_{jr}^*\,\phi_{kr}^*}{i\,\omega - s_r^*}\right) \tag{1.246}$$

The similarities between (1.246) and (1.66) are evident and indeed we could have arrived at the above receptance equation by applying the Laplace transform concept to the system equation (1.139).

The eigenvalues s_r are the poles (which occur in complex conjugate pairs) and the products $\phi_{jr}\,\phi_{kr}$ are the residues, for mode r. Once again, as in the case of hysteretic damping, we arrive at an expression which states that the total receptance is the result of the summation of the contributions of the different modes of vibration.

Considering (1.192) and denoting the residues by $_rA_{jk}$[†], equation (1.246) may be rewritten as

$$\alpha_{jk}(\omega) = \sum_{r=1}^{N} \left(\frac{_rA_{jk}}{\omega_r\,\xi_r + i\left(\omega - \omega_r\,\sqrt{1-\xi_r^2}\right)} + \frac{_rA_{jk}^*}{\omega_r\,\xi_r + i\left(\omega + \omega_r\,\sqrt{1-\xi_r^2}\right)} \right)$$

(1.247)

1.5 REPRESENTATION AND PROPERTIES OF MDOF FRFs
1.5.1 General considerations

In Section 1.3 we discussed the different representations and properties of the FRFs for SDOF systems. Though most of what has been said still applies, it is insufficient when the system under analysis is more complex than the simple SDOF system. MDOF systems present new features that must be taken into consideration. A good starting point is to consider, as an example, the system in figure 1.37 (a simple beam) for which two points (1 and 2) have been defined as those where motion coordinates in the y and θ directions are of interest for the analysis (bending of the beam). We are therefore assuming that, in this example, the system dynamic behaviour is described, with sufficient accuracy for our purposes, by four degrees of freedom, denoting the bending motion at the tips of the beam.

Fig. 1.37 Example of a 4 DOF system.

As we know, the dynamic behaviour of this system is described by:

$$\{q\} = \begin{Bmatrix} y_1 \\ \theta_1 \\ y_2 \\ \theta_2 \end{Bmatrix} = \begin{bmatrix} \alpha_{11} & \alpha_{12} & \alpha_{13} & \alpha_{14} \\ \alpha_{21} & \alpha_{22} & \alpha_{23} & \alpha_{24} \\ \alpha_{31} & \alpha_{32} & \alpha_{33} & \alpha_{34} \\ \alpha_{41} & \alpha_{42} & \alpha_{43} & \alpha_{44} \end{bmatrix} \begin{Bmatrix} f_{y_1} \\ f_{\theta_1} \\ f_{y_2} \\ f_{\theta_2} \end{Bmatrix} = [\alpha]\{f\}$$

(1.248)

where $\{q\}$ is a vector of generalised response coordinates and $\{f\}$ is a vector of generalised excitation forces. What is important to consider now is the fact that the

---

[†] It should be noted that, although we are using here a similar notation as for the hysteretically damped system, the mode shape elements ϕ_{jr} are not the same as before and therefore the residue $_rA_{jk}$ should not be confused with what we previously called modal constant. Nevertheless, residues and modal constants are naturally closely related.

symmetric receptance matrix $[\alpha]$ consists of different types of individual receptance elements which must be defined. For example, at point 1 of the example structure we can take the three (note that $\alpha_{12} = \alpha_{21}$) following different receptance elements:

$$\alpha_{11} = \alpha_{y_1 y_1} = \frac{y_1}{f_{y_1}} \quad ; \quad \alpha_{12} = \alpha_{y_1 \theta_1} = \frac{y_1}{f_{\theta_1}} \quad ; \quad \alpha_{22} = \alpha_{\theta_1 \theta_1} = \frac{\theta_1}{f_{\theta_1}} \qquad (1.249)$$

All of the above receptances are called point receptances because they are taken at the same point of the structure. However, α_{11} and α_{22} are direct point receptances, due to the fact that both response and excitation force are considered along the same coordinate, whereas α_{12} is called a cross point receptance. The same applies if we consider the other point of the example structure. Finally, we have receptances relating coordinates at different points of the structure, such as, for example

$$\alpha_{13} = \alpha_{y_1 y_2} = \frac{y_1}{f_{y_2}} \quad ; \quad \alpha_{14} = \alpha_{y_1 \theta_2} = \frac{y_1}{f_{\theta_2}} \qquad (1.250)$$

which are called transfer receptances as opposed to the point receptances previously defined. These definitions apply as well to any other form of the FRFs (mobility, accelerance, dynamic stiffness, mechanical impedance and apparent mass).

Let us now recall Section 1.3.2 where it was stated that the use of the inverse FRF forms, i.e., the FRFs denoting force/motion relationships, should be avoided[†]. Considering the example structure in figure 1.37 and the receptance characteristics described by (1.248), the response at any one coordinate, say y_1, is given by

$$y_1 = \alpha_{11} f_{y_1} + \alpha_{12} f_{\theta_1} + \alpha_{13} f_{y_2} + \alpha_{14} f_{\theta_2} \qquad (1.251)$$

Equation (1.251) shows clearly that measurement of any of the α_{jk} elements poses no problem. For example, suppose one wishes to measure α_{11}. The conditions to be fulfilled are very easy to achieve: it suffices that the only non-zero excitation force applied to the system be f_{y_1}. The same applies to any of the other α_{jk} elements. On the other hand, each α_{jk} element describes a dynamic characteristic of the structure, relating motion response and excitation force at coordinates j and k, respectively, and is totally independent of the number of coordinates one decides to consider.

For instance, let us take our example structure in figure 1.37 and assume that, after having measured all the 4x4 receptance matrix elements described by (1.248), it was found that the 4 DOF model had to be modified to include two more coordinates at a third point of the beam, as exemplified in figure 1.38. The system must now be described by a 6x6 receptance matrix, as follows:

[†] Actually, what should be avoided is the use of 'Free Impedances' as defined in the footnote of next page.

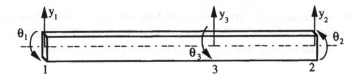

Fig. 1.38 System of figure 1.37 with two extra DOFs.

$$\{q\} = \begin{Bmatrix} y_1 \\ \theta_1 \\ y_2 \\ \theta_2 \\ y_3 \\ \theta_3 \end{Bmatrix} = \begin{bmatrix} \alpha_{11} & \alpha_{12} & \alpha_{13} & \alpha_{14} & \alpha_{15} & \alpha_{16} \\ \alpha_{21} & \alpha_{22} & \alpha_{23} & \alpha_{24} & \alpha_{25} & \alpha_{26} \\ \alpha_{31} & \alpha_{32} & \alpha_{33} & \alpha_{34} & \alpha_{35} & \alpha_{36} \\ \alpha_{41} & \alpha_{42} & \alpha_{43} & \alpha_{44} & \alpha_{45} & \alpha_{46} \\ \alpha_{51} & \alpha_{52} & \alpha_{53} & \alpha_{54} & \alpha_{55} & \alpha_{56} \\ \alpha_{61} & \alpha_{62} & \alpha_{63} & \alpha_{64} & \alpha_{65} & \alpha_{66} \end{bmatrix} \begin{Bmatrix} f_{y_1} \\ f_{\theta_1} \\ f_{y_2} \\ f_{\theta_2} \\ f_{y_3} \\ f_{\theta_3} \end{Bmatrix} = [\alpha]\{f\} \qquad (1.252)$$

and, for example, the response at coordinate y_1 is no longer given by (1.251) but by

$$y_1 = \alpha_{11} f_{y_1} + \alpha_{12} f_{\theta_1} + \alpha_{13} f_{y_2} + \alpha_{14} f_{\theta_2} + \alpha_{15} f_{y_3} + \alpha_{16} f_{\theta_3} \qquad (1.253)$$

Given the fact that the measurement of any α_{jk} is performed by making all the force vector elements equal to zero except f_k, it is immediately obvious that the elements on the initial 4x4 receptance matrix suffered no changes when the number of degrees of freedom was increased. Increasing the size of the receptance matrix and including (measuring) the new matrix elements is sufficient.

Let us now assume that we wanted to measure the elements of one of the inverse FRF relationships, say the dynamic stiffness matrix elements, for the system in figure 1.37. In this case we would have to consider the dynamic behaviour of our system in terms of

$$\{f\} = [\alpha]^{-1}\{q\} = [Z]\{q\} \qquad (1.254)$$

where [Z] is a 4x4 dynamic stiffness matrix. From (1.254) we may write, for example

$$f_{y_1} = z_{11} y_1 + z_{12} \theta_1 + z_{13} y_2 + z_{14} \theta_2 \qquad (1.255)$$

and therefore, if one wants to measure directly one of the dynamic stiffness elements, say z_{11}, one would have to prevent motion[†] in all coordinates but y_1,

---

[†] For this reason, the term 'Blocked Impedance' is becoming a common designation for the inverse forms of Mobility. Conversely, 'Free Impedance' is used to designate the pseudo-impedance elements resulting from a simple inversion of the mobility matrix elements ($f_k/x_j = 1/\alpha_{kj}$). The use of this term is not recommended.

thereby leaving only one non-zero term on the right-hand-side of (1.255). Preventing motion is a very difficult task from an experimental point of view. When the motions to be prevented refer to coordinates at the same point as the coordinate of interest, this condition becomes a practical impossibility.

In addition, if one thinks in terms of increasing the number of degrees of freedom, it is clear that any previously known element z_{jk} will no longer be valid. Therefore, an impedance matrix must almost invariably be obtained by inverting the corresponding mobility matrix. For the reasons given before, use of any inverse FRF form should be avoided.

1.5.2 MDOF FRF graphical representation

We have seen that the response model of a MDOF system consists of a set of different FRF functions and it has been shown that a system with N degrees of freedom is described by a modal model with N natural frequencies and N mode shapes. Also, it was shown that each FRF may be written under the form of a series of terms, each of which refers to the contribution to the total response of each mode of vibration, as stated by equations (1.223) and (1.247).

Bearing in mind the above features, let us look at the Bode representation of a receptance FRF for an example of an undamped system with 4 degrees of freedom. Figures 1.39 a) and b) display the magnitude and phase, respectively, using a linear scale, of a direct point receptance.

What is immediately obvious, from the magnitude plot, is that there are four peak amplitudes, corresponding to the four natural frequencies of the system. The meaning of this is that one is now confronted with four different resonances. In analogy with what we saw for SDOF systems, it is to be expected that, for each resonance, there will be a 180° phase shift.

However, looking at figure 1.39 b) it is clear that there are more than four such phase shifts. They not only occur at each resonance but also for intermediate frequency values that have no apparent special behaviour as far as the magnitude plot is concerned. This is just a consequence of using a linear scale for plotting the magnitude of the receptance, which hides the lower level behaviour. If we replace the linear plot in figure 1.39 a) by a logarithmic amplitude scale plot, we obtain what is shown in figure 1.40.

Now, we can also see detail at the lower levels of the response and the FRF shows that, in those regions, there are some 'inverted' peaks, each of which occurs in between the resonance peaks. These are the so-called antiresonances and they have an important feature which is a phase change just like the phase change associated with resonances.

For an undamped system, the antiresonance corresponds to no motion at all at the coordinate where the response is being considered. This situation can be explained if one recalls that the receptance FRF may be represented by a summation of terms, each of which corresponds to one of the modes of vibration of the system.

Taking, for example, equation (1.223) for zero damping,

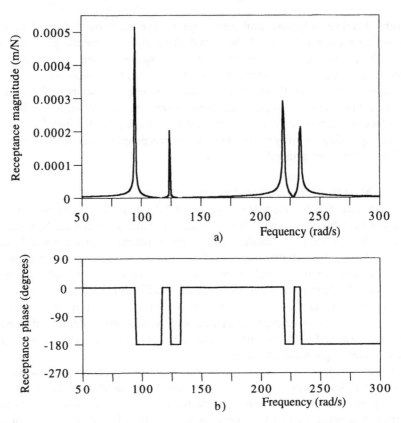

Fig. 1.39 Graphical display of a direct point receptance for a 4 DOF system: a) magnitude in linear scale; b) phase angle.

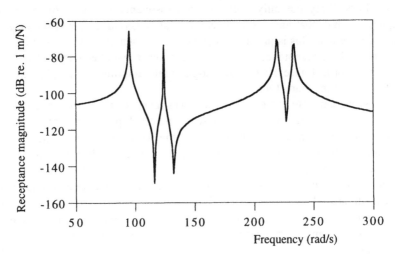

Fig. 1.40 Logarithmic plot of the receptance magnitude displayed in figure 1.39 a).

$$\alpha_{jk} = \sum_{r=1}^{N} \frac{_rA_{jk}}{\omega_r^2 - \omega^2} \qquad (1.256)$$

where the modal constant $_rA_{jk}$ is now a real quantity. If we consider a direct point measurement, say α_{kk}, the modal constant $_rA_{kk}$ is always positive due to it being the product (see (1.228)) of element k of the eigenvector for mode r, by itself.

What equation (1.256) states is that the total receptance FRF is the sum of the contributions of 'SDOF' terms corresponding to each of the system modes of vibration. For a direct point receptance:

$$\alpha_{kk} = \frac{_1A_{kk}}{\omega_1^2 - \omega^2} + \frac{_2A_{kk}}{\omega_2^2 - \omega^2} + \cdots + \frac{_NA_{kk}}{\omega_N^2 - \omega^2} \qquad (1.257)$$

Thus, in the lower frequency region, all the terms in the summation are positive and the receptance value is positive and dominated by the first mode ($r = 1$), for which the denominator $\omega_r^2 - \omega^2$ is smaller than for the other terms in the summation. After the first resonance, $\omega_1^2 - \omega^2$ becomes negative and therefore the first term in the series becomes negative, while still dominating the response, and therefore α_{kk} becomes negative. This change of sign corresponds to a phase shift from $0°$ to $-180°$.

As we approach ω_2, there will be a frequency value for which the first term of the series is cancelled out by the sum of all the other terms and, subsequently, there is a new change of sign of α_{kk} which becomes positive. Accordingly, the phase angle suffers a new shift and becomes zero. The frequency for which the cancelling out occurs is the antiresonance.

The same reasoning, for increasing values of the frequency, leads one to conclude that there will be an antiresonance between each pair of resonances. This particular feature of the direct point receptance (note that exactly the same applies for the other FRF forms) is very useful as a means of evaluating the validity of a measured FRF.

If one is considering transfer FRFs, the sign of the modal constants is no longer always positive and the occurrence of an antiresonance between two resonances is not certain, as exemplified in figure 1.41.

However, it may be concluded that, if the sign of the modal constant for two consecutive modes is the same, then there will be an antiresonance at some frequency between the natural frequencies of those two modes. When there is not an antiresonance, the FRF merely reaches a minimum non-zero value.

Another interesting feature associated with antiresonances is their physical meaning when considering direct point FRFs. In fact, each antiresonance is a natural frequency of the same system if the motion coordinate under consideration is fixed[†]. This property is useful in some experimental cases such as when using

---

[†] See also Chapter 6 where the physical meaning of antiresonances is discussed.

76

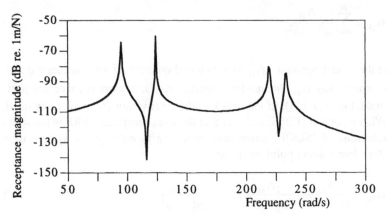

Fig. 1.41 Example of a transfer receptance.

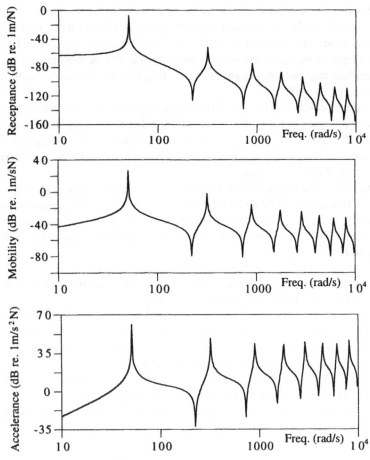

Fig. 1.42 Direct point FRF at the free tip of an undamped cantilevered beam.

seismic tables to test structures, where the excitation force and the motion response are measured at the seismic table. The antiresonance frequencies of the whole system (table plus structure) are the resonance frequencies of the structure under analysis (assuming that the table behaves as a rigid body, which is normally true for the low values of the frequency range of interest in such cases).

It is now interesting to see how the different FRF forms compare when displayed in log-log Bode plots. This is shown in figure 1.42 where a direct point FRF, at the free tip of an undamped cantilevered beam is displayed. It is clear that the receptance and accelerance plots make poor use of the available vertical space in the plot frame because they are generally displayed as falling (receptance) or rising (accelerance) curves. This is true for most beam or plate-like structures for which the mobility, over a wide range of frequencies, produces a roughly levelled plot. As a consequence of the above feature, FRFs Bode plots are usually produced using the mobility form. In fact, the three alternatives (receptance, mobility and accelerance) describe the same properties and each one has its own advantages. In general, receptance is convenient for analytical work whereas accelerance is used for direct plotting of measured data since it is customary to measure acceleration and force.

Now taking into consideration damped systems, the FRFs trace out Bode plots that are very similar to the ones just described. The differences are due to the resonances and antiresonances being less sharp and the phase angles being no longer exactly 0° or -180°. This is shown in figure 1.43 where the receptance for a

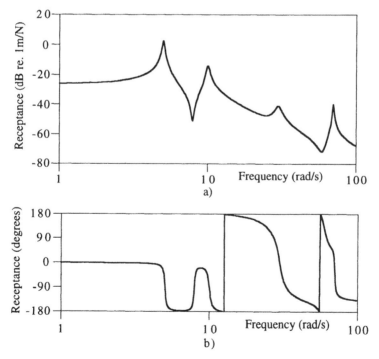

Fig. 1.43 Damped 4 DOF system (hysteretic model).

damped 4 DOF system is plotted. Note that, as shown in that figure, high damping values may hide the existence of an antiresonance, making the direct FRF look more like a transfer FRF.

As we have seen when studying the SDOF case, instead of plotting the magnitude and phase of the FRF, one may plot its real and imaginary parts. Figure 1.44 illustrates this type of display, using the same example as figure 1.43. What is immediately clear in figure 1.44 is that, due to the use of a linear scale and the fact that, in general, the receptance amplitude decays with frequency, the higher frequency modes tend not to show in the plots.

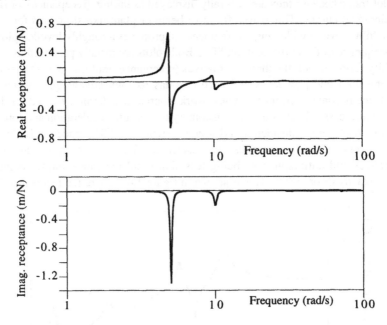

Fig. 1.44 Real and imaginary part of receptance plotted against frequency, for the 4 DOF damped system used in figure 1.43.

In order to avoid this problem, one could use several separated plots, each of which would cover a limited frequency range so that different amplitude scales could be used in each plot. As an alternative, the receptance plots may be replaced by accelerance plots, as shown in figure 1.45. Now, all modes are visible and it is clear that the real and imaginary parts do not exhibit the behaviour that would be expected from what has been seen in figure 1.28. The reason for this 'unexpected' behaviour is the fact that the 4 DOF example we used has highly non-proportional damping for the two last modes.

Let us turn our attention now to the use of Nyquist or Argand plane plots. The scaling problem we found when plotting the real and imaginary parts of receptance *versus* frequency will also be present here and make it difficult to 'read' a Nyquist plot of receptance covering the total frequency range of interest. The solution is to

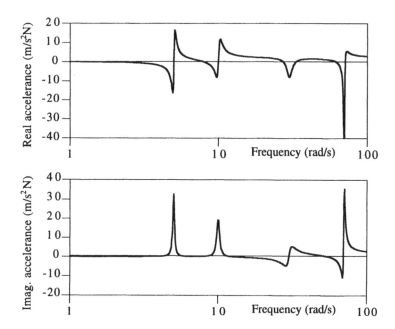

Fig. 1.45 Real and imaginary parts of the accelerance, plotted against frequency, for the example 4 DOF damped system used in figure 1.44.

use separate Nyquist plots, one for each natural frequency region. This is indeed performed when taking advantage of the particular features of Nyquist plots for the purpose of identifying system modal properties (see Section 4.4.1). Nevertheless, it will now be interesting to have a full representation of the FRF in one plot only. Thus, we are going to take an example where the modal constants have such values that all modes are visible. We shall start by plotting a direct point receptance of a 3 DOF system with proportional damping (figure 1.46).

As expected, the natural frequency regions plot as circular loops. However, it can be seen that the loops are not exactly centred with respect to the imaginary axis as in the case of a SDOF system. This can be easily explained if we recall equation (1.223) and rewrite it for a direct point receptance of a 3 DOF system:

$$\alpha_{kk}(\omega) = \frac{\overline{X}_k}{F_k} = \sum_{r=1}^{3} \frac{{}_rA_{kk}}{\omega_r^2 - \omega^2 + i\,\eta_r\,\omega_r^2}$$

$$= \frac{{}_1A_{kk}}{\omega_1^2 - \omega^2 + i\,\eta_1\,\omega_1^2} + \frac{{}_2A_{kk}}{\omega_2^2 - \omega^2 + i\,\eta_2\,\omega_2^2} + \frac{{}_3A_{kk}}{\omega_3^2 - \omega^2 + i\,\eta_3\,\omega_3^2}$$

$$(1.258)$$

where the modal constants ${}_rA_{kk}$ are real quantities due to the fact that damping was assumed to be proportional. Consider, for example, the first loop in our plot. Recalling that each loop occurs for a frequency region close to the corresponding

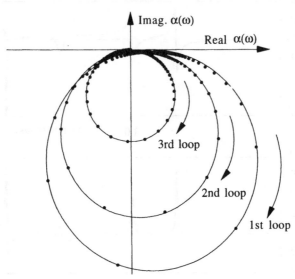

Fig. 1.46 Nyquist plot of a point receptance for a proportionally damped 3 DOF system.

natural frequency, then it may be assumed that, for that particular frequency range (1.258) can be approximated by

$$\alpha_{kk}(\omega) \approx \frac{{}_1A_{kk}}{\omega_1^2 - \omega^2 + i\,\eta_1\,\omega_1^2} + B_{kk} \qquad (1.259)$$

where B_{kk} is a constant complex quantity accounting for the contribution of the remaining modes to the total receptance value, which is dominated by the first mode. The first term of the summation plots as a circle with its centre on the imaginary axis, just like the receptance of a SDOF system. The only difference from a SDOF system is the fact that there is a real scaling factor (which alters the circle diameter), due to the existence of a modal constant ${}_1A_{kk}$ in the numerator. Summing a complex quantity B_{kk} will then produce a translation of the circle, displacing it from the original position.

Also, it can be seen in figure 1.46 that all circular loops are in the lower half of the complex plane. As explained above, the only difference from a SDOF Nyquist plot is the product by a scaling factor of each term in the summation. As we are considering a direct point receptance, the modal constants are all positive and, therefore, the loops remain in the lower half of the complex plane.

A different situation occurs if we are plotting a transfer receptance. In this case, the modal constants may be positive or negative and the opposing signs of these quantities may cause one or more loops to be in the upper half of the complex plane. This is exemplified in figure 1.47 where a transfer receptance of the same example system is plotted.

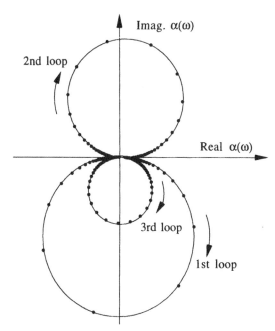

Fig. 1.47 Nyquist plot of a transfer receptance for a proportionally damped 3 DOF system.

If we consider the situation where damping is non-proportional, it is not difficult to predict what is going to happen. The difference now is the fact that the modal constants become complex quantities, i.e., they have a magnitude and a phase. Thus, the circular loop displacement and scaling effect remain and are due to the contribution of the off-resonant modes and to the magnitude of the modal constant, respectively.

In addition to the previous effects, the phases of the modal constants produce rotations of the modal loops which are no longer in the 'upright' position, as illustrated in figure 1.48.

1.6   COMPLETE AND INCOMPLETE MODELS
It has been seen that the dynamic properties of a system with N DOFs may be described by three different types of complete models: the Spatial model, the Modal model and the Response model.

In the first case, the system dynamic characteristics are contained in the spatial distribution of its mass, stiffness and damping properties, described by the NxN mass, stiffness, and damping matrices, respectively. The spatial model given by [M], [K] and [D] (or [C]) leads to an eigenproblem which, having been solved, yields the modal model constituted by the modal properties (N natural frequencies, N modal damping values and N mode shape vectors) contained in matrices $[\diagdown \lambda^2_r \diagdown]$ and $[\Phi]$ (as explained in Section 1.4.1).

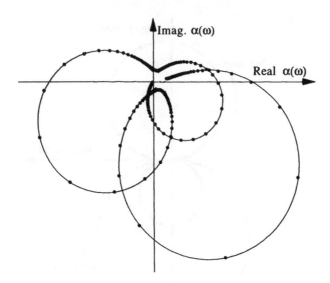

Fig. 1.48 Nyquist plot of a point receptance for a non-proportionally damped 3 DOF system.

If we take the modal model and recall the orthogonal properties of the modal matrix (let us assume hysteretic damping), we have

$$[\Phi]^T [M][\Phi] = [I]$$
$$[\Phi]^T [[K]+i[D]][\Phi] = \left[\diagdown \lambda_r^2 \diagdown \right] = \left[\diagdown \omega_r^2 (1+i\eta_r) \diagdown \right] \qquad (1.260)$$

and therefore

$$[\Phi]^{-T} [\Phi]^{-1} = [M]$$
$$[\Phi]^{-T} \left[\diagdown \omega_r^2 (1+i\eta_r) \diagdown \right][\Phi]^{-1} = [K]+i[D] \qquad (1.261)$$

i.e., it is possible, in principle, to obtain a spatial model from the knowledge of the modal model. This may be an important conclusion if one thinks that, experimentally, it is the modal model that is initially derived from the experimental FRF data. Furthermore, the modal model yields the response model as seen in Section 1.4.2. In fact, recalling the starting equation

$$\left[[K]-\omega^2 [M]+i[D] \right]\{\overline{X}\} = \{\overline{F}\} \qquad (1.262)$$

i.e.,

$$[\alpha(\omega)]^{-1} \{\overline{X}\} = \{\overline{F}\} \qquad (1.263)$$

and the inverse relationships (1.261), it may easily be concluded that

$$[\alpha(\omega)] = [\Phi]\left[\,^{\backprime}\,\omega_r^2\,(1+i\,\eta_r) - \omega^2\,_{\backprime}\,\right]^{-1}[\Phi]^T \tag{1.264}$$

Thus, starting with a spatial model we have ended with a response model after going through an intermediate modal model. This sequence is commonly performed when the starting point is a theoretical analysis.

However, if a system is too complex and therefore cannot be modelled analytically, one must revert to experimental analysis where the starting point is the measurement of the system FRFs, that is, the system is initially described by a response model $[\alpha(\omega)]$.

We shall see, in Chapter 4, that there are many techniques that allow derivation from the experimentally obtained response model, of the modal characteristics of a given system. The procedure is called Modal Identification or System Identification. From the modal model thus derived, one may apply (1.261) and obtain a spatial model. Figure 1.49 summarises the above discussed model interrelation, based on the undamped case.

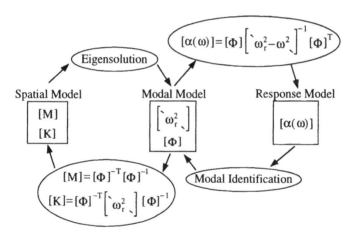

Fig. 1.49 Dynamic models interrelation (undamped case).

Up to now we have always assumed that our systems are described by complete models, i.e., that all their mass, stiffness and damping properties are known, or that all the eigenvalues and all the elements in the eigenvectors are known, or that all the elements in the FRF matrix are known.

Though this is a valid assumption from a theoretical point of view, in practice it corresponds to an impossibility. The knowledge that any real system has an infinite number of degrees of freedom is sufficient for understanding of the problem. However, when we refer to incomplete models we are considering further forms of incompleteness. In fact, even if we assume that when reducing a real system to a model with N DOFs we have a complete model, in practice and from

84

an experimental point of view, it is not usually possible to measure all the coordinate motion responses, or to apply all force excitations, or even to cover and analyse all modes.

The limitations introduced by the fact that we shall be using reduced or incomplete models, and the techniques to overcome the inherent problems, will be dealt with in subsequent chapters. We shall therefore only discuss the problem here in a very superficial way.

Let us consider an N DOFs response model and assume that, if completely known, it gives us a thorough description of the dynamic behaviour of our system. For instance, let us take the example in figure 1.38 where we discretised a simple beam as a 6 DOFs model described by (1.252).

The complete model receptance matrix is therefore of order 6x6 and includes both translation and rotation coordinates. Experimentally, measurement of rotational responses is a very difficult task. Furthermore, torque (or moment) excitation is almost impossible with current experimental means.

We may therefore have to decide to limit our model to include only the translation coordinates. The resulting reduced FRF model will be of order 3x3 and is obtained simply by extracting from (1.252) the relevant matrix elements:

$$\left\{q^R\right\} = \left\{\begin{matrix} y_1 \\ y_2 \\ y_3 \end{matrix}\right\} = \left[\alpha^R(\omega)\right]\left\{\begin{matrix} f_{y_1} \\ f_{y_2} \\ f_{y_3} \end{matrix}\right\} \qquad (1.265)$$

What is important to note is that, despite the fact that we limited our system description to a reduced number of coordinates, we have not altered the basic system which still has 6 degrees of freedom. We simply ceased to be able to describe all of the initial DOFs. Thus, when using a response model, reducing the coordinates of interest from N to p is a simple operation of elimination of the relevant N-p rows and columns of the frequency response matrix.

This is true if we are using any of the direct motion/force FRF relationships. However, if one is using any of the impedance type inverse FRFs, the elements in the reduced matrix will be completely different from the elements in the complete impedance matrix, i.e., while the elements in the pxp reduced matrix $[\alpha^R(\omega)]$ were directly extracted from the NxN matrix $[\alpha(\omega)]$, the reduced impedance matrix is given by

$$\left[Z^R(\omega)\right] = \left[\alpha^R(\omega)\right]^{-1} \qquad (1.266)$$

and it is clear that its elements are not the same quantities as the corresponding ones in the full impedance matrix[†].

[†] We are repeating what has already been explained in Section 1.5.1 but we think the repetition is justified by the importance of the statement.

Another important reduction that occurs in practical situations is related to the number of modes of vibration one can include in the analysis. The frequency range of an experimental analysis is limited, and therefore at least the high frequency modes will be omitted. Thus, despite the fact that the total number of coordinates N used in the model may remain without being reduced, one may have to consider a smaller number $m \le N$ of modes, that is, equation (1.223) would be of the form

$$\alpha_{jk}(\omega) = \sum_{r=1}^{m \le N} \frac{{}_r \overline{A}_{jk}}{\omega_r^2 - \omega^2 + i\, \eta_r\, \omega_r^2} \tag{1.267}$$

and, taking advantage of the orthogonality properties of the eigenvectors, the corresponding NxN receptance matrix containing a reduced modal information, can be written as

$$[\alpha(\omega)]_{NxN} = [\Phi]_{Nxm} \left[{}^\diagdown (\lambda_r^2 - \omega^2) {}_\diagdown \right]_{mxm}^{-1} [\Phi]_{mxN}^T \tag{1.268}$$

It is clear that each of the previous equations lacks information that may be of great importance. The fact that $m < N$ in (1.268) will lead to an NxN matrix $[\alpha(\omega)]$ that cannot be inverted due to its rows and columns being not linearly independent, i.e., the order of the matrix is N but the rank is m. If $m \ge N$, this inversion problem does not exist. This type of situation will be discussed in greater detail in Sections 5.2.4 to 5.2.6. Finally, the consequence of considering only a limited number of modes m is the fact that we end up with a square eigenvalue matrix of order mxm and the eigenvector matrix will be rectangular (Nxm).

Let us think now in terms of the response model and consider a system which is to be modelled with a finite number N of degrees of freedom. These DOFs will correspond to the coordinates of interest for the analysis. Let us assume as well that our modal model is to be derived from identification of the experimental data (as described in Chapter 4). As before, the analysis will have to be based on a necessarily limited experimental frequency range and, therefore, on a limited number of modes m.

Thus, the measured response model will consist of an NxN matrix of elements α_{jk} expressed in terms of experimental data. To replace these data by a mathematical model given by (1.267), with $m \ne N$, it is necessary to apply identification procedures such as the ones described in Chapter 4 and derive the values of the modal characteristics ω_r, η_r and ${}_r \overline{A}_{jk}$ for all measured modes r.

However, our response model described now by the identified NxN matrix $[\alpha(\omega)]$ will contain errors due to omitting all the out-of-range modes. These errors are usually visible when comparing the measured FRF data with the corresponding identified FRFs if represented only by (1.267). One way of minimising the consequences of using such a model is to introduce corrections on the identified FRFs, so that they approximate the measured data in the frequency range of interest, by including an extra term in the response equation

$$\alpha_{jk}(\omega) = \sum_{r=1}^{m} \frac{_r\overline{A}_{jk}}{\omega_r^2 - \omega^2 + i\,\eta_r\,\omega_r^2} + \overline{R}_{jk}(\omega) \tag{1.269}$$

where $\overline{R}_{jk}(\omega)$ is a complex residual term accounting for the contribution of the out-of-range modes.

The need to introduce a residual term brings forward another type of problem. In fact, if our FRF model was completely described by a matrix of elements α_{jk} given by (1.267), then it would be possible to derive the complete matrix from the knowledge of a single column (or a single row) of experimental data. This reasoning is based on the fact that, by definition, the modal properties ω_r and η_r are global properties, i.e., they constitute a fundamental characteristic of the system and remain the same no matter which FRF one takes. On the other hand, the modal constants, being products of the eigenvector elements, must obey the consistency relationships expressed by equations (1.228). Thus, in principle, the experimental procedure could be reduced to the measurement of a column of the frequency response matrix.

However, if there are modes that were not included in the analysis due to their being outside the frequency range of interest, one must use (1.269), and the previous modal constant derivation is limited to the modal terms, as the residuals $\overline{R}_{jk}(\omega)$ do not obey any specific relationship. Therefore, in this case, one must measure all the frequency response matrix elements.

Reduced models can also be obtained directly from the original NxN spatial model. In this case, reducing the number of coordinates entails reducing the mass, stiffness and damping matrices. Such a reduction cannot be performed by simply eliminating rows and columns as the consequence would be to obtain a system with completely different properties. In this case one must use specific techniques for redistribution of the system mass, stiffness and damping properties amongst the retained coordinates. It is therefore better to speak in terms of matrix condensation instead of matrix reduction. The implications of these situations will be more evident when coupling of substructures, structural modification and updating are discussed (Chapters 5, 6 and 7).

CHAPTER 2

Signal Processing for Modal Analysis

2.1 INTRODUCTION

Over the years, measurement techniques have been developed continuously to improve the accuracy of measured frequency response functions and to reduce the testing time, and therefore the cost, of modal testing. Historically, these measurement techniques have been categorised into tuned-sinusoidal methods and non-sinusoidal methods[†].

Tuned-sinusoidal methods consist of those procedures that attempt to establish natural modes of vibration by direct measurement of the test structure's forced vibration by multi-point excitation. To do so, several shakers are used to exert sinusoidally varying forces on the test structure. For a multi-shaker modal test, the frequency of excitation and the relative force level of the electrodynamic shakers are adjusted to isolate the target mode response from all other modes. The modal properties are then taken from direct measurement of the forced vibration response. A significant advantage of these methods is that they can be used to determine the modal properties of structures with high modal density, i.e., close modes. However, in tuning the shakers to excite the mode of interest, reasonably good knowledge about the nature of that particular mode is required. Furthermore, the tuning procedure can become quite complicated, time-consuming, and therefore costly. In reality, in tuned-sinusoidal methods there is not a clear distinction between measurement and analysis, which are somewhat mixed together. On the other hand, in non-sinusoidal methods, there is a clear separation between measurement and analysis. Occasionally, free response data are recorded and afterwards analysed. More often the time signals corresponding to an applied force and the responses are captured and processed in order to obtain frequency response

---

[†] Tuned-sinusoidal methods are also called Phase-Resonance or Force Appropriation methods. Non-sinusoidal methods are also widely known as Phase-Separation methods. These experimental techniques are covered in more detail in Chapter 3.

functions (FRFs) or impulse response functions (IRFs) which are better analysed by appropriate curve-fitting techniques to yield the modal properties of the structure. Details about these techniques will be covered in Chapter 4. Meanwhile, the treatment of the signals involves quite a number of precautions, techniques and procedures that justify a closer look at the area, known as Signal Processing.

In the past two decades, the introduction of the Fast Fourier Transform (FFT) algorithm, the availability of digital data processing equipment and powerful micro-computers have led to the development of test procedures that make no attempt to excite the test structure at discrete frequencies. Instead, all modes within the frequency range of interest are simultaneously excited with either a single broadband randomly varying force using an electrodynamic shaker, or multiple uncorrelated broadband randomly varying forces using multiple shakers, or an impulsive force using an instrumented hammer. The measured response data are then digitally processed to yield estimates of the FRFs or IRFs.

In this chapter the basic concepts of Fourier analysis, Fourier transform and FFT are presented. Their limitations, the pitfalls in using them and how they can be implemented are addressed.

After understanding the operation of Fourier transform and hence the capability of an FFT analyser, an FRF measurement for single-shaker modal testing using random excitation is presented. The traditional form of such an FRF measurement is an open-loop system. However, inevitable physical constraints cause such an FRF measurement to be inherently of a closed-loop form. A closed loop model for single-shaker modal testing will then be described. The effect of hidden feedback paths due to structure-shaker interaction on some current frequency response function estimators is covered. Attention is given to the effects of noise and leakage on these estimators. The generalised closed loop model is used to deal with multi-shaker modal testing using random excitation.

2.2 FOURIER ANALYSIS
When the response signal is a pure sine wave such as the one obtained from a modal test with sinusoidal excitation, the determination of the characteristic frequency is a simple process. However, in a modal test with random excitation, the displacement, velocity or acceleration of the test structure is normally a complicated function resulting from the interaction between various resonance frequencies of the structure and the frequency characteristics of external driving force(s).

Before approaching such a situation, let us start with the simpler case of a periodic excitation, leading to a periodic response. As the displacement x(t) is a periodic function with period T, x(t) can be expressed as the sum of a number of harmonically related sinusoidal waves and therefore may be represented by a Fourier series:

$$x(t) = a_0 + 2 \sum_{n=1}^{\infty} \left(a_n \cos \frac{2\pi n t}{T} + b_n \sin \frac{2\pi n t}{T} \right) \tag{2.1}$$

The constant a_0 is simply the mean value of x(t) over the period T (say from -T/2 to +T/2). The constants a_n and b_n can be evaluated from

$$a_n = \frac{1}{T} \int_{-T/2}^{T/2} x(t) \cos\frac{2\pi n t}{T} \, dt \tag{2.2}$$

$$b_n = \frac{1}{T} \int_{-T/2}^{T/2} x(t) \sin\frac{2\pi n t}{T} \, dt \tag{2.3}$$

The constants a_n and b_n can be computed in a straightforward manner for any periodic function. It should be noted that the above calculations are based on real functions and the frequency spectrum of x(t) is defined for positive frequencies only ($n = 1, 2, 3, \ldots$).

If $\cos 2\pi n t/T$ and $\sin 2\pi n t/T$ are expressed in terms of complex exponential functions as

$$\cos\frac{2\pi n t}{T} = \frac{e^{i 2\pi n t/T} + e^{-i 2\pi n t/T}}{2}$$
$$\sin\frac{2\pi n t}{T} = \frac{e^{i 2\pi n t/T} - e^{-i 2\pi n t/T}}{2i} \tag{2.4}$$

the resulting exponential form for the Fourier series can be written as

$$x(t) = \sum_{n=-\infty}^{+\infty} c_n e^{i 2\pi n t/T} \tag{2.5}$$

where

$$c_n = \frac{1}{T} \int_{-T/2}^{T/2} x(t) e^{-i 2\pi n t/T} \, dt \tag{2.6}$$

As c_n is defined for both positive and negative n's (frequencies), the following relationship holds between c_n and the real quantities a_n and b_n.

$$|c_n| = |c_{-n}| = \sqrt{a_n^2 + b_n^2} \tag{2.7}$$

If x(t) is not a periodic function but a single impulse, it is still possible to utilise Fourier's theorem, although in a somewhat modified version. Substituting (2.6) in (2.5) one obtains

$$x(t) = \sum_{n=-\infty}^{+\infty} \frac{1}{T} \left(\int_{-T/2}^{T/2} x(\tau) e^{-i 2\pi n \tau/T} \, d\tau \right) e^{i 2\pi n t/T} \tag{2.8}$$

It can be assumed that in this case the impulse will be repeated after an infinite period. As $T \to \infty$, $1/T = \Delta f \to df$ and $n/T = n\,\Delta f \to f$ which is a continuous variable. Hence, the sum in (2.8) turns into an integral:

$$x(t) = \int_{-\infty}^{+\infty} \left(\int_{-\infty}^{+\infty} x(\tau)\, e^{-i2\pi f \tau}\, d\tau \right) e^{i2\pi f t}\, df \tag{2.9}$$

or

$$x(t) = \int_{-\infty}^{+\infty} X(f)\, e^{i2\pi f t}\, df \tag{2.10}$$

where

$$X(f) = \int_{-\infty}^{+\infty} x(t)\, e^{-i2\pi f t}\, dt \tag{2.11}$$

Expressions (2.10) and (2.11) constitute the well-known pair of Fourier integrals (previously presented in Section 1.2.3), defined in complex form from $-\infty$ to $+\infty$. These integrals are very important because they allow a time domain signal to be transformed to and from a frequency domain signal.

So far, only periodic signals and a single impulse have been discussed. Often, in modal testing, other types of signal can be encountered. The various types of signals can be categorised as:

(a) harmonic   e.g. force and response signals of a structure excited by a sinusoidal force;

(b) periodic   e.g. vibration of a machine rotating at constant speed;

(c) transient   e.g. response of a structure excited by an impulsive force;

(d) random   e.g. response of a structure under white noise excitation.

Categories (a) and (b) are periodic signals. Therefore, a Fourier analysis can be carried out, as the period of the signals, T, is well defined. Category (c) is the result of an impulse signal and may be analysed using the Fourier integrals. Furthermore, if the transient signal has died away within the record length, it can be converted to a periodic signal by considering it to repeat itself at one record length, two-record length, three-record length ... etc. For Category (d), it is only possible in practice to analyse a finite record length of the random signal. Provided that the statistical properties do not vary either during the data record or between data records (i.e., the signal is stationary and ergodic), the signal is assumed to be periodic, the period being the record length T, and a Fourier analysis is carried out. Nevertheless, as will be seen later on, some additional and careful treatment is necessary, including some averaging of the results, which is always required.

The Fourier series given by (2.1) or (2.5) shows $x(t)$ to be represented by a series of harmonics, of frequencies $1/T$, $2/T$, $3/T$, ... The spacing of the frequency components and hence the resolution obtained is $1/T$ Hz. This is no problem if the signal being analysed is truly periodic in time T because there cannot be any components in the signal at frequencies between those calculated in the Fourier analysis. However, it is often necessary to analyse signals which are not truly periodic in time T (for example, the vibration of a rotating machine having a non-integer number of revolutions during the measurement period, or the vibration of a structure under random environmental loads).

The problem may be illustrated by comparing the Fourier series of a sine wave having n periods in time T with that of a sine wave having (2n+1)/2 periods within the measurement record as shown in figures 2.1 a) and 2.1 b). As one can see from figure 2.1 b), with non-integer periods of wave, if the signal in the recorded time T is repeated beyond that period, a slope discontinuity will occur. The resulting spectra are shown in figures 2.2 a) and 2.2 b). Figure 2.2 a) shows the frequency of the first sine wave to be correctly recorded while figure 2.2 b) shows the spectrum of the second sine wave to be relatively broadband with two peaks in the vicinity of the correct frequency. Moreover, the presence of other sinusoidal components in the second case may cause the original sine wave to be masked by the sidelobes of those components. This has important implications for modal testing where some modes may be less strongly excited than other close modes at some particular measurement positions, such as nodal points.

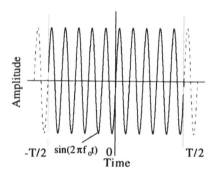

Fig. 2.1 a)   Sine wave with n periods in time T.

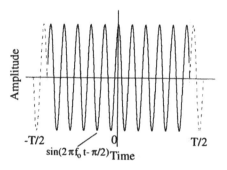

Fig. 2.1 b)   Sine wave with (2n+1)/2 periods in time T.

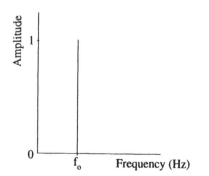

Fig. 2.2 a)   Power spectrum obtained when f_0T is an integer.

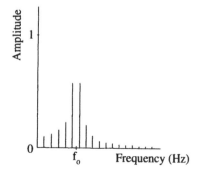

Fig. 2.2 b)   Power spectrum obtained when f_0T is a half integer.

The phenomenon of the spreading of the true spectrum components to other frequencies is named 'leakage'. Referring to figure 2.2 a), the interpretation is that the energy associated with the spectral line at f_0 has 'flown away', or 'leaked' to neighbouring frequencies, as in figure 2.2 b). Thus, the analysis of a finite time record can cause 'leakage' in the true spectrum. To cure or, at least, minimise such a problem, the time signal is usually multiplied by a function, known as window function or simply window. The objective is to obtain a smooth decay to zero at the limits of the recorded time period, so that the resulting signal approximates more closely an exact periodic one. From figure 2.1 b) to figure 2.2 b), no window was applied or else, one can say in general terms that a rectangular window was used, which is equivalent (see table 2.1).

The exact shape of the leaked spectrum depends on the frequency and the phase relationships between x(t) and the time window function. In order to minimise

Table 2.1 Window functions and their shapes.

	Window Function: w(t)		Shape
Rectangular (no window)	$w(t) = 1,$ $= 0,$	$\|t\| \leq \dfrac{T}{2}$ $\|t\| > \dfrac{T}{2}$	 -T/2 0 T/2
Barlett	$w(t) = 1 - \dfrac{2\|t\|}{T},$ $= 0,$	$\|t\| \leq \dfrac{T}{2}$ $\|t\| > \dfrac{T}{2}$	 -T/2 0 T/2
Hanning	$w(t) = \dfrac{1}{2}\left[1 + \cos\left(\dfrac{2\pi t}{T}\right)\right],$ $= 0,$	$\|t\| \leq \dfrac{T}{2}$ $\|t\| > \dfrac{T}{2}$	 -T/2 0 T/2
Hamming	$w(t) = 0.54 + 0.46 \cos\dfrac{2\pi t}{T},$ $= 0,$	$\|t\| \leq \dfrac{T}{2}$ $\|t\| > \dfrac{T}{2}$	 -T/2 0 T/2
Blackman	$w(t) = 0.42 + 0.5\cos\dfrac{2\pi t}{T} + 0.08\cos\dfrac{4\pi t}{T},$ $= 0,$	$\|t\| \leq \dfrac{T}{2}$ $\|t\| > \dfrac{T}{2}$	 -T/2 0 T/2
Kaiser-Bessel	$w(t) = \dfrac{I_o\left[\pi\beta\sqrt{1 - (2t/T)^2}\right]}{I_o(\pi\beta)},$ $= 0,$	$\|t\| \leq \dfrac{T}{2}$ $\|t\| > \dfrac{T}{2}$	 -T/2 0 T/2
	I_o = modified Bessel function		

leakage, a number of window functions have been proposed. Table 2.1 summarises some of the most important window functions normally incorporated in FFT analysers as built-in features.

The most commonly used window function is the Hanning one. Its name derives from von Hann, a scientist who applied an equivalent process to meteorological data. When the sine signals shown in figures 2.1 a) and 2.1 b) are multiplied by the Hanning function prior to the Fourier analysis, the resulting spectra of the signals are shown in figures 2.3 a) and 2.3 b). One can observe that the Hanning window spreads some of the energy of the original signals to the two adjoining spectral components while it suppresses the energy leaked to other spectral components which are far from the correct frequency. Therefore, application of the window generally limits the extent of leakage and reduces the chance of important components of the signal being masked.

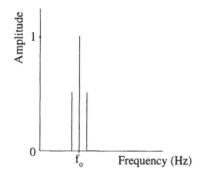

Fig. 2.3 a) Power spectrum obtained by the use of a Hanning window when $f_0 T$ is a integer.

Fig. 2.3 b) Power spectrum obtained by the use of a Hanning window when $f_0 T$ is a half integer.

It is recommended that a window should always be used except when the signal is truly periodic in time T or the signal is a transient which has died away within the record length. Moreover, the inevitable spreading of energy when Fourier analysis is used on signals which are not exactly periodic in the measurement time means that great care must be taken when obtaining damping values from the spectra, as these tend to appear as higher than they really are.

2.2.1 Discrete Fourier Transform (DFT)

The above applies equally to analogue and digital signals. In practice, Fourier analysis is almost always carried out using a digital processor even though most transducers generate an analogue output signal. For example, a piezoelectric force gauge generates an output voltage that is proportional to the driving force applied to the structure. As the force varies in a continuous manner, the transducer output shows a continuous variation. An analogue-to-digital (A/D) converter, which is an important part of the data acquisition system, is used to convert the analogue transducer signal into the digital code used by the processor.

In practice, the A/D converter records the level of the signal at a discrete set of times, say $1/f_s$, $2/f_s$, ... and N/f_s seconds where N is the total number of samples and f_s is the sampling frequency in Hz. Figure 2.4 shows the typical sampling process of a voltage measurement.

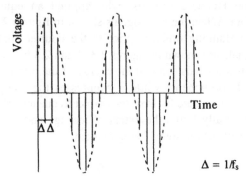

Fig. 2.4 A typical sampling process of voltage measurement.

Since there is no information for the time periods between the samples, incorrect selection of the sampling frequency can produce misleading results. Figure 2.5 a) shows a sine wave with a frequency f_o of 10 Hz. By taking the sampled values of this continuous time-dependent signal using a sampling frequency of 5 Hz, 12 Hz, and 20 Hz and connecting those values with straight-time segments, one can produce plots of amplitude against time as shown in figures 2.5 b) - 2.5 d). When examining the data in figure 2.5 b), it is reasonable to conclude that the sampled signal has a constant (DC) value. Such a conclusion is wrong because the original signal is a sine wave.

The amplitude of the sampled data, which depends on when the first sample was taken, is also misleading. This behaviour occurs if the wave is sampled at any rate that is an integer fraction of the frequency of the original signal, f_o (e.g., f_o, $f_o/2$, $f_o/3$, ...). Figure 2.5 c) shows the data to be a wave of two cycles per second. The frequency, 2 Hz, is the difference between the signal frequency, 10 Hz, and the sampling frequency, 12 Hz. The misinterpretation of a signal by a sinusoid at lower frequency is termed 'aliasing'. The minimum sampling rate or the lowest sampling frequency required for aliasing to be avoided is at least two times the signal frequency. This is known as the Nyquist-Shannon sampling theorem. Figure 2.5 d) shows the data to be a wave of 10 Hz, the same as the original data.

When sampling and digitisation are carried out by the A/D converter, the device generates a fixed number of possible discrete digital levels or quantisation levels. The value of the signal at the instant of sampling is rounded to the nearest digital level as shown in figure 2.6. The accuracy of the process depends on the number of quantisation levels available in the converter, which in turn depends on the number of bits in the converter (a n-bit converter has 2^n quantisation levels). Most FFT analysers use 12 bit A/D converters giving $2^{12} = 4096$ levels (or $2^{12} - 1 = 4095$ non-zero levels) and a $20\log_{10} 4095 = 72.25$ dB dynamic range for peak-to-peak

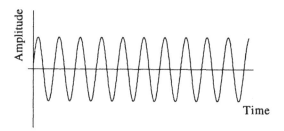

Fig. 2.5 a) A sine wave of 10 Hz.

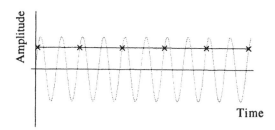

Fig. 2.5 b) Sampling of a 10 Hz sine wave at a rate of 5 Hz.

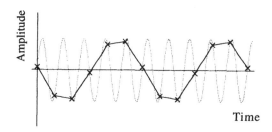

Fig. 2.5 c) Sampling of a 10 Hz sine wave at a rate of 12 Hz.

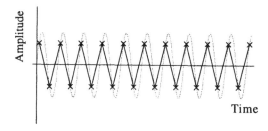

Fig. 2.5 d) Sampling of a 10 Hz sine wave at a rate of 20 Hz.

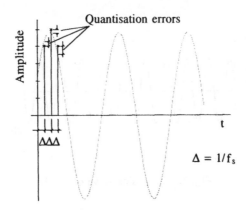

Fig. 2.6 Quantisation errors of a sampled signal.

measurements. They also provide a 69 dB dynamic range for root-mean-square (rms or RMS) measurements.

In practice, the signal should occupy as much of the range of the converter as possible. If the converter is a bipolar one having, for example, 4096 levels between -1V to +1V and the signal is always between -0.01 V and +0.01V, 40 dB of the measurement range will be lost. In that case, the signal should be amplified or the range of the converter changed. Provided that is done, quantisation errors are normally insignificant.

It was shown in Section 2.2 that if $x(t)$ is a periodic function with period T, then it is always possible to write $x(t)$ as the sum of two infinite cosine and sine series (see equation (2.1)) or that of one complex exponential series as follows

$$x(t) = \sum_{n=-\infty}^{+\infty} \frac{1}{T} \left(\int_{-T/2}^{T/2} x(\tau) \, e^{-i2\pi n\tau/T} \, d\tau \right) e^{i2\pi nt/T}$$
$$= \sum_{n=-\infty}^{+\infty} \frac{1}{T} \left(\int_{0}^{T} x(\tau) \, e^{-i2\pi n\tau/T} \, d\tau \right) e^{i2\pi nt/T}$$

(2.12)

which, in turn, is equal to

$$x(t) = \sum_{n=-\infty}^{+\infty} d_n \, e^{i2\pi nt/T}$$

(2.13)

where

$$d_n = \frac{1}{T} \int_{0}^{T} x(t) \, e^{-i2\pi nt/T} \, dt$$

(2.14)

In fact, d_n is the time-averaged value of $x(t)$ weighted by the function $e^{-i2\pi nt/T}$. Consider now that $x(t)$ is sampled at regular time intervals and

represented by the discrete series $\{x(k)\}$, $k = 0, 1, 2, ..., N\text{-}1$ where $t=kT/N$.
Denoting the coefficients d_n by the corresponding sampled coefficients $X(j)$, the
value of the weighted function is $x(k)e^{-i2\pi jk/T}$ at the time instant k.
The integral shown in equation (2.14) will be replaced by a summation:

$$X(j) = \frac{1}{N\,\Delta t} \sum_{k=0}^{N-1} x(k)\, e^{-i2\pi jk/N}\, \Delta t = \frac{1}{N} \sum_{k=0}^{N-1} x(k)\, e^{-i2\pi jk/N} \qquad (2.15)$$

and (2.13) takes the form

$$x(k) = \sum_{j=0}^{N-1} X(j)\, e^{i2\pi jk/N} \qquad (2.16)$$

for $j = 0, 1, 2, ..., N\text{-}1$; $k = 0, 1, 2, ..., N\text{-}1$. This discrete Fourier transform (DFT)
pair can also be obtained using equations (2.10) and (2.11).

A direct evaluation of equation (2.15) would require nearly N^2 complex
multiply-and-add operations. For large values of N, this can be prohibitive. A
sophisticated algorithm, the Fast Fourier Transform (FFT) proposed by Cooley and
Tukey in 1965 [1, 2], is described below, being - as the name suggests - much
faster.

2.2.2 Fast Fourier Transform (FFT)
When the expression $e^{-i2\pi/N}$ in (2.15) and (2.16) is replaced by the term W_N, the
DFT pair takes the form

$$X(j) = \frac{1}{N} \sum_{k=0}^{N-1} x(k)\, W_N^{jk} \qquad (2.17)$$

$$x(k) = \sum_{j=0}^{N-1} X(j)\, W_N^{-jk} \qquad (2.18)$$

Figure 2.7 shows the sampled displacement signal $x(kT/N)$ which is a real-
valued time series and its associated DFT $X(jf_s/N)$. If $x(kT/N)$ is periodic in time
T, the set of Fourier coefficients $X(jf_s/N)$ will be periodic over the sample
frequency f_s.

From figure 2.7 one observes the real part of $X(j)$ to be symmetric about the
folding frequency f_f where f_f equals $f_s/2$ and the imaginary part is antisymmetric.
These symmetries suggest that the real part of $X(j)$ is an even function while the
imaginary part of $X(j)$ is an odd function. This means that the Fourier coefficients
between $N/2$ and $N\text{-}1$ can be viewed as the 'negative frequency' harmonics between
$-N/2$ and -1. Likewise, the last half of the time series can be interpreted as negative
time which should be occurring before $t = 0$.

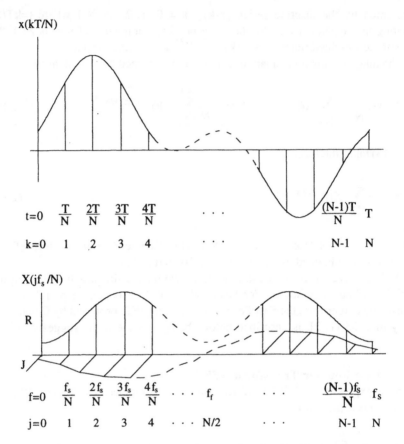

Fig. 2.7 A real time signal and its complex discrete Fourier transform.

In what follows, a derivation of the Cooley-Tukey FFT algorithm for evaluating (2.17) is presented, using an example of N=8. Using Cooley's notation, the FFT algorithm involves evaluating the expression

$$X(j) = \sum_{k=0}^{7} A(k) \, W^{jk} \tag{2.19}$$

where $j = 0, 1, 2, ..., 7$; $k = 0, 1, 2, ..., 7$; $A = x(k)/8$ and $W = e^{-i2\pi/8}$. One can express

$$\begin{aligned} j &= 4 \, j_2 + 2 \, j_1 + j_0 \\ k &= 4 \, k_2 + 2 \, k_1 + k_0 \end{aligned} \tag{2.20}$$

where j_0, j_1, j_2, k_0, k_1 and k_2 can have the value 0 or 1. Using this representation of j and k, (2.19) becomes

$$X(j_2, j_1, j_0) = \sum_{k_0=0}^{1} \sum_{k_1=0}^{1} \sum_{k_2=0}^{1} A(k_2, k_1, k_0)\, W^{(4j_2+2j_1+j_0)(4k_2+2k_1+k_0)} \tag{2.21}$$

Noting that $W^{m+n} = W^m \cdot W^n$, it can be shown that

$$W^{(4j_2+2j_1+j_0)(4k_2+2k_1+k_0)} = W^{(4j_2+2j_1+j_0)4k_2}\, W^{(4j_2+2j_1+j_0)2k_1}\, W^{(4j_2+2j_1+j_0)k_0} \tag{2.22}$$

Looking at these terms individually, one can write

$$W^{(4j_2+2j_1+j_0)4k_2} = \left[W^{8(2j_2+j_1)k_2} \right] W^{4j_0 k_2} \tag{2.23}$$

$$W^{(4j_2+2j_1+j_0)2k_1} = \left[W^{8j_2 k_1} \right] W^{(2j_1+j_0)2k_1} \tag{2.24}$$

$$W^{(4j_2+2j_1+j_0)k_0} = W^{(4j_2+2j_1+j_0)k_0} \tag{2.25}$$

As $W^8 = [e^{-i2\pi/8}]^8 = e^{-i2\pi} = 1$, the bracketed portions in (2.23) and (2.24) can be replaced by unity. This means that (2.21) can be written as

$$X(j_2, j_1, j_0) = \sum_{k_0=0}^{1} \sum_{k_1=0}^{1} \underbrace{\sum_{k_2=0}^{1} A(k_2, k_1, k_0)\, W^{4j_0 k_2}}_{A_1(j_0, k_1, k_0)}\, W^{(2j_1+j_0)2k_1}\, W^{(4j_2+2j_1+j_0)k_0}$$

$$A_2(j_0, j_1, k_0)$$

$$A_3(j_0, j_1, j_2) \tag{2.26}$$

In this form it is convenient to perform each of the summations separately and to label the intermediate results. It should be noted that each set consists of eight terms and only the latest set needs to be saved. Thus a set of equations can be derived as

$$A_1(j_0, k_1, k_0) = \sum_{k_2=0}^{1} A(k_2, k_1, k_0)\, W^{4j_0 k_2} \tag{2.27}$$

$$A_2(j_0, j_1, k_0) = \sum_{k_1=0}^{1} A_1(j_0, k_1, k_0)\, W^{(2j_1+j_0)2k_1} \tag{2.28}$$

$$A_3(j_0, j_1, j_2) = \sum_{k_0=0}^{1} A_2(j_0, j_1, k_0)\, W^{(4j_2+2j_1+j_0)k_0} \tag{2.29}$$

$$X(j_2, j_1, j_0) = A_3(j_0, j_1, j_2)$$ (2.30)

These FFT equations have 48 operations compared to the 64 complex multiply-and-add operations of the direct evaluation of Fourier coefficients using (2.15). As the first multiplication in each summation is actually a multiplication by +1, the number of operations involved in the above FFT can be reduced to 24. By noting that $W^0 = -W^4$, $W^1 = -W^5$, etc., the number of multiplications can be further reduced to 12. These kinds of reduction are applicable to the more general case of $N = 2^m$ where m is any positive integer, reducing the computation from nearly N^2 operations to $(N/2)\log_2 N$ multiplications, $(N/2)\log_2 N$ additions and $(N/2)\log_2 N$ subtractions. For $N = 1024$, this represents a computational reduction of more than 200 to 1.

The aliasing problem encountered in the DFT is also a problem in the FFT analysis. When the $\{x(kN/T)\}$ time series is processed, the corresponding Fourier coefficients range from -N/2T to N/2T which are in turn between $-f_s/2$ and $f_s/2$. Therefore, the maximum frequency at which unique spectral information is obtained is $f_s/2$, not f_s. Frequencies higher than $f_s/2$ merely reveal spurious frequency coefficients which are simply repetitions of those at lower frequencies. The cure for this problem involves sampling the signal at a rate at least twice as high as the highest frequency. This restriction on the sampling rate (the so-called Nyquist frequency) is known as Nyquist-Shannon's sampling theorem (see also Section 2.2.1). In most FFT analysers, this condition is automatically met by passing the signal through a low pass anti-aliasing filter prior to the A/D conversion. The cut-off frequency of the filter depends on the sampling rate selected. Furthermore, some analysers do not display the spectrum over the whole frequency range up to $f_s/2$. Frequently, only 256 or 400 out of 512 usable spectral lines for an analyser performing the FFT based on 1024 data points are shown, giving spectral information up to $0.25 f_s$ or $0.39 f_s$.

The problem of leakage described before is also inherent in FFT analysis because only a finite set of recorded data is analysed. To minimise leakage, a time window function must be used. The discrete form of various window functions and their DFT characteristics are given in table 2.2. The most popular window function is the Hanning. When it convolutes with the sampled data, sidelobes are produced which are at least 31 dB lower than the mainlobes of the measured spectrum.

In the following, we shall need to use the auto-spectrum and cross-spectrum concepts defined in Section 1.2.7. Thus, it is important now to note that the auto-spectrum of x(t) can be computed directly from the DFT of x(t) as:

$$S_{xx}(j f_s/N) = \frac{T}{N} \left| X(j f_s/N) \right|^2$$ (2.31)

where f_s/N is the frequency spacing. Equation (2.31) describes what is known as the periodogram. For the discrete version of two continuous signals, such as f(t) and x(t), one can obtain the cross-spectrum which is given by

$$S_{fx}(jf_s/N) = \frac{T}{N}F^*(jf_s/N)\,X(jf_s/N) \tag{2.32}$$

where the asterisk denotes the complex conjugate. S_{fx} is in general a complex quantity, as opposed to S_{xx} which is real. Finally, note that the following cross-spectrum inequality exists

$$\left|S_{fx}(jf_s/N)\right|^2 \le S_{ff}(jf_s/N)\,S_{xx}(jf_s/N) \tag{2.33}$$

Table 2.2 Discrete version of window functions and their characteristics.

Window function g(n)	Discrete Fourier Transform	
	Bandwidth of main-lobe in frequency	Peak sidelobe relative to main-lobe amplitude
Bartlett $g(n) = \begin{cases} \dfrac{2n}{N-1}, & 0 \le n \le \dfrac{N-1}{2} \\ 2-\dfrac{2n}{N-1}, & \dfrac{N-1}{2} \le n \le N-1 \end{cases}$	$\dfrac{4}{NT}$	0.056 (-25 dB)
Hanning $g(n) = \dfrac{1}{2}-\dfrac{1}{2}\cos\left(\dfrac{2\pi n}{N-1}\right)$ $0 \le n \le N-1$	$\dfrac{4}{NT}$	0.028 (-31 dB)
Hamming $g(n) = 0.54-0.46\cos\left(\dfrac{2\pi n}{N-1}\right)$ $0 \le n \le N-1$	$\dfrac{4}{NT}$	0.0089 (-41 dB)
Blackman $g(n) = 0.42-0.5\cos\left(\dfrac{2\pi n}{N-1}\right)+0.08\cos\left(\dfrac{4\pi n}{N-1}\right)$ $0 \le n \le N-1$	$\dfrac{6}{NT}$	0.0014 (-57 dB)
Kaiser-Bessel $g(n) = \dfrac{I_o\left[\pi\beta\sqrt{1-\left(\dfrac{2n}{N-1}-1\right)^2}\right]}{I_o(\pi\beta)}$ $0 \le n \le N-1$	$\dfrac{2(\beta^2+1)^{\frac{1}{2}}}{NT}$	$0.22\left[\dfrac{\sinh(\pi\beta)}{\pi\beta}\right]^{-1}$ which is a function of β

2.3   SINGLE-INPUT SINGLE-OUTPUT ANALYSIS OF MECHANICAL STRUCTURES

Most FRF measurement methods require the attachment of an electrodynamic shaker and transducers to the test structure. The transducers convert the force transmitted to the structure and the response into electric signals which, once filtered through signal conditioning equipment, are digitised and used to develop estimates of FRF in an FFT analyser. However, by its very nature, the excitation mechanism interferes with the test structure. Ewins [3], *inter alia*, has noticed that if that interference is not controlled to represent negligible proportions, it may result in the acquisition of poor data, with the consequence of loss of quality in the final model of the structure. This problem is also discussed in Section 3.5.

Shaker-structure interaction is always a concern in FRF measurements because it introduces the problem of 'force drop-out' at structural resonances. In the 1980s, there were a number of publications [4-6] which documented the force drop-out phenomenon at structural resonances and gave excellent models of both mechanical and electrical systems for an excitation mechanism. Also, there is a focus of interest in the estimation of FRFs of mechanical structures, and various alternative FRF estimators have been introduced and investigated [7-9] for single-shaker modal testing using random excitation.

However, the effect of shaker-structure interaction on those FRF estimates is not usually taken into account although it is obvious that the inevitable physical constraints cause such FRF measurements to be inherently of a closed-loop form rather than of the open-loop form usually supposed to apply. As will subsequently be made clear, the practical importance of the analysis of an FRF measurement system with feedback is quite significant.

2.3.1 Frequency Response Function estimators

Figure 2.8 shows a traditional measurement-system model used to describe FRF measurement when noise is present on both measured force and response signals. As modal practitioners normally express power spectra against radians per second, all spectra and FRF estimators in subsequent sections will be characterised by (ω) instead of (f).

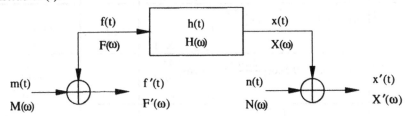

Fig. 2.8 A traditional measurement-system model.

The conventional frequency response function estimator $H_1(\omega)$ is determined by using the cross input-output spectrum and the input auto-spectrum (see also Section 1.2.7):

$$H_1(\omega) = \frac{S_{f'x'}(\omega)}{S_{f'f'}(\omega)} \tag{2.34}$$

Another version of the frequency response function estimator, $H_2(\omega)$, is obtained by normalising the output auto-spectrum by the cross input-output spectrum:

$$H_2(\omega) = \frac{S_{x'x'}}{S_{x'f'}}$$ (2.35)

As $H_1(\omega)$ and $H_2(\omega)$ should give the same result, an indicator of the quality of the analysis can be defined as the ratio of these two estimators. Thus,

$$\frac{H_1(\omega)}{H_2(\omega)} = \frac{S_{f'x'}(\omega)}{S_{f'f'}(\omega)} \frac{S_{x'f'}(\omega)}{S_{x'x'}(\omega)} = \frac{S_{f'x'}(\omega) \, S_{f'x'}^{\,*}(\omega)}{S_{f'f'}(\omega) \, S_{x'x'}(\omega)}$$

$$= \frac{\left|S_{f'x'}(\omega)\right|^2}{S_{f'f'}(\omega) \, S_{x'x'}(\omega)} = \gamma^2(\omega)$$ (2.36)

where $\gamma^2(\omega)$ is called the ordinary coherence function. It is a normalised coefficient of correlation between the measured force and response signals evaluated at each frequency. In practice, the ordinary coherence function is always greater than zero but less than unity.

When the ordinary coherence function is less than unity, one or more of the following four main conditions exist:

1. Extraneous noise is present in the FRF measurements;
2. Resolution bias errors are present in the spectral estimates;
3. The system relating f(t) and x(t) is not linear;
4. The measured response $x'(t)$ is due to other external inputs besides f(t).

Goyder [10] and Mitchell *et al* [9] pointed out that an estimator can extract the FRF ($H(\omega)$) with minimum noise-incurred errors by using the ratio of two complex cross-spectral estimates. The computation scheme of figure 2.8 can be extended to include the external input signal r(t), a broadband white noise, generated by a signal source generator (SSG), as shown in figure 2.9. The dynamic response of an electrodynamic shaker is s(t). K_{fg} is a constant derived from the stiffness of the force transducer.

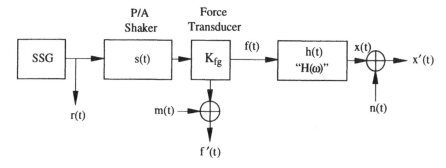

Fig. 2.9 An open-loop measurement-system model with external reference signal.

This measurement technique was used by Wellstead [11] long ago for determining transfer functions in control systems with a feedback path, and the estimator was called the 'instrumental frequency response function estimator', $H_3(\omega)$.

$$H_3(\omega) = \frac{S_{r'x'}(\omega)}{S_{r'f'}(\omega)} \tag{2.37}$$

The measurement-system models shown in figures 2.8 and 2.9 have been widely used in vibration testing. However, neither can be used to explain what causes the drop-out 'notches' in the input auto-spectrum at structural resonance frequencies. Since a realistic model is vital in mathematical analysis, an alternative model for the measurement-system is presented which seeks to reflect the true physical situation of an FRF measurement.

In order to establish a realistic measurement-system model it is necessary to characterise in detail the components of the excitation mechanism consisting of the force transducer and electrodynamic shaker. A force transducer (see Section 3.4.4) is the simplest type of piezoelectric transducer in which the relative displacement of the upper and lower plates of the cell generates a charge proportional to the transmitted force. When the force transducer is placed between the test structure and the shaker, it can be modelled as a spring with stiffness K_{fg}. By considering the shaker to have a suspension stiffness k_s, and an armature mass m_s, a greatly simplified diagram of an excitation mechanism with the test structure as used for a point FRF measurement may be shown as in figure 2.10.

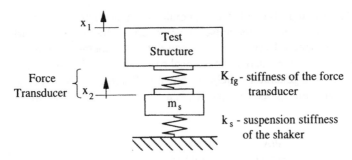

Fig. 2.10 Simplified diagram of a point FRF measurement.

It can be seen that by transforming the above figure into a block diagram, a closed-loop model is established for a point FRF measurement as shown in figure 2.11. The signal source generator provides a driving signal r(t), a broadband white noise, to a power amplifier. The dynamic response of the power amplifier $h_{em}(t)$ is assumed to be linear and thus to produce constant driving force p(t) on the armature of the shaker. The dynamic response of the shaker and its displacement output are represented by $h_s(t)$ and $x_2(t)$ respectively. The measured force $f'(t)$ is an internal force across the force transducer with uncorrelated measurement noise, m(t), and the

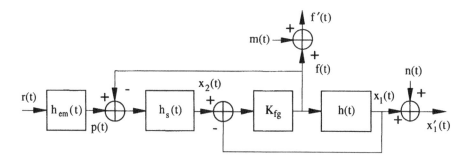

Fig. 2.11 Block diagram of a point FRF measurement with noise.

measured displacement is $x_1'(t)$ which is contaminated by uncorrelated measurement noise, $n(t)$.

The use of a reference test signal $r(t)$ for an estimate of FRF $H_3(\omega)$ has been established for many years in control engineering and the associated basic statistical properties were given by Wellstead [11, 12] for a single feedback control system. However, there are two interrelated feedback paths existing in a point FRF measurement, as shown in figure 2.11, and the following sections describe the effect of noise and the bias error due to leakage on FRF estimators.

2.3.2 Effects of noise on H_1, H_2 and H_3 estimators

With reference to figure 2.11, one can simplify the block diagram to produce an equivalent single-input two-output system as illustrated in figure 2.12. It is noted that the measured force signal is one of the output signals for a point FRF measurement.

Figure 2.12 shows that only the driving signal $r(t)$ generated from the signal source generator is not influenced by the excitation mechanism and hence is the only input signal for a point FRF measurement. Here, one can readily observe that $F'(\omega)$ contains the characteristics of the test structure as well as of the excitation mechanism itself.

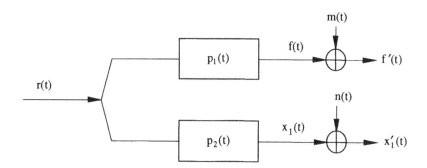

Fig. 2.12 An alternative model of a point FRF measurement.

$P_1(\omega)$ and $P_2(\omega)$ are the Fourier transforms of $p_1(t)$ and $p_2(t)$ and are expressed as

$$P_1(\omega) = \frac{H_{em}(\omega) \, H_s(\omega) \, K_{fg}}{1 + K_{fg} \left[H_s(\omega) + H(\omega) \right]} \tag{2.38}$$

$$P_2(\omega) = \frac{H_{em}(\omega) \, H_s(\omega) \, K_{fg} \, H(\omega)}{1 + K_{fg} \left[H_s(\omega) + H(\omega) \right]} \tag{2.39}$$

It is reasonable to assume that the noise terms $m(t)$ and $n(t)$ are uncorrelated with each other and with the input $r(t)$, and therefore, the following frequency domain equations apply to the model:

$$S_{rm}(\omega) = S_{fm}(\omega) = S_{rn}(\omega) = S_{x_1 n}(\omega) = S_{mn}(\omega) = 0 \tag{2.40}$$

$$S_{f'f'}(\omega) = S_{ff}(\omega) + S_{mm}(\omega) = \left| P_1(\omega) \right|^2 S_{rr}(\omega) + S_{mm}(\omega) \tag{2.41}$$

$$S_{x_1' x_1'}(\omega) = S_{x_1 x_1}(\omega) + S_{nn}(\omega) = \left| P_2(\omega) \right|^2 S_{rr}(\omega) + S_{nn}(\omega) \tag{2.42}$$

$$S_{rf'}(\omega) = S_{rf}(\omega) = P_1(\omega) \, S_{rr}(\omega) \tag{2.43}$$

$$S_{rx_1'}(\omega) = S_{rx_1}(\omega) = P_2(\omega) \, S_{rr}(\omega) \tag{2.44}$$

$$S_{f'x_1'}(\omega) = S_{fx_1}(\omega) = P_1(\omega)^* \, P_2(\omega) \, S_{rr}(\omega) \tag{2.45}$$

The conventional frequency response function estimator $H_1(\omega)$ is described by

$$H_1(\omega) = \frac{S_{f'x_1'}(\omega)}{S_{f'f'}(\omega)} = \frac{P_1(\omega)^* \, P_2(\omega) \, S_{rr}(\omega)}{\left| P_1(\omega) \right|^2 S_{rr}(\omega) + S_{mm}(\omega)} = \frac{H(\omega)}{1 + \dfrac{S_{mm}(\omega)}{S_{ff}(\omega)}} \tag{2.46}$$

Mitchell [7] showed the same result by using the model given in figure 2.8 and pointed out that the input force spectrum drops drastically near resonance, but without mathematical explanation because the model gave no information on the structure-shaker interaction. In contrast, the closed-loop model reveals that the measured force signal contains the characteristics of the test structure as well as of the excitation measurement. The auto-spectrum of the measured force is expressed by

$$S_{ff}(\omega) = \left| P_1(\omega) \right|^2 S_{rr}(\omega) = \left| \frac{H_{em}(\omega) \, H_s(\omega) \, K_{fg}}{1 + K_{fg} \left[H_s(\omega) + H(\omega) \right]} \right|^2 S_{rr}(\omega) \tag{2.47}$$

By representing the receptance of the test structure in terms of an effective stiffness k_e and an effective mass m_e with hysteretic loss factor η at a structural resonance, (2.47) can be rewritten as

$$S_{ff}(\omega) =$$

$$\left| \frac{H_{em}\left(k_e - m_e\,\omega^2 + i\,\eta\,k_e\right)}{\dfrac{1}{K_{fg}}\left(k_e - m_e\omega^2 + i\eta k_e\right)\!\left(k_s - m_s\omega^2\right) + \left(k_e - m_e\omega^2 + i\eta k_e\right) + \left(k_s - m_s\omega^2\right)} \right|^2 S_{rr}(\omega)$$

$$(2.48)$$

When the excitation frequency ω approaches the structural resonance frequency, $\omega_r = \sqrt{k_e / m_e}$, $S_{f'x_1'}(\omega)$ is noise-free, but the noise-to-signal ratio decreases due to the peak response of force spectrum $S_{ff}(\omega)$, if the structural resonance frequency is greater than the combined structure-shaker resonance frequency in (2.48). There is a drastic change in the small frequency range around the structure's resonance. The force drops to a minimum so that the noise-to-signal ratio increases from a minimum to a maximum, and when the excitation frequency is increased beyond the structural resonance frequency, the noise-to-signal ratio decreases again. A typical drop-out notch in the force auto-spectrum is shown in figure 2.13.

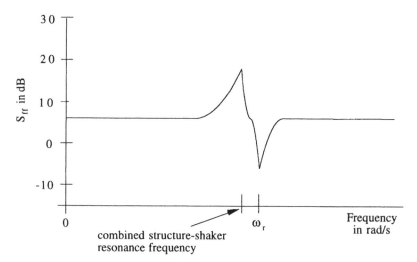

Fig. 2.13 Force auto-spectrum for a point FRF measurement.

An alternative frequency response function estimator $H_2(\omega)$ is described by

$$H_2(\omega) = \frac{S_{x_1'x_1'}(\omega)}{S_{x_1'f'}(\omega)} = \frac{\left|P_2(\omega)\right|^2 S_{rr}(\omega) + S_{nn}(\omega)}{P_2(\omega)^* P_1(\omega) S_{rr}(\omega)} = H(\omega)\left(1 + \frac{S_{nn}(\omega)}{S_{x_1x_1}(\omega)}\right) \quad (2.49)$$

One can observe that this estimator is contaminated only by the noise-to-signal ratio of the measured response. In the closed-loop model, it can be shown that the auto-spectrum of the measured response is given by

$$S_{x_1 x_1}(\omega) = |P_2(\omega)|^2 \, S_{rr}(\omega) = \left| \frac{H_{em}(\omega) \, H_s(\omega) \, K_{fg} \, H(\omega)}{1 + K_{fg} \left(H_s(\omega) + H(\omega) \right)} \right|^2 S_{rr}(\omega) \qquad (2.50)$$

When the excitation frequency approaches a structural antiresonance frequency, the noise-to-signal ratio increases to a maximum as the response auto-spectrum $S_{x_1 x_1}(\omega)$ becomes a minimum. Equation (2.50) shows a drop in the response auto-spectrum resulting from the nature of $H(\omega)$ at a structural antiresonance.

When the excitation frequency is increased beyond the structural antiresonance frequency, the noise-to-signal ratio decreases. At resonance one recalls that $S_{mm}(\omega)$, the input noise, can become significant. However, $H_2(\omega)$ is insensitive to such an effect, and, moreover, the response at resonance is still large compared to $S_{nn}(\omega)$, thus allowing (2.49) to provide a better estimate of $H(\omega)$ than (2.46).

Since $H_1(\omega)$ is a lower bound estimator and $H_2(\omega)$ is an upper bound estimator, one can calculate the average or the geometric mean of $H_1(\omega)$ and $H_2(\omega)$ to obtain another FRF estimate. The expression for the FRF estimator, $H_V(\omega)$, which is the geometric mean of $H_1(\omega)$ and $H_2(\omega)$ and lies between the upper and lower bound of the FRF estimate, is given as

$$H_V(\omega) = \sqrt{H_1(\omega) \, H_2(\omega)} = H(\omega) \sqrt{\left(1 + \frac{S_{nn}(\omega)}{S_{x_1 x_1}(\omega)} \right) \left(1 + \frac{S_{mm}(\omega)}{S_{ff}(\omega)} \right)^{-1}} \qquad (2.51)$$

This estimator takes the contamination in measured force and response auto-spectra into account. In the case when the noise-to-signal ratios for measured force and response signals are 'small' and approximately the same, $H_V(\omega)$ should provide a better picture of the true FRF.

However, practical experience and the aforementioned analysis show that those noise-to-signal ratios vary with the excitation frequency and their magnitude depends on the structure's properties as well as the excitation mechanism. Hence, this FRF estimator is unlikely to provide a meaningful result from a mathematical point of view.

The normal frequency response function estimators of $H(\omega)$ based upon (2.46) and (2.49) are systematically in error when the measured force and output signals are contaminated with noise. This difficulty can be avoided by introducing the independent external input signal $r(t)$, as shown in figure 2.11, where $r(t)$ is a broadband random signal. Because $r(t)$ is independent of $m(t)$ and $n(t)$, and is statistically uncorrelated to the disturbances $m(t)$ and $n(t)$, the cross-spectra $S_{rm}(\omega)$ and $S_{rn}(\omega)$ are identically zero. Hence, the instrumental frequency response estimator $H_3(\omega)$ is given by

$$H_3(\omega) = \frac{S_{rx_i'}(\omega)}{S_{rf'}(\omega)} = \frac{S_{rx_1}(\omega)}{S_{rf}(\omega)} = \frac{P_2(\omega)\,S_{rr}(\omega)}{P_1(\omega)\,S_{rr}(\omega)} = H(\omega) \qquad (2.52)$$

$H_3(\omega)$ is a noise-free FRF estimator provided that due attention is paid to the statistical error caused by windowing for a finite amount of data.

As mentioned in Section 2.2, when both measured force and response are random time signals, it is not sufficient to compute Fourier transforms because the signals are not deterministic in nature. Additional consideration must be given to the statistical reliability and accuracy of FRF estimates. Generally, it is necessary to perform an averaging of several individual time records, or samples, to reduce the random fluctuation in the estimation of FRFs.

These time records are normally independent, so that if m time records are averaged, the variability of the estimates will decrease inversely with m and will depend on the degree to which f(t) and x(t) are correlated. The variability of the magnitude and phase of the $H_1(\omega)$ estimator is given approximately by

$$\frac{\mathrm{var}\left[\left|\hat{H}_1(\omega)\right|\right]}{\left|H(\omega)\right|^2} = \mathrm{var}\left[\arg\hat{H}_1(\omega)\right] = \frac{1}{2\,m}\left(\frac{1}{\gamma_{f'x_i'}(\omega)^2} - 1\right) \qquad (2.53)$$

where the circumflex ($^\wedge$) indicates an estimated quantity.

The variability of the instrumental FRF estimator can, to a first approximation, be assessed by taking a Taylor series expansion about the expected values of $\hat{H}(\omega)$ [11]. In this manner it can be shown that around resonance

$$\frac{\mathrm{var}\left[\left|\hat{H}_3(\omega)\right|\right]}{\left|H(\omega)\right|^2} = \mathrm{var}\left[\arg\hat{H}_3(\omega)\right] \approx \frac{1}{2m}\left(\frac{S_{mm}(\omega)}{S_{rr}(\omega)\,|P_1(\omega)|}\right) \qquad (2.54)$$

and around antiresonance

$$\frac{\mathrm{var}\left[\left|\hat{H}_3(\omega)\right|\right]}{\left|H(\omega)\right|^2} = \mathrm{var}\left[\arg\hat{H}_3(\omega)\right] \approx \frac{1}{2\,m}\left(\frac{S_{nn}(\omega)}{S_{rr}(\omega)\,|P_2(\omega)|}\right) \qquad (2.55)$$

Here, the functions $P_1(\omega)$ and $P_2(\omega)$ are the equivalent closed-loop transfer functions used in figure 2.12. The desirable feature of the $H_3(\omega)$ estimate is that the variance of its magnitude and phase is inversely proportional to the number of averages m and the spectral power associated with r(t).

2.3.3   Effects of leakage on H_1, H_2 and H_3 estimators

Noise is not the only source of errors in an FRF measurement. The frequency resolution obtained from an FFT analyser is 1/T Hz, where T is the record length employed. The process is not equivalent to a frequency sweep using sinusoidal excitation where readings are taken at discrete frequencies. The spectral components obtained from an FFT analyser are average values across the band of the FFT 'filter'. Therefore, maxima (resonances) in the FRF tend to be underestimated while minima (antiresonances) tend to be overestimated. This inaccuracy is a bias error caused by the FFT of a finite record length and is known as leakage. A model for the measurement of a noise-free point FRF measurement is presented in figure 2.14, in which one mode of the system comprising an effective mass m_e, effective stiffness k_e, hysteretic loss factor η and its natural frequency ω_r dominates, under excitation by a broadband random reference signal of constant spectral power, $R(\omega)$.

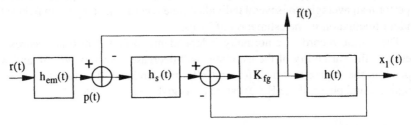

Fig. 2.14 A noise-free point FRF measurement.

The relationship between the displacement signal and the input signal produced by the random signal generator is shown as follows

$$\frac{X_1(\omega)}{R(\omega)} = \frac{H_{em}(\omega)\,H_s(\omega)\,K_{fg}\,H(\omega)}{1 + K_{fg}\left[H_s(\omega) + H(\omega)\right]} \tag{2.56}$$

The relationship between the force signal and the input signal is

$$\frac{F(\omega)}{R(\omega)} = \frac{H_{em}(\omega)\,H_s(\omega)\,K_{fg}}{1 + K_{fg}\left[H_s(\omega) + H(\omega)\right]} \tag{2.57}$$

For the system employing the symbols used in figure 2.14, (2.56) and (2.57) can be rewritten as

$$\frac{X_1(\omega)}{R(\omega)} = \frac{H_{em}(\omega)}{\dfrac{1}{K_{fg}}\left(k_e - m_e\omega^2 + i\eta k_e\right)\left(k_s - m_s\omega^2\right) + \left(k_e - m_e\omega^2 + i\eta k_e\right) + \left(k_s - m_s\omega^2\right)} \tag{2.58}$$

$$\frac{F(\omega)}{R(\omega)} = \frac{H_{em}(\omega)\,m_e\left(\omega_r^2 - \omega^2 + i\eta\omega_r^2\right)}{\dfrac{1}{K_{fg}}\left(k_e - m_e\omega^2 + i\eta k_e\right)\left(k_s - m_s\omega^2\right) + \left(k_e - m_e\omega^2 + i\eta k_e\right) + \left(k_s - m_s\omega^2\right)} \tag{2.59}$$

As the stiffness of the force transducer approaches infinity, the following approximate expression is derived

$$\frac{F(\omega)}{R(\omega)} \approx \frac{H_{em}(\omega)\, m_e \left(\omega_r^2 - \omega^2 + i\eta\omega_r^2\right)}{m_e \left(\omega_r^2 - \omega^2 + i\eta\omega_r^2\right) + \left(\dfrac{k_s}{m_e} - \dfrac{m_s}{m_e}\omega^2\right)} \tag{2.60}$$

Expression (2.60) is similar to the equation derived by Cawley [13]. The estimates of the spectra produced by a Fourier-based analysis may be derived as follows

$$S_{x_1 x_1}(\omega) = \int_{-\infty}^{+\infty} \frac{\left|H_{em}(\Omega)\right|^2}{\left|G(\Omega)\right|^2}\, R_0\, W(\Omega - \omega)\, d\Omega \tag{2.61}$$

$$S_{x_1 f}(\omega) = \int_{-\infty}^{+\infty} \frac{\left|H_{em}(\Omega)\right|^2}{\left|G(\Omega)\right|^2}\left(k_e - m_e\,\Omega^2 + i\eta k_e\right) R_0\, W(\Omega - \omega)\, d\Omega \tag{2.62}$$

where W is a function of the window, e.g., Hanning, and

$$G(\Omega) = \frac{1}{K_{fg}}\left(k_e - m_e\Omega^2 + i\eta k_e\right)\left(k_s - m_s\Omega^2\right) + \left(k_e - m_e\Omega^2 + i\eta k_e\right) + \left(k_s - m_s\Omega^2\right) \tag{2.63}$$

Substituting the estimates of the auto- and cross-spectra shown in (2.61) and (2.62) into (2.49), the expression of $H_2(\omega)$ for a point noise-free FRF measurement is derived:

$$H_2(\omega) = \frac{S_{x_1 x_1}(\omega)}{S_{x_1 f}(\omega)} = \frac{\displaystyle\int_{-\infty}^{+\infty} \frac{\left|H_{em}(\Omega)\right|^2}{\left|G(\Omega)\right|^2}\, R_0\, W(\Omega - \omega)\, d\Omega}{\displaystyle\int_{-\infty}^{+\infty} \frac{\left|H_{em}(\Omega)\right|^2}{\left|G(\Omega)\right|^2}\left(k_e - m_e\,\Omega^2 + i\eta k_e\right) R_0\, W(\Omega - \omega)\, d\Omega} \tag{2.64}$$

Setting

$$A = \int_{-\infty}^{+\infty} \frac{\left|H_{em}(\Omega)\right|^2}{\left|G(\Omega)\right|^2}\, R_0\, W(\Omega - \omega)\, d\Omega \tag{2.65}$$

$$B = \int_{-\infty}^{+\infty} \frac{\left|H_{em}(\Omega)\right|^2 \Omega^2}{\left|G(\Omega)\right|^2}\, R_0\, W(\Omega - \omega)\, d\Omega \tag{2.66}$$

112

gives

$$H_2(\omega) = \frac{A}{m_e \left(A\,\omega_r^2 - B + i\,A\,\eta\,\omega_r^2 \right)} = \frac{1}{m_e \left(\omega_r^2 - \frac{B}{A} + i\,\eta\,\omega_r^2 \right)} \tag{2.67}$$

while the true frequency response function is given by

$$H(\omega) = \frac{1}{m_e \left(\omega_r^2 - \omega^2 + i\,\eta\,\omega_r^2 \right)} \tag{2.68}$$

Comparison of (2.67) and (2.68) reveals that the expressions for the FRF estimate, $H_2(\omega)$, and the true value, $H(\omega)$, are of the same form and are identical if $B/A = \omega^2$. This condition is only satisfied at a very high frequency resolution ($\Omega \to \omega$). In general, the estimate $H_2(\omega)$ lies on the true modal circle but at different positions around the circle.

By using a similar analysis, one can derive the FRF estimate $H_1(\omega)$ based on the cross-spectrum and the measured force auto-spectrum, which are defined as follows

$$S_{fx_1}(\omega) = \int_{-\infty}^{+\infty} \frac{|H_{em}(\Omega)|^2}{|G(\Omega)|^2} \left(k_e - m_e\,\Omega^2 - i\,\eta\,k_e \right) R_o\, W(\Omega - \omega)\, d\Omega \tag{2.69}$$

$$S_{ff}(\omega) = \int_{-\infty}^{+\infty} \frac{|H_{em}(\Omega)|^2}{|G(\Omega)|^2} \left[\left(k_e - m_e\,\Omega^2 \right)^2 + \left(\eta\,k_e \right)^2 \right] R_o\, W(\Omega - \omega)\, d\Omega \tag{2.70}$$

Using the estimates of the cross- and auto-spectra derived in (2.69) and (2.70), the estimate of $H_1(\omega)$ is expressed by

$$H_1(\omega) = \frac{S_{fx_1}(\omega)}{S_{ff}(\omega)} = \frac{\displaystyle\int_{-\infty}^{+\infty} \frac{|H_{em}(\Omega)|^2}{|G(\Omega)|^2} \left(k_e - m_e\,\Omega^2 - i\,\eta\,k_e \right) R_o\, W(\Omega - \omega)\, d\Omega}{\displaystyle\int_{-\infty}^{+\infty} \frac{|H_{em}(\Omega)|^2}{|G(\Omega)|^2} \left[\left(k_e - m_e\,\Omega^2 \right)^2 + \left(\eta\,k_e \right)^2 \right] R_o\, W(\Omega - \omega)\, d\Omega} \tag{2.71}$$

Setting now

$$C = \int_{-\infty}^{+\infty} \frac{|H_{em}(\Omega)|^2\, \Omega^4}{|G(\Omega)|^2}\, R_o\, W(\Omega - \omega)\, d\Omega \tag{2.72}$$

gives

$$H_1(\omega) = \frac{A\,\omega_r^2 - B - iA\,\eta\,\omega_r^2}{m_e\left(A\,\omega_r^4 - 2B\,\omega_r^2 + C + A\,\eta^2\,\omega_r^4\right)}$$

$$= \frac{\omega_r^4 - 2\dfrac{B}{A}\,\omega_r^2 + \left(\dfrac{B}{A}\right)^2 + \eta^2\,\omega_r^4}{\omega_r^4 - 2\dfrac{B}{A}\,\omega_r^2 + \dfrac{C}{A} + \eta^2\,\omega_r^4}\,H_2(\omega) \qquad (2.73)$$

Hence, the estimates of $H_1(\omega)$ have the same phase as those of $H_2(\omega)$ but in general the magnitudes are smaller and, as a result, the estimate $H_1(\omega)$ always lies inside the true modal circle. If the frequency resolution is increased, thus reducing the leakage error, $H_1(\omega)$ tends towards $H_2(\omega)$ which tends towards the true FRF, and all three FRF estimates, $H_1(\omega)$, $H_2(\omega)$ and $H_V(\omega)$, give the correct magnitude and phase of the frequency response data.

The formulation of the instrumental FRF estimator $H_3(\omega)$ is different and requires the derivation of the estimates of two cross-spectra shown as follows:

$$S_{rx_1}(\omega) = \int_{-\infty}^{+\infty} \frac{H_{em}(\Omega)}{G(\Omega)}\, R_0\, W(\Omega - \omega)\, d\Omega \qquad (2.74)$$

$$S_{rf}(\omega) = \int_{-\infty}^{+\infty} \frac{H_{em}(\Omega)}{G(\Omega)} \left(k_e - m_e\,\Omega^2 + i\eta\,k_e\right) R_0\, W(\Omega - \omega)\, d\Omega \qquad (2.75)$$

Using these estimates of the cross-spectra, the estimate of $H_3(\omega)$ is expressed by

$$H_3(\omega) = \frac{S_{rx_1}(\omega)}{S_{rf}(\omega)} = \frac{\displaystyle\int_{-\infty}^{+\infty} \frac{H_{em}(\Omega)}{G(\Omega)}\, R_0\, W(\Omega - \omega)\, d\Omega}{\displaystyle\int_{-\infty}^{+\infty} \frac{H_{em}(\Omega)}{G(\Omega)} \left(k_e - m_e\,\Omega^2 + i\eta\,k_e\right) R_0\, W(\Omega - \omega)\, d\Omega}$$

$$(2.76)$$

As no simple or enlightening relation between $H_3(\omega)$ and the true $H(\omega)$ exists, computer simulation is employed to explore the bias error of the $H_3(\omega)$ FRF estimator. Let us consider a particular mode of a structure with effective mass m_e, effective stiffness k_e and hysteretic loss factor η, excited through a force gauge with stiffness K_{fg}, and a shaker with armature mass m_s and suspension stiffness k_s.

The predicted variations of the FRF estimators, $H_1(\omega)$, $H_2(\omega)$ and $H_3(\omega)$, for different armature suspension stiffness of the shaker are shown in figure 2.15. It should be noted that the value of $H_{em}(\omega)$ is unity so that the external input signal has a uniform frequency spectrum R_0. The integration of expressions (2.64), (2.73) and (2.76) can be carried out digitally using the FFT in which the bandwidth of the true spectra is limited. A Hanning window is chosen in all cases.

114

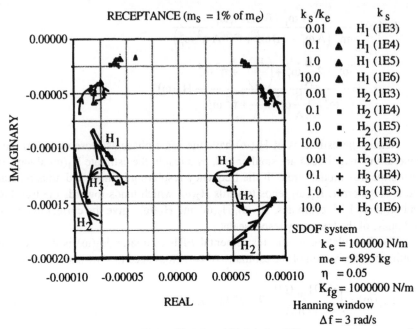

Fig. 2.15 The variations of $H_1(\omega)$, $H_2(\omega)$ and $H_3(\omega)$ for different armature suspension stiffness of the shaker.

The curves in figure 2.15 show how the estimates $H_1(\omega)$, $H_2(\omega)$ and $H_3(\omega)$ vary as the ratio of the armature suspension stiffness for the shaker to the effective stiffness of the system varies. When this stiffness ratio is small, $H_3(\omega)$ is close to $H_1(\omega)$. When the ratio is large, $H_3(\omega)$ is close to $H_2(\omega)$. The FRF data can be analysed by using a SDOF curve-fitting algorithm, and the bias errors for the resulting modal properties of this mode are shown in figures 2.16 a) - d).

Figure 2.16 a) shows a graph of the bias error on natural frequency estimate against the stiffness ratio. The natural frequency of the mode is overestimated when the stiffness ratio is below 0.1% for $H_1(\omega)$, 0.1% for $H_2(\omega)$ and 1% for $H_3(\omega)$ FRF estimates, respectively. The natural frequency is underestimated when the stiffness ratio increases. All estimates of natural frequency approach the correct analytical solution when the stiffness ratio is equal to or greater than 10.

Figure 2.16 b) shows a graph of the radius of the modal circle against the stiffness ratio from which it can be seen that only $H_2(\omega)$ always lies on the true modal circle. Both the curves $H_1(\omega)$ and $H_3(\omega)$ lie inside the true modal circle in most cases.

Figure 2.16 c) shows a graph of the bias error on the damping loss factor against the stiffness ratio. The damping loss factor is overestimated for all $H_1(\omega)$ FRF estimates. When the stiffness ratio is equal to or greater than 10, the $H_2(\omega)$ and $H_3(\omega)$ FRF estimates give accurate results for the identified damping loss factor. The bias error on the modal constant against the stiffness ratio is shown in figure 2.16 d) from which it can be seen that those estimates produce accurate results when the stiffness ratio is equal to or larger than 10. When the mass of the armature of the shaker increases from 1% to a higher percentage of the effective mass of the system, the computational results of various FRF estimators and their identified modal properties follow a trend similar to that mentioned earlier.

The effects of noise and leakage on $H_1(\omega)$, $H_2(\omega)$ and $H_3(\omega)$ estimators for any general FRF measurements are very similar.

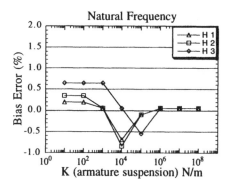

Fig. 2.16 a) The error on natural frequency using $H_1(\omega)$, $H_2(\omega)$ and $H_3(\omega)$.

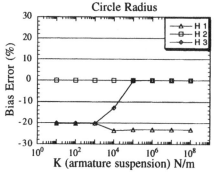

Fig. 2.16 b) The error on the radius of modal circle using $H_1(\omega)$, $H_2(\omega)$ and $H_3(\omega)$.

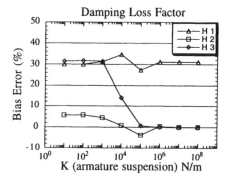

Fig. 2.16 c) The error on damping loss factor using $H_1(\omega)$, $H_2(\omega)$ and $H_3(\omega)$.

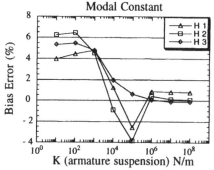

Fig. 2.16 d) The error on modal constant using $H_1(\omega)$, $H_2(\omega)$ and $H_3(\omega)$.

2.4 MULTI-INPUT MULTI-OUTPUT ANALYSIS OF MECHANICAL STRUCTURES

The preceding section was devoted to a proper understanding and description of the effect of feedback paths on currently-used FRF estimators for single-shaker modal testing. In some cases, a single shaker is unable to accomplish the requirements of a particular test, such as a need for application of large forces and linearisation of nonlinear structures. In such situations, one has to use two or more shakers in providing a better energy distribution in a test structure, resulting in a more consistent estimation of resonance frequencies, mode shapes and damping loss factors.

Another advantage of multi-shaker modal testing is the ability to excite all modes in a frequency range of interest. This is in contrast to single-shaker modal testing where (i) some modes cannot be observed if the force is applied to a location close to a nodal point for some particular modes, and (ii) inconsistencies may occur in the measured FRFs as the set of force and response transducers are moved to all the required locations on the structure.

Another deficiency of single-point excitation is that, in an effort to excite remote regions in a large structure, the single point forcing level is sometimes increased to excessive levels, thereby inducing nonlinear behaviour, especially in the region of the excitation point. With multiple-input excitation, the response amplitudes across the structure will be more uniform, with a consequent decrease in the effect of nonlinearity.

2.4.1 Partial and Multiple Coherence functions

For multi-shaker modal testing, the theoretical basis of existing FRF analysis has been well documented [14, 15]. The existing theories have been developed based on the general case of n inputs and m outputs measured during a modal test as shown in figure 2.17, from which an open-loop multi-input multi-output (MIMO) system is identified.

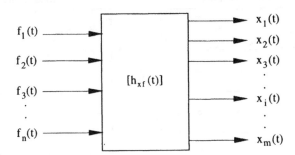

Fig. 2.17 Traditional multi-input multi-output model.

By considering the test structure to be a linear and time-invariant system, a measured output (response) signal $X_i'(\omega)$ in frequency domain, which is contaminated by an output noise, is expressed as:

$$X_i'(\omega) = \sum_{j=1}^{n} H_{ij}(\omega)\, F_j(\omega) + N_i(\omega) \tag{2.77}$$

Using matrix notation, the measured output vector of a *MIMO* system is written as

$$\{X'(\omega)\} = \left[H_{xf}(\omega)\right]\{F(\omega)\} + \{N(\omega)\} \tag{2.78}$$
$$\underset{(mx1)}{} \quad \underset{(mxn)}{} \quad \underset{(nx1)}{} \quad \underset{(mx1)}{}$$

The transpose of $\{X'(\omega)\}$ is

$$\{X'(\omega)\}^T = \{F(\omega)\}^T \left[H_{xf}(\omega)\right]^T + \{N(\omega)\}^T \tag{2.79}$$
$$\underset{(1xm)}{} \quad \underset{(1xn)}{} \quad \underset{(nxm)}{} \quad \underset{(1xm)}{}$$

Pre-multiplying both sides of (2.79) by $\{F(\omega)\}^*$ and taking expected values, gives the following matrix equation in terms of output-input and input cross-spectrum matrices, as the measured output noise vector and input vector are not correlated:

$$\left[S_{fx'}(\omega)\right] = \left[S_{ff}(\omega)\right]\left[H_{xf}(\omega)\right]^T \tag{2.80}$$
$$\underset{(nxm)}{} \quad \underset{(nxn)}{} \quad \underset{(nxm)}{}$$

As a result, the least-squares estimate of the FRF matrix can be computed by

$$\left[H_{xf}(\omega)\right] = \left(\left[S_{ff}(\omega)\right]^{-1}\left[S_{fx'}(\omega)\right]\right)^T \tag{2.81}$$
$$\underset{(nxm)}{} \quad \underset{(nxn)}{} \quad \underset{(nxm)}{}$$

Although most of the literature considers the effect of noise in measured output signals only, in practice both the measured input and measured output signals are contaminated by uncorrelated random noise. The measured input (force) vector can be considered as a summation of the input vector $\{F(\omega)\}$ and a measurement noise vector $\{M(\omega)\}$ as

$$\{F'(\omega)\} = \{F(\omega)\} + \{M(\omega)\} \tag{2.82}$$
$$\underset{(nx1)}{} \quad \underset{(nx1)}{} \quad \underset{(nx1)}{}$$

Substituting (2.82) into (2.79), gives

$$\{X'(\omega)\}^T = \left\{\{F'(\omega)\} - \{M(\omega)\}\right\}^T \left[H_{xf}(\omega)\right]^T + \{N(\omega)\}^T \tag{2.83}$$
$$\underset{(1xm)}{} \quad \underset{(nx1)}{} \quad \underset{(nx1)}{} \quad \underset{(nxm)}{} \quad \underset{(1xm)}{}$$

Pre-multiplying both sides of (2.83) by $\{F'(\omega)\}^*$ and taking expected values, one obtains:

$$\left[S_{f'x'}(\omega)\right] = \left[\left[S_{f'f'}(\omega)\right] - \left[S_{mm}(\omega)\right]\right]\left[H_{xf}(\omega)\right]^T \tag{2.84}$$
$$\underset{(nxm)}{} \quad \underset{(nxn)}{} \quad \underset{(nxm)}{}$$

where $[S_{mm}(\omega)]$ is a diagonal matrix whose elements represent noise auto-spectra at each transducer. Since it is not feasible to separate the input noise auto-spectrum matrix from the measured input (force) cross-spectrum matrix, the estimate of FRF matrix is written as

$$\left[{}_1H_{xf}(\omega)\right] = \left(\underset{(n\times n)}{\left[S_{f'f'}(\omega)\right]^{-1}} \underset{(n\times m)}{\left[S_{f'x'}(\omega)\right]} \right)^T \qquad (2.85)$$
$$\underset{(m\times n)}{}$$

It is noted that the measured input cross-spectrum matrix must be inverted at every frequency in the frequency range of interest. Equation (2.85) is valid when all the measured input signals are not fully correlated. Since the estimate of $[H_{xf}(\omega)]$ is contaminated by the measurement noise existing in the measured input (force) signals and the input cross-spectra are used, this $[{}_1H_{xf}(\omega)]$ estimator will converge to a biased, lower-bound estimate of the true $[H_{xf}(\omega)]$.

Moreover, practical experience shows that there are a number of situations where the input cross-spectrum matrix $[S_{ff}(\omega)]_{n\times n}$ is singular at specific frequencies or over specific frequency intervals, usually at and around the structural resonances. When this happens, the inverse of $[S_{ff}(\omega)]_{n\times n}$ will not exist and thus (2.85) cannot be used to determine the estimate of FRFs.

As there are a number of situations where the cross-spectrum matrix of measured force signals is singular at specific frequencies or frequency intervals, partial coherence functions [15] have been proposed for establishing the causal input-input, input-output and output-output relationships. Bendat [15] presented an iterative procedure for obtaining ordinary and partial coherence functions based on conditioned spectra.

The ordinary and partial coherence functions of the ith output are expressed as

$$\gamma^2_{x_i' f_1'}(\omega) = \frac{\left|S_{x_i' f_1'}(\omega)\right|^2}{S_{f_1' f_1'}(\omega) S_{x_i' x_i'}(\omega)} \qquad (2.86)$$

$$\gamma^2_{x_i' f_2' \bullet f_1'}(\omega) = \frac{\left|S_{x_i' f_2' \bullet f_1'}(\omega)\right|^2}{S_{f_2' f_2' \bullet f_1'}(\omega) S_{x_i' x_i' \bullet f_1'}(\omega)} \qquad (2.87)$$

$$\gamma^2_{x_i' f_3' \bullet f_2'!}(\omega) = \frac{\left|S_{x_i' f_3' \bullet f_2'!}(\omega)\right|^2}{S_{f_3' f_3' \bullet f_2'!}(\omega) S_{x_i' x_i' \bullet f_2'!}(\omega)} \qquad (2.88)$$

$$\vdots \qquad\qquad \vdots$$

$$\gamma^2_{x_i' f_n' \bullet f_{n-1}'!}(\omega) = \frac{\left|S_{x_i' f_n' \bullet f_{n-1}'!}(\omega)\right|^2}{S_{f_n' f_n' \bullet f_{n-1}'!}(\omega) S_{x_i' x_i' \bullet f_{n-1}'!}(\omega)} \qquad (2.89)$$

where $S_{f'_n f'_n \bullet f'_{n-1}!}(\omega)$, $S_{x'_i f'_n \bullet f'_{n-1}!}(\omega)$ and $S_{x'_i x'_i \bullet f'_{n-1}!}(\omega)$ are the conditioned spectra obtained by removing coherent contributions of the previous (n-1) inputs.

The operations involved in the calculation of conditioned spectra are identical to the ones used in the LU decomposition of the following matrix, using Gauss elimination:

$$
\begin{bmatrix}
\begin{bmatrix} S_{f'f'}(\omega) \end{bmatrix} & \begin{bmatrix} S_{f'x'}(\omega) \end{bmatrix} \\
\text{(nxn)} & \text{(nxm)} \\
\begin{bmatrix} S_{x'f'}(\omega) \end{bmatrix} & \begin{bmatrix} S_{x'x'}(\omega) \end{bmatrix} \\
\text{(mxn)} & \text{(mxm)}
\end{bmatrix}
\tag{2.90}
$$

The problem associated with this technique is the value of partial coherence functions being dependent on the order of the measured input signals. In practice, the usefulness of partial coherence functions is limited.

Multiple coherence is defined as the correlation coefficient describing the possible causal relationship between an output and all known inputs. Bendat [15] showed that for a n-input m-output model, the multiple coherence function between the ith output and all inputs satisfies the following relationship:

$$
\gamma^2_{x'_i : f'}(\omega) = 1 - \left(\left(1 - \gamma^2_{x'_i f'_1}(\omega)\right)\left(1 - \gamma^2_{x'_i f'_2 \bullet f'_1}(\omega)\right) \ldots \left(1 - \gamma^2_{x'_i f'_n \bullet f'_{n-1}!}(\omega)\right) \right)
\tag{2.91}
$$

where $\gamma^2_{x'_i : f'}(\omega)$ is the multiple coherence function of the ith output. For the special case when the inputs are mutually uncorrelated, this reduces to

$$
\gamma^2_{x'_i : f'}(\omega) = \sum_{k=1}^{n} \gamma^2_{x'_i f'_k}(\omega)
\tag{2.92}
$$

where $\gamma^2_{x'_i f'_k}(\omega)$ is the ordinary coherence function between the kth input and the ith output. The multiple coherence functions can also be obtained using the estimate of $[S_{x'x'}(\omega)]$ and the measured $[S_{x'x'}(\omega)]$. Equation (2.85) gives the estimate of the FRF matrix. Pre-multiplying the measured input-output cross-spectrum matrix by the complex conjugate of $[_1 H_{xf}(\omega)]$, one obtains

$$
\begin{bmatrix} \tilde{S}_{x'x'}(\omega) \end{bmatrix} = \begin{bmatrix} _1 H_{xf}(\omega) \end{bmatrix}^* \begin{bmatrix} S_{f'x'}(\omega) \end{bmatrix}
\tag{2.93}
$$
$$
\text{(mxm)} \qquad\qquad \text{(mxn)} \qquad\qquad \text{(nxm)}
$$

where $[\tilde{S}_{x'x'}(\omega)]$ is the estimate of the output cross-spectrum matrix. The multiple coherence function of the ith output is the ratio of the ith diagonal element of $[\tilde{S}_{x'x'}(\omega)]$ and $[S_{x'x'}(\omega)]$:

$$
\gamma^2_{x'_i : f'}(\omega) = \frac{\tilde{S}_{x'_i x'_i}(\omega)}{S_{x'_i x'_i}(\omega)}
\tag{2.94}
$$

If the multiple coherence functions for all output signals are close to unity, then the FRFs obtained using equation (2.85) are accurate.

Leuridan [16] noted that a unique solution for the *MIMO* problem can be worked out for the case where the measured inputs are not correlated. He also pointed out that most solution techniques applied to the *MIMO* situation can handle the case of some correlation between the measured inputs when the multiple coherence functions are not much less than unity, but fail if the measured inputs become highly correlated. This phenomenon is quite common at the frequency ranges around structural resonances. In those cases the input cross-spectrum matrix $[S_{f'f'}(\omega)]$ is rank deficient which is unavoidable due to structure-shakers interaction.

As will subsequently be shown, it is still possible to determine the dynamic characteristics of a test structure by using the cross-spectrum matrices even though the measured force signals are highly correlated. An FRF technique is presented below, which can give noise-free estimates of FRFs and requires minimum post-processing calculation.

2.4.2 Effects of Structure-Shakers interaction on FRFs

The closed-loop model used in Section 2.3 for single-shaker modal testing can be generalised to produce a model for multi-shaker modal testing as shown in figure 2.18.

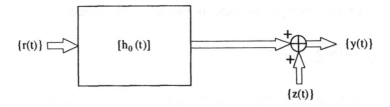

Fig. 2.18 A generalised model for a multi-shaker modal test which takes into account the effect of structure-shakers interaction on FRFs measurement (with measurement noise).

Figure 2.18 illustrates an equivalent open-loop system model for the multi-shaker modal test system with measurement noise in the time domain. It should be noted that $\{r(t)\}_{nx1}$ is a true nx1 external input vector while the output vector $\{y(t)\}_{(n+m)x1}$, is subdivided into two parts, $\{f'(t)\}_{nx1}$ and $\{x'(t)\}_{mx1}$, corresponding to the measured force and measured response vectors respectively.

The equivalent open-loop FRF matrix $[h_o(t)]_{(n+m)xn}$, is partitioned into two rectangular matrices $[h_f(t)]_{nxn}$ and $[h_x(t)]_{mxn}$, where $[h_f(t)]_{nxn}$ and $[h_x(t)]_{mxn}$ are the equivalent open-loop impulse response functions for the measured forces and measured responses respectively. The contaminating noise vector $\{z(t)\}_{(n+m)x1}$, is also subdivided into two parts, $\{m(t)\}_{nx1}$ and $\{n(t)\}_{mx1}$, and is defined to be uncorrelated with $\{r(t)\}_{nx1}$, as expressed in the frequency domain by

$$E\left[\{R(\omega)\}^*\{Z(\omega)\}^T\right]=[0]$$
$$(nx(n+m))$$

(2.95)

The output $\{Y(\omega)\}$ in the frequency domain is written as

$$\{Y(\omega)\}=\left[H_o(\omega)\right]\{R(\omega)\}+\{Z(\omega)\}$$
$$((n+m)x1)\quad((n+m)xn)\quad(nx1)\quad((n+m)x1)$$

(2.96)

or

$$\begin{Bmatrix}\{F'(\omega)\}\\(nx1)\\\{X'(\omega)\}\\(mx1)\end{Bmatrix}=\begin{bmatrix}\left[H_F(\omega)\right]\\(nxn)\\\left[H_X(\omega)\right]\\(mxn)\end{bmatrix}\{R(\omega)\}+\begin{Bmatrix}\{M(\omega)\}\\(nx1)\\\{N(\omega)\}\\(mx1)\end{Bmatrix}$$
$$((n+m)x1)\qquad\quad((n+m)xn)\qquad(nx1)\qquad((n+m)x1)$$

(2.97)

Pre-multiplying both sides of the transpose of (2.96) by $\{R(\omega)\}^*$ and taking expected values one gets

$$E\left[\{R(\omega)\}^*\{Y(\omega)\}^T\right]=E\left[\{R(\omega)\}^*\{R(\omega)\}^T\right]\left[H_o(\omega)\right]^T+E\left[\{R(\omega)\}^*\{Z(\omega)\}^T\right]$$
$$(nx(n+m))\qquad\qquad\qquad(nxn)\qquad\quad(nx(n+m))\qquad\qquad(nx(n+m))$$

(2.98)

Equation (2.98) can be expressed in terms of the auto- and cross-spectra:

$$\left[S_{ry}(\omega)\right]=\left[S_{rr}(\omega)\right]\left[H_o(\omega)\right]^T$$
$$(nx(n+m))\quad(nxn)\quad(nx(n+m))$$

(2.99)

from which, the FRF matrix $[H_o(\omega)]$ is obtained:

$$\left[H_o(\omega)\right]=\left(\left[S_{rr}(\omega)\right]^{-1}\left[S_{ry}(\omega)\right]\right)^T$$
$$((n+m)xn)\qquad(nxn)\qquad(nx(n+m))$$

(2.100)

or

$$\begin{bmatrix}\left[H_f(\omega)\right]\\(nxn)\\\left[H_x(\omega)\right]\\(mxn)\end{bmatrix}=\left(\left[S_{rr}(\omega)\right]^{-1}\left[\left[S_{rf'}(\omega)\right]\vdots\left[S_{rx'}(\omega)\right]\right]\right)^T$$
$$((n+m)xn)\qquad\quad(nxn)\qquad(nxn)\quad\ \ (nxm)\qquad\qquad(nx(n+m))$$

(2.101)

A necessary and sufficient condition for the inverse of $[S_{rr}(\omega)]$ to exist is that none of the external input signals are completely correlated with any other so that $[S_{rr}(\omega)]$ is non-singular. After the equivalent open-loop FRF matrix $[H_o(\omega)]$ has been obtained, the FRFs of the system are derived from:

$$\left[H_{xf}(\omega)\right] = \left[H_x(\omega)\right]\left[H_f(\omega)\right]^{-1} = \left(\left[S_{rf'}(\omega)\right]^{-1}\left[S_{rx'}(\omega)\right]\right)^T \qquad (2.102)$$
$$\underset{(mxn)}{} \qquad \underset{(mxn)}{} \qquad \underset{(nxn)}{} \qquad \underset{(nxn)}{} \qquad \underset{(nxm)}{}$$

Equation (2.102) provides a unique least-squares solution for the estimate of $[H_{xf}(\omega)]$. It can also be seen that the effect of measurement noise in the estimation of the FRFs for a multi-shaker random excitation test can be minimised by using (2.102), because cross-spectra are used. However, it should be noted that when (2.102) is applied to the estimation of FRFs in multi-shaker modal testing with n measured force signals and m measured response signals, 2n+m processing channels are needed, because n(n+m) additional cross-spectrum terms are needed for the $[S_{rf'}(\omega)]$ and $[S_{rx'}(\omega)]$ matrices, respectively.

2.5 CONCLUDING REMARKS

This chapter highlighted the basic concepts of Fourier analysis, discrete and fast Fourier transforms and their limitations. A closed-loop model for a single-shaker FRF measurement scheme with the potential of uncorrelated measurement noise was presented. That model explains the variation in the input force spectrum near structural resonances. It also gives the relationships between the measured force and response signals to the properties of a test structure and the excitation mechanism. When a specified amount of uncorrelated noise exists in the measured force and acceleration, the overall percentage of the noise to frequency response signal ratio for $H_1(\omega)$ and $H_2(\omega)$ FRF estimators is no longer a constant, because of the frequency dependent characteristics of the power spectra. It was shown that the instrumental FRF estimator $H_3(\omega)$ is a noise-free one. However, the instrumental FRF estimator $H_3(\omega)$ has a phase-dependent bias error due to leakage (windowing effect). The bias errors for different FRF estimators were discussed analytically. The computer-simulation results showed that the only estimate which will lie on the true modal circle is $H_2(\omega)$.

The generalised closed-loop model was also presented for multi-shaker modal testing. A frequency domain technique was described which would give a noise-free estimate of FRFs.

CHAPTER 3

Modal Testing

3.1 VIBRATION TESTING FOR MODAL ANALYSIS

3.1.1 Introduction

In many respects the practice of vibration testing is more of an art than a science. There is no single *right* way to perform a vibration test. In almost every case the support, the excitation equipment or the transducers will influence the dynamic behaviour of the article under test. The science in vibration testing is in realising that these influences exist, understanding them and then designing the test to minimise their effects on the dynamic behaviour of the structure. The art and science of vibration testing is to obtain results that are as close to *the correct answer* as required, at a cost that is within budget, and to achieve all this *first time*.

The purpose of this chapter is to provide an introductory overview of the extensive subject of vibration testing for modal analysis, commonly known as Modal Testing. The object of this form of vibration testing is to acquire sets of Frequency Response Functions (FRFs) that are sufficiently extensive and accurate, in both the frequency and spatial domains, to enable analysis and extraction of the properties for all the required modes of the structure. This chapter does not address explicitly other forms of vibration testing such as:

'Environmental Testing' - where a test article is subjected to vibration of a specified form and amplitude for a period of time in order to assess the operational integrity of the test article, or

'Operational Testing' - where the response of a structure to an unknown, or unquantifiable, operational excitation is measured,

although the selection and use of transducers apply in just the same way for these other forms of testing as well. In modal testing the excitation levels are frequently much lower than those found in operation, and the linearity of structures to different forcing levels is often assumed despite the fact that this may only apply over a limited range.

Every structure poses its own special problems for modal testing and there are many different ways of tackling the same problem - there is no definitive *right way* to test any given structure. All modal tests involve a degree of compromise and the test engineer is encouraged to consider all aspects of the test set-up, test equipment, data collection and assessment at an early stage. No recipes for a successful modal test are given because there aren't any. The reader is introduced to a selection of the ingredients and invited to consider whether they are appropriate for a proposed test. Thorough preparatory work is a vital prerequisite for successful modal testing. Unfortunately, this key stage is often rushed in the overwhelming desire to collect data.

The published literature concerning modal testing is extensive but the books by McConnell [17], the Dynamic Testing Agency in the UK [18] and Harris and Crede [19], are all very comprehensive and handy reference texts on the subject to have available. More specific references are given throughout this chapter.

Unless an analytical model or a Finite Element (FE) model is available, the vibration properties of a structure must be derived from experimental measurements. When there is a physical structure available, it is possible to derive a mathematical model of the dynamic behaviour based upon FRFs obtained from experimental measurements on the structure, as explained in Chapter 4. The advantage of this approach is that the actual structure, with all the manufacturing anomalies, is tested. No assumptions about the behaviour of joints are made; the joints simply contribute to the way in which the structure responds to the applied excitation. However, for practical reasons, the extent of a model derived in this way is more limited in the spatial and frequency domains than is the case for an FE model. In most cases, it is not practical to measure the response of the structure at anything like the number of points that may be used in the FE model. Also, the measured degrees-of-freedom may be different from those of the FE model in position and direction. Furthermore, it is rare for there to be any measurement of rotational response and, even rarer still, for there to be any torque excitation of the structure. The location of the excitation and response points is critical if all the modes are to be excited and identified uniquely.

The frequency domain extent of the model derived from experimental measurements will be dictated by the capabilities of the transducers and data processing equipment used in the measurements. Both the accuracy and correctness of the model can be influenced by the experimental set-up. Some form of excitation has to be applied to the structure (and measured) to generate the required responses. Almost all methods for applying the structural excitation will have some unwanted modification effect on the structure. Similarly, almost all the response measurement transducers and support fixtures (boundary conditions) will have an unwanted influence on the structure [20]. The experimentalist must be aware of these influences and strive to select methods of excitation and response measurement that minimise these effects. The acquisition of a measured database that is sufficiently extensive (in both the spatial and frequency domains) and of the high quality required for purposes of refining an FE model can be a very significant task. In the words of D. J. Ewins [21], " . . . *it is usually prudent and often necessary to measure and/or analyse at least some of the critical parameters several times, in order to attain the required consistency of results"*. Any measurement variability should be less than the effects of manufacturing variability inherent in the structures under test [22]. This is particularly relevant when the updated model (Chapter 7) is required to represent more than just a single example of the structure.

Many of the potential problems with modal testing only become apparent during the actual tests. Frequently, it is not possible to predict such problems beforehand either because there is no analytic model or because the model of the structure is unrepresentative. For this reason it is advocated strongly that a preliminary survey should be performed prior to the full measurement survey. In the preliminary measurement survey, the measurement frequency range, necessary frequency resolution and the likely positions for structural modification can all be defined. Furthermore, the preliminary survey offers the opportunity to assess the influences of the test equipment on the structure; the selections of structure support and exciter push-rods for instance. The main advantage of the preliminary survey is that more effective use can be made of subsequent testing time and resources.

As part of the preliminary survey, and during the full survey, critical assessments of the quality of the measured data should be made. Reciprocity and repeatability are two established methods for assessing the quality of measured data that have been used for many years. However, it has been found that much of the information that could be derived from these checks is lost because the respective FRFs are only compared by eye. Consequently, small differences between two large quantities are invariably overlooked despite the fact that the differences may be systematic and indicative of slight shifts in the structural resonance frequencies. To aid the assessment of data quality by these methods, the use of 'difference functions' alongside the actual FRFs has been found to be particularly beneficial.

3.1.2 Basic measurement chain

The previous introduction has made clear that understanding modal testing requires knowledge of a somewhat vast range of areas such as instrumentation, signal processing, parameter estimation (modal identification) and, of course, vibration analysis. Some of these areas are considered with sufficient detail in other chapters of this book. In this chapter, we are going to summarise the basic principles of modal testing and deal in greater detail with some specific topics of experimental FRF measurements.

Modal testing involves availability of various hardware components such as the ones schematically represented in figure 3.1, which shows a typical layout for a simple measurement system. Basically, there are three main measurement mechanisms:

- the excitation mechanism;
- the sensing mechanism; and

- the data acquisition and processing mechanism.

The excitation mechanism is constituted by a system which provides the input motion to the structure under analysis, generally under the form of a driving force f(t) applied at a given coordinate. There are many variants for this system, their choice depending on several factors such as the desired input, accessibility and physical properties of the test structure. The exciter, also known as the shaker, is usually an electromagnetic or electrohydraulic vibrator, driven by a power amplifier. The excitation signals, in these cases, are generated by a signal generator and can be chosen from a variety of different possibilities (stepped-sine, swept sine, impulse, random, etc.), to match the requirements of the structure under test.

This type of excitation mechanism may be easily controlled both in frequency and amplitude and therefore offers the best overall accuracy. However, it also has some disadvantages such as the need to have the exciter connected to the test structure. Despite the use of connecting devices (called push-rods, drive rods or stingers) designed to reduce the attachment influence, there are always some constraining effects and mass loading of the structure.

Conventional electromagnetic (or electrohydraulic) exciters vary in size and their choice depends on the structure under test. The objective is for them to provide inputs large enough to result in easily measured responses. The applied excitation force is commonly measured by means of a load cell (known as force transducer) which is located at the end of the push-rod and is rigidly connected to the test structure.

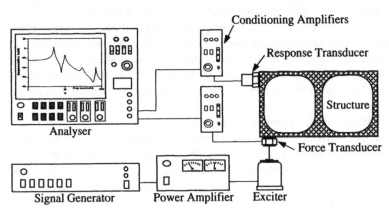

Fig. 3.1 Schematic representation of basic hardware for modal testing.

A very popular alternative as an excitation device is the impulse or impact hammer, which consists of a hammer with a force transducer attached to its head. This device does not need a signal generator and a power amplifier. The hammer, by itself, is the excitation mechanism and is used to impact the structure and thus excite a broad range of frequencies. On the other hand, as the impact hammer does not need a connecting device, its use avoids mass loading the test structure and it is faster than an exciter. The range of frequencies covered by a hammer depends on the hammer mass and on how hard its impacting head is (note that the mass and stiffness of the impacted structure also contribute to define this range). Furthermore, the mass - and therefore the size - of the hammer together with the velocity of the impact dictate the amplitude of the impact force.

The sensing mechanism is basically constituted by sensing devices known as transducers. There is a large variety of such devices (see Section 3.4) though the most commonly used in experimental modal analysis are the piezoelectric transducers either for measuring force excitation (force transducers) or for measuring acceleration response (accelerometers). The transducers generate electric signals that are proportional to the physical parameter one wants to measure. Most of the time, the electric signals generated by the transducers are not amenable to direct measurement and processing. This problem, usually related to the signals being very weak and to electric impedance mismatch, is solved by the conditioning amplifiers. These devices are usually considered as part of the transducers and therefore of the sensing mechanism (some transducers actually incorporate the basic conditioning electronics).

Finally, one has to consider the data acquisition and processing mechanism. Its basic objective is to measure the signals developed by the sensing mechanism and to ascertain the magnitudes and phases of the excitation forces and responses. There are very sophisticated devices for this purpose, called analysers, that incorporate many functions and even include the signal generation component. The most common analysers are based on the Fast Fourier Transform (FFT) algorithm (see Section 2.2.2) and provide direct measurement of the FRFs. They are known as Spectrum Analysers or FFT Analysers. Basically, they convert the analogue time domain signals generated by the transducers into digital frequency domain information that can be subsequently processed with digital computers. Multichannel FFT analysers are nowadays a normal component of any modal analysis laboratory. In short, an experimental modal analysis test set-up as described above is basically aimed at computing FRF accelerance data for a given structure, from measurements of time-varying excitation forces applied to the structure and corresponding dynamic acceleration responses.

Understanding the principles behind signal acquisition and processing (Chapter 2) is very important for anyone involved with the use of modal analysis test equipment. The validity and accuracy of the experimental results may strongly depend on the knowledge and experience of the equipment user.

The accuracy of experimental data is also very dependent on particular problems related to the experimental set-up and control of the different analysis steps. Some useful practical information is presented in this chapter.

3.2   TEST PLANNING AND OBJECTIVES IN MODAL TESTING

Before embarking on any modal test it is important to have a clear definition of the objectives of the test. The type of test, the extent of the test and the required quality of the results all follow from the defined objectives of the test. The generic requirement is usually of the form:

"to obtain a dynamic model of a structure that is suitable for a given purpose"

The purpose of a test may be:

- to obtain mode frequencies of the structure;
- to obtain mode shapes and damping information for the structure;
- to correlate an FE model of the structure with measured results from the real structure;
- to obtain a dynamic model of the structure that can then be used to assess the effects of a range of modifications to that structure; or
- to obtain a dynamic model that is suitable for updating a FE model of that structure such that the theoretical model is a better representation of the dynamic characteristics of the real structure than it was previously.

The quality and extent of the modal test required to achieve these aims increases as one progresses down the list. A very reasonable estimate of the mode frequencies of a structure may be made from a few impact test measurements. However, in order to update a FE model satisfactorily and reliably, very extensive and high quality measurements are necessary. The measurement of rotational degrees-of-freedom (Section 3.4.8) may be essential. For the purposes of model updating, it is important that the sizes of the measured and FE models are not too dissimilar. In practice, though, the measured model may be at least an order of magnitude smaller than the FE model.

Once the objectives of the test have been established, the practical details must be considered. The number and placement of exciters should be chosen so that all the modes of interest are excited properly. Similarly, the choice of response measurement locations should allow unique geometrical description of the mode shapes, avoiding problems of spatial aliasing.

Without prior knowledge of the dynamic characteristics of a structure the location of excitation and response measurement points is a matter of trial and error coupled with experience and engineering judgement. This trial and error approach to transducer placement is inefficient and a particular problem when a test structure may only be available for a limited period of time.

When a FE model of a structure is available this model can be used to determine 'optimum' locations for the excitation and response measurement points. Driving Point Residues [23, 24], calculated with data from the FE model, are used to select appropriate excitation locations. The magnitude of the driving point residue for a location is:

$$|r_{ik}| = \frac{v_{ik}^2}{2m_k \omega_{dk}} \tag{3.1}$$

where:

r_{ik} = driving point residue for degree-of-freedom i for mode k,
v_{ik} = mode shape coefficient for degree-of-freedom i for mode k,
m_k = modal mass for mode k,
ω_{dk} = damped natural frequency for mode k.

It is a measure of how much a point participates in the overall response of a mode and proportional to the resonance peak magnitude for that particular point FRF. The 'optimum' driving points are usually chosen as those having the largest driving point residues averaged and weighted for all the modes of interest. In a pre-test analysis such as this a mode is considered to be excited well if the excitation points are located at points with large amplitudes of motion. However, the practical consequences of this selection criterion should also be borne in mind. Although a mode may be excited well by locating the exciters at points with large amplitudes of motion, the mode will also be highly susceptible to modification or loading at these points - a most undesirable situation.

Suitable sets of response measurement locations are verified using the Modal Assurance Criterion (MAC)[†] correlation of the FE mode shapes defined only at the sets of reduced measurement locations. A good candidate set of response measurement locations will be one in which all the off-diagonal terms in the reduced MAC matrix are small. This indicates low cross-correlation between the 'measured' mode shapes and hence each mode shape will be sufficiently different from the other mode shapes to avoid spatial aliasing. Once again though, loading effects of the response transducers on the structure under test should be considered before settling on a set of locations for the actual test. The selection of transducers and influences of excitation equipment are dealt with later in this chapter.

It may be that, as a result of this assessment, the original objectives have to be modified to reflect what can be achieved with the resources available - a valid and well-justified compromise. Sufficient time should be allowed in the test plan, however, to calibrate all of the transducers (see Section 3.4.6) before the test (and possibly after as well). Calibration of equipment is a whole subject in its own right [25-27] and whatever procedures are adopted, documentation of every detail, no matter how small, is strongly advised. There should be no difference in the level of documentation used for a calibration or a structural test; a calibration is just another (special) form of test. In fact, the calibration test can highlight potential problems in the actual test.

3.3   TEST SET-UP
3.3.1 Support of the structure under test
Consideration of the support of the structure under test is an important part of the test set-up. The support conditions should be well defined and experimentally repeatable if the results of the dynamic measurements are to reflect the properties of

---

[†] See Section 7.4.

the structure without undue influence from the support. For test of components *in situ*, exact definition of the boundary conditions may be problematic but, nevertheless, tests should be considered to prove the repeatability of the installation.

For 'laboratory' testing of components or structures, grounded or free boundary conditions are the two extremes that are most frequently employed. It is almost impossible for either of these two conditions to be achieved in practice; a grounded structure will have some movement at the grounding point (usually rotation) and there will be some small restraint of a, nominally, free structure.

For a structure to be really free, it should be suspended (floating) in the air, free in space with no holding points whatsoever. Such a situation is commonly designated as 'free-free', 'freely supported' or 'ungrounded' and is clearly impossible. However, simulation of free-free conditions is very easy to achieve. It suffices to suspend or support the structure using very flexible (also designated as 'soft') springs so that the resonance frequencies of the mass of the structure on the stiffness of the supports or suspension devices are very low and far away from the frequency range of interest.

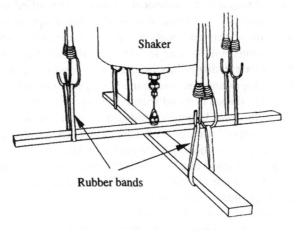

Fig. 3.2 Example of soft suspension.

Figure 3.2 exemplifies the use of rubber bands for the simulation of free-free conditions of a cross-beam assembly, tested in the vertical direction.

However, it must be noted that the method used to simulate free-free conditions can also interfere with the results. This is clearly shown in figure 3.3 where a simple steel beam was tested using different types of materials for the suspension. In the case illustrated, the first antiresonance suffers response changes due to damping being introduced by the suspension system.

Boundary conditions other than free may produce stronger interference with the test results. For example, a reasonably small structure was bolted to a large and massive steel bed-plate embedded in a concrete floor to produce a grounded test article. No rattles were evident during test but, to check the root condition the bolts were tightened still further and the test repeated. The two FRF results in figure 3.4 clearly show that, initially, the root of the structure was not properly grounded. Further tightening of the bolts had an insignificant effect.

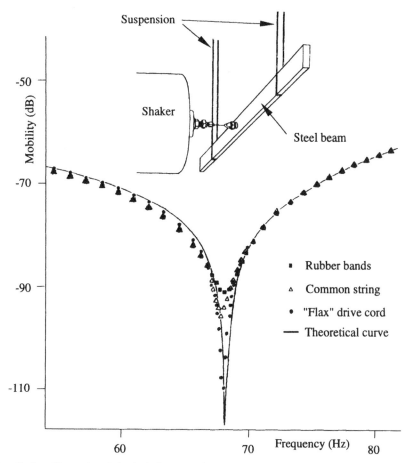

Fig. 3.3   Example of the interference of the suspension system on the response of a simple beam, at the first antiresonance.

3.3.2 Support of the excitation system

The support of the excitation system is not quite so important as the support of the structure under test, but it should still be considered. Rigid mounting of shakers on the floor or on stands is simple and straightforward, but care should be taken to avoid any ground transmission between the shakers.

The other common method of supporting a shaker is by suspension on some kind of hoist. Although this is a most convenient arrangement for positioning and aligning the shaker, it is important to remember that the shaker on the hoist is now a dynamic system in its own right. The shaker will move as a result of the internal force generated. The degree of movement will be related to the inertial mass of the shaker; the larger the mass, the smaller the movement. This movement of the shaker can cause problems at low frequencies when modes of the shaker suspension can be excited. This results in excessive motion and can lead to damage of the shaker, push-rod and force transducer.

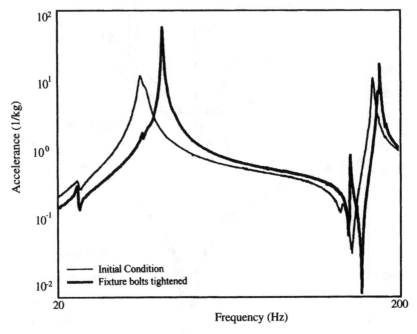

Fig. 3.4 The influence of root conditions on a cantilevered structure.

If the shaker suspension modes are below the frequency range of interest for the structure under test, the excitation signal may be filtered to remove the low frequency content and, thereby, avoid the problem. Increasing the shaker inertial mass with supplementary masses can be used to shift troublesome suspension mode frequencies below the frequency range of interest. Further discussion of the interaction between an exciter and the structure under test is given later.

3.3.3 Attachment of the excitation system to the structure under test

The majority of structural excitation techniques in common usage require some physical contact with the structure. The objective is to transmit controlled excitation to the structure in a given direction and, simultaneously, to impose as little restraint on the structure as possible in all other directions.

However, to achieve the previous purpose, the shaker must be rigidly attached to the structure and this connection is bound to introduce constraints that will affect the force transducer signal. In fact and except in particular cases of symmetry, the structure responds to the excitation by both translating and rotating, and therefore the shaker (and force transducer) will be affected by a reaction torque that will distort the force signal and introduce errors in the measurements. To avoid this, a flexible push-rod, or stinger, or drive rod, usually forms some part of the link between a shaker and the structure under test. The push-rod is designed so that it is relatively flexible to lateral and rotational motions between its ends, but very stiff axially. Such a device is schematically illustrated in figure 3.5. The paper on the subject of stinger design by Mitchell [28], can be of use in initial sizing of the stinger.

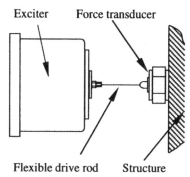

Fig. 3.5 Flexible drive rod between exciter and force transducer.

However, there is still no substitute for actually trying out several different push-rods during preliminary investigations to obtain one that is best suited. 'Piano wire', stranded steel cable, and plastic rods of various cross-sections are a few of the varied and diverse selection of materials to be found in use for push-rods. Push-rod extensions may also be used where access to part of a structure is limited.

A typical push-rod and extension link are shown in figure 3.6. The locking ball joint fixtures at each end allow for simple alignment of the excitation direction, even where the local structure surface is not perpendicular to the excitation direction.

The adjustable length extension rod also simplifies positioning of the shaker along the excitation direction. Once set up, the push-rod linkage is then easily removed and replaced, thereby avoiding damage to the shaker or structure either overnight, or while transducers are being repositioned.

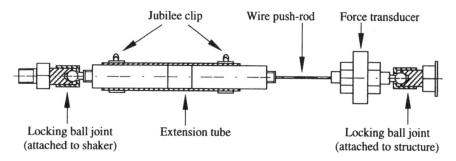

Fig. 3.6 A typical push-rod and extension link.

3.4   SELECTION AND USE OF TRANSDUCERS

The size and mass of the transducers have implications for their influence on the behaviour of the structure under test. It is very much better to use a signal directly from a transducer that has minimal effect on the structure than to try to compensate, at a later stage, for the effects of an inappropriate transducer. Careful selection of accelerometers such that their mass is small when compared with the local mass of the test structure can help to avoid frequency shift problems [29, 30]. Almost all transducers have some effect on the structure and it is a matter of engineering judgement as to when the influence can be considered to be negligible, and hence, which transducer is appropriate. The recent advances in laser based transducer systems now provide an extensive non-contact measurement capability.

For FRF type measurements, there must be at least one force input transducer and one response transducer as exemplified in figure 3.1. The force transducer is often overlooked in transducer selection considerations because it is quite usual for there to be just a single force to be measured, but a large number of responses. As the calculation of all the FRFs depends upon accurate measurement of the force, it is particularly important that the transducer responds to the true force input to the structure.

3.4.1 Response transducers

The mechanical response of a structure may be defined in terms of displacement, velocity or acceleration. Accelerometers are the most common form of response transducer used today although, as Laser Doppler systems become more readily available, velocity transducers are gaining ground rapidly. Devices using Moiré fringes, laser holography and dual-phase fibre optics may also be used for the measurement of vibration response although, due to their relative rarity, they are not described in this section.

Displacement transducers

The simplest form of displacement transducer is the potentiometer. Although they are available for measurement of linear and rotary displacements, they tend to be noisy and are only suitable for relatively low frequency, large displacement applications.

Linear Variable Differential Transformers (LVDTs), however, are another form of displacement transducer that have been used successfully for vibration measurements for many years. The principle of operation of an LVDT is that a free-moving magnetic core is used to link the magnetic flux between a surrounding primary coil and two secondary coils. A schematic cross-section of an LVDT is shown in figure 3.7. The primary coil is energised by an external ac source. The alternating magnetic flux induces voltages in each of the secondary coils. The two secondary coils are wound such that the voltages induced at the null position are of equal magnitude but opposite phase. When these two coils are connected together, the net output of the transducer at the central position is zero. As the magnetic core is moved away from the central position the induced voltage in one of the secondary coils increases. At the same time, the induced voltage in the other coil decreases, resulting in a differential voltage output that varies linearly with the position of the magnetic core. In moving from one side of the central position to the other, the polarity of the demodulated output changes instantaneously. The core has a small mass and is separated from the coil structure by a low-friction lining that produces an almost frictionless device that is insensitive to radial motion of the core.

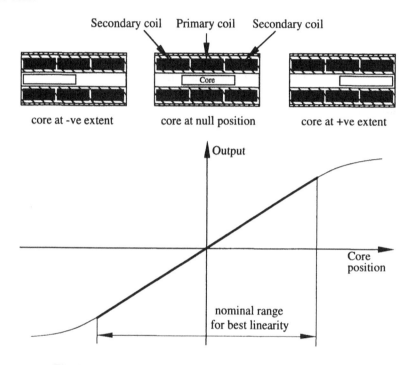

Fig. 3.7 A Linear Variable Differential Transformer (LVDT).

For vibration measurements, the core is usually connected to the structure via a push-rod. The push-rod has two functions: to decouple lateral motion and to provide a convenient method of attachment to the structure. To maintain the calibration of the device the push-rod should be non-magnetic.

Rotary Variable Differential Transformers (RVDTs) are also available for the measurement of angular position.

Laser optical range sensors are another form of displacement transducer (figure 3.8) that have recently become available. They operate on the principle of triangulation. A laser light beam reflected from the surface of a structure is focused onto an internal photo-sensitive device. As the structure moves, the position of the focused spot on the photo-sensitive device moves. The photo-sensitive device generates a signal according to the position of the focused spot. This output is then conditioned and linearised to give an analogue signal proportional to the range of the surface's motion.

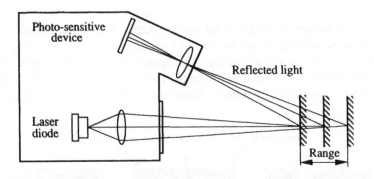

Fig. 3.8 A Laser Optical Range Sensor.

Velocity Transducer

A laser vibrometer is an optical system that can be used to measure the instantaneous velocity of a point (or points) on a structure. The instrument is a non-contact device in which the velocity measured is the velocity component in the direction of the incident laser beam. The velocity is measured by the detection of the Doppler frequency shift of light scattered from the moving surface. Sophisticated optics and signal processing mean that these devices are expensive. Scanning systems are now available in which the laser beam can be moved rapidly over a grid of measurement points on a structure. It is possible to make finely detailed measurements on complex structures that are not amenable to, or accessible for, conventional transducers. Further developments in laser measurement techniques now enable the measurement of rotational responses [31].

Accelerometers

Piezoelectric accelerometers

The cross-section of a typical piezoelectric accelerometer is shown in figure 3.9. There are four basic components; a base and case, a centre post, an annular section of piezoelectric ceramic and an annular seismic mass element. The piezoelectric crystal and seismic mass are arranged concentrically around the centre post. The base of the accelerometer moves with the motion of the structure

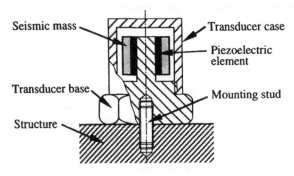

Fig. 3.9 Cross-section of a piezoelectric accelerometer.

to which it is attached, and to cause equivalent motion of the seismic mass, a force must be applied – Newton's 1st Law. This force is transmitted through the piezoelectric crystal that deforms slightly as a consequence. The deformation produces a charge in the piezoelectric crystal that is proportional to the deformation and hence, ultimately, to the acceleration of the seismic mass and the structure.

These devices operate well over a fairly wide frequency range, but they are not generally well suited to low-frequency applications.

Piezoresistive and Capacitive accelerometers
In a piezoresistive accelerometer (figure 3.10), semiconductor flexure elements, that support a seismic mass, form part or all of an active Wheatstone bridge (figure 3.11). As the elements are strained, the Wheatstone bridge is unbalanced and the differential output (proportional to the applied strain) is a measure of the acceleration.

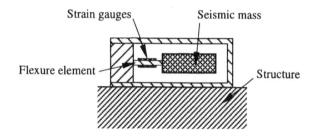

Fig. 3.10 Schematic cross-section of a piezoresistive accelerometer.

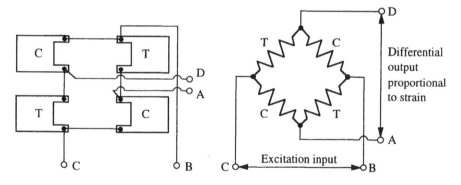

Fig. 3.11 Strain gauge schematic for a piezoresistive accelerometer.

In a capacitive-type accelerometer (figure 3.12), the measuring elements form a capacitive half-bridge. A seismic mass is supported on flexures between two electrodes. As the seismic mass moves from its central position under acceleration, the capacitive bridge is unbalanced. The differential output is then measured in a way analogous to that for the piezoresistive accelerometer. The advantage of the piezoresistive and capacitive types of accelerometer is their DC and low frequency response capability.

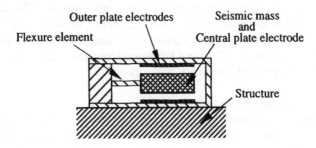

Fig. 3.12 Schematic cross-section of a capacitive-type accelerometer.

Force Balance accelerometers

Force balance accelerometers have been available for many years and were used in aircraft inertial navigation systems. Originally they tended to be relatively large but the requirement for reliable and sensitive yet rugged accelerometers for use in automobile applications has resulted in the manufacture of integrated circuit (IC) sensors in large quantities by micromachining techniques. A schematic diagram of the mechanical part of one of these sensors is shown in figure 3.13. The complete transducer has extensive signal conditioning circuitry built into the IC chip.

A central plate is supported on four suspension beams anchored at their tips. The central plate also forms the central electrode in many separate variable differential capacitors; shown as the unit cell in figure 3.13. So far, the principle of operation is identical to that for a capacitive type of accelerometer described above - although on a very much smaller scale. However, when the transducer experiences an acceleration the inertial force acting on the central plate causes it to move relative to the anchor points. This movement causes unequal capacitances between the central multi-plate electrode and the two fixed electrodes. These changes in capacitance are sensed, and conditioned and used to generate a DC voltage that is fed back to the capacitor plates and electrostatically forces the central plate back into the equilibrium position. The voltage required to hold the central plate in the equilibrium position provides the output signal that varies directly with the applied acceleration. In operation the response of the device is fast enough that there is virtually no relative motion between the central plate and the anchor points. This reduces substantially the effects of any nonlinearities in the mechanical spring rates and provides DC response capability.

The influence of cross-axis sensitivity

Transducers are designed to measure a particular quantity in a specific direction, e.g. acceleration in a direction perpendicular to the base of the transducer. Transducers are carefully designed and built such that motion of the transducer in directions other than the specified measuring direction has little effect on the output. Nevertheless, there is always some degree of cross-axis sensitivity that may influence the results adversely [32]. With care, this can be minimised by suitable alignment of the transducer. However, it is necessary to have a polar plot of the transducer cross-axis sensitivity that identifies the cross-axis sensitivity of the transducer as a function of the direction of cross-axis excitation.

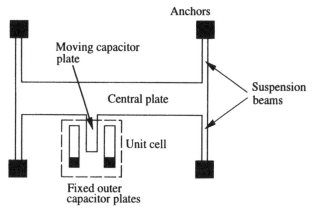

Accelerometer in the equilibrium position.

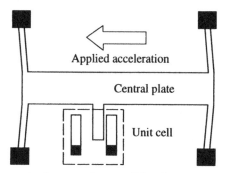

Accelerometer in a transient condition due to an external acceleration.

Fig. 3.13 Schematic view of an IC Force Balance type accelerometer.

Cross-axis sensitivity effects can be of particular importance in the calculation of rotational degree-of-freedom properties from two sets of closely spaced translational degree-of-freedom FRF measurements [32-37].

The calculation of the rotation/translation, translation/rotation, and rotation/rotation FRFs from the translation/translation data essentially involves differencing operations, as explained in Section 3.4.8. When the motion in one of the transverse directions is large, the difference between the measured FRFs can be of the same order of magnitude as the cross-axis sensitivity component. The calculated rotational degree-of-freedom properties will then contain significant errors. For minimum cross-axis effect, the direction of minimum cross-axis sensitivity should be aligned with the direction of the maximum cross-axis response of the structure. There are several difficulties with implementation of such a procedure:

(i) a polar plot of cross-axis sensitivity is seldom supplied with each transducer;

(ii) a preliminary measurement survey is necessary to deduce the directions of maximum cross-axis motion at the measurement points;

(iii) the directions of maximum cross-axis motion will change as different modes become more prominent; and

(iv) it can be quite difficult to set a particular angular orientation of the transducer because of the fixing methods used.

In the light of these difficulties, it is unusual for much consideration to be given to cross-axis sensitivity influences, except in the measurement of rotational responses (as mentioned above) and for measurements in rotating systems, e.g. out of plane vibration of a rotating disc.

3.4.2 Attachment of response transducers

The purpose of a transducer is to convert one physical quantity into another, usually electrical, that can be quantified by remote instrumentation. The signal produced is a representation of the motion of the transducer. The implicit assumption is made that the motion of the transducer is identical to that of the structure. This will be the case providing that the transducer is rigidly attached to the structure. The attachment stiffness must be sufficiently high that, throughout the frequency range of interest, the motion of the transducer is identical to that of the attachment point on the structure. For low-frequency work, it is quite acceptable to attach an accelerometer with beeswax or double-sided adhesive tape, but for high-frequency measurements, the accelerometer must be bolted or cemented firmly to the structure [3, 38].

Virtually all response measurement transducers, except those that are 'non-contact', will have some inertial loading on the structure [30]. The consequences of transducer inertial loading on a structure should be carefully considered at an early stage in the set-up phase of a measurement. In a large measurement survey, a further consideration is whether all the transducers are to be fixed to the structure before the start of the test, or whether a small number of transducers are to be moved around the structure to all of the test points in turn. This is largely dictated by the size of the test, the number of transducers and the number of simultaneous data acquisition channels that are available. In a test where all the transducers remain fixed in place throughout the complete measurement phase, the structure does not change as different response points on the structure are measured. The structure will be slightly altered from the base condition, but the important point here is that no further changes occur throughout the measurement phase. When the measurements are made in a series of tests where a small set of transducers is moved around, the structure is altered differently for each test – a systematic error.

The errors are most easily seen as different positions of the resonance peaks for each successive set of measured FRFs. Although the errors introduced in each test are potentially smaller than the overall error caused by attaching all the transducers at the start, the systematic errors make the task of analysis of the measured data significantly more complex.

3.4.3 The use of dummy response transducers

The systematic error incurred as a result of moving the transducers around on the structure during the course of the test can be turned into a single error by the use of dummy transducers. Instead of instrumenting the whole structure with transducers at the outset of the test program, dummy transducers are used. The dummy transducers are designed to have similar mass properties to the real transducers. The dummies are systematically replaced (temporarily) by the real transducers for the measurements, until all of the measurement locations have been covered. The results from this type of test should be the same as if a complete set of real transducers had been attached to the structure at the outset. It should be noted that the total time for a test will be greater than for a 'one-shot' test and the 'handling' of the structure provides an opportunity for small changes to take place during acquisition of the complete data set.

3.4.4 Approaches to force measurement
Direct approach, via a load cell or force transducer
The most common type of force transducer (figure 3.14), works on the principle that the deformation of a piezoelectric crystal produces a charge output proportional to the force acting on that crystal.

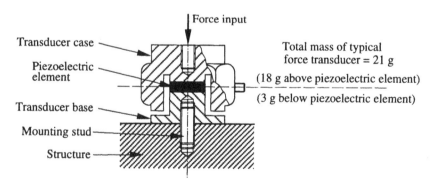

Fig. 3.14 Cross-section of a piezoelectric force transducer.

The shaker applies a force through the case of the force gauge to the top of the piezoelectric crystal. The lower end of the crystal is fixed to the base of the transducer that is, in turn, attached to the structure. In the axial direction, the force applied to the structure is that transmitted through the piezoelectric crystal (measured) minus the force that is required to accelerate the base of the force gauge.

A force transducer is constructed to have as little mass on the 'base-side' of the sensing element as possible. A conventional test set-up using a force transducer can be seen in figure 3.15 a). The force transducer is attached with its 'base' towards the structure and the 'top' is connected via a flexible push-rod to a shaker.

The 'live-side' mass of the force transducer is kept small to try to minimise the modification to the structure. However, what is minimised is the mass modification to the structure *in the sensing direction of the force transducer*. The full mass of the force transducer modifies the structure in directions perpendicular to the sensing direction.

There is a large discrepancy between the apparent mass of a force transducer, as seen by the structure, in the sensing direction and a perpendicular direction. For a typical commercial force transducer, the 'base-side' mass is only 3g and this is the mass *seen* by the structure in the sensing direction. The total transducer mass is 21g and this is the mass *seen* by the structure in directions perpendicular to the force sensing direction.

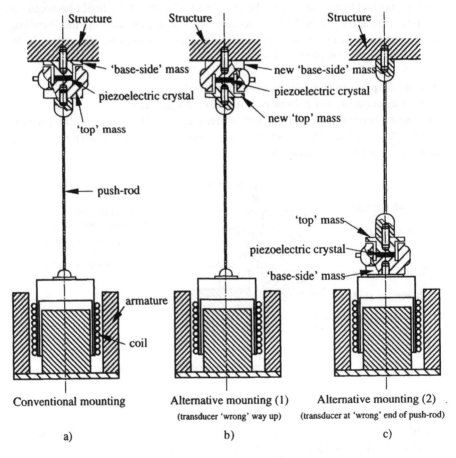

Fig. 3.15 Excitation set-ups for direct force measurements.

If the force transducer were to be mounted on the structure the 'wrong' way up (figure 3.15 b)), then there would be less difference in the apparent loadings on the structure in the sensing and perpendicular directions. For the typical commercial force transducer used before as an example, the 'live-side' mass as *seen* by the structure in the sensing direction would now be 18g. The mass *seen* by the structure in directions perpendicular to the force sensing direction is unchanged at 21g, the total mass of the force transducer.

Another way to reduce the discrepancy between the mass added in the sensing direction and that added in perpendicular directions is to mount the force transducer directly on the shaker platform, as shown in figure 3.15 c). The transducer 'base' is arranged towards the structure and the 'top' towards the shaker. A flexible push-rod then connects the force transducer to the structure. Only the small 'live-side' mass of the transducer plus the push-rod mass modify the structure in the sensing direction and only part of the push-rod mass modifies the structure in the perpendicular directions. In many instances the push-rod mass may be smaller than the total mass of the force transducer, so the overall modification effect will be less than for the conventional arrangement. Furthermore, there will be a much smaller difference between the modification effect in the sensing direction and that in the perpendicular directions. A supplementary advantage of this arrangement of components is that the force transducer cable (which can be relatively stiff and massive with indeterminate effect) may also have less influence on the measured characteristics of a relatively light structure.

If it is important to have results for the structure free from the loading effects of the measurement equipment, it is theoretically possible to use structural modification techniques to remove the effects of the transducers and excitation system to deduce the FRF characteristics of the underlying structure. In some commercial software packages allowing simple structural modifications to be made to a measured dynamic model of a structure, point mass modifications are assumed to act equally in all three translational degrees-of-freedom. This is a fairly reasonable assumption for a mass bolted to a structure. However, in the case of a force transducer mounted conventionally (figure 3.15 a)) the modification effects are not the same in all three translational degrees-of-freedom and care must be exercised when trying to perform the structural modifications. If, however, the component arrangements shown in figures 3.15 b) or c) are used, the discrepancies between the modification effects in the sensing and perpendicular directions may be sufficiently small for the transducer to be considered as an equal three degree-of-freedom mass modification.

In all the preceding discussion only the translational effects of the force transducer have been considered. In reality, the force transducer will modify the structure in all six degrees-of-freedom and the effects in the rotational degrees-of-freedom should also be considered.

The force measured with a conventional force transducer will not be in-phase with the excitation signal driving the shaker. This is caused by the dynamic characteristics of elements in the excitation drive-train. Normally it is not a problem but, if one wishes to control the relative phasing of several forcing inputs (as required in phase-separation or force appropriation tests - Section 3.7.2), the phase shift between the drive signal and the force must be taken into account. This can be done either iteratively or by use of the transfer function between each drive signal and its associated force. The simpler, but laborious, iterative approach is usually adopted in such situations.

Indirect approach, via the current flowing through the exciter coil
Instead of using a force transducer to measure the force input into a structure, the force can be calculated from measurements of the current fed to the coil of an electromagnetic exciter. The experimental arrangement is shown in figure 3.16. A flexible push-rod links the exciter to the structure directly.

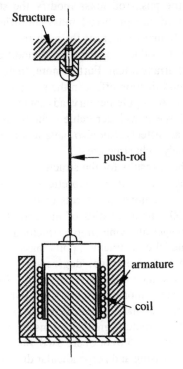

Fig. 3.16 Excitation set-up for indirect force measurement.

From basic electromagnetic principles it is known that a current-carrying conductor in a magnetic field experiences a force given by the expression

$$F_e = B \, I_e \, L \, N = \mu_F \, I_e \qquad (3.2)$$

where F_e = force (N)
  B = magnetic flux density (weber/m^2)

I_e = current (ampere)
L   = length of conductor in flux field (m)
N   = number of turns of length L in flux field
μ_F = the force/current factor for the coil. (newton/ampere)

It is assumed that the magnetic flux density, B, the length of conductor in the flux field, L, and the number of coil turns in the field, N, are constant throughout the full range of travel of an exciter. The force generated in the coil is, therefore, directly proportional to the current flowing in that coil. The linear relationship between the current and force generated in the coil can be derived from calibration measurements with subsidiary equipment.

By definition, the coil current and the force generated in the coil are directly proportional and are in-phase. Therefore, controlling the relative phasing of several excitation inputs for a force appropriation test is a trivial matter.

The force calculated is the electromagnetic force *applied to the coil* and not that actually transmitted to the structure. The moving components of the exciter and the push-rod have a dynamic modification effect on the force. With lightweight structures the errors can be significant, and corrections in the sensing and perpendicular directions will be required in just the same way as for a conventional force transducer. Further discussion on the direct or indirect measurement of force and the influence of an exciter on the behaviour of a structure is given later.

3.4.5 The use of dummy force transducers
The use of dummy transducers for response measurements is an established practice; what is rarely considered is the use of dummy force transducers at the excitation locations. These may be just as applicable as dummy response transducers when a series of single point tests is performed to measure several columns of the FRF matrix.

3.4.6 Transducer calibration
Transducer calibrations are often done on a relatively small dedicated calibration fixture where all the signal conditioning, acquisition and processing equipment is located nearby. In such circumstances lead lengths are not normally a problem. However, if a large structure is to be tested, then representative long lead lengths should be used during the calibration as well. Long transducer lead lengths can be detrimental to signal quality due to significant lead resistance, capacitance variation or noise pick-up. The location of signal conditioning and amplification equipment close to the transducers may have to be investigated for such situations.

The values measured by the test equipment represent electrical voltage and therefore it is necessary to obtain a calibration factor which translates these values into units of acceleration and force.

The use of the manufacturers' quoted transducers sensitivities may not be accurate enough since they can change with time and environmental conditions. In addition, the transducers may have suffered some kind of damage and, while still working, may have lost their response linearity. Finally, the remaining units in the measurement chain may change, albeit slightly, the overall sensitivity.

When measuring FRF data, one is concerned with the ratio of motion to force and not the individual values of any of these quantities. This fact allows the use of a simple and straightforward technique which provides an accurate calibration of the transducers, including the influence of the remaining units of the measurement chain. The technique requires only the use of a simple rigid structure, such as a steel block, together with the equipment that is going to be used for the accelerance measurements.

For each accelerometer to be used it is necessary to make a calibration test involving simultaneously the accelerometer and the force transducer. Applying a time-varying force (figure 3.17) to a solid block of known mass (which can be accurately measured), measuring the corresponding acceleration response and computing the accelerance through a specified frequency range, one obtains a value in units of volt/volt which corresponds to:

$$A(\omega) = \frac{\ddot{x}(t)}{f(t)} = \frac{1}{m} \qquad (3.3)$$

where m is the mass of the block (which may include the added transducers' masses). Thus, the measured accelerance will be a constant value proportional to the block mass, within a frequency range for which the block behaves as a rigid body.

This calibration technique ignores the individual transducers' sensitivity values and must be performed for each pair of accelerometers/force transducers. In the calibration curve exemplified in figure 3.17, the overall sensitivity of the measurement chain for the particular pair of transducers used was 1.1056 volt/volt per unit ms^{-2}/N.

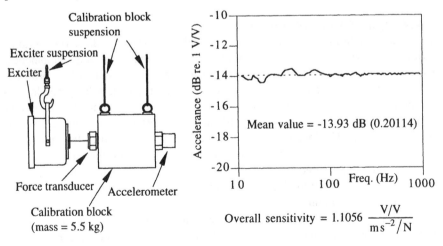

Fig. 3.17 Transducer calibration procedure.

3.4.7 Cancellation of added mass

Due to the transducers being attached to the test structure, there is a certain amount of mass which is added to the structure (accelerometer, mounting studs and washers and mass between the piezoelectric element of the force transducer and the structure). Though this mass may be small, there are many situations where the total added mass is not negligible when compared with the effective modal mass of a mode of vibration. In these circumstances and whenever possible, it is advisable to perform an operation known as mass cancellation.

Consider a direct point accelerance measurement where m_1 is the mass of the transducers' material added to the structure. The force measured by the force transducer (f_{meas}) is related to the additional mass and true force that would be applied if there was no added mass, through

$$f_{meas.} = f_{true} + m_1 \ddot{x} \tag{3.4}$$

and since $f_{meas.}$ and $\ddot{x}$ have been measured and m_1 is known, it is quite simple to derive the corrected FRF ratios

$$\text{Real}\left(\frac{f_{true}}{\ddot{x}}\right) = \text{Real}\left(\frac{f_{meas.}}{\ddot{x}}\right) - m_1$$

$$\text{Imag.}\left(\frac{f_{true}}{\ddot{x}}\right) = \text{Imag.}\left(\frac{f_{meas.}}{\ddot{x}}\right) \tag{3.5}$$

In terms of accelerance, it follows that

$$A_{true} = \frac{A_{meas.}}{1 - m_1 A_{meas.}} \tag{3.6}$$

The above equations are valid only for a direct point measurement. For transfer measurements and other inertial loading of the structure under test the problem becomes very complicated and it does not seem possible to perform mass cancellation[†].

3.4.8 Multi-directional measurements

One of the difficulties in experimental modal analysis is the measurement of moment excitation and rotational response. For many years, this problem has been accepted as having no easy solution (despite some attempts to develop suitable equipment) and measurements have been limited to conventional force excitation and translational response. The net result of this situation is an incomplete experimental model, the missing information easily representing more than 70% of the complete model (see, for example, the case of the 6 DOF clamped-free beam in

---

[†] However, following a different technique, the problem seems to have, at least theoretically, a general solution [39-41].

figure 1.38 that has to be represented by a 3 DOF model as in equation (1.265)). However, recent developments, namely in the fields of structural modification and updating (Chapters 6 and 7), show the increasing need for more accurate and thorough experimental dynamic characterisation; therefore, measurement of torque excitations and rotational responses is becoming increasingly more important.

Among other proposed solutions (namely special types of exciters and transducers), with more or less limited applications and success, there is one that is very simple and has been around for some time now, which is based on the conventional type of exciters and transducers.

The technique uses a pair of transducers placed a short distance apart on the structure or on an auxiliary rigid block attached to the structure as shown in figure 3.18. In both cases, it is assumed that the translational acceleration signals generated by the transducers may be converted into a translation and a rotation at a point P of the structure between their location:

$$\ddot{x}_P = \frac{\ddot{x}_B + \ddot{x}_A}{2}$$

$$\ddot{\theta}_P = \frac{\ddot{x}_B - \ddot{x}_A}{2s}$$

(3.7)

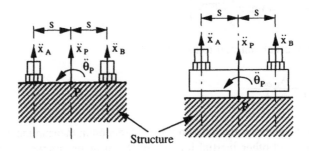

Fig. 3.18 Schematic representation of a technique to measure translation and rotation responses.

One of the problems associated with this technique derives from the fact that the acceleration difference in (3.7) may easily be of the same order as the value of the errors made when measuring the data. Despite this difficulty, there have been various applications of this method with reasonable success. It must also be noted that the technique is based on the assumption that the block is sufficiently stiff in order to behave as a rigid body within the test frequency range.

Measuring rotations only solves half of the problem. However, moment excitation may also be applied if one extends the above technique as shown in figure 3.19. Basically, one wants to relate the force f_x and the moment m_θ to the responses $\ddot{x}_P$ and $\ddot{\theta}_P$. So, performing a first test run with force f_1 applied at the right-hand side of the block, we may write the following relationships:

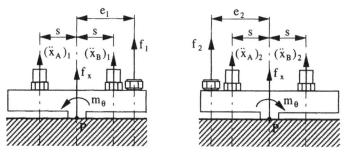

Fig. 3.19 Extension of the auxiliary block technique to include force and moment excitation measurements.

$$f_x = f_1 - m\left(\ddot{x}_P\right)_1$$
$$m_\theta = e_1 f_1 - I_P \left(\ddot{\theta}_P\right)_1 \tag{3.8}$$

where m is the mass and I_P is the inertia about P, respectively, of the block. The acceleration values $(\ddot{x}_P)_1$ and $(\ddot{\theta}_P)_1$ at point P may be simply related to the measured values through equations (3.7). The receptance equation relating these quantities is

$$\left\{ \begin{matrix} (x_P)_1 \\ (\theta_P)_1 \end{matrix} \right\} = \begin{bmatrix} \alpha_{xx} & \alpha_{x\theta} \\ \alpha_{\theta x} & \alpha_{\theta\theta} \end{bmatrix} \left\{ \begin{matrix} f_x \\ m_\theta \end{matrix} \right\} = [\alpha(\omega)] \left\{ \begin{matrix} f_1 + \omega^2 m (x_P)_1 \\ e_1 f_1 + \omega^2 I_P (\theta_P)_1 \end{matrix} \right\} \tag{3.9}$$

and therefore, dividing through by f_1, one obtains

$$\left\{ \begin{matrix} \dfrac{(x_P)_1}{f_1} \\[2mm] \dfrac{(\theta_P)_1}{f_1} \end{matrix} \right\} = [\alpha(\omega)] \left\{ \begin{matrix} 1 + \omega^2 m \dfrac{(x_P)_1}{f_1} \\[2mm] e_1 + \omega^2 I_P \dfrac{(\theta_P)_1}{f_1} \end{matrix} \right\} \tag{3.10}$$

Performing a second test run (within the same frequency range) with the force excitation f_2 applied at the other side of the block, similar reasoning will lead to

$$\left\{ \begin{matrix} \dfrac{(x_P)_2}{f_2} \\[2mm] \dfrac{(\theta_P)_2}{f_2} \end{matrix} \right\} = [\alpha(\omega)] \left\{ \begin{matrix} 1 + \omega^2 m \dfrac{(x_P)_2}{f_2} \\[2mm] -e_2 + \omega^2 I_P \dfrac{(\theta_P)_2}{f_2} \end{matrix} \right\} \tag{3.11}$$

By suitable combination of (3.10) and (3.11) and taking into account relationships (3.7) for $(x_P)_1$ and $(\theta_P)_1$ and equivalent ones for $(x_P)_2$ and $(\theta_P)_2$, the receptance matrix is found to be:

150

$$[\alpha(\omega)] = -\frac{1}{\omega^2}[T][G]\big[[\Pi] - [M][T][G]\big]^{-1} \tag{3.12}$$

where

$$[T] = \begin{bmatrix} 0.5 & 0.5 \\ (2s)^{-1} & -(2s)^{-1} \end{bmatrix} \tag{3.13}$$

$$[\Pi] = \begin{bmatrix} 1 & 1 \\ e_1 & -e_2 \end{bmatrix} \tag{3.14}$$

$$[G] = \begin{bmatrix} \left(\dfrac{\ddot{x}_A}{f}\right)_1 & \left(\dfrac{\ddot{x}_A}{f}\right)_2 \\ \left(\dfrac{\ddot{x}_B}{f}\right)_1 & \left(\dfrac{\ddot{x}_B}{f}\right)_2 \end{bmatrix} \tag{3.15}$$

and,

$$[M] = \begin{bmatrix} m & 0 \\ 0 & I_P \end{bmatrix} \tag{3.16}$$

Thus, the relevant elements of $[\alpha(\omega)]$ may be calculated from the knowledge of the exciting block geometry and after measuring the elements of [G] which are yielded directly by the measuring equipment.

When the inertia properties of the block are very small compared to the test structure they may be neglected, and (3.12) becomes

$$[\alpha(\omega)] = -\frac{1}{\omega^2}[T][G][\Pi]^{-1} \tag{3.17}$$

with

$$[\Pi]^{-1} = \begin{bmatrix} \dfrac{e_2}{e_1 + e_2} & \dfrac{1}{e_1 + e_2} \\ \dfrac{e_1}{e_1 + e_2} & \dfrac{-1}{e_1 + e_2} \end{bmatrix} \tag{3.18}$$

Although the equations previously derived are relative to point measurements, it is quite straightforward to derive similar expressions for transfer measurements. Note that, in this case, the inertia properties of the block cannot be taken into account[†].

---

† Unless a different theory is followed, as stated in a footnote for the case discussed in Section 3.4.7. In such a case, the problem seems to have, at least theoretically, a solution [41].

3.5   EXCITER-STRUCTURE INTERACTIONS

The excitation system can affect the dynamic behaviour of a structure in just the same was as a force or response transducer. The size of the excitation system is generally much larger than a force or response transducer and, hence, the influence will be greater. In addition to the effects of the exciter itself, the exciter suspension system also has a contribution. By use of a force transducer in the push-rod link, the primary effects of the excitation system on the structure may be avoided. However, excessive response of the exciter on its suspension may give rise to unwanted and unmeasured excitation (moment and transverse) of the structure. At all times it should be remembered that the system in motion includes *all* the structural components as well as components of the excitation and response measurement equipment.

The relative ratios of modal properties for the exciter system to those of the structure dictate the dynamic range of the force measured in the push-rod link, assuming a constant amplitude voltage input to the amplifier. The magnitude and phase relationships between the voltage input to an amplifier and the force measured in a push-rod link or derived from the current flowing in the coil of an electromagnetic exciter are considered further. Extension of these issues to multiple input and multiple output vibration tests is discussed with particular reference to the use of indirect force measurement for force appropriation test techniques.

3.5.1 Theory
Grounded exciter body
Consider a simple model of an electrodynamic vibration exciter attached via a rigid link to a representative single degree-of-freedom system as shown in figure 3.20. The exciter coil has mass m and is supported on a flexure of stiffness k and damping c. The force acting on the exciter coil due to the electromagnetic effect is f_e and the force actually applied to the system is f_a. The system has effective mass M, stiffness K and damping C. The exciter body is grounded.

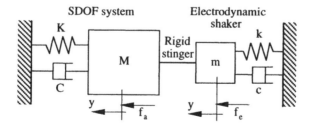

Fig. 3.20 SDOF system with exciter body grounded.

The equation of motion of the masses is:

$$M\ddot{y} + C\dot{y} + Ky = f_a \tag{3.19}$$

$$m\ddot{y} + c\dot{y} + ky + f_a = f_e = \mu_F\, i_e \tag{3.20}$$

152

where i_e is the current in the coil and μ_F is the force/current factor.

Consider a harmonic electromagnetic force $f_e = F_e\,e^{i\omega t}$ giving rise to a force $f_a = F_a\,e^{i\omega t}$ applied to the system and response $y = Y\,e^{i\omega t}$. From (3.19), the FRF of the SDOF system alone is:

$$H_a = \frac{Y}{F_a} = \frac{1}{K - \omega^2 M + i\omega C} \tag{3.21}$$

This is the FRF that would be obtained if a massless force transducer were used to measure the force in the rigid stinger link.

From (3.19) and (3.20), the FRF of the complete system is:

$$H_e = \frac{Y}{F_e} = \frac{1}{(K+k) - \omega^2(M+m) + i\omega(C+c)} \tag{3.22}$$

This is the FRF that would be obtained if the force were derived from the current flowing in the exciter coil: the electromagnetic force. This FRF is for the SDOF system plus exciter.

The electrical behaviour of the exciter is governed by the equation

$$G v = L\frac{di_e}{dt} + R\,i_e + \mu_B\,\dot{y} \tag{3.23}$$

where v is the applied voltage, G is the amplifier voltage gain, L is the coil inductance, R is the coil resistance and μ_B is the back-emf factor (Note: the back-emf factor μ_B in V/ms^{-1}, and the force/current factor μ_F in N/A, are numerically equal).

Now, assume the harmonic voltage $v = V\,e^{i\omega t}$ leads to the current $i_e = I_e\,e^{i\omega t}$ and response $y = Y\,e^{i\omega t}$, etc., and if L is small and the frequency is low, the $L\,di_e/dt$ term can be neglected. Then, using (3.20), (3.22) and (3.23), the FRF based upon the input voltage to the amplifier is:

$$\frac{Y}{V} = \frac{\mu_F\,G}{R}\left[\frac{1}{(K+k) - \omega^2(M+m) + i\omega(C+c+C_B)}\right] \tag{3.24}$$

where $C_B = \mu_F\mu_B/R$ is an electrical 'damping factor'.

The transfer function between the applied force and the input voltage is:

$$\frac{F_a}{V} = \frac{\mu_F\,G}{R}\left[\frac{K - \omega^2 M + i\omega C}{(K+k) - \omega^2(M+m) + i\omega(C+c+C_B)}\right] \tag{3.25}$$

It may be seen from these last two equations that there is an electrical damping term C_B present due to the back-emf that causes a reduction in the coil current as the response velocity increases. This reduction in coil current leads to a corresponding reduction in the electromagnetic force generated within the exciter coil. For a constant voltage amplifier, the gain G is essentially constant with frequency.

If a constant current amplifier is employed, the amplifier output voltage changes to overcome the back-emf generated in the coil as it moves through the magnetic field and maintains the current flowing. The force generated in the coil therefore remains directly proportional to, and in-phase with, the amplifier input voltage.

It should be noted that even when a constant current amplifier is used, the actual force applied to the structure for a constant amplifier input voltage will experience a pole (i.e. peak) at the system-plus-exciter natural frequency and a zero (i.e. trough) at the system natural frequency, equation (3.25). The trough in the actual force applied to the system is well known and often referred to as force 'drop-out' [4, 5, 42, 43][†].

The force drop-out is reduced by use of a constant current amplifier, because the back-emf effect is prevented. Nevertheless, there will always be some force drop-out due to the added mass, stiffness and damping of the exciter. If the 'structure' is re-defined to include the moving parts of the exciter ('structure' mass as (M+m), 'structure' stiffness as (K+k) and 'structure' damping as (C+c)) then it can be seen that H_e becomes synonymous with H_a. Providing the constant current amplifier has sufficient capacity there will be no force drop-out at resonance for this re-defined system.

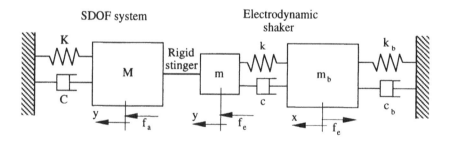

Fig. 3.21   SDOF system with exciter body flexibly mounted.

Exciter body flexibly mounted

Now, consider the case where the exciter body of mass m_b is flexibly mounted to ground via a stiffness k_b and damper c_b, as shown in figure 3.21. Note that there is an equal and opposite force f_e applied to the exciter body by the electromagnetic effect.

If the body motion is given by x and the stinger is rigid, then the equations of motion are:

[†] See also Section 2.3.2.

$$M\ddot{y}+C\dot{y}+Ky=f_a \tag{3.26}$$

$$m\ddot{y}+c(\dot{y}-\dot{x})+k(y-x)+f_a-f_e=0 \tag{3.27}$$

$$m_b\ddot{x}+c_b\dot{x}+k_b x-c(\dot{y}-\dot{x})-k(y-x)+f_e=0 \tag{3.28}$$

The true FRF, H_a, is the same as before, but the FRF based on the electromagnetic force is:

$$H_e=\frac{Y}{F_e}=$$

$$\frac{k_b-\omega^2 m_b+i\omega c_b}{\left[(K+k)-\omega^2(M+m)+i\omega(C+c)\right]\left[(k+k_b)-\omega^2 m_b+i\omega(c+c_b)\right]-(k+i\omega c)^2}$$

$$\tag{3.29}$$

which does not lend itself to obvious simplification. The difference between H_a and H_e will best be seen using simulated data.

3.5.2 Simulations

In order to illustrate the interaction of the exciter with the structure and to compare the different approaches for force measurement the following set-up will be considered:

Structure:
>Effective mass (M) = 10 kg
>(natural frequency = 20 Hz)
>Effective stiffness (K) = 157914 N/m
>Effective damping (C) = 25.13 Ns/m
>(damping factor = 1 %)

Exciter Coil:
>m = 0.2 kg
>k = 14100 N/m
>(natural frequency ≈ 42 Hz)
>c = 10.62 Ns/m (damping factor = 10 %)
>μ_F = 22.3 N/A, μ_B = 22.3 V/ms⁻¹
>R = 4.8 Ω, and G = 1

Exciter Body:
>m_b = 13.9 kg
>k_b = 137.2 N/m (pendulum support with 0.5 Hz natural frequency)
>c_b = 0.87 Ns/m (damping factor = 1 %)

The values for the exciter coil and body are those for a commercial shaker and the structure values are for a small aircraft model structure. The results are only intended to give an idea of the effects that can occur.

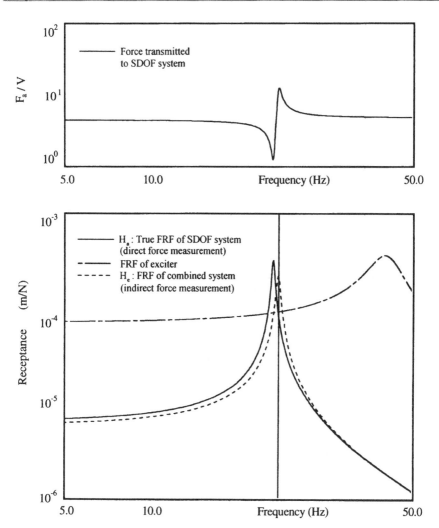

Fig. 3.22 Various FRFs for SDOF system with a grounded exciter.

Grounded Exciter Body

The FRFs H_a and H_e are shown for a frequency range 5-50 Hz in figure 3.22. H_a represents the 'true' behaviour of the structure found using a force transducer whose added mass effect is negligible. H_e is the FRF of the combined system, or that which would be obtained using the electromagnetic force. The FRF of the exciter alone is also shown in this plot.

The function H_e may be seen to overestimate both the natural frequency and the damping ratio. The natural frequency is overestimated because the structural mode frequency is less than the exciter natural frequency. Below its natural frequency, the exciter acts like a stiffness and increases the apparent structural mode frequency. It can be seen that where the point FRFs of the SDOF structure and the

156

exciter are equal in magnitude but opposite in phase, there is a resonance of the combined system. This follows directly from the impedance coupling equation for a single degree-of-freedom:

$$\frac{1}{A} + \frac{1}{B} = \frac{1}{C}$$

(3.30)

where, A and B are receptances of the constituent systems and C is the receptance of the combined system.

The magnitude of these effects depends upon the relative ratios of the properties of the exciter to those of the structure: m/M, k/K and c/C in (3.22). For a different structure or exciter in which the system natural frequency was greater than that of the exciter, the exciter mass effect would be dominant and the system natural frequency would be underestimated. The use of 'Salter Skeletons' [44], can help to simplify the interpretation of these plots. The variation of the force transmitted to the SDOF system per unit voltage input, F_a/V, is shown in figure 3.22. The classical 'drop-out' in the force applied to the structure occurs at the natural frequency of the SDOF system. If a constant voltage amplifier were used, the force measured by a massless force transducer in the rigid stinger link would have an identical form to this transfer function. In this instance, the applied force drops to approximately 1/5 of the level at frequencies away from resonance.

Exciter body flexibly mounted

The support chosen for this illustration was a pendulum arrangement with a 0.5 Hz natural frequency. The FRF for the 'true' behaviour of the structure found using a massless force transducer, H_a, and that for the combined system (or that obtained using the electromagnetic force), H_e, are shown for a frequency range 0.1-100 Hz in figure 3.23. Again, the FRF of the exciter is also shown in this plot.

More detailed plots around the resonance frequencies of the combined system are given in figures 3.24 and 3.25. It can be seen that the resonances of the combined system occur where the point FRFs of the constituent systems have equal magnitude but opposite phase.

There is also an antiresonance of the combined system at the exciter suspension frequency (0.5 Hz). At this frequency, the exciter body on its suspension acts as a dynamic absorber: the point of attachment to the SDOF system is almost stationary and there is virtually no force transmitted through the stinger link. The FRF of the SDOF system remains unchanged by the attachment of a different exciter. However, the combined system now has two resonance peaks: one at approximately 5 Hz and another just above the resonance frequency of the SDOF system alone (20 Hz).

Figure 3.25 (10-100 Hz) shows FRF characteristics almost identical to those for the grounded exciter except that the peak in transmitted force is attenuated as a result of the flexibly mounted exciter body. Increasing the reaction mass of the exciter body will reduce the frequency of the 5 Hz mode.

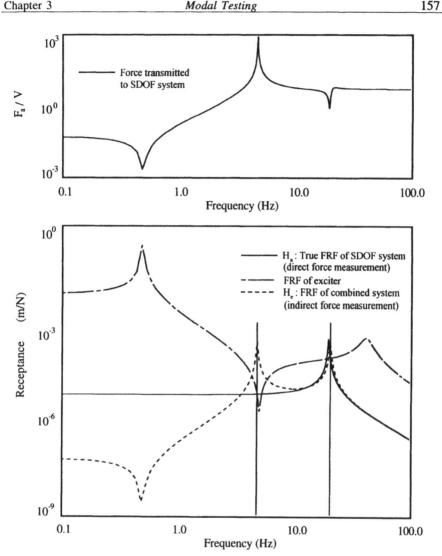

Fig. 3.23 Various FRFs for SDOF system with exciter body flexibly mounted, 0.1-100 Hz.

3.5.3 Extension to Multi-DOF/Multi-Reference

In practical modal testing, the system under test will have many modes and often multiple exciters are employed simultaneously. Clearly the relevant equations of motion could be expressed but they are inevitably rather involved. Therefore, the extension to multi-DOF and multi-references will be described only in concept. To and Ewins explore a closed-loop model for multiple-input testing in [45].

For multi-DOF, the simple FRF H_a will involve a summation of modes. The exciter effects add into all the system modes and will be seen in H_e. The severity of the exciter effects will depend upon the effective contribution of the exciter coil mass, etc., to the modal mass of the mode of interest, e.g. they will depend upon the ratio of $\varphi_j^2 m_c / M_j$ where M_j is the modal mass of the jth mode and φ_j is the mode shape value at the exciter attachment point. Thus, any error in the estimated FRF will increase as the exciter is moved into regions of larger modal displacement.

When multiple exciters are used, the effects are additive. For classical FRF estimation using uncorrelated random excitation, then H_a will be accurate, apart from any effects due to the mass of the force gauge. However, H_e will differ from the true FRF depending upon the accumulated effects of the exciter parameters.

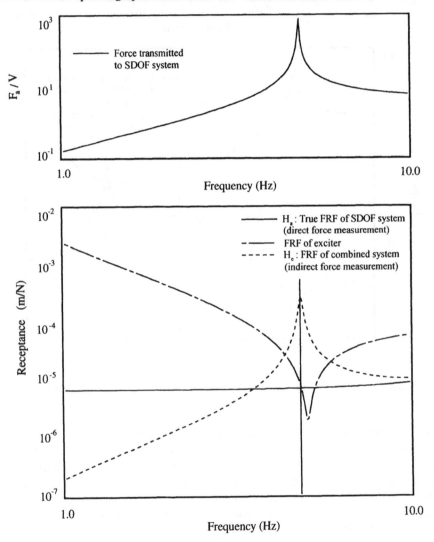

Fig. 3.24 Various FRFs for SDOF system with exciter body flexibly mounted, 1-10 Hz.

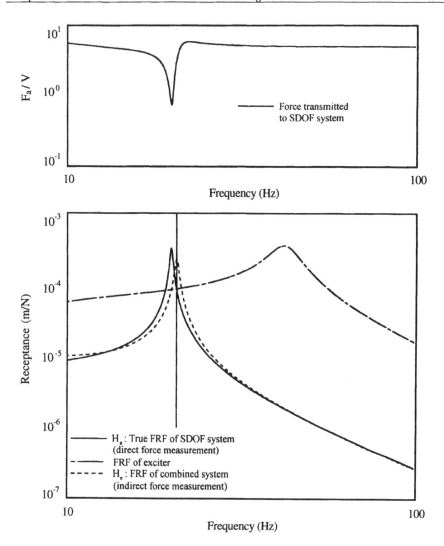

Fig. 3.25 Various FRFs for SDOF system with exciter body flexibly mounted, 10-100 Hz.

Thus it is arguable that force gauges should always be employed for multiple input testing. However, the gain and phasing of applied forces and voltage drive signals will be different for each excitation channel and for each frequency. For normal mode force appropriation testing a method of force control has to be employed by estimating the FRF matrix between the input voltages and the applied forces.

Therefore, in order to obtain mono-phased applied forces, the voltages supplied to the exciters become complex. If the indirect estimation of forces from the coil currents is employed together with constant current amplifiers then there is no need for complex force control. The coil currents, and hence electromagnetic forces, are constant with frequency and in-phase with the input voltages. In this way, the normal mode tuning process is greatly simplified, leading to improved efficiency in force appropriation test methods.

3.6 PRELIMINARY DATA COLLECTION AND ASSESSMENT (PRELIMINARY SURVEY)

In a comprehensive paper by Mitchell [46], various techniques are discussed that may be used to obtain a greater quantity of higher quality experimental data for complex structures. Smallwood [47] addresses the problem of qualifying the accuracy of FRFs by study of rigid-body response shapes. Common errors such as the location and measurement direction of transducers are easily detected by the visual inspection of these rigid-body shapes. Problems affecting the faithful generation, transmission and quantification of the measurement signals are considered in a paper by Stein [48]. In this paper it is stressed that thorough, careful and systematic validation of measured data must be undertaken and documented at every stage and that assumptions are unacceptable. The ideas and suggestions contained in these papers could also be implemented during a preliminary survey to ascertain their applicability in a particular test. In this section, however, only a few very simple checks for data quality are introduced and discussed.

3.6.1 Measurement frequency range

Usually, in structural modification exercises, the region of interest is focused on a relatively small frequency range, situated about a problem frequency at the operating condition of the machine. This small frequency range should be considered in careful detail both before and after the modification prediction, so that the precise effect of the modification in that region can be properly assessed.

Although the main interest is concentrated in a narrow frequency range, it is still vital to have a much wider picture of the situation. It may be that as a consequence of improving the dynamic characteristics in one particular frequency range detrimental effects occur elsewhere. This should be known, so that the overall suitability of a modification can be assessed.

It has already been mentioned that the final results will be judged over a small frequency range around a particular frequency of interest. However, it is important to remember that the input data may have to be collected accurately for a very much wider frequency range. This is almost certain to be the case if a process of modal analysis and subsequent synthesis of the FRFs is to be employed. In principle, if the raw measured FRF data were to be used directly in a coupling analysis, then it would be necessary to measure data only over the small frequency range of interest because these data include the effects of *all* the modes of vibration.

It has been found [49] that slight inconsistencies in the raw data (perhaps brought about by taking several measurements sequentially rather than all simultaneously) leads to predictions for the FRFs of the modified structure that are contaminated by 'breakthroughs' from the original data. This problem can be avoided if all the data are measured simultaneously, or if the FRFs are synthesised from a modal database. Practically, it is not possible to collect all the data for the complete FRF matrix in one simultaneous multi-point excitation measurement. Such a measurement would necessitate an excitation source and a response transducer at every degree-of-freedom of interest, including rotational degrees-of-freedom.

If the FRFs used in a coupling analysis are to be synthesised from a modal database, then either:

(i)   sufficient modes must be included in the measured frequency range to account for the effect of out-of-range modes in the narrow frequency band of interest; or

(ii) residual terms must be accurately defined and incorporated for each FRF to be used in the modification procedure.

The answer to the question - "What is a sufficient number of modes?" - depends upon factors such as the local modal density and modal dampings, and whether there are any dominant modes nearby. It will be highly specific to the particular structure, and the modification sites chosen.

Once a sufficient number of modes has been measured and analysed for one row or column of the FRF matrix, then all the terms in the matrix may be synthesised accurately from the modal database. While the agreement of measured and synthesised FRFs serves as a check on the results for the unmodified structure, there is still no guarantee that the model will be adequate for accurate prediction of the effects of any given modification.

If an impedance coupling method is used, the effects of out-of-range modes can be included by the addition of residuals to the synthesised FRFs. The residual terms approximate the effects of the out-of-range modes in the frequency range of interest (see Section 1.6, equation 1.269). However, if a modal coupling method is used, it is not easy to add in such correction factors.

3.6.2 Frequency resolution

Adequate definition of the resonance features of an FRF is most important since these data points are highly influential in the modal analysis process. Circle-fitting of isolated modes in the Nyquist plane presentation of an FRF provides a clear example of the need for sufficient data points between the half-power points of the Nyquist circle (figure 1.30 in Chapter 1 illustrates this problem).

From preliminary measurements and the following analysis, it is a relatively simple task to calculate the frequency resolution necessary to define the circle adequately for the purposes of modal analysis. Consider the Nyquist circle in figure 3.26. If we require a minimum of three points between the half-power points, ω_1, ω_r and ω_2, to define the circle for analysis purposes, then the frequency spacing $\delta\omega$ is given by

$$\delta\omega = \frac{\omega_2 - \omega_1}{2} \tag{3.31}$$

It can be shown that for small levels of viscous damping[†],

$$\xi_r \approx \frac{\omega_2 - \omega_1}{2\omega_r} \tag{3.32}$$

where

ξ_r is the viscous damping factor for r^{th} mode;
ω_r is the resonance frequency;
ω_1 and ω_2 are the half-power points.

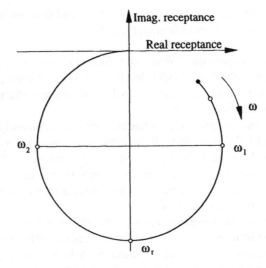

Fig. 3.26 Derivation of a criterion for frequency point spacing.

Therefore, by substituting (3.31) into (3.32), the point frequency spacing $\delta\omega$ is found to be

$$\delta\omega = \xi_r \omega_r \tag{3.33}$$

For broadband random testing, the frequency resolution is generally dictated by the properties of the lowest frequency mode of interest. Elsewhere in the frequency range, the resolution is usually more than adequate – data points are used inefficiently. The maximum number of data points for a test is limited by the data block size available in the analyser. Thus, in combination, the necessary resolution (for the low frequency modes) and the restricted number of data points limit the

---

[†] See also Section 1.3.3.

frequency range that can be measured in any one test. If the frequency range of interest is larger than this, there are two options available:

(i)   measure the additional frequency range using zoom measurements [50];

(ii) reduce the frequency resolution and re-measure over a larger baseband frequency range.

The choice is dependent largely upon the features of the acquisition system and the number of channels measured simultaneously. Implementation of the zoom method requires significantly greater computational power and it is usual for either the number of frequency points or the number of channels to be reduced by a factor of 2. Zoom measurement has the advantage over extended baseband testing in that the data points are used more efficiently. Each mode is measured only once. However, zoom measurements take longer than standard baseband measurements and there are some computational difficulties (frequently zoom FRFs have a 'spike' at the centre-frequency).

When using extended baseband testing methods, the low frequency modes are measured several times over, each with gradually increasing frequency spacing as the frequency range is extended. In the subsequent modal analysis phase, only the highest frequency resolution data for each mode are used – the lower resolution data are superfluous.

Any noise on the force signal is likely to influence the resonances whereas noise on the response signal will affect the antiresonances. To minimise these effects of noise, different estimators of the FRF can be used for the best estimate of resonances. Where there is noise on the response signal, H_1 should be used, and H_2 should be used where there is noise on the force signal. The reverse applies for the antiresonances. Derivation of these results can be found in Chapter 2.

3.6.3   Quantity of data – number of degrees-of-freedom to measure

The quantity of data to measure depends largely upon the purposes for which the data are required - the defined objectives of the tests. If the data are only required to identify the resonance frequencies of the structure, a small number of FRFs is all that is necessary. To be able to characterise the mode shapes uniquely requires significantly more data. If the purpose of measuring the structure is to obtain data for use in a coupling analysis then the measurements *must* include, as a minimum, FRF data for all the coupling freedoms. The minimum quantity of data will enable the new resonance frequencies for the modified structure to be identified but it is probably insufficient to define the new mode shapes adequately.

3.6.4   Quality of measured data

Measurement accuracy is the ability to quantify the behaviour of the structure precisely. A lack of accuracy in test measurements may arise from two sources:

(i)   the inability to measure the motion of the structure without affecting its behaviour; and

164

(ii) the inability to quantify the transducer signals.

Once the transducer and conditioning equipment have converted the force or response quantities into electrical signals, the signals have to be quantified. Nowadays, the measurement instrumentation is primarily digital in nature and a large proportion of the errors in quantification of these signals are associated with details of digital signal processing.

The force and response signals returning to the analyser from the transducers are usually analogue signals. The analysers work with digital representations of these analogue signals. The interface process is called Analogue-to-Digital conversion (ADC).

The signals are sampled at regular intervals and the instantaneous value of the analogue signal is compared with fixed, discrete, digital levels in the analogue-to-digital converter. These are the so-called quantisation levels already discussed in Section 2.2.1. The digital level that most closely approximates the instantaneous value of the analogue signal is selected. The dynamic range of the ADC is expressed as the ratio of the maximum to the minimum signal level that can be resolved, in dB. A 12 bit ADC[†], typical of the type found in many analysers today, has a dynamic range of 72.25 dB. Before the start of a measurement, the input range for each signal is scaled to make the best possible use of the dynamic range available. The ranges are set using the peaks of the input signals. Provided that the dynamic range of the analogue input signal is less than 72 dB, the digital representation will be quite adequate. However, if the analogue signal has a dynamic range greater than 72 dB, very small levels in the signal cannot be resolved and they are all represented as zero.

During the preliminary survey, and indeed throughout the full modal test, some checks should be made to assess the quality of the measured data. There are several techniques already in general use that can provide an indication of the quality of the measured data, e.g. repeatability, reciprocity and coherence. However, application of some of these techniques in a much more systematic, rigorous and critical manner will give a better performance than is presently achieved.

Almost always, repeatability and reciprocity checks are done by 'eyeballing' sets of FRF curves to see if there are any major differences. The comparisons are made significantly easier if difference function curves (ΔFRF) are plotted for the sets of data.

3.6.5 Difference functions

The ΔFRF is defined as the following *vector* difference:

$$\Delta FRF(\omega) = FRF(\omega) \text{ for comparison - Reference } FRF(\omega)$$

or

$$\Delta H(\omega) = H'(\omega) - H(\omega) \tag{3.34}$$

[†] For which there are $2^{12} - 1 = 4095$ possible non-zero discrete levels.

The ΔFRF has the same units as the original FRF, and the same 'form', i.e., Receptance, Mobility, Accelerance, etc., and indeed, exhibits the properties and characteristics of a multi-degree-of-freedom linear system. Once the various features of the difference curves have been characterised and are recognised, a single ΔFRF curve is easier to read and interpret than visually looking for differences between two FRF curves.

The ΔFRF is the *vector* difference of the two FRFs at each frequency ω. This is shown clearly in the Argand plane (figure 3.27). The vector joining the two response function vectors is the difference, and the magnitude of this difference vector can often exceed the magnitude of one of the constituent FRF vectors.

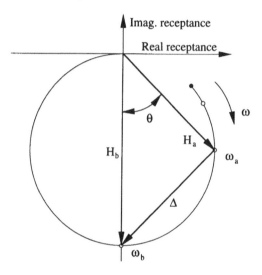

Fig. 3.27 Derivation of the Vector Difference Function.

The difference is given by the vector difference

$$\Delta = H_b - H_a \tag{3.35}$$

Now, if we assume that $|\Delta| = |H_a|$, then

$$\theta = \pi/4 \text{ rads } (45°) \quad \text{and} \quad |H_b| = \sqrt{2}\,|H_a| \tag{3.36}$$

On a log plot, the difference in the magnitudes of H_a and H_b would be 3 dB in a total range of the function of possibly 60 dB and, therefore, not easily seen. The difference between a half-power point frequency and the resonance frequency is given by $\xi_r\omega_r$ (from equation (3.33)). For a system with a viscous damping factor of 1% there would only need to be a 1% shift in the measured resonance frequency for the magnitude of the ΔFRF to be the same as one of the constituent FRFs. This serves to illustrate that the difference function is a very sensitive function.

Without very careful inspection of the FRF Bode plots, even differences as large as this can be difficult to see.

3.6.6 Use of the ΔFRF in Repeatability and Reciprocity

Repeatability checks are a means of assessing the stability of the structural characteristics over a period of time. It is usually assumed that the structure does not change with time or as a result of the excitation itself but there are a number of practical effects that can alter the characteristics of a structure:

> e.g. Bolts slackening;
>
> Release of pre-loads;
>
> Fretting;
>
> Environmental factors, such as temperature and humidity.

For a linear conservative system, Maxwell's Rule of Reciprocity applies: the measured FRF for a force at location j and a response at location i should correspond directly with the measured FRF for a force at location i and response at location j. The FRF matrix is symmetric and this property can be used as a check on the quality of the measured data.

Where a multiple single-input test[†] strategy is used, the reciprocity check can give an indication of shaker and accelerometer loading effects on the structure. The positions of the shaker and accelerometer are reversed in multiple single-input reciprocity checks. If the shaker and accelerometer have negligible effect on the structure, then there should be good reciprocity. If the shaker and accelerometer do have a significant loading effect on the structure, then the effects in the two configurations will be different and the reciprocity check will reveal any differences between the FRFs.

With a simultaneous multiple-input test strategy, the loading effects of the shakers and accelerometers will be constant (although not necessarily negligible) and this will not show up in the reciprocity check.

Repeatability and reciprocity checks are usually carried out over a frequency range incorporating several modes, with the emphasis placed on the magnitudes of the functions. In keeping with this practice, the magnitude of the ΔFRF is plotted alongside the two FRFs of the repeatability or reciprocity check for comparison. It is important that the ΔFRF is interpreted with reference to the FRFs from which it was derived, to understand the possible causes of any large discrepancies indicated.

If a large peak on the ΔFRF corresponds with a resonance region of the constituent FRFs, then it is likely that the difference is due to a slight shift in the natural frequency. When a peak occurs away from resonance, it is usually caused by differences in the magnitudes of the FRFs.

---

† Meaning various separate single-input tests with the shaker located at a different position for each test.

ΔFRF examples

The following examples show the benefits to be gained from the use of the ΔFRF in the presentation of repeatability and reciprocity data. Most of the measurements are taken from a survey in which the repeatability measurements were made over a period of one week. The same level of input excitation was used for all the measurements and, in the case of the reciprocity measurements, no attempt was made to keep the response at a particular point the same for each excitation. Therefore, there could be the influences of nonlinearities included in these results.

The ΔFRF in Repeatability checks

The two point FRF measurements, shown in figure 3.28, were made at different times as a check of the measurement repeatability. The vector difference between these two functions has been calculated and forms the ΔFRF that shows clearly the large differences between the FRF measurements around the resonance regions. Without the ΔFRF it may be easy to dismiss the small differences in the magnitude plots of the FRFs. The ΔFRF is of little value on its own; only when it is plotted alongside the FRFs from which it was derived can meaningful interpretations be made.

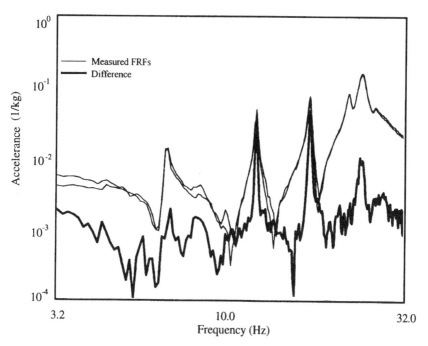

Fig. 3.28 Repeatability check using the ΔFRF.

The ΔFRF and Reciprocity checks

The use of the ΔFRF in assessing the reciprocity of measured FRFs is shown in figure 3.29: two measured reciprocal FRFs are plotted together with the ΔFRF.

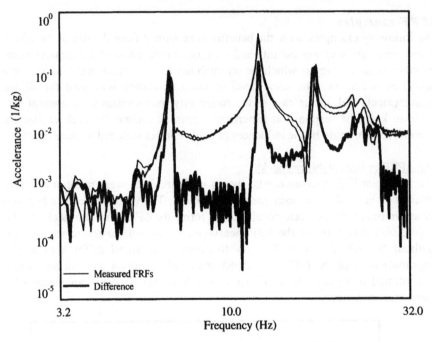

Fig. 3.29 Reciprocity check using the ΔFRF.

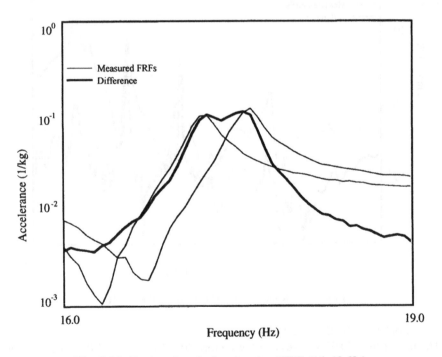

Fig. 3.30 Reciprocity check using the ΔFRF (16–19 Hz).

These FRFs were measured in separate single input surveys, the excitation system being moved from one input point to the other for the second survey. It will be noticed that the largest differences occur in the resonance regions. Once again, this is due primarily to slight shifts in the resonance frequencies rather than to changes in the magnitudes of the FRFs. The mode at 17 Hz illustrates this clearly. An enlarged portion (16 to 19 Hz) of the plot is shown in figure 3.30 where the shift in the resonance frequency can be seen clearly. Frequency shifts such as these are not always quite so obvious but, nevertheless, they can still give rise to problems at the modal analysis stage.

Figure 3.31 provides an example of the level of reciprocity that it is possible to achieve (albeit for a different structure) using a carefully set up multiple input test. The ΔFRF is at least an order of magnitude smaller than the constituent FRFs throughout the whole of the frequency range, including the resonance regions. This highlights the consistency of the measured data that can be obtained with simultaneous multiple input testing strategies.

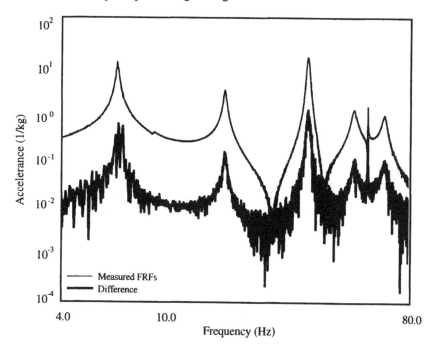

Fig. 3.31 An example of the Reciprocity attainable with a multiple input test.

Note: throughout the complete assessment, it should be remembered that the quantity and quality of the data are being assessed for their suitability for a particular purpose. The data may not have to be highly accurate in an absolute sense, just sufficiently accurate for the required purpose, a fact that can contribute to significant savings in cost and time.

3.7 AN OVERVIEW OF TESTING METHODS

There are several different approaches for measurement of the dynamic characteristics of a structure. Historically, the approaches have been divided into Phase-Resonance methods (Normal Mode testing or Sine Dwell testing) and Phase-Separation methods (Stepped-Sine and Broadband testing). The Hewlett Packard 'Application Note' on the fundamentals of modal testing [38] provides a very good basic introduction to many aspects of testing, and analysis. Olsen [51], has conducted a survey of excitation functions for structural FRF measurements for both phase-resonance and phase-separation test strategies.

Stepped-sine testing is explained in detail in references [52, 53]. It is interesting to note that some issues considered in an earlier section regarding the frequency resolution in measured FRFs, the accuracy of data, and the number of response and excitation points, were discussed by Craig and Su [54] many years ago in relation to multiple-shaker resonance testing. The same problems are still with us today.

Evaluations of phase-resonance test techniques may be found in references [55, 56] together with some advanced applications of normal mode testing described in reference [57].

Although each method has specific advantages and disadvantages, the selection of a testing technique is often based (unfortunately) on the type of equipment and expertise available rather than its suitability for a particular job.

3.7.1 Phase-Resonance testing

Phase-resonance methods[†] rely on the ability to excite a single mode of vibration by use of multiple shakers with independently variable force levels. The shakers each produce sinusoidal excitation at the same frequency and are either in-phase or out-of-phase with a reference source.

The 'normal-mode' excitation set-up effectively cancels the damping in the structure – the exciting forces are distributed such that each energy sink is cancelled by a corresponding energy source – and single real modes can be excited. In this condition of normal-mode vibration, the excitation frequency is the undamped natural frequency of the mode. The response at all the points on the structure is in quadrature with the excitation forces, and the structural responses relate directly to the mode shape vector.

The previous method can be very powerful and is popular for the ground vibration testing of aircraft structures because of the ability to measure real normal modes (for direct comparison with FE results) and the ability to investigate nonlinear behaviour.

The main difficulties with normal-mode testing are the selection of excitation locations, the tuning of the force pattern and the choice of excitation frequency. The complete process has to be repeated for each different mode and consequently the testing time can be lengthy.

[†] Also known as tuned-sinusoidal or force appropriation methods.

3.7.2 Phase-Separation testing

Phase-separation[†] methods rely on the fact that the forced response of a linear structure is a weighted linear summation of all the uncoupled modes of the structure. The forced response to a known excitation is measured and then the dynamic properties are extracted by means of mathematical curve fitting techniques (Chapter 4). Phase-separation methods rely upon the mathematical assumption that the actual responses are formed from a linear combination of the modes. In fact, the measured data may be inconsistent and contain the effects of nonlinearities. Nevertheless, with more and more sophisticated analysis techniques and coupled with the fact that the phase-separation methods are much cheaper, easier and quicker to implement than phase-resonance techniques, the phase-separation methods are now the predominant form of modal testing.

Almost by definition, phase-resonance testing methods involve multiple simultaneous excitations applied to the structure. This is not the case for phase-separation testing methods: 'single-point' test strategies may be applied, including the simplest form of modal testing, an impact test.

Single-Point testing

Within the phase-separation category of modal testing, the single-input type of test is the most straightforward. It is simple and quick to implement and requires a minimum of equipment. The extraction of FRFs from the measured data is also a simple matter. One row or column of the FRF matrix is measured in a single input vibration test. However, information from this single row or column of the FRF matrix may be insufficient to define all the required modes. To overcome this deficiency, several single-input tests may be performed one after another, thereby obtaining progressively more rows or columns of the FRF matrix. Unfortunately it is highly likely that the FRFs from these multiple single-point excitation tests will be inconsistent and therefore the resonance peaks obtained using different excitation points will not be exactly the same.

A deficiency with single-point testing is that the energy input is very localised. In large heavily-damped structures the excitation energy is quickly dissipated before it has propagated far within the structure. In an effort to excite remote regions, the single-point forcing level is sometimes increased to excessive levels. Local nonlinearities are thereby exercised to a much greater extent than would otherwise be the case if the excitation energy were distributed more evenly throughout the structure.

Probably the simplest form of single-point excitation is hammer, or impact, testing. A specially-instrumented hammer is used to impact the structure at the desired excitation point. A very short sharp excitation pulse is produced – approximating a Dirac delta function – which has a flat spectrum over a wide frequency range. The amount of energy contained in the impact pulse is small. By using hammer tips with different resiliences, different pulse widths can be obtained.

[†] Also known as non-sinusoidal methods.

Used with a two channel data acquisition system, it is customary for the response measurement point to remain the same as the hammer is moved around the structure to different excitation points. In this way, a row of the FRF matrix is measured rather than a column which is usually the case when a shaker is used for excitation and a roving response transducer is moved to measurement points around the structure.

The main advantage of hammer testing is that the excitation equipment (the hammer) is small, light and cheap. It is used mainly for diagnostic purposes rather than for precise measurement of FRF properties. Disadvantages of the hammer testing method are related to the inconsistency of the excitation. The impact pulse is difficult to control accurately in size, in shape and in direction, and the duration of the pulse is very small compared with the measurement time frame. Furthermore, impulse excitation is a major disadvantage when the structure contains nonlinearities.

Multi-Point testing

In tests on large complicated structures, with numerous joints and nonlinearities, the vibration energy is quickly dissipated within the structure. This may indicate that the use of multi-point testing is preferable to a series of single-point tests. By the use of multi-point excitation, the energy can be fed into the structure more uniformly than with single-point excitation and the response amplitudes at various locations can be kept much closer to those found in operation. Energy is supplied to the structure by several shakers, and so smaller and cheaper shakers can be used than would be necessary for single-point testing. Furthermore, the effects of nonlinearities, which may occur with excessive single-point forcing, can be substantially reduced. Since all the shakers are connected to the structure before the start of the test, systematic errors, resulting from repositioning the shaker during a series of single-point tests, do not occur. The influence of the shakers on the structure is not removed, it just remains constant throughout the whole of the test programme: a static base-condition error. Simultaneous measurement of multiple columns of the FRF matrix means that the overall test time is shortened. There is less opportunity for structural changes (with time, temperature or humidity) to affect the measured results.

Suitable test control hardware and software are also required for multiple-input tests, and the process for extraction of FRFs is more complicated than for a single-input test. The most commonly used multiple shaker technique is known as Multi-Point Random (MPR). For all multi-point excitation techniques, more sophisticated computer programs are necessary to extract the standard FRFs from the measured data.

To comply with the mathematical assumptions made in the analysis it is important that the excitation inputs to the structure are purely random and uncorrelated. Although it is possible to ensure that the excitation signals driving the shakers are uncorrelated, it does not necessarily follow that the excitation force signals are uncorrelated. At the resonance frequencies, in particular, it is found that the motion of the structure tends to correlate the actual multiple forcing inputs. This can lead to degradation in the quality of the derived FRFs at the resonance frequencies. Stepped-sine testing and broadband testing are two types of phase-separation methods of modal testing that will be discussed further in this brief overview. The requirements for extending the simple single-point testing methods to multi-point methods are also presented together with discussion of some of the multi-point testing techniques.

Stepped-Sine testing

In the stepped-sine testing technique, a shaker is used to excite the structure sinusoidally at a single, precisely controlled, frequency [52]. The structure is allowed to settle under this excitation. Steady-state measurements are then made of the magnitude and phase relationship between the input force and the response at the precise excitation frequency for any desired response location. Division of the response by the force input gives the value of the FRF at that particular frequency. The excitation frequency is then changed by a small increment and the measurement process is repeated once the structure has settled at the new frequency. By this means, the FRF can be constructed for any frequency range. There are no restrictions on the frequency spacing imposed by the testing method, although there may be restrictions due to the analysis methods to be used.

At each excitation frequency the structure's response is allowed to settle to remove any transient effects associated with the step change in excitation frequency from the previous measurement. The transient effects will be included in the measurement if insufficient time is allowed for them to die away [53]. A clear indication of whether the settling period is long enough can be obtained by performing the measurement twice around a resonance peak; once with an increasing frequency sweep, and once with a decreasing frequency sweep.

An example is presented in figure 3.32 where the resonance peaks occur at different frequencies for the two measurements. The transient effects have not had sufficient time to decay and the settling time should be increased. An alternative strategy is to monitor the response and only record data when successive acquisitions at that frequency have less than a pre-defined variation. In this way the acquisition becomes adaptive; the settling time varies to allow the structure to reach a steady-state at each frequency point across the frequency range of interest.

When the structure has had time to settle, a single measurement of force and response should be sufficient to define the point on the FRF accurately. However, in practice, noise is invariably superimposed on the signals to be measured and it is prudent to make several measurements of the data and to calculate an average value. Fortunately, one of the properties of random noise is that when averaged over a long enough period of time the errors on the data reduce to zero. Averaging will only reduce the effects of random errors; it will not reduce the effects of any systematic errors.

A prime advantage of the stepped-sine testing technique is the large signal-to-noise ratio for all the force and response measurements. This is a consequence of the single excitation frequency for each measurement – there are no other significant sources of excitation to contaminate the results. Furthermore, the input ranges for the force and response signals can be adjusted automatically for each excitation frequency point, which allows the best possible use of the measuring equipment. The ability to concentrate data points in regions of greatest rates of change on FRFs is also particularly advantageous for efficient and accurate definition of resonance peaks and antiresonance troughs.

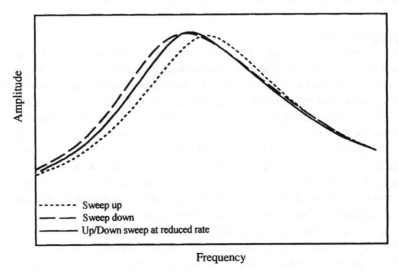

Fig. 3.32 Example of sine sweep rate influences on measured FRFs.

Close to a resonance frequency, a structure will exhibit large amplitude responses that are particularly dependent upon the excitation pattern applied. Unless great care is taken to monitor and control the response level for each force pattern at a given frequency, response dependent nonlinearities will be exercised completely differently for each force pattern. This will accentuate the nonlinear behaviour of the structure and give rise to inconsistent measurements that present difficulties for analysis.

Broadband testing
For broadband testing, the structure is excited with a signal containing energy over a wide range of frequencies simultaneously. The time domain force and response signals are filtered, digitised and then passed through a Fourier analysis process to transform the time domain information to frequency domain spectra. By appropriate combination of the force and response spectra, the required FRFs for the structure can be derived.

Because of constraints imposed by the Fourier analysis process, the frequency point spacing is constant across the whole measurement frequency range. This represents an inefficient use of the limited number of frequency points available because they cannot be concentrated in the regions of greatest change on the FRFs – around the resonances and antiresonances.

The most popular form of excitation is by use of an electromagnetic shaker driven with a signal from the data acquisition system. The mechanical set-up is identical to that used for sine testing. There are very many different types of excitation that can be used for broadband testing. Two common types of broadband excitation signals – random and periodic random – will be discussed here.

Random

A random excitation signal has a continuous spectrum that is flat across the frequency range of interest. The random time domain excitation signal is a reasonable approximation to the type of excitation found in many typical installations. The probability distribution of the excitation is approximately Gaussian (Normal) and hence the peak-to-RMS ratio is about 3:1. For a linear system, a Gaussian input will produce a Gaussian output. The structural response does not tend to 'peak-up' at resonance frequencies to the same extent as it does with some other forms of excitation, e.g. swept sine. Since the peak-to-RMS ratio of the output is relatively small, the random type of excitation does not cause undue problems with excitation of nonlinear structures. However, because the force and response signals are random, a weighting function (e.g. Hanning window) must be applied to these signals before the Discrete Fourier Transform (DFT) is performed, otherwise the mathematical assumptions of periodicity in the measurement time frame are invalid (see Section 2.2).

The force and response signals from the transducers contain energy at all frequencies. Because the measurement time frame is of a finite length, however, the Fourier Analysis produces spectra that are discrete, rather than continuous, functions. Therefore, there is a 'spreading' of energy from the actual continuous spectra into adjacent spectral lines – this phenomenon is known as leakage (see Section 2.2).

Periodic Random

Periodic random excitation is a special form of periodic excitation that has several benefits for signal processing. Although called random, it is not random in the true sense of the word. Within a time period equal to the analysis time frame, the signal is random, but the *same* signal is then repeated continuously. Usually, the signals are generated inside the data acquisition equipment in the frequency domain. Once the measurement frequency range and number of frequency lines have been selected, a flat excitation frequency spectrum may be generated by setting the magnitude of all spectral lines to the same value. The phases of the spectral components are then randomised. By use of the Inverse Fourier Transform (IFT) this frequency domain spectrum is transformed into the time domain to produce a random-like excitation signal in the analysis time frame. As a consequence of generating the signal in this way, it is exactly periodic in the analysis time frame, and both the force and response signals will be periodic in the analysis time frame also.

Provided that the excitation is applied to the structure more than once so that any start-up transients have had sufficient time to decay, there is no requirement for any window function to be applied to this type of data. In fact, it is detrimental to the quality of the results to apply *any* form of window function to this type of data, other than a uniform (rectangular) window. These signals only contain energy at the precise spectral component frequencies of the analysis and, therefore, there are no leakage problems when periodic random excitation signals are used. Furthermore, the signal-to-noise ratios for the measured force and response signals are much better than they would be with pure random excitation because there is no superfluous energy contained in the excitation.

Every different randomised arrangement of the spectral component phases will give a different time domain signal. If a large number of these signals are inspected, it is found that the peak-to-RMS ratio is often greater than 3:1. However, it is possible to arrange the phase relationships of the spectral components such that the time domain signal closely approximates the form of a swept sine:

$$f(t) = \sin\ (at)^2 \tag{3.37}$$

which has the instantaneous frequency given by

$$\frac{d}{dt}\left\{(at)^2\right\} = 2a^2t \tag{3.38}$$

For this signal the spectral components have the following relationship:

$$F\left(\Omega_r\right) = \frac{1+i}{a}\left(\frac{\pi}{2}\right)^{1/2} e^{-i\Omega_r^2/4a^2} \tag{3.39}$$

Provided that the sweep rate is not too rapid, each cycle is approximately a sine wave of constant frequency. The peak-to-RMS ratio of the signal will then be very similar to that of a pure sine wave. For measurement of the input signal, this low peak-to-RMS ratio is advantageous. Unfortunately, by the nature of the excitation signal, the structural response builds up very significantly in the resonance regions, leading to high peak-to-RMS ratios in the measured responses.

Furthermore, an excessive response range, such as this, can cause problems when the structure contains nonlinearities because these nonlinearities will be exercised to a greater extent than with a less well ordered excitation. Detailed derivations of the equations relating to the sine sweep excitation functions may be found in the paper by White and Pinnington [58].

Multi-Phased Stepped-Sine testing (MPSS)
A new multi-point testing technique that is gaining in popularity is the Multi-Phased Stepped-Sine technique (MPSS) developed by SDRC [59]. In essence, the technique is a combination of the phase-resonance and phase-separation methods of modal testing.

By using multiple inputs with specific mono-phased force patterns (forces in-phase or out-of-phase with each other), sweeps are made over the complete frequency range of interest. The data are analysed and by use of the Multi-variate Mode Indicator Function (MvMIF) [60][†] it is possible to identify the specific force patterns required to excite each normal mode. Narrow frequency sweeps are performed around each resonance, using the appropriate force patterns, and the results obtained can be analysed easily by circle fitting methods. At the undamped natural frequency the displacements of the system correspond to the undamped normal mode shape – the basis of the phase-resonance technique of modal testing.

The benefits of the MPSS technique are the high signal-to-noise ratio, the variable frequency point spacing, combined with a multi-point test configuration that allows full identification of modes with a multiplicity greater than one. The FRF data obtained by this method are generally smoother and easier to analyse than data from MPR type tests.

As its name implies, the 'Stepped-Sine' vibration test method involves the accumulation of a complete cross-spectrum matrix for all the responses and excitation forces for each frequency of interest in turn. At each frequency, time history measurements are made for all the responses and excitation forces. The fundamental components corresponding to the excitation frequency are extracted by means of a single frequency Discrete Fourier Transform (DFT). In traditional broadband test methods, the spectra are obtained in a single acquisition by means of the FFT.

The frequency-by-frequency approach of stepped-sine testing is clearly much slower than the single acquisition approach of broadband test methods. The major advantage is that a much greater noise reduction is obtained due to the high signal-to-noise ratio that is characteristic for the measurement of sinusoidal signals. Furthermore, it is possible to set the most appropriate ADC gain for each measurement channel for each measurement frequency to ensure the most accurate digitisation of the analogue signals.

The ability to keep the applied excitation forces, or a response, constant throughout the test is useful for identifying the presence of, and possibly identifying, structural nonlinearities. As with all sine test techniques though, the structural response at the resonance frequencies can be very large. Great care must be taken to prevent damage to the structure or the excitation equipment.

Stepped-sine testing is not intended to be a replacement for other test methods but an addition. An initial test on a structure will undoubtedly be a broadband random type of test that will be used to identify any structural mode frequencies and damping values. Information from the broadband test will then help in the definition of subsequent stepped-sine tests.

It is vital to remember that stepped-sine testing requires that the whole system should be in a steady-state when the data are measured at each frequency of interest; all transient effects should have had sufficient time to decay before any data are

---

[†] See also Section 4.4.4.

measured. For lightly damped modes at low frequencies, this settling time can be quite considerable. A useful guide is the number of cycles taken for a single degree-of-freedom system to decay to half its initial amplitude (figure 3.33).

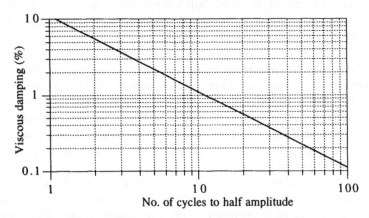

Fig. 3.33 Viscous damping factor *versus* number of cycles to half amplitude for the free decay of a SDOF system.

For a mode with a viscous damping factor of 0.5%, the number of cycles to half amplitude is 22. If this mode has a frequency of 6 Hz, then the time taken to reach half amplitude is 3.67 seconds. The actual settling time should be at least 3 times the time taken to reach half amplitude, e.g. 11 seconds for the example mode. Because a change of frequency involves a transient, this settling time is required after each frequency step before any data are acquired.

Software for stepped-sine testing often allows various strategies for control of the excitation levels and phase relationships, e.g. to keep the force input levels constant or a single measure of response level constant. Binary phase patterns (components either in-phase or out-of-phase with each other) for the excitation drive signals may be used, or other phase relationships may be specified explicitly. For binary phase patterns, there are $2^{(n-1)}$ different patterns that can be generated with n shakers.

Some of the algorithms for control of the excitation signals involve additional measurements with the magnitudes and phases of the drive signals slightly perturbed at each measurement frequency. These additional measurements enable the relationships between the drive signals and the applied forces to be determined. Settling times are required after each perturbation and the total test time can be increased very substantially if these control strategies are adopted.

The investigation of nonlinear effects may dictate that the excitation or the response levels should be kept constant. However, very precise control of the phase relationships of the actual force inputs is not necessary for recovery of FRFs. The actual force inputs do not have to be exactly in-phase or 180° out-of-phase; all that is required is that there should be an orthogonal set. The minimum number of different excitation patterns that *must* be applied in any multiple-input stepped-sine test is equal to the number of shakers, but this may be insufficient if there is significant correlation between some of the excitation forces.

3.8 MULTIPLE–INPUT MULTIPLE–OUTPUT DATA CHECK

3.8.1 General considerations

The theory of multiple-input testing is based upon the assumption that the inputs are not fully correlated. Some correlation of inputs is acceptable and inevitable if force transducers are used to measure the force input to the structure at the excitation points, even if the drive signals to the shakers are completely uncorrelated. The correlation of the force signals is due to the interaction of the excitation systems with the structure and will tend to be greater at the structural resonance frequencies.

In vibration tests it is quite often the case that a selection of (force) input and response channels will be monitored using an oscilloscope. If random excitation is being used, an oscilloscope only allows the dynamicist to observe that the signals are present and their approximate amplitude range. It is not usually possible to deduce from the traces whether the signals are fully correlated - and hence that the data collected are of little use for modal analysis. Because it is the forces between the structure and the shakers that are recorded as the inputs, it may not even be possible to tell whether all of the shakers are being driven correctly. The structure will still react against a 'dead' shaker, giving a measurable input force. This force will, however, be fully correlated with the other excitation forces.

It is to this problem of how to ascertain the degree of correlation of the inputs that the 'principal component analysis check' is directed. Performed at a preliminary stage in the set-up of a large multiple–input multiple–output (MIMO) test programme, this check will enable the integrity and suitability of the chosen excitation system to be assessed.

The basic theory behind the use of principal component analysis is given by Leuridan [16], but some extensions have been introduced here to aid its use in practical situations.

3.8.2 Principal Component Analysis theory

In a preliminary test, the cross spectrum matrix for the force inputs is measured. The Principal Component Analysis theory is based on the spectral decomposition (or eigenvalue decomposition) of a matrix which, in turn, is a particular case of the Singular Value Decomposition (see Appendix A), when applied to a square, Hermitian matrix. The spectral or eigenvalue decomposition of a Hermitian matrix [A] is therefore written as:

$$[A] = [U] \ [\Sigma] \ [U]^H \qquad (3.40)$$
$$\underset{(NxN)}{} \quad \underset{(NxN)(NxN)}{} \ \underset{(NxN)}{}$$

where [U] is a unitary matrix ($[U]^H[U]=[I]$). Post-multiplication of (3.40) by [U] reveals the classic eigenvalue equation, in which the diagonal matrix $[\Sigma]$

consists of the eigenvalues arranged down the leading diagonal and matrix [U] contains the corresponding eigenvectors:

$$[A][U] = [U][\Sigma] \tag{3.41}$$

The eigenvalues are effectively weighting factors applied to the eigenvectors to reconstruct the original matrix [A]. The non-zero eigenvalues and corresponding eigenvectors are known as the principal components.

The eigenvalue decomposition is applied to the $N \times N$ cross spectrum matrix. If there are p distinct non-zero eigenvalues for this matrix, then there are only p independent and uncorrelated columns of the cross spectrum matrix. In practice though, none of the eigenvalues will be identically zero and the problem is to decide the threshold value above which an eigenvalue will be considered to be non-zero. This is most easily done by plotting the eigenvalues and then looking for the most significant change between adjacent eigenvalues. The eigenvalues are calculated and plotted at each frequency point in the range of interest.

Example
A triangular truss structure, figure 3.34, was tested using three shakers simultaneously using a standard *MIMO* procedure with burst random excitation. With all three shakers driven correctly the principal components plot for this test is given in figure 3.35.

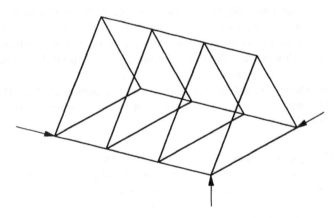

Fig. 3.34 Triangular truss structure.

One of the shakers was then disconnected from its power supply. A principal components check of the force cross spectrum matrix now shows that there are only two independent forces. A plot of the three principal components across the frequency range of interest is given in figure 3.36. One of the components in figure 3.36 is significantly smaller than the other two, whereas all three components are grouped together much more closely in figure 3.35.

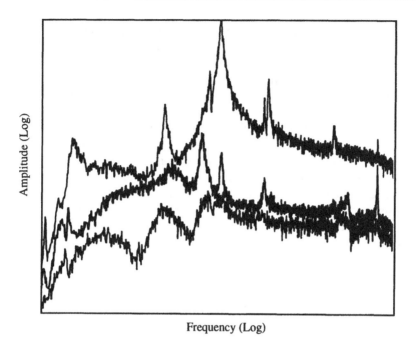

Fig. 3.35 Principal Components Plot; all 3 shakers driven.

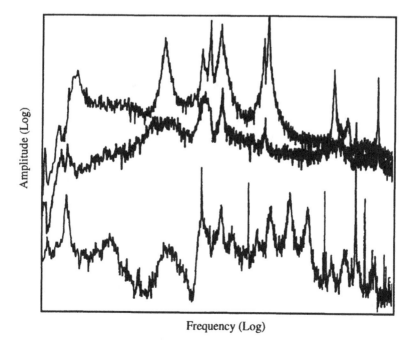

Fig. 3.36 Principal Components Plot: only 2 of the 3 shakers driven.

However, it can sometimes be difficult to judge whether one of the components is sufficiently small in comparison with the rest to indicate a dependent input. This is when it can be helpful to calculate and plot the principal components for more of the full cross spectrum matrix than just the force submatrix. Inclusion of one response in the calculations ensures that there will be at least one 'zero' eigenvalue, because a response must be dependent upon the force inputs. This can be used as a reference to help judge whether any of the other principal components are 'zero', hence indicating a dependent force input. A certain amount of caution is required here because the units of the eigenvalues depend upon whether they relate to a response column or a force column of the cross spectrum matrix; i.e. units of (response)^2, or (force)^2. Plots of four principal components for the tests described above are shown in figures 3.37 and 3.38, respectively. The presence of a dependent force input is shown much more clearly in figure 3.38 than in figure 3.36.

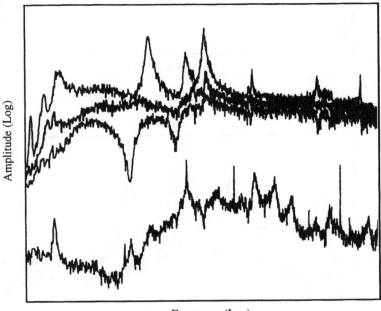

Frequency (Log)

Fig. 3.37 Principal Components Plot: all 3 shakers driven (1 response DOF included as a reference).

3.9 FINAL REMARKS

As was mentioned at the start of this chapter ..."There is no single *right* way to perform a vibration test". The aim has been to encourage the reader to question all aspects of the testing procedure from pre-test analysis through to checks on the quality and consistency of the measured FRFs. The availability of an FE model of the structure for pre-test analysis can be a very great benefit for locating excitation

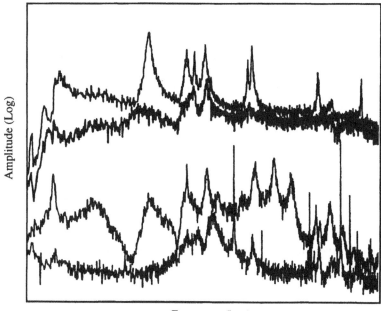

Frequency (Log)

Fig. 3.38 Principal Components Plot: only 2 of the 3 shakers driven (1 response DOF included as a reference).

and response positions without trial and error. But, regardless of whether or not an FE model is available, a thorough process of test planning is considered to be absolutely vital to achieve a successful outcome for the test. Time spent in planning a test is time well spent. Decisions in the early stages of test planning will undoubtedly have repercussions later. The test engineer should understand the reasons for selection of a particular transducer, for instance, and be aware of the possible consequences with regard to mass or inertia loading on the structure. A knowledge of how a transducer functions and how the electrical signals are conditioned and processed will also enable transducer leads and cables to be routed in such a way as to minimise the chance of signal corruption by external electrical or mechanical means. Similar care should be exercised in choosing how to support a structure during a test and how to support and connect the exciter(s).

Once the time and effort has been expended in setting up the structure for test, there is a great temptation to proceed immediately with the collection of vast quantities of data to justify all the 'non-productive' work up to this point. It is preferable at this stage to acquire and process relatively small quantities of data as part of a preliminary survey. This is also the time when full use should be made of the test engineer's senses. Visual observation of various test signals on an oscilloscope can be valuable in checking for signal clipping or determining whether the responses have decayed to zero before the end of the acquisition block in burst random testing, for example. Visual observation of the exciters will show whether there is excessive motion leading to unwanted bending of the push-rods. Listening to the sounds produced during the test can be an effective way of locating rattles and unlocked push-rod linkages, for instance. Transient 'clicks' in the excitation may point to buffer switching problems in digital-to-analogue converters. Surprisingly, smell and touch are other senses that may be brought into play, albeit subconsciously. An over-driven power amplifier or shaker will probably get hot, resulting in a characteristic (and expensive!) smell.

Results from different test strategies, different levels of excitation, reciprocity and repeatability checks made as part of a preliminary survey should be scrutinised very carefully before finally proceeding to the major data acquisition task. Even at this stage, the quest for complete understanding of the test is not over because expansion to the full set of data acquisition channels may bring its own problems.

Thorough documentation throughout the whole procedure is a good habit to adopt; it may prove to be invaluable for resolving problems that come to light after the test equipment has been dismantled. The process of systematically documenting every stage also encourages questions to be asked and answered - the key to successful testing.

CHAPTER 4

Modal Analysis Identification Methods

4.1 INTRODUCTION

Experimental Modal Analysis (EMA) comprises nowadays a vast range of different areas, like Modal Correlation and Updating, Modal Testing methods, Nonlinear Modal Analysis, Substructuring, Structural Modification, to name a few. However, one of the fundamental areas is, precisely, the identification of the dynamic properties of a structure from the measured data. In the seventies and more so in the eighties one has witnessed the development of a vast range of identification methods and techniques. Some of the best-performing have been integrated into commercial software, and nowadays one can find entire modal analysis packages running in computers linked to modal testing systems and doing not only modal parameter estimation but also other pre- and post-processing calculations, like geometry definition, mode animation, structural modification prediction, coupling of substructures, etc.

The objective of the present chapter is to give a general classification of modal analysis identification methods, together with a relatively detailed description of the most significant and most used. The authors believe this chapter may be important in the sense that it may help the users of modal analysis software to acquire a deeper understanding of the way the various methods work, so that they can, on a stronger basis, take decisions on the most suitable technique to adopt for their specific application, use it with more confidence, and to better judge the obtained results and choose an alternative method if that is appropriate. The detailed description given is also intended to make life easier for the interested reader who wants to write a computer program on one or several methods. Moreover, it may help the reader to understand more readily the similarities and differences between the various approaches. Modal Analysis packages tend to be mostly black boxes, very much automatic; however, it is always nice to have at least a minimum idea of how a black box works.

4.2 GENERAL CLASSIFICATION OF MODAL ANALYSIS IDENTIFICATION METHODS

During the last three decades or so, many researchers have devoted their efforts to the development of techniques that aim to produce a reliable identification of the dynamic properties of structures. Those efforts have been fruitful due largely to the introduction of the Fast Fourier Transform (FFT) [1] and to the development in recent years of very powerful multi-channel spectrum analysers, computers and instrumentation in general that permit the acquisition and treatment of large quantities of data. In this way, it was possible to evolve from very simple techniques (for example, [61-64]) where analyses were based on data from single-input excitation and single-output response to highly sophisticated ones (for example, [65, 66]) where data from multi-input excitation and multi-output responses are treated simultaneously. The turning point in this evolution was made in 1971 by Klosterman [67], introducing on-line testing controlled by mini-computers.

The current trend of experimental modal analysis seems to be a progressive abandonment of interactive analysis programs and a development of completely automated systems of acquisition and analysis of data, to accompany the developments in other areas of structural design, like the Finite Element Method (FEM) and Computer-Aided Design (CAD). A perspective of the evolution and future of EMA is given in [68].

Nowadays, the number of technical publications on EMA is such that the task of classifying the available methods of analysis (itself only part of EMA's vast field of study) represents a great effort. Previous surveys, [69-79], are a good starting point for such a work.

The major grouping concerns the domain in which the data are treated numerically. There are time domain and frequency domain methods. Tuned-sinusoidal methods are a special category and will be considered separately. Early methods used to work in the frequency domain, but problems associated with frequency resolution, leakage and high modal densities led people to start looking at time domain methods as a promising alternative.

The calculation of the impulse response function (IRF) corresponding to an FRF involves the calculation of the inverse FFT, a standard feature in spectral analysers. In this case, however, leakage can still be a problem, and to avoid this some methods use the force and response histories directly. In very general terms, time domain models tend to provide the best results when a large frequency range or large number of modes exist in the data, whereas frequency domain models tend to provide the best results when the frequency range of interest is limited and the number of modes is relatively small. However, time domain methods have a major disadvantage in that they can only estimate modes inside the frequency range of analysis, and take no account of the residual effects of modes that lie outside that range. This is why, some years ago, people returned to frequency domain techniques, which can improve the accuracy of the results by accounting for residual terms or by increasing the order of the model.

Time domain and frequency domain methods can be divided into indirect (or modal) and direct methods. The former designation (indirect) means that the identification of the FRF(s) is based on the modal model, i.e., on the modal parameters (natural frequencies, damping ratios, modal constants and their phases). The latter designation (direct) means that the identification is directly based on the spatial model, i.e., on the general matrix equation of dynamic equilibrium, the primitive equation from which all the methods are derived. In some of the methods in this last category, the system matrices of the referred equation can be evaluated and the corresponding eigenproblem solved in order to calculate the modal parameters.

A further division concerns the number of modes that can be analysed. In this respect, we can have single degree-of-freedom (SDOF) and multiple degree-of-freedom (MDOF) analyses. In the time domain we have only MDOF analysis, while in the frequency domain we can have SDOF or MDOF analyses with the indirect methods. Direct methods only apply to MDOF analysis.

When a structure is tested in order to collect the measured data, we usually have at our disposal a set of FRFs. These FRFs are the result of exciting the structure at each selected point and measuring the response at several locations along that structure. Some modal analysis methods can be applied only to a single FRF at a time. These are called single FRF or single-input-single-output (SISO) methods. Other methods allow for several FRFs to be analysed simultaneously, with responses taken at various points on the structure, but using one excitation point. These are called global or single-input-multi-output (SIMO) methods. The philosophy behind this category of methods is that the natural frequencies and damping ratios do not vary (theoretically) from FRF to FRF (they are global properties of a structure) and, thus, it should be possible to obtain a consistent and unique set of those properties by processing several FRFs at the same time. This procedure would automatically average out small variations in those modal properties that will necessarily occur when analysing one FRF at a time and it would, in principle, be preferable to a simple or weighted average of the results from several single analyses. Finally, there are methods that can process all the available FRFs simultaneously, from various excitation and response locations. Those methods are usually called polyreference or multi-input-multi-output (MIMO) methods. Situations of multi-input-single-output (MISO) are also possible, but are used to a much lesser extent [80]. Figure 4.1 shows a diagram with the various possible categories of methods.

In the description of modal analysis identification methods that follows, we have tried to follow a certain chronological order, usually accompanied by a growth in the complication and sophistication of the algorithms used. Based on the classification given in figure 4.1, a wide range of modal analysis identification methods will be described. The most relevant methods in the time and frequency domains are presented and a historical note on the tuned-sinusoidal methods is also given. The reader will be given a comprehensive review of the various methods of identification, usually distributed over a large number of publications and with a wide diversity in the notation used. We have tried to keep the notation as consistent and standard as possible.

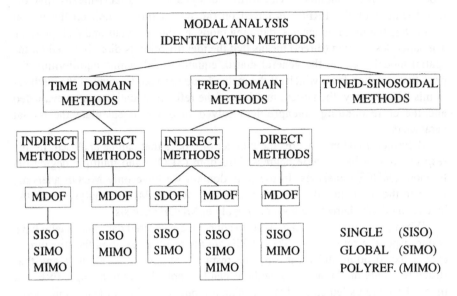

Fig. 4.1 Classification of modal analysis methods.

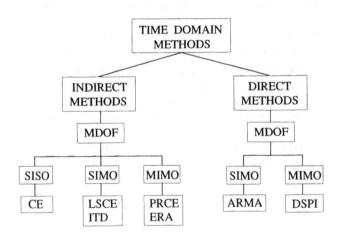

Fig. 4.2 Classification of time domain methods.

4.3   TIME DOMAIN METHODS

4.3.1 Indirect methods in the time domain

In this category, the most widely known methods are the Complex Exponential (CE), the Least-Squares Complex Exponential (LSCE), the Polyreference Complex Exponential (PRCE), the Ibrahim Time Domain (ITD) and the Eigensystem Realisation Algorithm (ERA). We shall describe each of these, providing an explanation of the theory and the most relevant references, and seeking to highlight the common features and the differences between them. In figure 4.2, a diagram of the classification of time domain methods is shown.

The Complex Exponential method (CE)

In the frequency domain, the FRF in terms of receptance α_{jk} (displacement at point j due to a force at point k) for a linear, viscously damped system with N degrees-of-freedom (DOF) can be given by (see equation (1.247)):

$$\alpha_{jk}(\omega) = \sum_{r=1}^{N} \left(\frac{{}_rA_{jk}}{\omega_r\xi_r + i\left(\omega - \omega_r\sqrt{\left(1-\xi_r^2\right)}\right)} + \frac{{}_rA_{jk}^*}{\omega_r\xi_r + i\left(\omega + \omega_r\sqrt{\left(1-\xi_r^2\right)}\right)} \right) \quad (4.1)$$

where ω_r is the natural frequency, ξ_r is the viscous damping factor and ${}_rA_{jk}$ is the residue corresponding to each mode r; * denotes complex conjugate. Another way of writing (4.1) is

$$\alpha_{jk}(\omega) = \sum_{r=1}^{2N} \frac{{}_rA_{jk}}{\omega_r\xi_r + i\left(\omega - \omega_r'\right)} \quad (4.2)$$

where

$$\omega_r' = \omega_r\sqrt{\left(1-\xi_r^2\right)}$$
$$\omega_{r+N}' = -\omega_r' \quad (4.3)$$
$$_{r+N}A_{jk} = {}_rA_{jk}^*$$

The CE method [77, 81, 82] works with the corresponding impulse response function (IRF), obtained from (4.2) by an inverse Fourier transform:

$$h_{jk}(t) = \sum_{r=1}^{2N} {}_rA_{jk}\, e^{s_r t} \quad (4.4)$$

or, simply

$$h(t) = \sum_{r=1}^{2N} A_r'\, e^{s_r t} \quad (4.5)$$

where $s_r = -\omega_r\xi_r + i\omega_r'$ and the properties (4.3) hold. The time response h(t) (real-valued) at a series of L equally spaced time intervals Δt, is

$$h_0 = h(0) = \sum_{r=1}^{2N} A'_r$$

$$h_1 = h(\Delta t) = \sum_{r=1}^{2N} A'_r e^{s_r(\Delta t)}$$

$$\vdots \qquad \vdots \qquad \vdots$$

$$h_L = h(L\Delta t) = \sum_{r=1}^{2N} A'_r e^{s_r(L\Delta t)}$$

(4.6)

or, simply

$$h_0 = \sum_{r=1}^{2N} A'_r$$

$$h_1 = \sum_{r=1}^{2N} A'_r V_r$$

$$\vdots \qquad \vdots$$

$$h_L = \sum_{r=1}^{2N} A'_r V_r^L$$

(4.7)

with

$$V_r = e^{s_r \Delta t}$$

(4.8)

It must be noted that in (4.7) the values of A'_r and V_r are not known. How can these values be calculated? The solution is an ingenious technique devised in 1795 by Prony [83] and known as Prony's method. Because the roots s_r for a sub-critically damped (underdamped) system always occur in complex conjugate pairs, so do the modified variables V_r.

Thus, there always exists a polynomial in V_r of order L with real coefficients β (called the autoregressive coefficients) such that the following relation is verified:

$$\beta_0 + \beta_1 V_r + \beta_2 V_r^2 + \ldots + \beta_L V_r^L = 0$$

(4.9)

In order to calculate the coefficients β to evaluate V_r, it is necessary just to multiply both sides of (4.7) by β_0 to β_L and to sum the result. This gives:

$$\sum_{j=0}^{L} \beta_j h_j = \sum_{j=0}^{L} \left(\beta_j \sum_{r=1}^{2N} A'_r V_r^j \right) = \sum_{r=1}^{2N} \left(A'_r \sum_{j=0}^{L} \beta_j V_r^j \right)$$

(4.10)

The inner summation in (4.10) is exactly the polynomial (4.9). As this polynomial vanishes for each value of V_r, it follows that

$$\sum_{j=0}^{L} \beta_j \, h_j = 0 \quad , \text{ for each } V_r \qquad (4.11)$$

From (4.11) it will be possible to calculate the coefficients β_j which will yield the solution of the polynomial (4.9) for its roots, V_r. To calculate β_j, we proceed as follows: L will be taken as equal to 2N for convenience and so there will be 2N sets of data points h_j, each one shifted one time interval, and β_{2N} will be set to 1. The result is:

$$\begin{bmatrix} h_0 & h_1 & h_2 & \cdots & h_{2N-1} \\ h_1 & h_2 & h_3 & \cdots & h_{2N} \\ \vdots & \vdots & \vdots & \cdots & \vdots \\ h_{2N-1} & h_{2N} & h_{2N+1} & \cdots & h_{4N-2} \end{bmatrix} \begin{Bmatrix} \beta_0 \\ \beta_1 \\ \vdots \\ \beta_{2N-1} \end{Bmatrix} = - \begin{Bmatrix} h_{2N} \\ h_{2N+1} \\ \vdots \\ h_{4N-1} \end{Bmatrix} \qquad (4.12)$$

or simply

$$\underset{(2Nx2N)}{[h]} \ \underset{(2Nx1)}{\{\beta\}} = \underset{(2Nx1)}{\{h'\}} \qquad (4.13)$$

Knowing the coefficients β_j, a polynomial solver can be used to calculate the roots V_r. Using (4.8) and its corresponding complex conjugate value, we can determine the natural frequencies and damping factors. With the values of V_r, we can use (4.7) to calculate the residues and consequently the modal constants and phase angles. The residues A'_r are easy to calculate if (4.7) is written as:

$$\begin{bmatrix} 1 & 1 & \cdots & 1 \\ V_1 & V_2 & \cdots & V_{2N} \\ V_1^2 & V_2^2 & \cdots & V_{2N}^2 \\ \vdots & \vdots & \cdots & \vdots \\ V_1^{2N-1} & V_2^{2N-1} & \cdots & V_{2N}^{2N-1} \end{bmatrix} \begin{Bmatrix} A'_1 \\ A'_2 \\ A'_3 \\ \vdots \\ A'_{2N} \end{Bmatrix} = - \begin{Bmatrix} h_0 \\ h_1 \\ h_2 \\ \vdots \\ h_{2N-1} \end{Bmatrix} \qquad (4.14)$$

where, for convenience (to have a square matrix), we have only taken the first 2N-1 values of h_j. In fact, it is enough to take N-1 values, as V_r and A'_r appear in complex conjugate pairs.

The CE method is a MDOF indirect method that falls in the category of *SISO* methods, as it is designed to analyse a single IRF at a time. It is a simple method that does not require initial estimates for the modal parameters, and the only unknown is the number of modes that must be considered in the analysis. An overspecified number of modes is usually given, and thus it will be necessary to distinguish later between genuine and computational modes. Another way of determining the correct number of modes is to repeat the analysis several times, decreasing each time the specified number of modes. A plot of the error between the original and regenerated curves for each number of modes gives an indication of the correct order of the model. A drop in this error curve corresponding to the correct number of modes is expected to be visible. An alternative to the calculation of the effective number of modes is the calculation of the rank of the coefficient matrix used to calculate the eigenvalues. Another way could be to use different sets of data and analyse the consistency (or the variation) of the solutions for the modal parameters. One of the biggest disadvantages of the CE method appears to be its sensitivity to noise [77].

The Least-Squares Complex Exponential method (LSCE)

The LSCE, introduced in 1979 [77], is the extension of the CE to a global procedure. It is therefore a *SIMO* method, processing simultaneously several IRFs obtained by exciting a structure at one single point and measuring the responses at several locations. With such a procedure, a consistent set of global parameters (natural frequencies and damping factors) is obtained, thus overcoming the variations obtained in the results for those parameters when applying the CE method on different IRFs.

The extension from the CE to the LSCE method is quite straightforward. Referring to (4.13), it can be seen that the coefficients β that provide the solution of the characteristic polynomial (4.9) are global quantities, i.e., they must be the same for every IRF used. Therefore, if we write (4.13) for p IRFs, we obtain

$$
\begin{bmatrix} [h]_1 \\ [h]_2 \\ \vdots \\ [h]_p \end{bmatrix} \{\beta\} = \begin{Bmatrix} \{h'\}_1 \\ \{h'\}_2 \\ \vdots \\ \{h'\}_p \end{Bmatrix} \tag{4.15}
$$

or

$$
\underset{(2Npx2N)}{[h_G]} \; \underset{(2Nx1)}{\{\beta\}} = \underset{(2Npx1)}{\{h'_G\}} \tag{4.16}
$$

The least-squares solution can be found via the pseudo-inverse technique:

$$
\{\beta\} = \left([h_G]^T [h_G] \right)^{-1} [h_G]^T \{h'_G\} \tag{4.17}
$$

A solution of this type could already have been applied in the CE method for (4.12). Considering more than 2N sets of points would already have been an improvement in the averaging out of noise disturbances in the data. Therefore, in (4.16) we can also have more than 2N sets of data points. Knowing the coefficients $\{\beta\}$, we obtain the values of V_r as before (solving (4.9)) and then, for each IRF, the residues A_r' can be calculated using again (4.14) and, consequently, the modal constants and phase angles. For these parameters, a frequency domain algorithm could alternatively be used.

The problems associated with the estimation of the correct number of modes still remain, as for the CE method. The calculation of the rank of matrix $[h_G]$ in (4.16) can be used as an indication of that quantity [77].

The Polyreference Complex Exponential method (PRCE)

The PRCE method, developed in 1982 by Vold *et al* [84, 85], constitutes the extension of the LSCE method to a *MIMO* version, i.e., including information not only from several output locations, but also from several input reference points on the structure. Apart from being a more general and automatic way of analysing dynamically a structure, this overcomes the problem that sometimes occurs when using a *SIMO* method, where one mode of vibration may not be excited because the excitation may be located close to a node of the structure.

In the explanation that follows, we keep a similar notation as used in the previous two methods, and a similar philosophy, to show as clearly as possible how the PRCE is an extension of the LSCE method. We shall follow closely the development given by Deblauwe and Allemang [86].

As seen in (4.4), the IRF at a point j due to an input at point k is given by

$$h_{jk}(t) = \sum_{r=1}^{2N} {}_rA_{jk}\, e^{s_r t} \tag{4.4}$$

Considering q input reference points, it follows that

$$h_{j1}(t) = \sum_{r=1}^{2N} {}_rA_{j1}\, e^{s_r t}$$
$$h_{j2}(t) = \sum_{r=1}^{2N} {}_rA_{j2}\, e^{s_r t}$$
$$\vdots \qquad \vdots \tag{4.18}$$
$$h_{jq}(t) = \sum_{r=1}^{2N} {}_rA_{jq}\, e^{s_r t}$$

But, for each mode r, the residues ${}_rA_{jk}$ are related through a scaling factor Q_r to the eigenvector elements ψ_{jr} and ψ_{kr} of the system by the relationship:

$${}_rA_{jk} = Q_r\, \psi_{jr}\, \psi_{kr} \tag{4.19}$$

So, if the first residue is given by

$$_rA_{j1} = Q_r \, \psi_{jr} \, \psi_{1r} \tag{4.20}$$

we can write for the residue at point k,

$$_rA_{jk} = {}_rW_{k1} \, {}_rA_{j1} \tag{4.21}$$

where $_rW_{k1}$ is a weighting factor or a modal participation factor:

$$_rW_{k1} = \frac{\psi_{kr}}{\psi_{1r}} \tag{4.22}$$

Thus, (4.18) can be written as:

$$
\begin{aligned}
h_{j1}(t) &= \sum_{r=1}^{2N} {}_rA_{j1} \, e^{s_r t} \\
h_{j2}(t) &= \sum_{r=1}^{2N} {}_rW_{21} \, {}_rA_{j1} \, e^{s_r t} \\
&\vdots \qquad\qquad \vdots \\
h_{jq}(t) &= \sum_{r=1}^{2N} {}_rW_{q1} \, {}_rA_{j1} \, e^{s_r t}
\end{aligned}
\tag{4.23}
$$

In matrix form, (4.23) becomes

$$
\begin{Bmatrix} h_{j1}(t) \\ h_{j2}(t) \\ \vdots \\ h_{jq}(t) \end{Bmatrix}
=
\begin{bmatrix}
1 & 1 & \cdots & 1 \\
{}_1W_{21} & {}_2W_{21} & \cdots & {}_{2N}W_{21} \\
\vdots & \vdots & \cdots & \vdots \\
{}_1W_{q1} & {}_2W_{q1} & \cdots & {}_{2N}W_{q1}
\end{bmatrix}
\begin{bmatrix}
e^{s_1 t} & 0 & \cdots & 0 \\
0 & e^{s_2 t} & \cdots & 0 \\
\vdots & \vdots & \ddots & \vdots \\
0 & 0 & \cdots & e^{s_{2N} t}
\end{bmatrix}
\begin{Bmatrix} {}_1A_{j1} \\ {}_2A_{j1} \\ \vdots \\ {}_{2N}A_{j1} \end{Bmatrix}
$$

$$\quad (q \times 1) \qquad\qquad (q \times 2N) \qquad\qquad (2N \times 2N) \qquad\qquad (2N \times 1)$$

$$\tag{4.24}$$

or simply

$$
\{h_j(t)\} = [W]
\begin{bmatrix}
e^{s_1 t} & 0 & \cdots & 0 \\
0 & e^{s_2 t} & \cdots & 0 \\
\vdots & \vdots & \ddots & \vdots \\
0 & 0 & \cdots & e^{s_{2N} t}
\end{bmatrix}
\{A_{j1}\}
\tag{4.25}
$$

where [W] is the modal participation matrix. Considering L+1 time intervals,

$$\left\{h_j(0)\right\} = [W]\left\{A_{jl}\right\}$$

$$\left\{h_j(\Delta t)\right\} = [W]\left[\,\check{}\, V \,\check{}\,\right]\left\{A_{jl}\right\}$$

$$\vdots \qquad\qquad \vdots \tag{4.26}$$

$$\left\{h_j(L\Delta t)\right\} = [W]\left[\,\check{}\, V \,\check{}\,\right]^L\left\{A_{jl}\right\}$$

where

$$\left[\,\check{}\, V \,\check{}\,\right] = \begin{bmatrix} e^{s_1\Delta t} & 0 & \cdots & 0 \\ 0 & e^{s_2\Delta t} & \cdots & 0 \\ \vdots & \vdots & \ddots & \vdots \\ 0 & 0 & \cdots & e^{s_{2N}\Delta t} \end{bmatrix} \tag{4.27}$$

It must be remembered that in (4.26) the elements of $\{h(t)\}$ are the only known quantities. As for the CE method (see (4.9)), the 2N eigenvalues can now be seen as solutions of a matrix polynomial, given by:

$$[\beta_0] + [\beta_1][W]\left[\,\check{}\, V \,\check{}\,\right] + [\beta_2][W]\left[\,\check{}\, V \,\check{}\,\right]^2 + \ldots + [\beta_L][W]\left[\,\check{}\, V \,\check{}\,\right]^L = [0] \tag{4.28}$$

where $[\beta_0]$, ..., $[\beta_L]$ are real, square coefficient matrices of order q (the number of input references). How big must the order of the matrix polynomial (L) be in order to obtain the 2N eigenvalue solutions? The number of eigenvalues of a matrix polynomial is equal to the order of the polynomial times the dimension of the matrix coefficients, i.e., in this case, 2N = Lq. This means that in order to calculate at least 2N eigenvalues we must have Lq ≥ 2N, i.e., L ≥ 2N/q. If Lq > 2N, then there will be computational modes. As before, we multiply each side of (4.26) by $[\beta_0]$, ..., $[\beta_L]$:

$$[\beta_0]\left\{h_j(0)\right\} = [\beta_0][W]\left\{A_{jl}\right\}$$

$$[\beta_1]\left\{h_j(\Delta t)\right\} = [\beta_1][W]\left[\,\check{}\, V \,\check{}\,\right]\left\{A_{jl}\right\}$$

$$[\beta_2]\left\{h_j(\Delta t)\right\} = [\beta_2][W]\left[\,\check{}\, V \,\check{}\,\right]^2\left\{A_{jl}\right\}$$

$$\vdots \qquad\qquad \vdots \tag{4.29}$$

$$[\beta_L]\left\{h_j(L\Delta t)\right\} = [\beta_L][W]\left[\,\check{}\, V \,\check{}\,\right]^L\left\{A_{jl}\right\}$$

Summing each side of (4.29), it follows that

$$\sum_{k=0}^{L} [\beta_k] \{h_j(k\Delta t)\} = \sum_{k=0}^{L} [\beta_k][W][\,\diagdown\, V \diagdown\,]^k \{A_{j1}\} \tag{4.30}$$

From (4.28), it is apparent that the right-hand side of (4.30) is zero, and thus,

$$\sum_{k=0}^{L} [\beta_k] \{h_j(k\Delta t)\} = \{0\} \tag{4.31}$$

Assuming $[\beta_L]$ to be the unit matrix $[I]$, we have

$$\sum_{k=0}^{L-1} [\beta_k] \{h_j(k\Delta t)\} = -\{h_j(L\Delta t)\} \tag{4.32}$$

We can consider now several sets of time data points (N_t sets of L points each: $N_t \geq L$), each set shifted by one interval Δt, leading to:

$$[[\beta_0]\,[\beta_1]\,...\,[\beta_{L-1}]]
\begin{bmatrix}
\{h_j(0)\} & \{h_j(\Delta t)\} & ... & \{h_j(N_t-1)\Delta t)\} \\
\{h_j(\Delta t)\} & \{h_j(2\Delta t)\} & ... & \{h_j(N_t\Delta t)\} \\
\vdots & \vdots & ... & \vdots \\
\{h_j((L-1)\Delta t)\} & \{h_j(L\Delta t)\} & ... & \{h_j((L+N_t-2)\Delta t)\}
\end{bmatrix} =$$

$$-[\{h_j(L\Delta t)\}\,\{h_j((L+1)\Delta t)\}\,...\,\{h_j((L+N_t-1)\Delta t)\}] \tag{4.33}$$

or simply

$$\underset{(q \times Lq)}{[B_T]}\ \underset{(Lq \times N_t)}{[h_j]}\ =\ \underset{(q \times N_t)}{[h'_j]} \tag{4.34}$$

Considering now (4.34) for each response location, with $j = 1, ..., p$,

$$\underset{(q \times Lq)}{[B_T]}\underset{(Lq \times N_t p)}{[[h_1]\,[h_2]\,...\,[h_p]]} = \underset{(q \times N_t p)}{[[h'_1]\,[h'_2]\,...\,[h'_p]]} \tag{4.35}$$

or, in short form,

$$[B_T][h_T] = [h'_T] \tag{4.36}$$

from which

$$[B_T] = [h'_T][h_T]^T \left([h_T][h_T]^T\right)^{-1} \qquad (4.37)$$
$$\underset{(qxLq)}{} \quad \underset{(qxLq)}{} \quad \underset{(LqxLq)}{}$$

It should be noted that in (4.35) we must have $N_t p \geq Lq$ in order that in (4.37) the resulting matrices are of full rank. Knowing the coefficient matrix [B], we can return to (4.28), and must now solve it for the eigenvalues $\lceil \ V \ \rceil$. Let us rewrite (4.28) as

$$\sum_{k=0}^{L} [\beta_k][W] \lceil \ V \ \rceil^k = [0] \qquad (4.38)$$

Post-multiplying each side of this equation by a unit vector of dimensions 2Nx1, composed of zeros except for unity at the position corresponding to the eigenvalue to be calculated, would give

$$\sum_{k=0}^{L} [\beta_k][W] \lceil \ V \ \rceil^k \begin{Bmatrix} 1 \\ 0 \\ \vdots \\ 0 \end{Bmatrix} = \sum_{k=0}^{L} [\beta_k] \left(e^{s_1 \Delta t}\right)^k \{W_1\} \quad = \{0\}$$

$$\sum_{k=0}^{L} [\beta_k][W] \lceil \ V \ \rceil^k \begin{Bmatrix} 0 \\ 1 \\ \vdots \\ 0 \end{Bmatrix} = \sum_{k=0}^{L} [\beta_k] \left(e^{s_2 \Delta t}\right)^k \{W_2\} \quad = \{0\}$$
$$(4.39)$$

$$\sum_{k=0}^{L} [\beta_k][W] \lceil \ V \ \rceil^k \begin{Bmatrix} 0 \\ 0 \\ \vdots \\ 1 \end{Bmatrix} = \sum_{k=0}^{L} [\beta_k] \left(e^{s_{2N} \Delta t}\right)^k \{W_{2N}\} = \{0\}$$

where $\{W_1\}, ..., \{W_{2N}\}$ are the columns of the modal participation matrix. For each eigenvalue r, and using the definition (4.8),

$$\left[\sum_{k=0}^{L} [\beta_k] V_r^k\right]\{W_r\} = \{0\} \qquad r = 1, ..., 2N \qquad (4.40)$$

{W_r} is a non-zero vector, independent of the summation in k. Each one of the 2N possible equations (4.40) represents the same eigenvalue problem, providing Lq solutions. Expanding (4.40) and remembering that $[\beta_L] = [I]$, it follows that

$$\left[[\beta_0]+[\beta_1]V_r+[\beta_2]V_r^2+...+[\beta_{L-1}]V_r^{L-1}\right]\{W_r\}=-V_r^L\{W_r\} \qquad (4.41)$$

If one defines

$$
\begin{aligned}
\{z_0\} &= \{W_r\} \\
\{z_1\} &= V_r\{W_r\} &&= V_r\{z_0\} \\
\{z_2\} &= V_r^2\{W_r\} &&= V_r\{z_1\} \\
&\vdots &&\vdots \\
\{z_{L-1}\} &= V_r^{L-1}\{W_r\} = V_r\{z_{L-2}\} \\
\{z_L\} &= V_r^L\{W_r\} &&= V_r\{z_{L-1}\}
\end{aligned}
\qquad (4.42)
$$

then,

$$[\beta_0]\{z_0\}+[\beta_1]\{z_1\}+...+[\beta_{L-1}]\{z_{L-1}\}=-V_r\{z_{L-1}\} \qquad (4.43)$$

or

$$
\begin{bmatrix}
-[\beta_{L-1}] & -[\beta_{L-2}] & \cdots & -[\beta_1] & -[\beta_0] \\
[I] & [0] & \cdots & [0] & [0] \\
\vdots & \vdots & \cdots & \vdots & \vdots \\
[0] & [0] & \cdots & [I] & [0]
\end{bmatrix}
\begin{Bmatrix}
\{z_{L-1}\} \\
\{z_{L-2}\} \\
\vdots \\
\{z_0\}
\end{Bmatrix}
= V_r
\begin{Bmatrix}
\{z_{L-1}\} \\
\{z_{L-2}\} \\
\vdots \\
\{z_0\}
\end{Bmatrix}
\qquad (4.44)
$$

$$(LqxLq) \qquad\qquad (Lqx1) \qquad (Lqx1)$$

Equation (4.44) is known as the companion matrix equation of the eigenproblem (4.41). This represents a standard eigenvalue problem of the type $[[A]-\lambda[I]]\{x\}=\{0\}$. Knowing the Lq eigenvalues V_r, it is easy to calculate the natural frequencies and damping factors, using (4.8). We can also calculate the Lq eigenvectors of (4.44) for each value of V_r. The corresponding values of $\{z_0\}$ are the values of $\{W_r\}$ in (4.41) (see (4.42)). Therefore, at this stage the modal participation matrix [W] is known and we now have to calculate the residues. Returning to (4.26),

$$\{h_j(k\Delta t)\}=[W]\left[\,\ddots V\,\ddots\,\right]^k\{A_{jl}\} \qquad k = 0, 1, ..., L \qquad (4.45)$$

where

$$\{h_j(k\Delta t)\} = \begin{cases} h_{j1}(k\Delta t) \\ h_{j2}(k\Delta t) \\ \vdots \\ h_{jq}(k\Delta t) \end{cases} \tag{4.46}$$

and $[\ ^{\backprime} V \backsim]$ is given by (4.27). While in the CE or LSCE methods each residue vector was calculated based upon one location of the input, now there is information from several inputs for each time interval. Varying k in (4.45), we obtain:

$$\begin{cases} \{h_j(0)\} \\ \{h_j(\Delta t)\} \\ \vdots \\ \{h_j(L\Delta t)\} \end{cases} = \begin{bmatrix} [W][\ ^{\backprime} V \backsim]^0 \\ [W][\ ^{\backprime} V \backsim]^1 \\ \vdots \\ [W][\ ^{\backprime} V \backsim]^L \end{bmatrix} \{A_{j1}\} \tag{4.47}$$

or

$$\{H_j\} = [W_V] \{A_{j1}\} \tag{4.48}$$
$$\scriptstyle ((L+1)qx1) \qquad ((L+1)qx2N) \ (2Nx1)$$

from which

$$\{A_{j1}\} = \left([W_V]^H[W_V]\right)^{-1} [W_V]^H\{H_j\} \tag{4.49}$$

where the superscript ^{H} denotes Hermitian transpose. This calculation is repeated for all the response locations, i.e., j = 1, ..., p. Knowing all the $\{A_{j1}\}$, (4.21) can be used to calculate all the residues. In summary, the necessary steps to use the present method are as follows: first, to take all the time records of the IRFs to enter in (4.34) and solve for the coefficients [B] using (4.37); second, to calculate the eigenvalues and eigenvectors of (4.44) in order to find the natural frequencies and damping factors, and build the matrices [W] and $[\ ^{\backprime} V \backsim]$; and finally, to calculate the residues by making use of (4.49) and (4.21).

Besides the fact of providing a more accurate modal representation of the test structure, this method can determine multiple roots or closely spaced modes of a structure. The time required for the analysis is reduced and the accuracy in the results increased. The major disadvantages seem to be the sensitivity to nonlinearities and to any lack of reciprocity in the frequency response matrix. It has also shown some difficulties in analysing satisfactorily structures with more than 5% of equivalent viscous damping [87]. The problems associated with the judgement of genuine and computational modes remain. Finally, it requires a considerable computer capacity.

The Ibrahim Time Domain method (ITD)

This method was introduced by Ibrahim in the 1970's [88, 89]. The formulation of the method included state vectors, where displacement and velocity responses were needed and were calculated by integration of the free acceleration response. Further improvements were given by the same author in 1977 [90], where only free acceleration responses were used. This is a *SIMO* method that uses free decay responses instead of IRFs, as in the previously described methods (the IRFs may benefit from the fact that they are inverse FFTs of FRFs that could already have been averaged, reducing the noise). For a system with N DOF, the free response of the structure at a point i and for the instant of time t_j is expressed as a summation of the individual responses of each mode:

$$x_i(t_j) = \sum_{r=1}^{2N} \psi_{ir}\, e^{s_r t_j} \tag{4.50}$$

where ψ_{ir} is the ith component of the eigenvector $\{\psi_r\}$ (complex, in general). Considering q response locations and L time instants, we can write:

$$\begin{bmatrix} x_1(t_1) & x_1(t_2) & \cdots & x_1(t_L) \\ x_2(t_1) & x_2(t_2) & \cdots & x_2(t_L) \\ \vdots & \vdots & \cdots & \vdots \\ x_q(t_1) & x_q(t_2) & \cdots & x_q(t_L) \end{bmatrix} = \begin{bmatrix} \psi_{11} & \psi_{12} & \cdots & \psi_{12N} \\ \psi_{21} & \psi_{22} & \cdots & \psi_{22N} \\ \vdots & \vdots & \cdots & \vdots \\ \psi_{q1} & \psi_{q2} & \cdots & \psi_{q2N} \end{bmatrix} \begin{bmatrix} e^{s_1 t_1} & e^{s_1 t_2} & \cdots & e^{s_1 t_L} \\ e^{s_2 t_1} & e^{s_2 t_2} & \cdots & e^{s_2 t_L} \\ \vdots & \vdots & \cdots & \vdots \\ e^{s_{2N} t_1} & e^{s_{2N} t_2} & \cdots & e^{s_{2N} t_L} \end{bmatrix}$$

$$\tag{4.51}$$

or just

$$[X] = [\Psi]\ [\Lambda] \tag{4.52}$$
$$\scriptstyle (q \times L) \quad (q \times 2N)(2N \times L)$$

where it is reasonable to admit $L \geq q \geq 2N$. Considering a second set of L data points, shifted one interval Δt with respect to the first, it follows that

$$x_i(t_j + \Delta t) = \sum_{r=1}^{2N} \psi_{ir}\, e^{s_r(t_j + \Delta t)} = \sum_{r=1}^{2N} \left(\psi_{ir}\, e^{s_r \Delta t} \right) e^{s_r t_j} \tag{4.53}$$

Defining $\hat{x}_i(t_j) = x_i(t_j + \Delta t)$ and $\hat{\psi}_{ir} = \psi_{ir}\, e^{s_r \Delta t}$,

$$\hat{x}_i(t_j) = \sum_{r=1}^{2N} \hat{\psi}_{ir}\, e^{s_r t_j} \tag{4.54}$$

and therefore, we can write an expression similar to (4.52):

$$\left[\hat{X}\right]_{(qxL)} = \left[\hat{\Psi}\right]_{(qx2N)} \left[\Lambda\right]_{(2NxL)}$$

(4.55)

A square matrix $[A_S]$ of order q is now defined (usually called the 'system matrix' and, in general, complex), such that

$$\left[A_S\right]_{(qxq)} \left[\Psi\right]_{(qx2N)} = \left[\hat{\Psi}\right]_{(qx2N)}$$

(4.56)

Pre-multiplying (4.52) by $[A_S]$ leads to

$$\left[A_S\right]_{(qxq)} \left[X\right]_{(qxL)} = \left[A_S\right]_{(qxq)} \left[\Psi\right]_{(qx2N)} \left[\Lambda\right]_{(2NxL)}$$

(4.57)

Substituting (4.56) in (4.57),

$$\left[A_S\right]_{(qxq)} \left[X\right]_{(qxL)} = \left[\hat{\Psi}\right]_{(qx2N)} \left[\Lambda\right]_{(2NxL)}$$

(4.58)

and substituting (4.55) in (4.58), we obtain

$$\left[A_S\right]_{(qxq)} \left[X\right]_{(qxL)} = \left[\hat{X}\right]_{(qxL)}$$

(4.59)

From this equation it is possible to calculate $[A_S]$, as $[X]$ and $\left[\hat{X}\right]$ are known matrices. This can be done via the pseudo-inverse technique, either by post-multiplying (4.59) by $[X]^T$ or by $\left[\hat{X}\right]^T$. In the first case, it gives

$$\left[A_S\right] = \left(\left[\hat{X}\right]\left[\hat{X}\right]^T\right)\left(\left[X\right]\left[\hat{X}\right]^T\right)^{-1}$$

(4.60)

In the second case, it gives

$$\left[A_S\right] = \left(\left[\hat{X}\right]\left[X\right]^T\right)\left(\left[X\right]\left[X\right]^T\right)^{-1}$$

(4.61)

Which of these two expressions should be used? A combination of both (4.60) and (4.61), known as Double Least-Squares (DLS), seems to be preferable, as it leads to better estimates of the damping factors [91]. Thus, the following expression is used:

$$[A_S] = \frac{1}{2}\left[\left([\hat{X}][\hat{X}]^T\right)\left([X][\hat{X}]^T\right)^{-1} + \left([\hat{X}][X]^T\right)\left([X][X]^T\right)^{-1}\right] \qquad (4.62)$$

As $\hat{\psi}_{ir} = \psi_{ir}\, e^{s_r \Delta t}$, each eigenvector $\{\hat{\psi}_r\}$ can be written as

$$\{\hat{\psi}_r\} = \{\psi_r\}\, e^{s_r \Delta t} \qquad (4.63)$$

and from (4.56), we have

$$[A_S]\{\psi_r\} = \{\psi_r\}\, e^{s_r \Delta t} \qquad (4.64)$$

or

$$\left[[A_S] - e^{s_r \Delta t}\,[I]\right]\{\psi_r\} = \{0\} \qquad (4.65)$$

which is a standard eigenvalue problem. Since $[A_S]$ is of order q, there will be q eigenvalues and eigenvectors and if $q > 2N$, there will be computational modes. From the eigenvalues, it is easy to calculate the natural frequencies and damping factors. As noted by Ewins [3], the eigenvectors cannot be mass-normalised, as we have only recorded free-response data. This fact may or may not constitute a disadvantage, depending on the purpose of the study.

There will generally be computational modes. One way of distinguishing these modes from the genuine ones can be established by means of the relationship (4.63). If the calculations are repeated taking a different time interval shift, it is possible, for each mode, to assess its authenticity by means of a Modal Confidence Factor (MCF) [92]; this factor compares the expected value $\{\hat{\psi}_r\}$ (from the calculated $\{\psi_r\}$) for one time interval with the calculated value of $\{\hat{\psi}_r\}$ for the following time interval. Thus, the philosophy of this method provides a very useful and automatic check on the calculated modes of vibration.

An extensive evaluation of the ITD method is provided in [93]. Among the advantages of this method are the need for little interaction, effective calculation of closely spaced modes and the possibility of verification of the quality of the results, via the MCF. The main disadvantage seems to be the tendency to give non-conservative damping estimates with noisy data.

A variation of the ITD method, the Sparse Time Domain algorithm (STD), has been proposed [94], making use of a sparse upper Hessenberg matrix. The main advantages are the reduction in computer storage and time, and higher identification accuracy.

The Eigensystem Realisation Algorithm (ERA)
This method is due to Juang and Pappa [95]. It is a *MIMO* method and we shall try to give a concise but hopefully clear explanation of the philosophy and main steps taken in this technique. Some alterations to the notation usually given by the

authors of this method (following mainly [96]) will be made, in order to retain as much coherence as possible with the general notation used in this and other chapters of the book.

Let the dynamic equations of equilibrium for an N DOF viscously damped system be expressed as:

$$[M]\{\ddot{x}(t)\} + [C]\{\dot{x}(t)\} + [K]\{x(t)\} = \{f(x(t),t)\} \tag{4.66}$$

Defining a state vector of dimensions 2Nx1, we have

$$\{u(t)\}_{(2Nx1)} = \left\{ \begin{array}{c} \{x(t)\} \\ \{\dot{x}(t)\} \end{array} \right\} \tag{4.67}$$

Defining also

$$[A']_{(2Nx2N)} = \begin{bmatrix} [0] & [I] \\ -[M]^{-1}[K] & -[M]^{-1}[C] \end{bmatrix} \tag{4.68}$$

$$\{f(x,t)\}_{(Nx1)} = [F]_{(Nxq)} \{\delta(t)\}_{(qx1)} \tag{4.69}$$

$$[B']_{(2Nxq)} = \begin{bmatrix} [0] \\ [M]^{-1}[F] \end{bmatrix} \tag{4.70}$$

where $\{\delta(t)\}$ is the input vector at q locations and [F] is a matrix of input coefficients, (4.66) is written in the state-space, as:

$$\{\dot{u}(t)\}_{(2Nx1)} = [A']_{(2Nx2N)} \{u(t)\}_{(2Nx1)} + [B']_{(2Nxq)} \{\delta(t)\}_{(qx1)} \tag{4.71}$$

It is possible to relate $\{u(t)\}$ to the measured responses at p physical coordinates $\{x(t)\}$ through a transformation matrix [R]:

$$\{x(t)\}_{(px1)} = [R]_{(px2N)} \{u(t)\}_{(2Nx1)} \tag{4.72}$$

The solution of (4.71) to an input $\{\delta(t)\}$ is given by

$$\{u(t)\} = e^{[A'](t-t_0)}\{u(t_0)\} + \int_{t_0}^{t} e^{[A'](t-\tau)}[B']\{\delta(\tau)\}d\tau \tag{4.73}$$

for any time t after an initial time t_0 ($t \geq t_0$). To give a discrete representation of (4.73), several equally spaced time intervals, 0, Δt, ..., $k\Delta t$ are considered. We can then take $t = (k+1)\Delta t$ and $t_0 = k\Delta t$:

$$\{u((k+1)\Delta t)\} = e^{[A']\Delta t}\{u(k\Delta t)\} + \int_{k\Delta t}^{(k+1)\Delta t} e^{[A']((k+1)\Delta t-\tau)}[B']\{\delta(\tau)\}d\tau \qquad (4.74)$$

Assuming that the input $\{\delta(\tau)\}$ is constant during the time interval $k\Delta t \le \tau \le (k+1)\Delta t$, given by $\{\delta(k\Delta t)\}$, and making the change of variable $\tau' = (k+1)\Delta t - \tau$, it follows that:

$$\{u((k+1)\Delta t)\} = e^{[A']\Delta t}\{u(k\Delta t)\} - \int_0^{\Delta t} e^{[A']\tau'}d\tau'[B']\{\delta(k\Delta t)\} \qquad (4.75)$$

Let

$$[A] = e^{[A']\Delta t} \qquad (4.76)$$

$$[B] = -\int_0^{\Delta t} e^{[A']\tau'}d\tau'[B'] \qquad (4.77)$$

$$\{u(k+1)\} = \{u((k+1)\Delta t)\} \qquad (4.78)$$

$$\{\delta(k)\} = \{\delta(k\Delta t)\} \qquad (4.79)$$

Thus, (4.75) is written as

$$\{u(k+1)\} = [A]\{u(k)\} + [B]\{\delta(k)\} \qquad \text{for} \quad k = 0, 1, 2, ... \qquad (4.80)$$

Equation (4.72) then becomes

$$\{x(k)\} = [R]\{u(k)\} \qquad (4.81)$$

Let us consider the response to an impulse at $k = 0$ and at one of the first of the input variables. Like this, $\{\delta(0)\} = \{1, 0, ...0\}^T$ and $\{\delta(k)\} = \{0\}$ for $k > 0$.
Substituting in (4.80),

$$\{u(1)\} = [A]\{u(0)\} + [B] \qquad (4.82)$$

and hence,

$$\{x(1)\} = [R]\{u(1)\} \qquad (4.83)$$

Substitution of (4.82) in (4.83) leads to

$$\{x(1)\} = [R][A]\{u(0)\} + [R][B] \qquad (4.84)$$

Considering, for simplicity, $\{u(0)\} = \{0\}$,

$$\{u(1)\} = \{B\} \tag{4.85}$$

$$\{x(1)\} = \underset{(px1)}{[R]} \underset{(px2N)(2Nx1)}{\{B\}} \tag{4.86}$$

For the other time intervals, $\{\delta(k)\} = \{0\}$. Thus,

$$\{u(2)\} = [A]\{u(1)\} \tag{4.87}$$

Hence,

$$\{x(2)\} = [R][A]\{B\} \tag{4.88}$$

Likewise,

$$\{x(3)\} = [R]\{u(3)\} = [R][A]\{u(2)\} = [R][A]^2\{B\} \tag{4.89}$$

and, in general,

$$\underset{(px1)}{\{x(k)\}} = \underset{(px2N)}{[R]} \underset{(2Nx2N)}{[A]^{k-1}} \underset{(2Nx1)}{\{B\}} \tag{4.90}$$

If we consider the impulse at all the q input locations, it follows that

$$\underset{(pxq)}{[X(k)]} = \underset{(px2N)}{[R]} \underset{(2Nx2N)}{[A]^{k-1}} \underset{(2Nxq)}{[B]} \tag{4.91}$$

Matrices $[X(k)]$ are usually called the Markov parameters. These are used to form the generalised Hankel matrices, given by:

$$\underset{(prxqs)}{[H(k-1)]} = \begin{bmatrix} [X(k)] & [X(k+1)] & \ldots & [X(k+j)] \\ [X(k+1)] & [X(k+2)] & \ldots & [X(k+j+1)] \\ \vdots & \vdots & \ldots & \vdots \\ [X(k+i)] & [X(k+i+1)] & \ldots & [X(k+i+j)] \end{bmatrix} \tag{4.92}$$

where $i = 1, ..., r\text{-}1$ and $j = 1, ..., s\text{-}1$, with r and s as integers. If there is an initial state response measurement, $[H(k\text{-}1)]$ is simply replaced by $[H(k)]$. In (4.92), k will be greater or equal to 1. Substitution of (4.91) in (4.92) gives

$$[H(k)] = [Q][A]^k[W] \qquad k \geq 0 \tag{4.93}$$

where

$$[Q] \atop (pr \times 2N) = \begin{bmatrix} [R] \\ [R][A] \\ \vdots \\ [R][A]^{r-1} \end{bmatrix}$$

(4.94)

$$[W] \atop (2N \times qs) = \begin{bmatrix} [B] & [A][B] & \cdots & [A]^{s-1}[B] \end{bmatrix}$$

[Q] and [W] are called the observability and controllability matrices, respectively. It must be remembered that [H(k)] is a known matrix of the responses. One of the advantages of this formulation is that in [H(k)] we may include only good responses, i.e., responses with low levels of noise. The objective is to reconstruct (4.91) from the experimental data. This process is known as realisation and implies the determination of matrices [R], [A] and [B]. There are an infinite number of sets of these three matrices satisfying (4.91), i.e., there are an infinite number of realisations. The objective is to obtain a minimum realisation, i.e., the realisation corresponding to the minimum order of the state-space formulation that can still represent the dynamic behaviour of the structure.

In the first place, we look for a matrix $[H]'$ such that

$$[W] \quad [H]' \quad [Q] \quad = \quad [I]$$
$$\scriptstyle (2N \times qs)(qs \times pr)(pr \times 2N) \quad (2N \times 2N)$$

(4.95)

Let us pre-multiply and post-multiply (4.95) by [Q] and [W], respectively:

$$[Q][W][H]'[Q][W] = [Q][W]$$

(4.96)

But, from (4.93), it is clear that

$$[Q][W] = [H(0)]$$

(4.97)

Thus,

$$[H(0)][H]'[H(0)] = [H(0)]$$

(4.98)

Therefore, $[H]'$ is the pseudo-inverse of [H(0)]:

$$[H]' = [H(0)]^+$$

(4.99)

The pseudo-inverse of [H(0)] can be calculated via the Singular Value Decomposition - SVD - (see Appendix A):

$$[H(0)] \atop (pr \times qs) = [U] \quad [\Sigma] \quad [V]^T$$
$$\scriptstyle (pr \times pr)(pr \times ps)(ps \times qs)$$

(4.100)

Matrix [H(0)] has 2N non-zero singular values (rank = 2N), equivalent to the order of the state-space system. [H(0)] can therefore be recomputed using only the first 2N columns of [U] and [V]:

$$[H(0)] = [U_{2N}][\Sigma_{2N}][V_{2N}]^T \qquad (4.101)$$

$$\text{(prxqs)} \quad \text{(prx2N)(2Nx2N)} \ (2Nxqs)$$

with

$$[U_{2N}]^T[U_{2N}] = [V_{2N}]^T[V_{2N}] = [I] \qquad (4.102)$$

Matrix [H]′ (= [H(0)]⁺) is therefore given by

$$[H]' = [V_{2N}][\Sigma_{2N}]^{-1}[U_{2N}]^T \qquad (4.103)$$

To obtain the desired realisation, we start from (4.91), which can be written for k≥0 as

$$[X(k+1)] = [R][A]^k[B] \qquad (4.104)$$

and use the identity

$$[X(k+1)] = [E_p]^T[H(k)][E_q] \qquad (4.105)$$

$$\text{(pxq)} \quad \text{(pxpr)} \ \text{(prxqs)} \ \text{(qsxq)}$$

where

$$[E_p]^T = [[I] \quad [0] \quad \dots \quad [0]] \qquad (4.106\ a)$$

$$\text{(pxpr)} \quad \text{(pxp)(pxp)} \quad \text{(pxp)}$$

$$[E_q] = \begin{bmatrix} [I] \\ [0] \\ \vdots \\ [0] \end{bmatrix} \qquad (4.106\ b)$$

$$\text{(qsxq)}$$

[I] being the identity matrix. Using (4.93), (4.95), (4.97), (4.101), (4.102) and (4.103), it is possible to show that

$$[X(k+1)] = \left[[E_p]^T[U_{2N}][\Sigma_{2N}]^{1/2} \right]\left[[\Sigma_{2N}]^{-1/2}[U_{2N}]^T [[Q][A]^k[W] \right]$$

$$[V_{2N}][\Sigma_{2N}]^{1/2} \left][[\Sigma_{2N}]^{1/2}[V_{2N}]^T [E_q] \right] \qquad (4.107)$$

Similarities with (4.104) (that we wish to recover) are already apparent. Some modifications, though, need to be made in (4.107), in the second block of matrices.

After some mathematical manipulations, it follows that

$$[X(k+1)] = \left[\left[E_p\right]^T \left[U_{2N}\right]\left[\Sigma_{2N}\right]^{1/2}\right]\left[\left[\Sigma_{2N}\right]^{-1/2}\left[U_{2N}\right]^T \left[H(1)\right]\left[V_{2N}\right]\left[\Sigma_{2N}\right]^{1/2}\right]^k$$

$$\left[\left[\Sigma_{2N}\right]^{1/2}\left[V_{2N}\right]^T \left[E_q\right]\right] \tag{4.108}$$

Comparing (4.108) with (4.104), it is clear that the desired realisation has been achieved, where

$$[R] = \left[E_p\right]^T \left[U_{2N}\right]\left[\Sigma_{2N}\right]^{1/2}$$

$$[A] = \left[\Sigma_{2N}\right]^{-1/2}\left[U_{2N}\right]^T \left[H(1)\right]\left[V_{2N}\right]\left[\Sigma_{2N}\right]^{1/2} \tag{4.109}$$

$$[B] = \left[\Sigma_{2N}\right]^{1/2}\left[V_{2N}\right]^T \left[E_q\right]$$

In order to determine the modal parameters of the system, we must solve an eigenproblem based on the 'realised' matrix [A], of the form

$$[A]\{\psi_u\} = \lambda\{\psi_u\} \tag{4.110}$$

To obtain the mode shapes in terms of the physical coordinates of the system, the following transformation is used (see (4.81)):

$$\{\psi_x\} = [R] \{\psi_u\}$$
$$\text{(p\times1)} \quad \text{(p\times2N)} \text{ (2N\times1)} \tag{4.111}$$

The modal parameters are easily calculated from the eigenproblem results. In summary, the necessary steps to perform an analysis with the ERA are as follows:

(1) choice of the measured data to construct matrix [H(0)];

(2) calculation of the SVD of [H(0)], to calculate $[U_{2N}]$, $[V_{2N}]$ and $[\Sigma_{2N}]$ and to recalculate [H(0)] based on the value of the rank found (order of the system);

(3) construction of matrix [H(1)] and calculation of the 'realised' matrices [R], [A] and [B], using (4.109); and

(4) calculation of the eigenvalues and eigenvectors of matrix [A] and the eigenvectors corresponding to the physical coordinates ((4.110) and (4.111)) and calculation of the modal parameters.

This method (like the ITD), also provides checks on the calculated modes, to distinguish between genuine and computational modes, as the results from the SVD, in some cases, may not be correct due to noise or nonlinearities. The first check is known as the Modal Amplitude Coherence and is defined as the coherence between each modal amplitude history and an ideal one, formed by extrapolating the initial value of the history to other points, using the identified eigenvalue. Another check is the Modal Phase Collinearity, for lightly damped structures, where real mode behaviour is expected. This indicator measures the strength of the linear functional relationship between the real and imaginary parts of the mode shape, for each mode. Model reduction is then possible, by truncating the modes with low accuracy indicators. The final model can be assessed by comparing the initial free responses with the ones calculated by (4.108).

An investigation on the effects of noise on the identified modal parameters using the ERA is given in [97]. In [96], the ERA is shown to be a more general formulation for modal analysis identification, as some other methods can be understood as particular cases of a unified approach. In the same reference, an extensive bibliography on System-Realisation Theory can be found.

4.3.2 Direct methods in the time domain

According to the classification given in Section 4.2, there are in this category essentially two methods, the Autoregressive Moving-Average method (ARMA) and the Direct System Parameter Identification method (DSPI).

The Autoregressive Moving-Average method (ARMA)

This is a *SISO* method that is mainly based on the works of Gersch [98-107]. Here, the basic ideas are given and the interested reader in invited to study the mentioned references, where the theory is explained in detail and different variations of the algorithm are given. References to other papers and fundamental texts can be found in those works. Let us consider the behaviour of a linear system with a single input $f(t)$ and a single output $y(t)$ as described by the following linear differential equation of constant coefficients:

$$a_n \frac{d^n y(t)}{dt^n} + a_{n-1} \frac{d^{n-1} y(t)}{dt^{n-1}} + \ldots + a_1 \frac{dy(t)}{dt} + a_0 y(t) = b_m \frac{d^m f(t)}{dt^m} + \ldots$$
$$+ b_1 \frac{df(t)}{dt} + b_0 f(t)$$

$$(4.112)$$

Taking the Laplace transform of (4.112) and considering the initial conditions as zero, it follows that

$$\left(a_n s^n + \ldots + a_1 s + a_0 \right) Y(s) = \left(b_m s^m + \ldots + b_1 s + b_0 \right) F(s) \qquad (4.113)$$

where s is the Laplace variable. The transfer function, defined as $H(s) = Y(s)/F(s)$, is

$$H(s) = \frac{b_m s^m + \ldots + b_1 s + b_0}{a_n s^n + \ldots + a_1 s + a_0} \qquad (4.114)$$

In the frequency domain, putting $s = i\omega$, the frequency response function is obtained:

$$H(\omega) = \frac{b_m(i\omega)^m + \ldots + b_1(i\omega) + b_0}{a_n(i\omega)^n + \ldots + a_1(i\omega) + a_0} \tag{4.115}$$

It is also possible to establish a linear difference equation corresponding to (4.112), when there are equally spaced time samples, as the following time series:

$$\alpha_n y(t - n) + \alpha_{n-1} y(t - n + 1) + \ldots + \alpha_1 y(t - 1) + \alpha_0 y(t) = \beta_m f(t - m) + \ldots$$

$$+ \beta_1 f(t - 1) + \beta_0 f(t) \tag{4.116}$$

or, in a more concise form,

$$\sum_{k=0}^{n} \alpha_k y(t - k) = \sum_{k=0}^{m} \beta_k f(t - k) \tag{4.117}$$

where α_k and β_k are known as the autoregressive and moving-average parameters, respectively; α_0 and β_0 are taken as 1. This model assumes that the output $y(t)$ is contaminated with a zero mean sequence of additive noise and that the time series input set $f(t)$ is a zero mean uncorrelated history. The input and output histories are supposed to be known, and α_k and β_k unknown. If the sampled time interval is Δt, then by introducing the z transform, where $z = e^{s\Delta t}$, we can write (4.116) as

$$\left(\alpha_n z^{-n} + \alpha_{n-1} z^{-n+1} + \ldots + \alpha_1 z^{-1} + \alpha_0\right) Y(z) = \left(\beta_m z^{-m} + \ldots + \beta_1 z^{-1} + \beta_0\right) F(z)$$

$$\tag{4.118}$$

Because (4.118) is the equivalent sampled representation of (4.112), the transfer function in terms of the z variable is given as:

$$H(z) = \frac{Y(z)}{F(z)} = \frac{\displaystyle\sum_{k=0}^{m} \beta_k z^{-k}}{\displaystyle\sum_{k=0}^{n} \alpha_k z^{-k}} \tag{4.119}$$

where the roots (poles) of the denominator polynomial are related to the natural frequencies and damping factors of the system. For a system with N DOF, those roots will be given by the solution of the characteristic polynomial:

$$\sum_{k=0}^{2N} \alpha_k u^{2N-k} = 0 \tag{4.120}$$

where α_k are the same as in (4.119). The problem now arises as to how to compute the autoregressive parameters α_k. To do this, (4.117) is written for an N DOF system as

$$\alpha_0 y(t) + \sum_{k=1}^{2N} \alpha_k y(t-k) = \sum_{k=0}^{2N-1} \beta_k f(t-k) \tag{4.121}$$

and, because $\alpha_0 = 1$, we can write the current observation $y(t)$ as the sum of its own past (the autoregressive part) plus a linear combination of uncorrelated terms (the moving-average):

$$y(t) = -\sum_{k=1}^{2N} \alpha_k y(t-k) + \sum_{k=0}^{2N-1} \beta_k f(t-k) \tag{4.122}$$

More realistically, by considering a prediction error $e(t)$,

$$y(t) = \left\{ -y(t-1) \; \ldots \; -y(t-2N) \quad f(t) \; \ldots \; f(t-2N+1) \right\} \begin{Bmatrix} \alpha_1 \\ \vdots \\ \alpha_{2N} \\ \beta_1 \\ \vdots \\ \beta_{2N-1} \end{Bmatrix} + e(t) \tag{4.123}$$

Collecting terms for $t = 2N+1, 2N+2, \ldots, 2N+L$, it follows that

$$\underset{(Lx1)}{\begin{Bmatrix} y(2N+1) \\ y(2N+2) \\ \vdots \\ y(2N+L) \end{Bmatrix}} = \underset{(Lx4N)}{\begin{bmatrix} -y(2N) & \ldots & -y(1) & f(2N+1) & \ldots & f(2) \\ -y(2N+1) & \ldots & -y(2) & f(2N+2) & \ldots & f(3) \\ \vdots & \ldots & \vdots & \vdots & \ldots & \vdots \\ -y(2N+L-1) & \ldots & -y(L) & f(2N+L) & \ldots & f(L+1) \end{bmatrix}} \underset{(4Nx1)}{\begin{Bmatrix} \alpha_1 \\ \vdots \\ \alpha_{2N} \\ \beta_1 \\ \vdots \\ \beta_{2N-1} \end{Bmatrix}}$$

$$+ \underset{(Lx1)}{\begin{Bmatrix} e(2N+1) \\ e(2N+2) \\ \vdots \\ e(2N+L) \end{Bmatrix}} \tag{4.124}$$

or simply,

$$\{y\} = [X]\{\theta\} + \{e\} \tag{4.125}$$

Minimising the squared error $\{e\}^T\{e\}$, leads to:

$$\{\theta\} = \left([X]^T [X]\right)^{-1} [X]^T \{y\} \tag{4.126}$$

Knowing $\{\theta\}$ and therefore the 2N values of α_k, we can return to (4.120), and write it as

$$\sum_{k=0}^{2N} \alpha_k u^{2N-k} = \prod_{k=1}^{N} (u - u_k)(u - u_k^*) = 0 \tag{4.127}$$

where $u_k = e^{s_r \Delta t}$ and $u_k^* = e^{s_r^* \Delta t}$ are the roots (poles) of the polynomial, from which it is possible to calculate the natural frequencies and damping factors. Knowing the 2N values of β_k, it is also possible to calculate the residues, using (4.119). This method can also provide statistical confidence factors (coefficients of variation). Variations of the ARMA method to accommodate multi-input multi-output situations are the so-called Autoregressive Moving-Average with exogenous variables, ARMAX [80], and the Autoregressive Moving-Average Vector, ARMAV [108-111].

The basic principles are the same as for the ARMA method, but the time series is now expressed in terms of matrix difference equations, where the autoregressive and moving-average coefficients are assembled in matrices. These types of methods are particularly advantageous whenever the excitation is not known in a deterministic sense and often taken statistically as white noise. It is possible to analyse systems where only the output is measured, while the input is unknown, as - for instance - in the case of bridges, subjected to traffic and/or wind excitation [111].

The Direct System Parameter Identification method (DSPI)
This method is due to Leuridan [65, 112], and is based on the kind of approach given above for the ARMA method. Let us consider the dynamic equilibrium equation for N_0 DOF, where N_0 is part of the N expected number of DOF of the structure, as

$$[M]\{\ddot{y}(t)\} + [C]\{\dot{y}(t)\} + [K]\{y(t)\} = [\Gamma]\{f(t)\} \tag{4.128}$$

where [M], [C] and [K] are real matrices of order N_0, $\{f\}$ is the input vector at q locations, and $[\Gamma]$ is an $N_0 \times q$ matrix whose elements e_{ij} are 1 when the input location j corresponds to the response location i and zero elsewhere. Pre-multiplying by $[M]^{-1}$, (4.128) can be written alternatively as

$$\{\ddot{y}(t)\} + [C]'\{\dot{y}(t)\} + [K]'\{y(t)\} = [\Gamma]'\{f(t)\} \tag{4.129}$$

As for (4.123), (4.129) can be written as an autoregressive moving-average model:

$$\{y(t)\} = [A_1]\{y(t-1)\} + ... + [A_p]\{y(t-p)\} + [B_0]\{f(t)\} + [B_1]\{f(t-1)\} + ...$$

$$+ [B_{p-1}]\{f(t-p+1)\} + \{e(t)\} \tag{4.130}$$

where the coefficient matrices [A] and [B] are of dimensions $N_0 x N_0$ and $N_0 x q$, respectively, and p is chosen so that $pN_0 \geq 2N$. Considering m sets of inputs and responses, we have

$$[y(t)] = [A_1][y(t-1)] + ... + [A_p][y(t-p)] + [B_0][f(t)] + ...$$

$$+ [B_{p-1}][f(t-p+1)] + [e(t)] \tag{4.131}$$

where

$$\underset{(N_0 xm)}{[y(t)]} = \left[\{y(t)\}_1 ... \{y(t)\}_m \right]$$

$$\underset{(qxm)}{[f(t)]} = \left[\{f(t)\}_1 ... \{f(t)\}_m \right] \tag{4.132}$$

$$\underset{(N_0 xm)}{[e(t)]} = \left[\{e(t)\}_1 ... \{e(t)\}_m \right]$$

Equation (4.131) can still be written as

$$[y(t)] = \left[[A_1]\ [A_2]\ ...\ [A_p]\ [B_0]\ [B_1]\ ...\ [B_{p-1}] \right] \begin{bmatrix} [y(t-1)] \\ [y(t-2)] \\ \vdots \\ [y(t-p)] \\ [f(t)] \\ [f(t-1)] \\ \vdots \\ [f(t-p+1)] \end{bmatrix} + [e(t)] \tag{4.133}$$

Considering all the L sampled time intervals, t=p+1, ..., p+L,

$$[[y(p+1)] \ldots [y(p+L)]] =$$
$$(N_0 \times Lm)$$

$$[[A_1] \ldots [A_p] \ [B_0] \ldots [B_{p-1}]] \begin{bmatrix} [y(p)] & \ldots & [y(p+L-1)] \\ \vdots & \vdots & \vdots \\ [y(1)] & \ldots & [y(L)] \\ [f(p+1)] & \ldots & [f(p+L)] \\ \vdots & \vdots & \vdots \\ [f(2)] & \ldots & [f(L+1)] \end{bmatrix}$$
$$(N_0 \times (N_0 p + qp)) \qquad\qquad\qquad ((N_0 p + qp) \times Lm)$$

$$+ [[e(p+1)] \ldots [e(p+L)]] \qquad\qquad (4.134)$$
$$(N_0 \times Lm)$$

In a compact form, (4.134) can be written as

$$[y] = [\theta][X] + [e] \qquad\qquad (4.135)$$

The minimisation of $[e]^T [e]$ leads to

$$[\theta] = [y][X]^T ([X][X]^T)^{-1} \qquad\qquad (4.136)$$

Considering impulse response functions instead of free decay responses, and using the z transform, with $z = e^{s\Delta t}$, we have a similar expression to (4.119) for the transfer function $H(z)$:

$$[[I] - [A_1]z^{p-1} - \ldots - [A_p]z^{-p}][H(z)] = (\Delta t)^{-1}[[B_0] + [B_1]z^{-1} + \ldots$$
$$+ [B_{p-1}]z^{-p+1}] \qquad (4.137)$$

or,

$$[[I]z^p - [A_1]z^{p-1} - \ldots - [A_p]][H(z)] = (\Delta t)^{-1}[[B_0]z^p + [B_1]z^{p-1} + \ldots$$
$$+ [B_{p-1}]z] \qquad (4.138)$$

where matrices $[A_1], \ldots, [A_p]$ and $[B_0], \ldots, [B_{p-1}]$ are already known. This equation can be written as

$$[z[I] - [A]][I(z)][H(z)] = [B(z)] \qquad\qquad (4.139)$$

with

$$\underset{(N_0p \times N_0p)}{[A]} = \begin{bmatrix} [A_1] & [A_2] & \cdots & \cdots & [A_p] \\ [I] & [0] & \cdots & \cdots & [0] \\ [0] & [I] & \ddots & \cdots & [0] \\ \vdots & \vdots & \ddots & \ddots & \vdots \\ [0] & [0] & \cdots & [I] & [0] \end{bmatrix} \qquad \underset{(N_0p \times N_0)}{[I(z)]} = \begin{bmatrix} z^{p-1}[I] \\ \vdots \\ [I] \end{bmatrix}$$

$$\underset{(N_0p \times q)}{[B(z)]} = (\Delta t)^{-1} \begin{bmatrix} z^p[B_0] + \ldots + z[B_{p-1}] \\ [0] \\ \vdots \\ [0] \end{bmatrix} \tag{4.140}$$

The N_0p eigenvalues and eigenvectors are obtained from

$$\underset{(N_0p \times N_0p)}{[z[I] - [A]]} \underset{(N_0p \times 1)}{\{\tilde{\psi}\}} = \underset{(N_0p \times 1)}{\{0\}} \tag{4.141}$$

with

$$\underset{(N_0p \times 1)}{\{\tilde{\psi}\}} = \underset{(N_0p \times N_0)}{[I(z)]} \underset{(N_0p \times 1)}{\{\psi\}} \tag{4.142}$$

From the eigenvalues of matrix [A], the natural frequencies and damping factors are easily calculated. $\{\psi\}$ are the eigenvectors we are looking for. From (4.140) and (4.142), and for all the N_0p eigenvectors $[\tilde{\Psi}]$,

$$\underset{(N_0p \times N_0p)}{[\tilde{\Psi}]} = \begin{bmatrix} [\Psi] \begin{bmatrix} `z_{r`} \end{bmatrix}^{2N-1} \\ \vdots \\ [\Psi] \end{bmatrix} \tag{4.143}$$

where $[`z_{r`}] = [`e^{s_r \Delta t}`]$. The last block of matrix $[\tilde{\Psi}]$ in (4.143) is precisely $[\Psi]$. It is known that the impulse response function matrix [H(t)] can be given by

$$\underset{(N_0p \times q)}{[H(t)]} = \underset{(N_0 \times N_0p)}{[\Psi]} \underset{(N_0p \times N_0p)}{[`e^{s_r t`}]} \underset{(N_0p \times q)}{[W]^T} \tag{4.144}$$

where [W] is the matrix of participation factors, as defined in (4.24) and (4.25). This is the matrix that remains to be calculated to complete the method.

In the z domain, (4.144) becomes:

$$[H(z)] = \underset{(N_0 p \times q)}{[\Psi]} \underset{(N_0 \times N_0 p)}{\left[[I] - z^{-1} \left[\diagdown z_{r \diagdown} \right] \right]^{-1}}_{(N_0 p \times N_0 p)} \underset{(N_0 p \times q)}{[W]^T} \tag{4.145}$$

Pre-multiplying (4.139) by $\left[\tilde{\Psi} \right]^{-1}$, we obtain

$$\left[\tilde{\Psi} \right]^{-1} [z[I] - [A]] [I(z)] [H(z)] = \left[\tilde{\Psi} \right]^{-1} [B(z)] \tag{4.146}$$

Substituting (4.145) in (4.146), it follows that

$$\left[\tilde{\Psi} \right]^{-1} (z[I] - [A]) [I(z)] [\Psi] \left[[I] - z^{-1} \left[\diagdown z_{r \diagdown} \right] \right]^{-1} [W]^T = \left[\tilde{\Psi} \right]^{-1} [B(z)] \tag{4.147}$$

From (4.141), we can write

$$[A] = \left[\tilde{\Psi} \right] \left[\diagdown z_{r \diagdown} \right] \left[\tilde{\Psi} \right]^{-1} \tag{4.148}$$

Substitution of (4.148) in (4.147) leads to

$$\left[z[I] - \left[\diagdown z_{r \diagdown} \right] \right] \left[\tilde{\Psi} \right]^{-1} [I(z)] [\Psi] \left[[I] - z^{-1} \left[\diagdown z_{r \diagdown} \right] \right]^{-1} [W]^T = \left[\tilde{\Psi} \right]^{-1} [B(z)] \tag{4.149}$$

where everything is known except the modal participation matrix [W]. By taking limits on both sides of (4.149) when $z \to z_r$ [65], and assuming that the 2N modes have already been sorted out from the $N_0 p$ modes, we obtain

$$\left\{ W^T \right\}_r = z_r \left\{ \left[\tilde{\Psi} \right]^{-1} \left\{ B(z_r) \right\} \right\}_r \qquad \text{for} \quad r = 1, ..., 2N \tag{4.150}$$

The model of expression (4.144) is therefore complete. The residue matrix for each mode r is given by

$$\underset{(N_0 q)}{[A]_r} = \underset{(N_0 \times 1)}{\{\psi\}_r} \underset{(1 \times q)}{\left\{ W^T \right\}_r} \tag{4.151}$$

Knowing the residues, it is easy to evaluate the modal constants and phase angles. This method is said to give very good results, even for almost repeated modes. In [65] it is shown that this method can be seen as a generalisation of other methods, like the CE, PRCE and ITD.

4.3.3 Other methods in the time domain
In 1980, Zaghlool [113] presented the Single-Station Time Domain method (SSTD), which is a *SISO* version of the ITD method. Matrices [X] and $[\hat{X}]$ in (4.52) and (4.55) are formed by shifting several times the response from one single station. It will not be developed here, but the corresponding version in the frequency domain will be explained in Section 4.4.3 (CEFD method). The similarities between the two domains are immediately obvious.

In [114], Swevers *et al* present another *SISO* method, which builds an over-determined set of linear equations, containing stepped-sine measurements, that is solved in a least-squares sense. The resulting error covariance matrix is used to construct the noise covariance matrix for a second set of linear equations, where all data are noisy, being solved by the quotient SVD, a generalisation of the common SVD (Appendix A). It seems adequate for very low signal-to-noise ratios.

Another method [115] uses the complex envelope signal of the impulse response. The parameters are extracted by mode decomposition. It is convenient for closely spaced modes.

In [116] a *MIMO* method is presented as a unifying approach, showing that many identification methods in the time domain, like the ARMAV, ITD, ERA, etc., can be derived from such a general approach.

In [117], several time domain methods are compared, to estimate natural frequencies and damping ratios from the response caused by an unknown random input. Among those methods are the Random Decrement Technique (Randomdec or RDD) and the Maximum Entropy Method (MEM). These are particularly suited for flight flutter testing, where an accurate estimation of the damping ratio is crucial. More recently, the RDD method has been the object of active research [118-122].

4.4   FREQUENCY DOMAIN METHODS
Frequency domain methods are presented in this section. Due to the enormous quantity of available methods, it is not possible to explain them all in great detail, so a selection had to be made *a priori*.

The chosen ones are those which seemed to be the most important, considering their historical relevance as well as their current impact in terms of practical use and implementation.

Figure 4.3 presents a diagram of the fundamental classification of the various methods referred to here, within this category.

4.4.1 SDOF methods
In this category, we shall discuss the Peak Amplitude, the Quadrature Response, the Maximum Quadrature Component, the Kennedy-Pancu, the Circle-Fitting, the Inverse and Dobson's methods. All these are *SISO* methods. Some *SIMO* variants are referred to later on in this section.

The Peak Amplitude method
This is the simplest known method for identifying the modal parameters of a structure [61]. The natural frequencies are simply taken from the observation of the

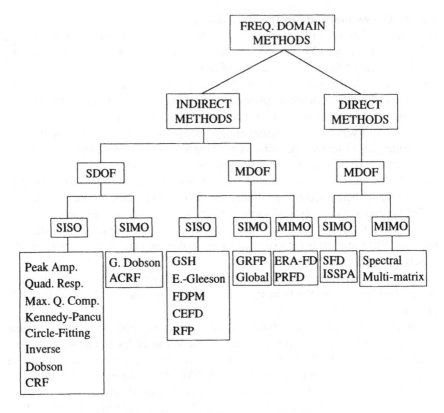

Fig. 4.3 Classification of frequency domain methods.

peaks on the graphs of the magnitude of the response. The damping ratios are calculated from the sharpness of the peaks and the mode shapes are calculated from the ratios of the peak amplitudes at various points in the structure. In order to take into account the amplitude of the excitation force, the use of the receptance represented an improvement to this method [123]. This method assumes that the modes are real and, albeit being quite 'crude', it may provide reasonable results if the modes are well separated and if the damping is not very high. A critical comparison between this and the Kennedy-Pancu method is given in [123], where some alternatives are suggested for the calculation of the damping ratios.

The Quadrature Response
and Maximum Quadrature Component methods

These methods [124] differ from the Peak Amplitude method in the location of the natural frequencies of the structure. The Quadrature Response locates the natural frequencies at the points where the in-phase component of the response (the real part) is zero. This corresponds to a 90 degree phase difference between the forcing function and the response. The Maximum Quadrature Component considers that the natural frequencies occur at the points where the quadrature component of the response (the imaginary part) has a maximum (or minimum). This component is 90 degrees out-of-phase with the forcing function.

The Kennedy-Pancu (or Maximum Frequency Spacing) method

This method, first introduced by Kennedy and Pancu [63] uses the Argand plane to display the real and the imaginary parts of the receptance. Around each natural frequency the curve approaches a circle (see Section 1.5) and the natural frequency is located at the point where the rate of change of arc length with frequency attains a maximum. The model assumed for the damping is the hysteretic one and the damping factor is evaluated from a simplified half-power points calculation and the mode shapes are calculated from the ratios of the diameters of the circles, fitted around each natural frequency for the various output responses. It is still assumed that the modes are real and that the damping is small. Some problems concerning the existence of close modes were already encountered by these authors.

Pendered [124] made a critical comparison between this and the previous methods cited above, concluding that the Kennedy-Pancu method was the one that could resolve more accurately two close modes and that the Quadrature Response Method was the worst one, concerning this aspect. Woodcock [125] extended this method to systems with viscous, non-proportional damping and without the restriction of small amounts of damping. Klosterman [67] continued the investigation on the Kennedy-Pancu method, establishing more efficient techniques for the determination of the modal parameters for systems with general non-proportional damping, using either the viscous or the hysteretic model. In 1973, Marples [126] gave a new formula for the calculation of the hysteretic damping factor. A systematic use of this formula [127] around the resonance region allows for the calculation of the mean value for the damping factor.

The Circle-Fitting method

Basically, after the cited work of Klosterman, the Kennedy-Pancu method became known as the Circle-Fitting method, easy to implement in small computers. A comprehensive study of the Circle-Fitting method has been presented by Ewins [3]. Nevertheless, a chapter on identification methods would not be complete without an explanation of this technique, so we decided to include it here.

As already seen in Section 1.4.2 (equation (1.233)), the receptance of an N DOF system with hysteretic damping is given by the following expression:

$$\alpha_{jk}(\omega) = \sum_{r=1}^{N} \frac{{}_rC_{jk}}{\omega_r^2 - \omega^2 + i\eta_r\omega_r^2} \tag{4.152}$$

where η_r and ${}_rC_{jk}$ are the hysteretic damping ratio and the complex modal constant ${}_rC_{jk} = (C_r e^{i\phi_r})_{jk}$, respectively, associated with each mode r.

In practice, there is a limited frequency range for which experimental data are collected. The contribution to the total response of the terms situated outside the experimental frequency range may be taken into account by means of 'residuals', a subject already discussed in Section 1.6. The circle-fitting method relies on the assumption that the contribution of the out-of-range modes to the particular one under study is a constant; (4.152) is therefore approximated by:

$$\alpha_{jk}(\omega) = \frac{{}_rC_{jk}}{\omega_r^2 - \omega^2 + i\eta_r\omega_r^2} + {}_rD_{jk} \qquad (4.153)$$

where ${}_rD_{jk}$ is a complex constant associated with mode r. As was also discussed in Section 1.3.3, the Nyquist plot of $1/(\omega_r^2 - \omega^2 + i\eta_r\omega_r^2)$ is a circle. Looking at (4.153), we see that the multiplication by the complex constant ${}_rC_{jk}$ means a magnification or reduction of the circle radius, as well as a certain rotation. ${}_rD_{jk}$ corresponds to a simple translation. As, in fact, we are representing (4.152) in the Nyquist plot, the complete curve will not be exactly a circle around each natural frequency, but the approach (4.153) tells us that the curve will have sections of near-circular arcs around those frequencies as illustrated in figure 4.4.

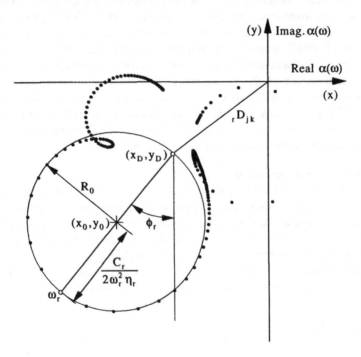

Fig. 4.4 Nyquist plot of the receptance, showing the SDOF circle-fit approach.

The derivation of the modal parameters associated with mode r relies on the fitting of a circle to the frequency response curve near the natural frequency ω_r. This first objective is usually achieved through the use of a least-squares technique. Let us assume that our data points are in the Argand plane represented by coordinates x_j, y_j. One is looking for a circle that best fits these data, i.e., for which the error e_1 is minimised:

$$e_1 = \sum_{j=1}^{L} \left[R_0 - \sqrt{\left(x_j - x_0\right)^2 + \left(y_j - y_0\right)^2} \right]^2 \tag{4.154}$$

However, such an error definition does not lead to an explicit calculation of x_0, y_0 and R_0. As a consequence, an iterative procedure is necessary based on a given initial estimate of the values of the previous parameters, which can be quite time-consuming. This problem can be solved if one defines a new error function [128] and takes the circle that best fits the L data points as being the one for which

$$e_2 = \sum_{j=1}^{L} \left\{ R_0^2 - \left[\left(x_j - x_0\right)^2 + \left(y_j - y_0\right)^2 \right] \right\}^2 \tag{4.155}$$

is minimised. Rearranging (4.155), one can write

$$e_2 = \sum_{j=1}^{L} \left[c - \left(x_j^2 + a\,x_j + b\,y_j + y_j^2 \right) \right]^2 \tag{4.156}$$

where

$$a = -2x_0 \quad\quad b = -2y_0 \quad\quad c = R_0^2 - x_0^2 - y_0^2 \tag{4.157}$$

Minimisation of e_2 relative to the new set of parameters a, b and c, leads to:

$$\begin{bmatrix} \sum x_j^2 & \sum x_j y_j & -\sum x_j \\ \sum x_j y_j & \sum y_j^2 & -\sum y_j \\ -\sum x_j & -\sum y_j & L \end{bmatrix} \begin{bmatrix} a \\ b \\ c \end{bmatrix} = \left\{ \begin{array}{c} -\left(\sum x_j^3 + \sum x_j y_j^2 \right) \\ -\left(\sum y_j^3 + \sum y_j x_j^2 \right) \\ \sum x_j^2 + \sum y_j^2 \end{array} \right\} \tag{4.158}$$

where the summations extend to the L points considered for the fit. This expression allows for an explicit calculation of a, b and c, and hence of x_0, y_0 and R_0, through relationships (4.157). So, an iterative process has been replaced by a direct and fast calculation. Nevertheless, it is important to note that the error function e_2 does not lead exactly to the same results as obtained when using e_1. A straightforward mathematical manipulation shows that if one neglects higher order terms,

$$e_2 \approx 4R_0^2 e_1 \tag{4.159}$$

or, upon comparing the minimisation procedure relative to the parameters x_0, y_0 and R_0,

$$\frac{\partial e_1}{\partial x_0} = 0 \implies \frac{\partial e_2}{\partial x_0} \approx 0, \quad \frac{\partial e_1}{\partial y_0} = 0 \implies \frac{\partial e_2}{\partial y_0} \approx 0,$$

$$\frac{\partial e_1}{\partial R_0} = 0 \implies \frac{\partial e_2}{\partial R_0} \approx 8R_0 e_1$$

(4.160)

When one takes into consideration the fact that the standard procedure based on e_1 also relies on a truncation of higher order terms in a Taylor series expansion, it becomes apparent that the linearised procedure can be reliably used at a very small cost. Practice shows extremely good results, and therefore the advantages of using such an approach greatly outweigh the inconvenience of the introduction of an additional small error.

The location and determination of the natural frequency are usually based on a frequency spacing technique [67]. For a given mode, and apart from the effect of the complex modal constant, the phase angle θ_r associated with the dynamic response is given by:

$$\theta_r = \tan^{-1}\left\{\eta_r / \left[1 - \left(\omega / \omega_r\right)^2\right]\right\}$$

(4.161)

and it is easy to show that $d(\omega^2)/d\theta_r$ is minimum when $\omega = \omega_r$, i.e.,

$$\frac{d}{d\omega}\left[\frac{d(\omega^2)}{d\theta_r}\right] = 0 \implies \omega = \omega_r$$

(4.162)

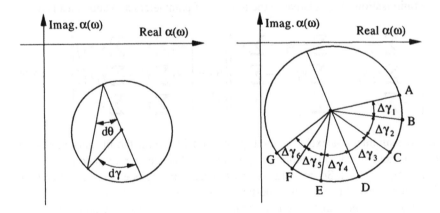

Fig. 4.5 Basis for the natural frequency determination.

From figure 4.5, it is evident that calculating the minimum of $d(\omega^2)/d\theta$ is the same as calculating the minimum of $d(\omega^2)/d\gamma$. For a set of experimental data values with equally spaced frequency increments, a finite-difference table can be constructed (table 4.1):

Table 4.1 Finite difference table.

Points	Frequency	γ	$\Delta\gamma$	$\Delta^2\gamma$	...
A	ω_A	γ_1			
B	ω_B	γ_2	$\Delta\gamma_1$		
C	ω_C	γ_3	$\Delta\gamma_2$	$\Delta^2\gamma_1$	
D	ω_D	γ_4	$\Delta\gamma_3$	$\Delta^2\gamma_2$	
E	ω_E	γ_5	$\Delta\gamma_4$	$\Delta^2\gamma_3$	...
F	ω_F	γ_6	$\Delta\gamma_5$	$\Delta^2\gamma_4$	
G	ω_G	γ_7	$\Delta\gamma_6$	$\Delta^2\gamma_5$	

There will be a change of sign of $\Delta^2\gamma$ as soon as one goes through data points immediately before and after the natural frequency. A more accurate location of the natural frequency and determination of its value are achieved by means of Newton's divided differences formula:

$$f(v) = f(v_0) + (v - v_0)f(v_0, v_1) + (v - v_0)(v - v_1)f(v_0, v_1, v_2) + \ldots$$
$$+ (v - v_0)(v - v_1)\ldots(v - v_L)f(v_0, v_1, \ldots, v_L) \tag{4.163}$$

where

$$f(v_0, v_1, \ldots, v_L) = \frac{f(v_0, v_1, \ldots, v_{L-1}) - f(v_1, v_2, \ldots, v_L)}{v_0 - v_L} \tag{4.164}$$

For our purposes, v represents the square of the frequency and $f(v)$ the phase angle γ, or θ. Taking into consideration four known points, two immediately before the natural frequency and two after, and neglecting the residual error term, (4.163) reduces to:

$$\theta = \theta_0 + (\omega^2 - \omega_0^2)(\theta_0, \theta_1) + (\omega^2 - \omega_0^2)(\omega^2 - \omega_1^2)(\theta_0, \theta_1, \theta_2)$$
$$+ (\omega^2 - \omega_0^2)(\omega^2 - \omega_1^2)(\omega^2 - \omega_2^2)(\theta_0, \theta_1, \theta_2, \theta_3) \tag{4.165}$$

with

$$(\theta_0, \theta_1) = \frac{\theta_0 - \theta_1}{\omega_0^2 - \omega_1^2}$$

$$(\theta_0, \theta_1, \theta_2) = \frac{(\theta_0, \theta_1) - (\theta_1, \theta_2)}{\omega_0^2 - \omega_2^2} \tag{4.166}$$

$$(\theta_0, \theta_1, \theta_2, \theta_3) = \frac{(\theta_0, \theta_1, \theta_2) - (\theta_1, \theta_2, \theta_3)}{\omega_0^2 - \omega_3^2}$$

As the minimum of $d(\omega^2)/d\theta$ corresponds to the maximum of $d\theta/d(\omega^2)$, the natural frequency is obtained by differentiating twice (4.165) and setting it equal to zero, which gives:

$$\omega_r^2 = \frac{1}{3}\left(\omega_0^2 + \omega_1^2 + \omega_2^2 - \frac{(\theta_0, \theta_1, \theta_2)}{(\theta_0, \theta_1, \theta_2, \theta_3)}\right) \qquad (4.167)$$

Substituting back in (4.165), we calculate the value of θ_r. This technique is very simple to implement and leads to very accurate values of the natural frequency and of its 'exact' location on the response curve. Although other techniques may also provide a precise derivation of the value of the natural frequency, they may not be so accurate in terms of its location on the response curve. The importance of the accuracy of that location is only apparent when one needs to derive the value of the phase angle ϕ_r of the complex modal constant, calculated by 'drawing' a line joining the natural frequency point and the circle centre (see figure 4.4).

Estimation of the damping factor η_r is now an easy task. From (4.161), taking two points on the circle, one corresponding to a frequency below the natural frequency (ω_b) and another to a frequency above the natural frequency (ω_a), one can write:

$$\tan\theta_a = \frac{\eta_r}{1 - \left(\dfrac{\omega_a^2}{\omega_r^2}\right)}$$

$$\tan\theta_b = \frac{\eta_r}{1 - \left(\dfrac{\omega_b^2}{\omega_r^2}\right)} \qquad (4.168)$$

Let us assume that $\theta_r = \pi/2$ and define:

$$\Delta\theta_a = \frac{\Delta\gamma_a}{2} = \theta_a - \theta_r$$

$$\Delta\theta_b = \frac{\Delta\gamma_b}{2} = \theta_r - \theta_b \qquad (4.169)$$

Thus,

$$\tan\left(\frac{\Delta\gamma_a}{2}\right) = \tan\left(\theta_a - \frac{\pi}{2}\right) = -\frac{1}{\tan\theta_a} = -\frac{1}{\eta_r}\left[1 - \left(\frac{\omega_a^2}{\omega_r^2}\right)\right]$$

$$\tan\left(\frac{\Delta\gamma_b}{2}\right) = \tan\left(\frac{\pi}{2} - \theta_b\right) = \frac{1}{\tan\theta_b} = \frac{1}{\eta_r}\left[1 - \left(\frac{\omega_b^2}{\omega_r^2}\right)\right] \qquad (4.170)$$

and, therefore,

$$\eta_r = \frac{\omega_a^2 - \omega_b^2}{\omega_r^2} \frac{1}{\tan(\Delta\theta_a) + \tan(\Delta\theta_b)} \tag{4.171}$$

In fact, this expression is valid, even when $\theta_r \neq \pi/2$, as we are dealing with differences of angles (see figure 4.6).

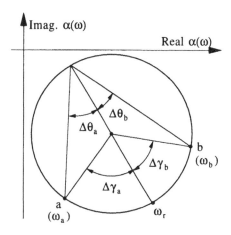

Fig. 4.6 Use of the natural frequency and two data points to derive the damping factor.

As can be seen, (4.171) gives the familiar half-power points relationship when $\Delta\gamma_a = \Delta\gamma_b = \pi/2$. For a given set of results, it is possible to derive several values for η_r, depending on the pair of data points that are used in (4.171). Ewins [127] showed that this could be a very useful way of determining the existence of nonlinear effects. In principle, for a linear system, the values of η_r should all be identical. However, in practice, this is not the case. The deviations in the estimates of η_r are therefore a useful means of assessing the validity of the analysis. If the variations are random then the scatter is probably due to measurement errors, but if they are systematic then it could be caused by various effects, namely a nonlinear behaviour.

On the other hand, one should be cautious and not jump to quick conclusions, as some variations in the damping estimates will always occur to some extent, even in an ideal and linear system. The reason for this is inherent to the computations associated with expression (4.171) and mainly due to the nature of the denominator. The end result can be as shown in figure 4.7, where the plane instead of being horizontal is somewhat tilted. The best results are obtained when angles $\Delta\gamma_a$ and $\Delta\gamma_b$ are not very small and have similar values. Thus, it is advisable to use, in practice, the combinations of data points corresponding to the diagonal of the η_r surface represented in figure 4.7, and calculate the average value.

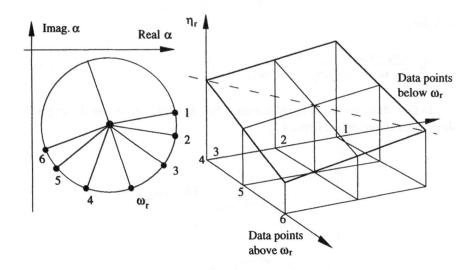

Fig. 4.7 Determination of η_r, using several combinations of data points.

Once ω_r and η_r are known, the derivation of the modulus C_r and phase angle ϕ_r of the modal constant is a straightforward calculation. In fact, and according to (4.153), the diameter of the circle is given by:

$$\text{Diameter} = \frac{C_r}{\omega_r^2 \eta_r} \tag{4.172}$$

as shown in figure 4.4. Also, ϕ_r is given by

$$\phi_r = \tan^{-1}\left(\frac{x_0 - x_D}{y_0 - y_D}\right) \tag{4.173}$$

where x_D and y_D are the coordinates of the displaced origin (figure 4.4) and their values are known as soon as the natural frequency location is determined.

Despite being well known, the Circle-Fitting method is often disregarded and said to only work well for widely separated modes and lightly damped structures. This is probably because in some modal analysis packages and analysers the versions used are the very basic ones, that coincide nearly with the Kennedy-Pancu method. It is our opinion (and experience) that the Circle-Fitting method works very well for the majority of situations and even for highly complicated structures. It is, naturally, very time consuming. One of the most important improvements associated with the Circle-Fitting method is the possibility to subtract the effect of modes already analysed before analysing the one we are interested in [129]. The idea is very simple: after a first identification of each of the modes individually, the analysis is repeated for each mode, this time subtracting from the original FRF data the contribution of the modes (besides the one under study) that have already been identified. Mathematically, this can be expressed as

$$\tilde{\alpha}_r = \tilde{\alpha} - \sum_{\substack{s=1 \\ s \neq r}}^{N} \alpha_s \qquad (4.174)$$

where $\tilde{\alpha}$ is the initially measured FRF data, $\tilde{\alpha}_r$ is the resulting FRF of the mode under consideration and α_s is the regenerated FRF contribution of each mode already analysed. This technique is very convenient for two close modes in particular, and an iterative procedure can be established between the two modes until convergence is obtained. This iterative procedure is, in the majority of cases, convergent, although it can be quite slow.

Some efforts to obtain consistent sets of natural frequencies and damping factors have been tried. The objective is to avoid variations on the estimates of those parameters when several FRFs are analysed individually. The simplest way of all is to average all the estimates obtained for each FRF. This was applied with success by Talapatra [130]. A weighted average of the natural frequencies and damping factors was also proposed and applied successfully by Kirshenboim [131]. The weighting factors are given by the RMS errors obtained in the circle-fitting procedure divided by the diameter, for each mode. The initial estimates of the modal constants are then corrected, using the averaged values.

The Inverse method
The Inverse method was presented by Dobson [132] and relies on the fact that the imaginary and real parts of the inverse of the receptance (dynamic stiffness) are straight lines in the frequency and frequency squared, respectively. Because it is based on the assumption that the modes are real and relies on well spaced natural frequencies, we can say that it represents the 'inverse' of the Kennedy-Pancu method.

The extension of the method to the more general case of complex modes is quite straightforward. Writing the receptance of a SDOF system with hysteretic damping as

$$\alpha_r = \frac{C_r e^{i\phi_r}}{\omega_r^2 - \omega^2 + i\eta_r\omega_r^2} \qquad (4.175)$$

the inverse is

$$\frac{1}{\alpha_r} = \frac{\omega_r^2 - \omega^2 + i\eta_r\omega_r^2}{C_r e^{i\phi_r}} \qquad (4.176)$$

Writing (4.176) as

$$\frac{1}{\alpha_r} = \frac{\omega_r^2 - \omega^2 + i\eta_r\omega_r^2}{A_r + iB_r} \tag{4.177}$$

where $A_r = C_r \cos\phi_r$ and $B_r = C_r \sin\phi_r$, it follows that

$$\mathrm{Re}\left(\frac{1}{\alpha_r}\right) = \frac{(A_r + B_r\eta_r)\omega_r^2 - A_r\omega^2}{A_r^2 + B_r^2}$$

$$\mathrm{Im}\left(\frac{1}{\alpha_r}\right) = \frac{(A_r\eta_r - B_r)\omega_r^2 + B_r\omega^2}{A_r^2 + B_r^2} \tag{4.178}$$

Both real and imaginary parts are straight lines in ω^2, of the form

$$\mathrm{Re}\left(\frac{1}{\alpha_r}\right) = m_R + n_R\omega^2$$

$$\mathrm{Im}\left(\frac{1}{\alpha_r}\right) = m_I + n_I\omega^2 \tag{4.179}$$

with

$$m_R = \frac{(A_r + B_r\eta_r)\omega_r^2}{A_r^2 + B_r^2} \qquad n_R = -\frac{A_r}{A_r^2 + B_r^2}$$

$$m_I = \frac{(A_r\eta_r - B_r)\omega_r^2}{A_r^2 + B_r^2} \qquad n_I = \frac{B_r}{A_r^2 + B_r^2} \tag{4.180}$$

After fitting the graphs of $\mathrm{Re}(1/\alpha_r)$ and $\mathrm{Im}(1/\alpha_r)$ to straight lines, m_R, m_I, n_R and n_I are known and by convenient manipulation of (4.180), we can determine the four modal parameters:

$$\omega_r = \sqrt{\frac{-m_R n_R - m_I n_I}{n_R^2 + n_I^2}} \qquad \eta_r = \frac{m_R n_I - m_I n_R}{-m_R n_R - m_I n_I}$$

$$C_r = \frac{1}{\sqrt{n_R^2 + n_I^2}} \qquad \phi_r = \mathrm{tg}^{-1}\left(-\frac{n_I}{n_R}\right) \tag{4.181}$$

In the case of real modes, $n_I = 0$ and these expressions coincide with the ones for the real modes case. Both slopes and intercepts of the real and imaginary parts contribute to the calculation of the modal parameters and so the imaginary part is no longer uniquely linked with the damping factor, as happens in the real modes assumption. Also, the natural frequency does not correspond to the intercept with the zero horizontal line, as in the real modes case. The Inverse method may have advantages in some cases over the Circle-Fitting, namely when the damping is very small and/or there are significant measurement errors in the areas close to resonances, as it is easier to fit a straight line rather than a circle, when one only has a good definition of the response away from the resonance peaks. Of course, if the modes are close together, the iterative procedure of removing the effect of the modes already identified has to be implemented, as for the Circle-Fitting.

Dobson's method

Also developed by Dobson [133] this method is an extension of the inverse method, considering complex modes and automatically compensating for the effects of neighbouring modes. Let us consider the receptance FRF of a system with hysteretic damping given by

$$\alpha(\omega) = \frac{A_r + iB_r}{\omega_r^2 - \omega^2 + i\eta_r\omega_r^2} + \text{Residual term} \qquad (4.182)$$

For a particular value $\omega = \Omega$ close to the resonance, one has

$$\alpha(\Omega) = \frac{A_r + iB_r}{\omega_r^2 - \Omega^2 + i\eta_r\omega_r^2} + \text{Residual term} \qquad (4.183)$$

The residual term, considered constant over the chosen frequency range, can be eliminated by subtracting each side of (4.182) and (4.183):

$$\alpha(\omega) - \alpha(\Omega) =$$

$$\left(A_r + iB_r\right)\left[\frac{\omega^2 - \Omega^2}{\left(\omega_r^2 - \omega^2\right)\left(\omega_r^2 - \Omega^2\right) - \eta_r^2\omega_r^4 + i\eta_r\omega_r^2\left(2\omega_r^2 - \omega^2 - \Omega^2\right)}\right] \qquad (4.184)$$

Defining a function Δ as

$$\begin{aligned}
\Delta &= \frac{\omega^2 - \Omega^2}{\alpha(\omega) - \alpha(\Omega)} \\
&= \frac{A_r - iB_r}{A_r^2 + B_r^2}\left[\left(\omega_r^2 - \omega^2\right)\left(\omega_r^2 - \Omega^2\right) - \eta_r^2\omega_r^4 + i\eta_r\omega_r^2\left(2\omega_r^2 - \omega^2 - \Omega^2\right)\right]
\end{aligned} \qquad (4.185)$$

it follows that

$$\begin{aligned}
\text{Re}(\Delta) &= c_R + t_R\omega^2 \\
\text{Im}(\Delta) &= c_I + t_I\omega^2
\end{aligned} \qquad (4.186)$$

which are linear functions in ω^2, with slopes given by

$$t_R = -\frac{1}{A_r^2 + B_r^2}\left[A_r\left(\omega_r^2 - \Omega^2\right) + B_r\eta_r\omega_r^2\right]$$

$$t_I = -\frac{1}{A_r^2 + B_r^2}\left[A_r\eta_r\omega_r^2 - B_r\left(\omega_r^2 - \Omega^2\right)\right]$$

(4.187)

By varying Ω around ω_r, with $\Omega \neq \omega$, we obtain a family of straight lines for Re(Δ) and Im(Δ). *A priori*, it is not easy to derive the modal parameters from these expressions. However, (4.187) represent straight lines in Ω^2, such that

$$t_R = d_R + u_R\Omega^2$$

$$t_I = d_I + u_I\Omega^2$$

(4.188)

where

$$d_R = -\frac{\left(A_r + B_r\eta_r\right)\omega_r^2}{A_r^2 + B_r^2} \qquad d_I = -\frac{\left(A_r\eta_r - B_r\right)\omega_r^2}{A_r^2 + B_r^2}$$

$$u_R = \frac{A_r}{A_r^2 + B_r^2} \qquad u_I = -\frac{B_r}{A_r^2 + B_r^2}$$

(4.189)

The modal parameters can be derived from (4.189). Figure 4.8 is a theoretical example of the type of plots we obtain in this method, where the data correspond to a real mode ($\phi_r = 0°$). If the mode is complex, the graphs on the right have a similar shape to the ones on the left.

Summarising the procedure: we have the display of Re(Δ) and Im(Δ) from the measurements (top graphs of figure 4.8), which constitute, theoretically, families of straight lines in ω^2, with slopes t_R and t_I. These parameters can, therefore, be computed by fitting Re(Δ) and Im(Δ) to straight lines. On the other hand, t_R, t_I are themselves, in theory, straight lines in Ω^2 and by a second straight-line fitting (bottom graphs of figure 4.8) it is possible to calculate the values on the left-hand sides of (4.189) and then, the modal parameters.

In addition, by comparing (4.189) with (4.180), it may be observed that these are the same, apart from a minus sign in all of them. This also means that expressions (4.187) are the same as the inverse of receptance (equations (4.178)), again apart from a minus sign.

Apparently, it seems that the modal parameters derived from Dobson's method should be the same as the ones derived from the Inverse method. However, when comparing the results of both methods, they are slightly different. This is due to two reasons. The first is because the Inverse method does not take into account the effect of other modes, while Dobson's method does. The second reason is because in the Inverse method the values of Re($1/\alpha_r$) and Im($1/\alpha_r$) are used directly from the measurement, while in Dobson's method the parameters t_R and t_I are already the result of a straight line fitting.

However, with noiseless theoretical data and for one single DOF, Dobson's method coincides with the Inverse method. For widely spaced modes the residual effect is not very large and the results from both methods are very similar.

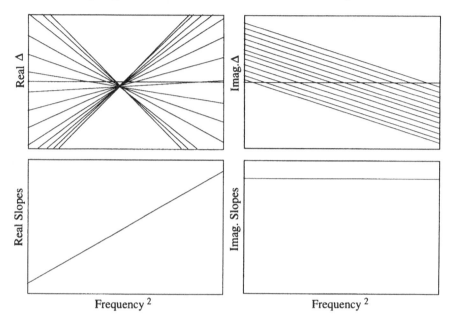

Frequency ^{2} Frequency ^{2}

Fig. 4.8 Example of analysis using Dobson's method (theoretical data).

4.4.2 Other SDOF methods
Global Dobson's and the Characteristic Response Function (CRF) methods

An extension of Dobson's method, converting it into a *SIMO* method, has been presented by Maia [134]. It was called Global Dobson's method.

Maia *et al* [135] have also developed the concept of Characteristic Response Function (CRF), another *SISO* method. It is based on the Inverse and Dobson's methods, but it goes further, obtaining a function that, besides being independent of the residual terms, is also independent of the modal constants, i.e., it depends only on the global parameters and that is the reason for its name.

No matter where the measurements are taken on the structure, the CRF should always provide the same results around the natural frequencies. It is primarily an indicator of the number of modes, as the CRF enhances them, but it can be used to identify the modal parameters as well.

Other extensions of this method have also been developed, namely the Advanced Characteristic Response Function (ACRF) [136], where the residual term is frequency dependent rather than constant, and a *SIMO* version of the CRF that also contemplates the consistency relationships (equations 1.228) among the modal constants [137].

Goyder's method

The method proposed by Goyder [138] is a variation of the GSH method that we shall refer to in Section 4.4.4, applied for the SDOF case. It considers hysteretic damping instead, and the extraction of the modal parameters for each mode is also based on the minimisation of an error function in a least-squares sense, but the linearisation of the error is made through a weighting function instead of by truncation of the Taylor expansion.

This method was also formulated for the case of one force input and several response locations, i.e., a global formulation but for one mode at a time. It is therefore a *SIMO* method. As with the GSH method, it also requires initial estimates and as the process progresses from mode to mode, the ones already identified are subtracted from the initial FRFs, in order to take out the influence of the residual terms. This process is repeated until convergence is attained.

Because this method is a particular case of a MDOF method, applied to analyse one DOF at a time, it is not considered as a true SDOF method. SDOF methods are usually designed specifically for one DOF and therefore have particular characteristics. Otherwise, all MDOF methods could be considered as SDOF ones. The same applies, for instance, to the method sometimes called the SDOF polynomial, which is the RFP method applied for one DOF [139]. In [139], the author applies the SDOF polynomial to several FRFs, one at a time, and calculates average values for the modal parameters as in [131]. The approach assumes, however, real modes.

4.4.3 Indirect MDOF methods in the frequency domain

Indirect MDOF methods include the Ewins-Gleeson, the Frequency Domain Prony (FDPM), the Complex Exponential Frequency Domain (CEFD), the Eigensystem Realisation Algorithm in the frequency domain (ERA-FD), the Rational Fraction Polynomial (RFP), the Global Rational Fraction Polynomial (GRFP), the Global method and the Polyreference Frequency Domain (PRFD).

The Ewins-Gleeson method

Introduced in 1982 [140], this *SISO* method is dedicated to the identification of lightly damped structures, assuming the hysteretic model. In [140], the problem is presented in terms of accelerance, but we shall derive it here for an FRF in terms of receptance.

Because the system is assumed to be lightly damped, the modes are taken as real, i.e., with phase angles of 0 or 180 degrees. The mathematical model is given by:

$$\alpha(\omega) = \sum_{r=1}^{N} \frac{C_r}{\omega_r^2 - \omega^2 + i\eta_r\omega_r^2} \qquad (4.190)$$

where C_r is therefore a real quantity.

In a first stage, it is considered that there is no damping. Hence,

$$\text{Re}(\alpha(\omega)) = \sum_{r=1}^{N} \frac{C_r}{\omega_r^2 - \omega^2} \tag{4.191}$$

If, in addition, the natural frequencies are taken as the frequencies corresponding to the peaks in the FRF receptance curve (which are supposed, in such a structure, to be well separated and sharp and, thus, easily visible), the only unknowns are the modal constants C_r.

To calculate the N unknowns C_r, it is necessary to have N equations. Thus, we take N frequencies Ω_1, Ω_2, ..., Ω_N and the corresponding real part of the measured receptances, which we write, shortly, as $\text{Re}(\tilde{\alpha}_1)$, $\text{Re}(\tilde{\alpha}_2)$, ..., $\text{Re}(\tilde{\alpha}_N)$. Thus, the modal constants are given by:

$$\begin{Bmatrix} C_1 \\ C_2 \\ \vdots \\ C_N \end{Bmatrix} = \begin{bmatrix} \dfrac{1}{\omega_1^2 - \Omega_1^2} & \dfrac{1}{\omega_2^2 - \Omega_1^2} & \cdots & \dfrac{1}{\omega_N^2 - \Omega_1^2} \\ \dfrac{1}{\omega_1^2 - \Omega_2^2} & \dfrac{1}{\omega_2^2 - \Omega_2^2} & \cdots & \dfrac{1}{\omega_N^2 - \Omega_2^2} \\ \vdots & \vdots & \cdots & \vdots \\ \dfrac{1}{\omega_1^2 - \Omega_N^2} & \dfrac{1}{\omega_2^2 - \Omega_N^2} & \cdots & \dfrac{1}{\omega_N^2 - \Omega_N^2} \end{bmatrix}^{-1} \begin{Bmatrix} \text{Re}(\tilde{\alpha}_1) \\ \text{Re}(\tilde{\alpha}_2) \\ \vdots \\ \text{Re}(\tilde{\alpha}_N) \end{Bmatrix} \tag{4.192}$$

or, in short form,

$$\{C\} = [R]\ \text{Re}\{\tilde{\alpha}\} \tag{4.193}$$
$$\text{(Nx1)} \quad \text{(NxN)} \quad \text{(Nx1)}$$

The sign of each C_r tells us whether a response is in-phase or out-of-phase. In a second stage, the damping factors are calculated from the peak amplitudes. From (4.190), at each resonance, we have

$$\left| \tilde{\alpha}(\omega_r) \right| \approx \frac{|C_r|}{\eta_r \omega_r^2} \tag{4.194}$$

from which,

$$\eta_r \approx \frac{|C_r|}{\left| \tilde{\alpha}(\omega_r) \right| \omega_r^2} \tag{4.195}$$

The effects of residual modes can be introduced later and details are given in [140]. An alternative to this is to consider N+2 points and calculate N+2 modal constants. The matrix [R] in (4.193) would be of order N+2.

This method works very well if in fact the structure is lightly damped. The method is extremely simple and fast, and very easy to incorporate in small computers. The disadvantage is that it is sensitive to the points chosen from the FRF, mainly with noisy measurements. Some experience is required to make the proper choice of points. To avoid this drawback, a different approach was developed by Maia and Ewins [141].

The Complex Exponential Frequency Domain method (CEFD)
This method, developed by Schmerr [142], is the corresponding frequency domain version of the SSTD, which in turn is, as mentioned previously in Section 4.3.3, the ITD method for the particular case of only one output measurement (perhaps it should more appropriately be called SSFD). The CEFD is therefore, like the SSTD, a *SISO* method. From expression (4.4), and considering only one input and one output location, the IRF is given by:

$$h(t) = \sum_{r=1}^{2N} A'_r e^{s_r t} \tag{4.196}$$

Applying a Fourier transform, the FRF is

$$\alpha(\omega) = \sum_{r=1}^{2N} \frac{A'_r}{i\omega - s_r} \tag{4.197}$$

Considering a series of N shifted responses from the same IRF,

$$h_k(t) = h(t + k\,T) \qquad k = 1, 2, ..., N \tag{4.198}$$

where T is an arbitrary time lag. The corresponding FRF is given by

$$\alpha_k(\omega) = \sum_{r=1}^{2N} \frac{A'_r e^{s_r kT}}{i\omega - s_r} \qquad k = 1, 2, ..., N \tag{4.199}$$

Expanding (4.199) for all k values, we obtain

$$\{\alpha(\omega)\}_{(N\times 1)} = \begin{bmatrix} A'_1 e^{s_1 T} & A'_2 e^{s_2 T} & ... & A'_{2N} e^{s_{2N} T} \\ A'_1 e^{s_1 2T} & A'_2 e^{s_2 2T} & ... & A'_{2N} e^{s_{2N} 2T} \\ \vdots & \vdots & ... & \vdots \\ A'_1 e^{s_1 NT} & A'_2 e^{s_2 NT} & ... & A'_{2N} e^{s_{2N} NT} \end{bmatrix}_{(N\times 2N)} \left\{ \begin{array}{c} \dfrac{1}{i\omega - s_1} \\ \dfrac{1}{i\omega - s_2} \\ \vdots \\ \dfrac{1}{i\omega - s_{2N}} \end{array} \right\}_{(2N\times 1)} \tag{4.200}$$

If each shifted sample contains L frequency points, then

$$\left[\{\alpha(\omega_1)\}\{\alpha(\omega_2)\}...\{\alpha(\omega_L)\}\right] =$$

$$
\begin{bmatrix}
A_1' e^{s_1 T} & A_2' e^{s_2 T} & \cdots & A_{2N}' e^{s_{2N} T} \\
A_1' e^{s_1 2T} & A_2' e^{s_2 2T} & \cdots & A_{2N}' e^{s_{2N} 2T} \\
\vdots & \vdots & \cdots & \vdots \\
A_1' e^{s_1 NT} & A_2' e^{s_2 NT} & \cdots & A_{2N}' e^{s_{2N} NT}
\end{bmatrix}
\begin{bmatrix}
\dfrac{1}{i\omega_1 - s_1} & \dfrac{1}{i\omega_2 - s_1} & \cdots & \dfrac{1}{i\omega_L - s_1} \\
\dfrac{1}{i\omega_1 - s_2} & \dfrac{1}{i\omega_2 - s_2} & \cdots & \dfrac{1}{i\omega_L - s_2} \\
\vdots & \vdots & \cdots & \vdots \\
\dfrac{1}{i\omega_1 - s_{2N}} & \dfrac{1}{i\omega_2 - s_{2N}} & \cdots & \dfrac{1}{i\omega_L - s_{2N}}
\end{bmatrix}
$$

$$\text{(4.201)}$$

or

$$\underset{\text{(NxL)}}{[\alpha]} = \underset{\text{(Nx2N)}}{[A_T]} \; \underset{\text{(2NxL)}}{[\Lambda]} \qquad\qquad\qquad \text{(4.202)}$$

If we repeat this whole procedure taking the responses shifted one interval of time Δt, (4.199) becomes:

$$\alpha_k'(\omega) = \sum_{r=1}^{2N} \frac{A_r' e^{s_r kT} e^{s_r \Delta t}}{i\omega - s_r} \qquad k = 1, 2, ..., N \qquad\qquad \text{(4.203)}$$

Shifting two time intervals, we obtain

$$\alpha_k''(\omega) = \sum_{r=1}^{2N} \frac{A_r' e^{s_r kT} e^{s_r 2\Delta t}}{i\omega - s_r} \qquad k = 1, 2, ..., N \qquad\qquad \text{(4.204)}$$

For the series of N shifted responses and for the L frequency points, the following matrix corresponding to (4.203) is obtained:

$$\underset{\text{(NxL)}}{[\alpha']} = \underset{\text{(Nx2N)}}{[A_T]}
\underset{\text{(2Nx2N)}}{\begin{bmatrix}
e^{s_1 \Delta t} & 0 & \cdots & 0 \\
0 & e^{s_2 \Delta t} & \cdots & 0 \\
\vdots & \vdots & \ddots & \vdots \\
0 & 0 & \cdots & e^{s_{2N} \Delta t}
\end{bmatrix}}
\underset{\text{(2NxL)}}{[\Lambda]} = [A_T'][\Lambda] \qquad \text{(4.205)}$$

Similarly, for (4.204),

$$[\alpha''] = [A_T]
\begin{bmatrix}
e^{s_1 2\Delta t} & 0 & \cdots & 0 \\
0 & e^{s_2 2\Delta t} & \cdots & 0 \\
\vdots & \vdots & \ddots & \vdots \\
0 & 0 & \cdots & e^{s_{2N} 2\Delta t}
\end{bmatrix}
[\Lambda] = [A_T''][\Lambda] \qquad \text{(4.206)}$$

Combining (4.202) and (4.205), and assuming L=2N, to simplify the exposition, we can write

$$\begin{bmatrix} [\alpha] \\ [\alpha'] \end{bmatrix}_{(2N \times 2N)} = \begin{bmatrix} [A_T] \\ [A_T'] \end{bmatrix}_{(2N \times 2N)} [\Lambda]_{(2N \times 2N)} \qquad \text{or} \qquad [\alpha_A]_{(2N \times 2N)} = [A_A]_{(2N \times 2N)} [\Lambda]_{(2N \times 2N)} \qquad (4.207)$$

Combining (4.205) and (4.206), it follows that

$$\begin{bmatrix} [\alpha'] \\ [\alpha''] \end{bmatrix}_{(2N \times 2N)} = \begin{bmatrix} [A_T'] \\ [A_T''] \end{bmatrix}_{(2N \times 2N)} [\Lambda]_{(2N \times 2N)} \qquad \text{or} \qquad [\hat{\alpha}_A]_{(2N \times 2N)} = [\hat{A}_A]_{(2N \times 2N)} [\Lambda]_{(2N \times 2N)} \qquad (4.208)$$

From (4.207),

$$[A_A]^{-1}[\alpha_A] = [\Lambda] \qquad (4.209)$$

Substitution of (4.209) in (4.208) leads to

$$[\hat{\alpha}_A] = [\hat{A}_A][A_A]^{-1}[\alpha_A] \qquad (4.210)$$

Post-multiplying by $[\alpha_A]^{-1}$, it follows that

$$[A_S] = [\hat{\alpha}_A][\alpha_A]^{-1} = [\hat{A}_A][A_A]^{-1} \qquad (4.211)$$

where $[A_S]$ is known as the system matrix. Post-multiplying (4.211) by $[A_A]$, gives

$$[A_S][A_A] = [\hat{A}_A] \qquad (4.212)$$

But each column r of $[\hat{A}_A]$ is related to the corresponding one of matrix $[A_A]$ by

$$\{\hat{A}_A\}_r = \{A_A\}_r e^{s_r \Delta t} \qquad r = 1, 2, ..., 2N \qquad (4.213)$$

Thus, one has the following eigenproblem to solve:

$$[[A_S] - e^{s_r \Delta t}[I]]\{A_A\}_r = \{0\} \qquad (4.214)$$

from which the modal parameters can be calculated. The calculation of mass-normalised mode shapes is not a problem in this case. A *SIMO* version of this method could also be developed and this would be the equivalent to the ITD method. Even as a *SISO* method, we can easily establish a parallel with the ITD method.

The Rational Fraction Polynomial method (RFP)

This is a *SISO* method that first appeared in 1982 [143] and is now one of the most popular and widely used MDOF frequency domain methods, employed by many commercial packages of modal analysis software. The formulation of the FRF is expressed in rational fraction form instead of the partial fraction form, and the error function to be minimised is established in such a way that the resulting system of equations is linear, without requiring initial estimates for the modal parameters. Because the resulting linear system of equations involves matrices that are ill-conditioned, the rational fraction form of the FRF is expressed in terms of orthogonal Forsythe polynomials. For this reason, this method is also known as the Rational Fraction Orthogonal Polynomial method (RFOP).

The FRF, in terms of receptance, for a linear system with N DOF and viscous damping can be given by the partial fraction form:

$$\alpha(\omega) = \sum_{r=1}^{N} \frac{A_r + i\omega B_r}{\omega_r^2 - \omega^2 + i2\xi_r\omega_r\omega} \tag{4.215}$$

where A_r and B_r are constants. It is also easy to show that expression (4.215) can be written as a ratio of two polynomials in $i\omega$:

$$\alpha(\omega) = \frac{\displaystyle\sum_{k=0}^{2N-1} a_k(i\omega)^k}{\displaystyle\sum_{k=0}^{2N} b_k(i\omega)^k} \tag{4.216}$$

This is the rational fraction form of $\alpha(\omega)$. An error function between the analytical FRF $\alpha(\omega)$ and the experimental values $\tilde{\alpha}(\omega)$ at each frequency ω_j, is defined as:

$$e_j = \frac{\displaystyle\sum_{k=0}^{2N-1} a_k(i\omega_j)^k}{\displaystyle\sum_{k=0}^{2N} b_k(i\omega_j)^k} - \tilde{\alpha}(\omega_j) \tag{4.217}$$

Working with a modified error function $e_j' = e_j \displaystyle\sum_{k=0}^{2N} b_k(i\omega_j)^k$ and making $b_{2N} = 1$,

$$e'_j = \sum_{k=0}^{2N-1} a_k (i\omega_j)^k - \tilde{\alpha}(\omega_j) \left[\sum_{k=0}^{2N-1} b_k (i\omega_j)^k + (i\omega_j)^{2N} \right] \tag{4.218}$$

This formulation leads to a linear system of equations, avoiding the necessity of initial estimates for the modal parameters. Let us define an error vector including all the L measured frequencies:

$$\{E\} = \begin{Bmatrix} e'_1 \\ e'_2 \\ \vdots \\ e'_L \end{Bmatrix} \tag{4.219}$$

Equation (4.218) becomes, in full,

$$\{E\} = \begin{bmatrix} 1 & (i\omega_1) & (i\omega_1)^2 & \cdots & (i\omega_1)^{2N-1} \\ 1 & (i\omega_2) & (i\omega_2)^2 & \cdots & (i\omega_2)^{2N-1} \\ \vdots & \vdots & \vdots & & \vdots \\ 1 & (i\omega_L) & (i\omega_L)^2 & \cdots & (i\omega_L)^{2N-1} \end{bmatrix} \begin{Bmatrix} a_0 \\ a_1 \\ \vdots \\ a_{2N-1} \end{Bmatrix} -$$

$$\begin{bmatrix} \tilde{\alpha}(\omega_1) & \tilde{\alpha}(\omega_1)(i\omega_1) & \cdots & \tilde{\alpha}(\omega_1)(i\omega_1)^{2N-1} \\ \tilde{\alpha}(\omega_2) & \tilde{\alpha}(\omega_2)(i\omega_2) & \cdots & \tilde{\alpha}(\omega_2)(i\omega_2)^{2N-1} \\ \vdots & \vdots & & \vdots \\ \tilde{\alpha}(\omega_L) & \tilde{\alpha}(\omega_L)(i\omega_L) & \cdots & \tilde{\alpha}(\omega_L)(i\omega_L)^{2N-1} \end{bmatrix} \begin{Bmatrix} b_0 \\ b_1 \\ \vdots \\ b_{2N-1} \end{Bmatrix} - \begin{Bmatrix} \tilde{\alpha}(\omega_1)(i\omega_1)^{2N} \\ \tilde{\alpha}(\omega_2)(i\omega_2)^{2N} \\ \vdots \\ \tilde{\alpha}(\omega_L)(i\omega_L)^{2N} \end{Bmatrix}$$

$$\tag{4.220}$$

or, in a more compact form:

$$\underset{(L \times 1)}{\{E\}} = \underset{(L \times 2N)}{[P]} \underset{(2N \times 1)}{\{a\}} - \underset{(L \times 2N)}{[T]} \underset{(2N \times 1)}{\{b\}} - \underset{(L \times 1)}{\{W\}} \tag{4.221}$$

To calculate the polynomial coefficients {a} and {b} a least-squares procedure is used, minimising a squared error function J, defined as

$$J = \{E*\}^T \{E\} \tag{4.222}$$

where * denotes complex conjugate. Substituting (4.221) in (4.222), and after some manipulations, we obtain

$$J = \{a\}^T \, \text{Re}\big([P*]^T[P]\big)\{a\} + \{b\}^T \, \text{Re}\big([T*]^T[T]\big)\{b\} +$$

$$\{W*\}^T\{W\} - 2\{a\}^T \, \text{Re}\big([P*]^T[T]\big)\{b\} - \qquad (4.223)$$

$$2\{a\}^T \, \text{Re}\big([P*]^T\{W\}\big) + 2\{b\}^T \, \text{Re}\big([T*]^T\{W\}\big)$$

J is minimised taking its derivatives with respect to $\{a\}$ and $\{b\}$ and equalling them to zero:

$$\text{Re}\big([P*]^T[P]\big)\{a\} - \text{Re}\big([P*]^T[T]\big)\{b\} - \text{Re}\big([P*]^T\{W\}\big) = \{0\}$$

$$\text{Re}\big([T*]^T[T]\big)\{b\} - \text{Re}\big([T*]^T[P]\big)\{a\} + \text{Re}\big([T*]^T\{W\}\big) = \{0\} \qquad (4.224)$$

or, in matrix form:

$$\begin{bmatrix} [Y] & [X] \\ [X]^T & [Z] \end{bmatrix} \begin{Bmatrix} \{a\} \\ \{b\} \end{Bmatrix} = \begin{Bmatrix} \{G\} \\ \{F\} \end{Bmatrix} \qquad (4.225)$$

where

$$
\begin{aligned}
[Y] &= \text{Re}\big([P*]^T[P]\big) \\
[X] &= -\text{Re}\big([P*]^T[T]\big) \\
[Z] &= \text{Re}\big([T*]^T[T]\big) \qquad\qquad (4.226) \\
\{G\} &= \text{Re}\big([P*]^T\{W\}\big) \\
\{F\} &= -\text{Re}\big([T*]^T\{W\}\big)
\end{aligned}
$$

Expression (4.225) contains the normal equations of the least-squares problem. It was found [143] that the numerical problems (ill-conditioning) encountered in the resolution of (4.225) could be overcome if matrices [Y] and [Z] were each an identity matrix. So, instead of $[Y] = \text{Re}([P*]^T[P])$ and $[Z] = \text{Re}([T*]^T[T])$, these matrices will have to be the product of two orthonormal matrices. So, [P] must be replaced by

$$[P] = \begin{bmatrix} \varphi_{1,0} & \varphi_{1,1} & \cdots & \varphi_{1,2N-1} \\ \varphi_{2,0} & \varphi_{2,1} & \cdots & \varphi_{2,2N-1} \\ \vdots & \vdots & & \vdots \\ \varphi_{L,0} & \varphi_{L,1} & \cdots & \varphi_{L,2N-1} \end{bmatrix} = [\vartheta] \qquad (4.227)$$

where $\varphi_{j,i}$ means a polynomial of order i evaluated at frequency ω_j. Likewise, [T] will now be given by

$$[T] = \begin{bmatrix} \tilde{\alpha}(\omega_1)\theta_{1,0} & \tilde{\alpha}(\omega_1)\theta_{1,1} & \cdots & \tilde{\alpha}(\omega_1)\theta_{1,2N-1} \\ \tilde{\alpha}(\omega_2)\theta_{2,0} & \tilde{\alpha}(\omega_2)\theta_{2,1} & \cdots & \tilde{\alpha}(\omega_2)\theta_{2,2N-1} \\ \vdots & \vdots & & \vdots \\ \tilde{\alpha}(\omega_L)\theta_{L,0} & \tilde{\alpha}(\omega_L)\theta_{L,1} & \cdots & \tilde{\alpha}(\omega_L)\theta_{L,2N-1} \end{bmatrix} \quad (4.228)$$

or

$$[T] = \begin{bmatrix} \tilde{\alpha}(\omega_1) & & & \\ & \tilde{\alpha}(\omega_2) & & 0 \\ & & \ddots & \\ 0 & & & \tilde{\alpha}(\omega_L) \end{bmatrix} \begin{bmatrix} \theta_{1,0} & \theta_{1,1} & \cdots & \theta_{1,2N-1} \\ \theta_{2,0} & \theta_{2,1} & \cdots & \theta_{2,2N-1} \\ \vdots & \vdots & & \vdots \\ \theta_{L,0} & \theta_{L,1} & \cdots & \theta_{L,2N-1} \end{bmatrix} \quad (4.229)$$

$$= \begin{bmatrix} \ddots & \tilde{\alpha} & \ddots \end{bmatrix}[\Theta]$$

and

$$\{W\} = \begin{Bmatrix} \tilde{\alpha}(\omega_1)\theta_{1,2N} \\ \tilde{\alpha}(\omega_2)\theta_{2,2N} \\ \vdots \\ \tilde{\alpha}(\omega_L)\theta_{L,2N} \end{Bmatrix} \quad (4.230)$$

So, it is necessary to find the complex functions φ and θ such that $\mathrm{Re}([P^*]^T[P])$ and $\mathrm{Re}([T^*]^T[T])$ are unit matrices. This is possible if φ and θ are complex orthonormal polynomials (see Appendix B).

The FRF is then written in terms of these polynomials, with c_k and d_k as the unknowns, instead of a_k and b_k:

$$\alpha(\omega) = \frac{\displaystyle\sum_{k=0}^{2N-1} c_k \varphi_k}{\displaystyle\sum_{k=0}^{2N} d_k \theta_k} \quad (4.231)$$

After finding $\{c\}$ and $\{d\}$ (populated by coefficients c_k and d_k respectively), $\{a\}$ and $\{b\}$ can be recovered to calculate the modal parameters.

Calculation of orthogonal polynomials φ and θ

As explained before, it is necessary to have

$$\text{Re}\left([\vartheta *]^T[\vartheta]\right) = [I]$$

$$\text{Re}\left([\Theta *]^T\left[\tilde{\alpha} *_\searrow\right]\left[\tilde{\alpha}_\searrow\right][\Theta]\right) = [I] \tag{4.232}$$

According to Appendix B, this means that polynomials φ must be calculated using a unit weighting function and θ must be calculated using a weighting function given by $|\tilde{\alpha}^2|$. Going back to (4.216) and developing it for a 2 DOF case (for example), it follows that

$$\alpha(\omega) = \frac{a_0 + a_1(i\omega) + a_2(i\omega)^2 + a_3(i\omega)^3}{b_0 + b_1(i\omega) + b_2(i\omega)^2 + b_3(i\omega)^3 + b_4(i\omega)^4} \tag{4.233}$$

For this case it is clear that each polynomial can be regarded as a linear combination of real and imaginary orthogonal polynomials, the real ones being even functions and the imaginary ones odd functions, as follows (for θ the procedure is similar):

Real (even)   Imaginary (odd)

$\varphi_0 = a_0'$ $\varphi_1 = a_1'(i\omega)$

$\varphi_2 = a_2' + a_3'(i\omega)^2$ $\varphi_3 = a_4'(i\omega) + a_5'(i\omega)^3$

$\varphi_4 = a_6' + a_7'(i\omega)^2 + a_8'(i\omega)^4$ $\varphi_5 = a_9'(i\omega) + a_{10}'(i\omega)^3 + a_{11}'(i\omega)^5$

$\vdots$ $\vdots$

$$\tag{4.234}$$

So, the numerator of (4.233) will be:

$$a_0 + a_1(i\omega) + a_2(i\omega)^2 + a_3(i\omega)^3$$

$$= a_0' + a_1'(i\omega) + a_2' + a_3'(i\omega)^2 + a_4'(i\omega) + a_5'(i\omega)^3 \tag{4.235}$$

$$= a_0' + a_2' + \left(a_1' + a_4'\right)(i\omega) + a_3'(i\omega)^2 + a_5'(i\omega)^3$$

This definition for the polynomials is not unique, but defining them as even and odd functions, besides being directly related to the 'Hermitian nature' of an FRF, simplifies the calculations, as will be shown later.

Once we are talking about even and odd functions, and if we want somehow to take advantage of them, it is necessary to consider both positive and negative frequencies. If there are L points to be fitted, the orthogonal property will be:

$$\text{Re}\left(\sum_{j=-L}^{L} \varphi_{j,k}^* \, \varphi_{j,i}\right) = \begin{cases} 0 & k \neq i \\ 1 & k = i \end{cases} \tag{4.236}$$

Writing φ as $\text{Re}(\varphi) + i\text{Im}(\varphi)$, it follows that

$$\sum_{j=-L}^{L}\left(\text{Re}\big(\varphi_{j,k}\big)\text{Re}\big(\varphi_{j,i}\big) + \text{Im}\big(\varphi_{j,k}\big)\text{Im}\big(\varphi_{j,i}\big)\right) = \begin{cases} 0 & k \neq i \\ 1 & k = i \end{cases} \tag{4.237}$$

In fact, the FRF is only defined for positive frequencies and so (4.237) must be transformed so that only the positive functions for points 1 to L are considered. Extracting the half-functions, negative and positive, from (4.237) we obtain:

$$\sum_{j=-L}^{-1}\left(\text{Re}\big(\varphi_{j,k}^-\big)\text{Re}\big(\varphi_{j,i}^-\big) + \text{Im}\big(\varphi_{j,k}^-\big)\text{Im}\big(\varphi_{j,i}^-\big)\right) +$$
$$\sum_{j=1}^{L}\left(\text{Re}\big(\varphi_{j,k}^+\big)\text{Re}\big(\varphi_{j,i}^+\big) + \text{Im}\big(\varphi_{j,k}^+\big)\text{Im}\big(\varphi_{j,i}^+\big)\right) = \begin{cases} 0 & k \neq i \\ 1 & k = i \end{cases} \tag{4.238}$$

According to the definition of even and odd functions, $\text{Re}(\varphi^-) = \text{Re}(\varphi^+)$ and $\text{Im}(\varphi^-) = -\text{Im}(\varphi^+)$, and therefore,

$$2\sum_{j=1}^{L}\left(\text{Re}\big(\varphi_{j,k}^+\big)\text{Re}\big(\varphi_{j,i}^+\big) + \text{Im}\big(\varphi_{j,k}^+\big)\text{Im}\big(\varphi_{j,i}^+\big)\right) = \begin{cases} 0 & k \neq i \\ 1 & k = i \end{cases} \tag{4.239}$$

or

$$\text{Re}\left(\sum_{j=1}^{L}\big(\varphi_{j,k}^+\big)^*\big(\varphi_{j,i}^+\big)\right) = \begin{cases} 0 & k \neq i \\ 0.5 & k = i \end{cases} \tag{4.240}$$

For polynomials θ^+ (corresponding to polynomials θ_k in (4.231)) everything is similar but, as seen before, there will be a weighting function equal to $|\tilde{\alpha}(\omega_j)|^2$. Hence,

$$\text{Re}\left(\sum_{j=1}^{L}\big(\theta_{j,k}^+\big)^*\big(\theta_{j,i}^+\big)\big|\tilde{\alpha}(\omega_j)\big|^2\right) = \begin{cases} 0 & k \neq i \\ 0.5 & k = i \end{cases} \tag{4.241}$$

The problem now is to generate φ^+ and θ^+ automatically so that they satisfy (4.240) and (4.241). Out of many possible types of orthogonal polynomials [144], the Forsythe ones are chosen due to their computational advantages. The elementary theory of orthogonal polynomials is well known and can be found in many textbooks (e.g., [145]). For the Forsythe polynomials specifically, we suggest [146-148]. The Forsythe recursion *formulæ* given in [147] and [148] can be used very conveniently as they are easy to program. Firstly, the *formulæ* for the general case (not only for positive half-functions) will be presented, using a common notation of γ for either φ or θ. Let q_j be the weighting function and m the degree of the polynomial. The recursion *formulæ* are given by:

$$\gamma_0(\omega) = 1$$
$$\gamma_1(\omega) = (\omega - u_1)\gamma_0(\omega)$$
$$\gamma_2(\omega) = (\omega - u_2)\gamma_1(\omega) - v_1\gamma_0(\omega)$$
$$\vdots \qquad\qquad \vdots \qquad\qquad\qquad (4.242)$$
$$\gamma_k(\omega) = (\omega - u_k)\gamma_{k-1}(\omega) - v_{k-1}\gamma_{k-2}(\omega)$$
$$\vdots \qquad\qquad \vdots$$
$$\gamma_m(\omega) = (\omega - u_m)\gamma_{m-1}(\omega) - v_{m-1}\gamma_{m-2}(\omega)$$

where

$$u_k = \frac{\displaystyle\sum_{j=1}^{L} \omega_j \left(\gamma_{k-1}(\omega_j)\right)^2 q_j}{D_{k-1}} \qquad\qquad (4.243\ a)$$

$$v_k = \frac{\displaystyle\sum_{j=1}^{L} \omega_j\, \gamma_k(\omega_j)\gamma_{k-1}(\omega_j)q_j}{D_{k-1}} \qquad\qquad (4.243\ b)$$

$$D_k = \sum_{j=1}^{L} \left(\gamma_k(\omega_j)\right)^2 q_j \qquad\qquad (4.243\ c)$$

If, in (4.243), the summation is extended from -L to +L, it is easily recognised that u_k is an odd function and v_k an even function of ω and so u_k will be zero and the summation in v_k and D_k will be $2\sum_{j=1}^{L}$ instead of $\sum_{j=-L}^{L}$. Then, expressions (4.242) become:

$$\gamma_0^+(\omega) = 1$$
$$\gamma_1^+(\omega) = \omega\,\gamma_0^+(\omega)$$
$$\gamma_2^+(\omega) = \omega\,\gamma_1^+(\omega) - v_1^+\gamma_0^+(\omega)$$
$$\vdots \qquad\qquad \vdots \qquad\qquad\qquad (4.244)$$
$$\gamma_k^+(\omega) = \omega\,\gamma_{k-1}^+(\omega) - v_{k-1}^+\gamma_{k-2}^+(\omega)$$
$$\vdots \qquad\qquad \vdots$$
$$\gamma_m^+(\omega) = \omega\,\gamma_{m-1}^+(\omega) - v_{m-1}^+\gamma_{m-2}^+(\omega)$$

with

$$v_k^+ = \frac{2\sum_{j=1}^{L} \omega_j \gamma_k^+(\omega_j)\gamma_{k-1}^+(\omega_j)q_j}{D_{k-1}^+} \tag{4.245 a}$$

$$D_k^+ = 2\sum_{j=1}^{L} \left(\gamma_k^+(\omega_j)\right)^2 q_j \tag{4.245 b}$$

This represents a major simplification (given in [143]) and is the reason why the half-functions, defined for positive and negative frequencies, have been considered.

After calculating all the polynomials γ_k^+, they must be normalised by dividing by $\sqrt{D_k^+}$, and to obtain the complex polynomials, they must be multiplied by i^k. To multiply the resulting polynomials γ_k^+ by i^k produces the same result as calculating them for $i\omega$, instead of ω, in (4.244) and (4.245). To avoid further numerical problems, the frequency range of interest is scaled by dividing all the frequencies by their maximum value in the considered interval, so that the maximum frequency value is 1.

Applying (4.244), polynomials $\varphi^+(\omega_j)$ and $\theta^+(\omega_j)$ can be calculated by making $q_j = 1$ or $q_j = |\bar{\alpha}(\omega_j)|^2$, respectively. Knowing $\varphi^+(\omega_j)$ and $\theta^+(\omega_j)$, it is possible to calculate the new [P] and [T] matrices, given by (4.227) and (4.229), where φ and θ must be understood as φ^+ and θ^+. Now, (4.225) will look like

$$\begin{bmatrix} \begin{bmatrix} `0.5` \end{bmatrix} & [X] \\ [X]^T & \begin{bmatrix} `0.5` \end{bmatrix} \end{bmatrix} \begin{Bmatrix} \{c\} \\ \{d\} \end{Bmatrix} = \begin{Bmatrix} \{G\} \\ \{0\} \end{Bmatrix} \tag{4.246}$$

with [X] and {G} defined as before in (4.226), but with [P] and [T] now in terms of φ^+ and θ^+. From (4.241) it is obvious that the former vector $\{F\} = -\operatorname{Re}([T^*]^T\{W\})$ is now a zero vector. Multiplying both sides of (4.246) by 2, we obtain

$$\begin{bmatrix} [I] & [X'] \\ [X']^T & [I] \end{bmatrix} \begin{Bmatrix} \{c\} \\ \{d\} \end{Bmatrix} = \begin{Bmatrix} \{G'\} \\ \{0\} \end{Bmatrix} \tag{4.247}$$

where $[X'] = 2[X]$ and $\{G'\} = 2\{G\}$. Solving (4.247), it follows that

$$\{d\} = -\left[[I] - [X']^T[X']\right]^{-1}[X']^T\{G'\} \tag{4.248 a}$$

$$\{c\} = \{G'\} - [X']\{d\} \tag{4.248 b}$$

Calculation of the modal parameters

We may now return to (4.231) (where φ and θ must be replaced by φ^+ and θ^+), with the FRF expressed in terms of the coefficients $\{c\}$ and $\{d\}$ and orthogonal polynomials φ^+ and θ^+.

In order to calculate the modal parameters, (4.231) must be rewritten in terms of coefficients $\{a\}$ and $\{b\}$, as in (4.216). If the coefficients of polynomials φ^+ and θ^+ have been stored, it is possible to find linear transformation relationships between $\{a\}$ and $\{c\}$ and between $\{b\}$ and $\{d\}$ in the following form:

$$\underset{(2Nx1)}{\{a\}} = \underset{(2Nx2N)}{\left[T_{ac}\right]} \underset{(2Nx1)}{\{c\}} \tag{4.249 a}$$

$$\underset{(2Nx1)}{\{b\}} = \underset{(2Nx2N)}{\left[T_{bd}\right]} \underset{(2Nx1)}{\{d\}} + \underset{(2Nx1)}{\{R\}} \tag{4.249 b}$$

Knowing $\{b\}$, these values can be introduced into a complex polynomial solver routine to calculate the roots, which are directly related to the resonant frequencies and damping ratios. With these results, the other modal parameters can be calculated, relating expressions (4.215) and (4.216).

A variation of this method, the Complex Orthogonal Polynomial Functions method (COPF), was described in [149], but the theory was not presented. Another variation was proposed in [150], using Chebyshev polynomials instead of the Forsythe ones, with the main objective of obtaining a faster algorithm.

The Global Rational Fraction Polynomial method (GRFP)
This method is an extension of the RFP method to analyse globally a set of FRFs, using one single input reference. It is, therefore, a *SIMO* method. This extension of the RFP, already mentioned in [143], was developed further in [151] and [152].

If other FRFs measured on the same structure are also considered, one might expect (theoretically) exactly the same results for the resonant frequencies and damping ratios, as these are characteristic properties of the structure, although different modal constants and phase angles. The former are called 'global' properties and the latter 'local' properties. Using several FRFs to calculate the global properties implies overspecification of the number of equations in relation to the number of unknowns, and so we shall make use again of a least-squares procedure. Knowing the global properties, the local ones can be calculated from each FRF.

As seen in the RFP method, coefficients b_k (expression (4.216)) are the ones which yield the global properties of the system. These are related to coefficients d_k (expression (4.231)) through the relationship (4.249 b), where $[T_{bd}]$ and $\{R\}$ are direct functions of the orthogonal polynomials coefficients. Taking $\{d\}$ from (4.249 b) and substituting in (4.248 a), results in:

$$\left[[I] - [X']^T[X']\right]\left[T_{bd}\right]^{-1}\left\{\{b\} - \{R\}\right\} = -[X']^T\{G'\} \tag{4.250}$$

or

$$[U_G]\{b\} = \{V_G\} \qquad (4.251)$$

where

$$[U_G] = \left[[I] - [X']^T[X']\right][T_{bd}]^{-1} \qquad (4.252\ a)$$
$$\text{(2Nx2N)}$$

$$\{V_G\} = [U_G]\{R\} - [X']^T\{G'\} \qquad (4.252\ b)$$
$$\text{(2Nx1)}$$

For each FRF there is an equation of the form of (4.251). Considering a total of p FRFs,

$$\begin{bmatrix} [U_G]_1 \\ [U_G]_2 \\ \vdots \\ [U_G]_p \end{bmatrix} \underset{\text{(2Nx1)}}{\{b\}} = \begin{Bmatrix} \{V_G\}_1 \\ \{V_G\}_2 \\ \vdots \\ \{V_G\}_p \end{Bmatrix} \qquad (4.253)$$
$$\text{((2Nxp)x2N)} \qquad \text{((2Nxp)x1)}$$

or

$$[U_T]\{b\} = \{V_T\} \qquad (4.254)$$

Solving (4.254) in a least-squares sense gives

$$\{b\} = \left([U_T]^T[U_T]\right)^{-1}[U_T]^T\{V_T\} \qquad (4.255)$$

With $\{b\}$, a polynomial solver can be used to find the roots and thence to calculate the natural frequencies and damping factors.

Calculation of the local properties

To calculate $\{a\}$ (vector of coefficients a_k), there are two options: either to use (4.249 a) with $\{c\}$ (vector of coefficients c_k) given by (4.248 b), using $\{d\}$ calculated for each FRF, or, once the resonant frequencies and damping ratios have been calculated, to curve-fit *each* FRF again but now only to calculate $\{a\}$ and then the modal constants.

The second option seems to be much more rational and consistent with the global curve-fitting philosophy, and will be explained next. Let us rewrite (4.216), for each measured frequency, as

$$\alpha(\omega_j) = \sum_{k=0}^{2N-1} \left(\frac{(i\,\omega_j)^k}{\sum\limits_{k=0}^{2N} b_k (i\,\omega_j)^k} \right) a_k \tag{4.256}$$

or simply

$$\alpha(\omega_j) = \sum_{k=0}^{2N-1} t_{jk}\, a_k \tag{4.257}$$

The error between the analytical and experimental values at each frequency ω_j, is

$$e_j = \alpha(\omega_j) - \tilde{\alpha}(\omega_j) = \sum_{k=0}^{2N-1} t_{jk}\, a_k - \tilde{\alpha}(\omega_j) \tag{4.258}$$

For all L measured frequencies, the error vector is given by:

$$\{E\}_{(Lx1)} = [T_G]_{(Lx2N)} \{a\}_{(2Nx1)} - \{\tilde{\alpha}\}_{(Lx1)} \tag{4.259}$$

where

$$[T_G] = \begin{bmatrix} t_{1,0} & t_{1,1} & \cdots & t_{1,2N-1} \\ t_{2,0} & t_{2,1} & \cdots & t_{2,2N-1} \\ \vdots & \vdots & & \vdots \\ t_{L,0} & t_{L,1} & \cdots & t_{L,2N-1} \end{bmatrix} \tag{4.260}$$

The squared error J is

$$J = \{E*\}^T \{E\}$$
$$= \{a\}^T [T_G^*]^T [T_G]\{a\} - \{\tilde{\alpha}*\}^T [T_G]\{a\} - \{a\}^T [T_G^*]^T \{\tilde{\alpha}\} + \{\tilde{\alpha}*\}^T \{\tilde{\alpha}\} \tag{4.261}$$

Minimising J by making $\partial J / \partial \{a\} = \{0\}$ and after some algebraic manipulations, it follows that

$$\text{Re}\left([T_G^*]^T [T_G] \right) \{a\} = \text{Re}\left([T_G^*]^T \{\tilde{\alpha}\} \right) \tag{4.262}$$

$$\{a\} = \text{Re}\left([T_G^*]^T [T_G] \right)^{-1} \text{Re}\left([T_G^*]^T \{\tilde{\alpha}\} \right) \tag{4.263}$$

However, once again, some numerical problems may be encountered in the inversion of $Re([T_G^*]^T[T_G])$ although these may again be avoided by using orthogonal polynomials replacing that term by the unit matrix. This means, following what has been done for the RFP method, changing (4.256) or (4.257) into:

$$\alpha(\omega_j) = \sum_{k=0}^{2N-1} \left(\frac{\varphi_{j,k}^+}{g_j} \right) c_k = \sum_{k=0}^{2N-1} z_{j,k} c_k \tag{4.264}$$

where g_j is the denominator of (4.256), which is known, as the coefficients b_k have already been calculated, and $\varphi_{j,k}^+$ is the half-positive orthogonal polynomial defined in (4.244) and (4.245) as γ^+ and for $q_j = | 1/g_j |^2$. For all L measured points, (4.264) can still be written as

$$\{\alpha(\omega)\} = [Z_G]\{c\} \tag{4.265}$$

where $[Z_G]$ can be expressed as $[`1/g_`][\vartheta^+]$, $[\vartheta^+]$ being the matrix composed by $\varphi_{j,k}^+$. Thus, to have $Re([T_G^*]^T[T_G]) = [I]$ means to have

$$Re\left(\left[\vartheta^{+*} \right]^T \left[`(1/g)^* \right] \left[ `1/g_` \right] \left[\vartheta^+ \right] \right) = [I] \tag{4.266}$$

Similarly to what was found in (4.241), we have to generate orthogonal polynomials with respect to a weighting function $| 1/g_j |^2$ and the orthogonality condition is (using the positive half-functions):

$$Re\left(\sum_{j=1}^{L} \left(\varphi_{j,k}^+ \right)^* \left(\varphi_{j,i}^+ \right) | 1/g_j |^2 \right) = \begin{cases} 0 & k \neq i \\ 0.5 & k = i \end{cases} \tag{4.267}$$

Hence, (4.262) becomes

$$\left[`0.5_` \right]\{c\} = Re\left(\left[Z_G^* \right]^T \{\tilde{\alpha}\} \right) \quad \Rightarrow \quad \{c\} = 2\,Re\left(\left[Z_G^* \right]^T \{\tilde{\alpha}\} \right) \tag{4.268}$$

$\{a\}$ can be calculated using a transformation as in (4.249 a), from which the local modal parameters can be evaluated.

The Polyreference Frequency Domain method (PRFD)
This is a *MIMO* method, due to the works of Zhang *et al* [153-155]. Considering p response locations and q input references, the impulse response matrix in terms of receptance, for an N DOF system, can be given by (see (4.144)):

$$[H(t)] = [\Psi] \left[\diagdown e^{s_r t} \diagdown \right] [W]^T \tag{4.269}$$
$$\underset{(pxq)}{} \quad \underset{(px2N)}{} \underset{(2Nx2N)}{} \underset{(2Nxq)}{}$$

where $[\Psi]$ is the mode shape matrix and $[W]$ is the modal participation matrix, as defined in (4.24) and (4.25). The corresponding Laplace transform of (4.232), is

$$[H(s)] = [\Psi] \left[s[I] - \left[\diagdown s_r \diagdown \right] \right]^{-1} [W]^T \tag{4.270}$$

Let

$$\underset{(2Nxq)}{[G(s)]} = \left[s[I] - \left[\diagdown s_r \diagdown \right] \right]^{-1}_{\underset{(2Nx2N)}{}} \underset{(2Nxq)}{[W]^T} \tag{4.271}$$

Thus, (4.270) is written as

$$\underset{(pxq)}{[H(s)]} = \underset{(px2N)}{[\Psi]} \underset{(2Nxq)}{[G(s)]} \tag{4.272}$$

Applying the Laplace transform to the first time derivative of (4.269), i.e., to the impulse response matrix in terms of mobility, we obtain

$$[H(s)]_M = \mathscr{L}\left[\dot{H}(t) \right] = s\left[H(s) \right] - \left[H(t) \right]_{t=0} \tag{4.273}$$

where the subscript $_M$ denotes mobility. From (4.269) for $t=0$,

$$[H(s)]_M = s[H(s)] - [\Psi][W]^T \tag{4.274}$$

But it is also true that

$$[H(s)]_M = [\Psi]\left[\diagdown s_r \diagdown \right][G(s)] \tag{4.275}$$

and thus,

$$s[H(s)] - [\Psi][W]^T = [\Psi]\left[\diagdown s_r \diagdown \right][G(s)] \tag{4.276}$$

Combining (4.272) and (4.276), it follows that

$$\begin{bmatrix} [H(s)] \\ s[H(s)] \end{bmatrix} - \begin{bmatrix} [0] \\ [\Psi][W]^T \end{bmatrix} = \begin{bmatrix} [\Psi] \\ [\Psi]\left[\diagdown s_r \diagdown \right] \end{bmatrix} [G(s)] \tag{4.277}$$

From (4.272), the value of $[G(s)]$ can be extracted:

$$[G(s)] = \left([\Psi]^T[\Psi]\right)^{-1}[\Psi]^T[H(s)] \tag{4.278}$$

Substituting (4.278) into (4.275) leads to

$$[H(s)]_M = [\Psi]\left[\,\text{`}s_r\,\text{`}\right]\left([\Psi]^T[\Psi]\right)^{-1}[\Psi]^T[H(s)] \tag{4.279}$$

or

$$[H(s)]_M = [A_S][H(s)] \tag{4.280}$$
$$\underset{(pxq)}{} \quad \underset{(pxp)}{} \underset{(pxq)}{}$$

where $[A_s] = [\Psi]\left[\,\text{`}s_r\,\text{`}\right]([\Psi]^T[\Psi])^{-1}[\Psi]^T$ is called the system matrix. Post-multiplying this by $[\Psi]$, gives

$$[A_S][\Psi] = [\Psi]\left[\,\text{`}s_r\,\text{`}\right] \tag{4.281}$$

which represents an eigenproblem that will give us the mode shapes $[\Psi]$ and the natural frequencies and damping factors from $[\,\text{`}s_r\,\text{`}]$. The problem now is how to evaluate the system matrix $[A_S]$. Equation (4.281) can be written as

$$\left[-[A_S] \quad [I]\right]\begin{bmatrix} [\Psi] \\ [\Psi]\left[\,\text{`}s_r\,\text{`}\right] \end{bmatrix} = [0] \tag{4.282}$$

Post-multiplying both sides by $[G(s)]$ gives

$$\left[-[A_S] \quad [I]\right]\begin{bmatrix} [\Psi] \\ [\Psi]\left[\,\text{`}s_r\,\text{`}\right] \end{bmatrix}[G(s)] = [0] \tag{4.283}$$

Substituting (4.277) in (4.283), we obtain

$$\left[-[A_S] \quad [I]\right]\begin{bmatrix} [H(s)] \\ s[H(s)]-[\Psi][W]^T \end{bmatrix} = [0] \tag{4.284}$$

or

$$-[A_S][H(s)] + s[H(s)] - [\Psi][W]^T = [0] \tag{4.285}$$

from which

$$\left[[A_S] \quad [\Psi][W]^T\right]\begin{bmatrix} [H(s)] \\ [I] \end{bmatrix} = s[H(s)] \tag{4.286}$$
$$\underset{(px(p+q))}{} \quad \underset{((p+q)xq)}{} \quad \underset{(pxq)}{}$$

Passing to the frequency domain by putting $s = i\omega$ and considering L measured frequencies, the following expression is obtained:

$$\left[\left[A_S \right] \underset{(px(p+q))}{\left[\Psi \right] \left[W \right]^T} \right] \underset{((p+q)xLq)}{\begin{bmatrix} \left[H(\omega_1) \right] \cdots \left[H(\omega_L) \right] \\ \left[I \right] \cdots \left[I \right] \end{bmatrix}}$$

$$= \underset{(pxLq)}{\left[\left[H(\omega_1) \right] \cdots \left[H(\omega_L) \right] \right]} \underset{(LqxLq)}{\begin{bmatrix} i\omega_1[I] & & [0] \\ & \ddots & \\ [0] & & i\omega_L[I] \end{bmatrix}} \qquad (4.287)$$

From (4.287), it is easy to extract $[A_S]$ and $[\Psi][W]^T$. After solving the eigenproblem (4.281), $[\Psi]$ will be known and therefore $[W]^T$ will be evaluated, completing the mathematical model. The residues can be calculated as in (4.151), from which it is easy to calculate the modal constants and phase angles. The correct number of modes can be estimated by the SVD.

The PRFD method has given evidence of good performance when dealing with close modes; it appears also to be less sensitive to computational modes when compared to time domain methods and allows for the use of unequal frequency steps. A more general version of this method, including also information in terms of accelerance, is presented in [156].

4.4.4 Other indirect MDOF methods in the frequency domain

An enormous number of methods can nowadays be found, the majority of them being just slight variations of the best known ones. We shall refer briefly to some of these methods.

The Gaukroger-Skingle-Heron method (GSH)

Presented in 1973 [157], this method is based on a least-squares fit of the receptance, considering several modes at a time, for one input and one output location (SISO). The model assumes viscous damping, and is an attempt to circumvent possible difficulties associated with the initial graphical techniques used in the Circle-Fitting method. It is, however, interactive, giving the option for the user to take decisions. As the procedure needs initial estimates for the modal parameters, the Circle-Fitting procedure is suggested at the first stage of the analysis. The method considers the receptance response of an N DOF system as:

$$\alpha(\omega) = \alpha_0 + \left(\sum_{r=1}^{N} \frac{A_r + i\omega B_r}{\omega_r^2 - \omega^2 + i2\xi_r \omega_r \omega} \right) e^{i\phi} \qquad (4.288)$$

where α_0 is a complex constant and ϕ is a rotation angle that seek to reflect the influence of out-of-range modes. An error function is defined between the theoretical values $\alpha(\omega)$ and the measured values $\tilde{\alpha}(\omega)$, for all the measured points L:

$$e = \sum_{j=1}^{L} \left(\tilde{\alpha}(\omega_j) - \alpha(\omega_j)\right)^* \left(\tilde{\alpha}(\omega_j) - \alpha(\omega_j)\right) \qquad (4.289)$$

The minimisation of this error leads to a nonlinear system to be solved. Using a Taylor expansion and keeping just the first order terms, the problem is linearised. Good initial estimates must be provided, in order that the process converges. It was found that initial zero values for α_0 and ϕ were satisfactory. Initial estimates for the natural frequencies and damping factors can be provided by a quick circle-fitting analysis. Then, A_r and B_r can be obtained by a non-iterative least-squares analysis. All the modal parameters will then be known and the iterative process can begin.

Although this method gives very satisfactory results, the whole process and in particular the iterative part is, generally, very slow.

The Frequency Domain Prony method (FDPM)

This is also a *SISO* method and it corresponds to the complex exponential method in the time domain. It was introduced in 1980 [158]. For an N DOF system with viscous damping, we saw that it is possible to write the receptance FRF as the ratio of two polynomials in $i\omega$ (see (4.216)). That expression is then separated into its real and imaginary parts. The 4N unknowns (coefficients a_k and b_k) can be obtained by considering at least 2N data points (because each one has real and imaginary information), by solving the resulting linear system of equations.

After finding those coefficients, it is possible to solve the denominator polynomial to calculate the poles and consequently the natural frequencies and damping factors. The residues and the modal constants and phase angles are calculated afterwards, as explained for the RFP method. The authors of this method also addressed an alternative procedure, when only the magnitude of the FRF is available from the measurements. In such a case, the squared magnitude of the FRF can be given as

$$|\alpha(\omega)|^2 = \frac{\displaystyle\sum_{k=0}^{2N-1} a_k'(i\omega)^{2k}}{\displaystyle\sum_{k=0}^{2N} b_k'(i\omega)^{2k}} \qquad (4.290)$$

Taking $b_{2N}' = 1$, one can write

$$\sum_{k=0}^{2N-1} (-1)^k \omega^{2k} \left(b_k' |\alpha(\omega)|^2 - a_k'\right) = (-1)^{N+1} \omega^{2N} |\alpha(\omega)|^2 \qquad (4.291)$$

Here, there are again 4N unknowns (a'_k and b'_k) and this time we need at least 4N data points to solve the linear system of equations. With a'_k and b'_k, $|\alpha(\omega)|^2$ is reconstituted using (4.290), and by solving the denominator polynomial the poles can be calculated.

It should be noted that in this case there are 4N poles, instead of 2N in the previous case. The 2N poles in excess are reflections of the other ones, with respect to the imaginary axis, i.e., they have positive real parts, while the ones we are interested in have negative real parts. This fact allows for the sorting of the poles we want. The calculation of the natural frequencies and damping factors is therefore possible.

For the calculation of the residues that would lead us to the modal constants and phase angles, a degree of ambiguity is expected, as we have only measured the magnitude of the FRF.

Thus, this alternative procedure is not convenient for the calculation of those parameters. Comparison of this method with the CE method showed a reasonable agreement [158].

The Eigensystem Realisation Algorithm in the frequency domain (ERA-FD)

A different version of the ERA, in the frequency domain, was presented in [159]. The procedure follows the same steps as for the time domain version and the mathematical development is entirely similar. Therefore we shall not develop it here.

The frequency domain version can also provide checks to distinguish between genuine and computational modes. It features a reduced computational time and storage and the examples tried revealed its good performance, both on theoretical and experimental cases.

The Global method

The Global method [160, 161] is a *SIMO* method that is based on the construction of receptance and mobility difference matrices, in order to avoid the effects of neighbouring modes, the same philosophy followed in Dobson's and CRF methods. In those matrices, the number of FRFs is overspecified with respect to the expected number of modes existing in the selected frequency range.

Application of the SVD technique (Appendix A) enables the determination of the genuine number of existing modes. A variation of the method, allowing the structure to be excited simultaneously by several input forces (with the condition that the input force vectors are linearly independent), is proposed in [162], with the objective of having a better excitation of all the modes that are to be identified.

The Constrained Global Nonlinear method (CGN)

In this *SIMO* method [163], the curve-fitting is reduced to a minimisation problem with constraint equations, related to the orthogonal properties of the mode shapes. An objective function is then constructed, where the constraint equations are incorporated in the form of Lagrange multipliers. All the modal parameters are derived from the minimisation procedure, and therefore in this method there is not the usual two stage calculation, where the natural frequencies and damping factors are calculated first and the mode shapes in a second stage. The procedure is, however, restricted to real modes.

The Matrix Decomposition Method (MDM)

The Matrix Decomposition Method (MDM) [164] is proposed for time-invariant and proportionally damped systems. The FRF matrix is treated by eigenvalue decomposition, rather than by SVD. The proportional damping assumption is, however, a limitation for real cases.

Extensions of the GRFP method

An extension of the GRFP method to incorporate multiple input locations, i.e., its *MIMO* version, is developed in [165]. This development follows exactly the same steps as explained for the RFP and GRFP methods, with the corresponding extensions to incorporate the additional information from several inputs.

In [166], another improvement to the GRFP method was introduced with the objective of determining the correct number of modes present in a given frequency range.

The analysis is repeated with an increasing number of assumed modes and a statistical parameter is introduced with the property that it stabilises when the correct number of modes is achieved; further increments of the number of modes will not alter the value of that parameter. This parameter is a direct function of the computed least-squares error of the fit.

In [167], the GRFP method is also extended to a *MIMO* version, but making use of a matrix autoregressive moving-average model in the Laplace domain, with the advantages of being appropriate for high modal density cases. A complex mode indicator function (CMIF), referred to below, which is a plot of eigenvalues as a function of the frequency, helps in determining the order of the orthogonal polynomials.

Other improvements to the GRFP method are also given in [168] and [169].

Mode indicators

In [170], the already mentioned Complex Mode Indicator Function (CMIF) is introduced. Besides being a *MIMO* method, it is also an indicator to identify the number of modes, by displaying the physical magnitude of each one and the damped natural frequency. Repeated modes are also detected. It is based on the calculation of the SVD of the FRF matrix at each spectral line.

While another well-known indicator, The Multivariate Mode Indicator Function (MvMIF) [60], indicates the existence of real normal modes, the CMIF indicates the existence of real or complex modes, as well as their relative magnitude. A comparison of and discussion on the performance of several functions that indicate the number of modes present in a frequency range are given in [171].

The Transfer Function Real Condensation method (TFRC)

This method, presented in 1995, [172] is based on the fitting of the squared amplitudes of the FRFs. It condenses the magnitude information included in all the measurements, no matter where the forces are located and identifies the global parameters from such a unique response.

The Direct Simultaneous Modal Approximation method (DSMA)

Although its name suggests that it is a direct method, the DSMA [173] is an indirect one. It minimises a weighted error function, evaluating simultaneously eigenvalues and eigenvectors, avoiding the numerical problems associated with the usual two-step calculations in most identification methods: first, the estimation of the eigenvalues (or natural frequencies and damping ratios) and second, the estimation of the eigenvectors (or modal constants and phase angles).

The disadvantage is that a considerable effort is necessary to obtain initial estimates, so that the optimisation process converges to a global minimum. It seems appropriate for high noise level and close modes.

4.4.5 Direct methods in the frequency domain

In this category, several methods can be found. Here, we selected the Identification of Structural System Parameters method (ISSPA) for a detailed explanation. Other methods are referred to briefly.

The Identification of Structural System Parameters method (ISSPA)

Presented in 1978 [174], this *SIMO* direct method is based on the identification of the system matrices to solve an eigenproblem.

Let us begin by considering the differential equation of equilibrium for a SDOF system:

$$m\ddot{y} + c\dot{y} + ky = f \tag{4.292}$$

This method considers the more general case where there is also a moving base exciting the system, as in figure 4.9. For this case, (4.292) becomes:

$$m\ddot{y} + c(\dot{y} - \dot{x}) + k(y - x) = f \tag{4.293}$$

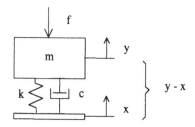

Fig. 4.9 Moving base SDOF system.

Defining the relative displacement between the mass and the base, as $y_r = y - x$, we have, in terms of y_r, the following equation:

$$m\ddot{y}_r + c\dot{y}_r + ky_r = -m\ddot{x} + f \tag{4.294}$$

Let us consider now that we have a MDOF system. Instead of (4.294) we shall have a system of equations which may be written as

$$[M]\{\ddot{y}_r\} + [C]\{\dot{y}_r\} + [K]\{y_r\} = -[M]\{\ddot{x}\} + \{F\} \tag{4.295}$$

For harmonic base and force excitation, (4.295) becomes

$$\left[-\omega^2[M] + i\omega[C] + [K]\right]\{\bar{y}_r\} = -\omega^2[M]\{x\} + \{F\} \tag{4.296}$$

or, pre-multiplying by $[M]^{-1}$,

$$\left[-\omega^2[I] + i\omega[C]' + [K]'\right]\{\bar{y}_r\} = -\omega^2\{x\} + [M]^{-1}\{F\} \tag{4.297}$$

where $\{\bar{y}_r\}$ is the complex amplitude of $\{y_r\}$. Equation (4.297) can still be written as

$$\left[-\omega^2[I] + i\omega[C]' + [K]'\right]\{\bar{y}_r\} = \omega^2\{F\}' \tag{4.298}$$

where

$$\{F\}' = -\{x\} + \omega^{-2}[M]^{-1}\{F\} \tag{4.299}$$

Supposing an N DOF system, the matrices in equation (4.298) have dimensions NxN and the vectors Nx1. From (4.299), for pure excitation of the base it is not necessary to know [M], but for an applied force, [M] must be known; it is usually introduced from a theoretical evaluation via a finite element analysis. Considering L measured frequency points, from (4.298),

$$-\omega_1^2\{\bar{y}_r\}_1 + i\omega_1[C]'\{\bar{y}_r\}_1 + [K]'\{\bar{y}_r\}_1 = \omega_1^2\{F\}'$$
$$-\omega_2^2\{\bar{y}_r\}_2 + i\omega_2[C]'\{\bar{y}_r\}_2 + [K]'\{\bar{y}_r\}_2 = \omega_2^2\{F\}'$$
$$\vdots \qquad\qquad\qquad \vdots \tag{4.300}$$
$$-\omega_L^2\{\bar{y}_r\}_L + i\omega_L[C]'\{\bar{y}_r\}_L + [K]'\{\bar{y}_r\}_L = \omega_L^2\{F\}'$$

or

$$\left[\{\bar{y}_r\}_1 \; \{\bar{y}_r\}_2 \; \cdots \; \{\bar{y}_r\}_L\right]_{(N\times L)} \begin{bmatrix} -\omega_1^2 & 0 & \cdots & 0 \\ 0 & -\omega_2^2 & \cdots & 0 \\ \vdots & \vdots & \ddots & \vdots \\ 0 & 0 & \cdots & -\omega_L^2 \end{bmatrix}_{(L\times L)} +$$

$$i\,[C]'_{(N\times N)} \left[\{\bar{y}_r\}_1 \; \{\bar{y}_r\}_2 \; \cdots \; \{\bar{y}_r\}_L\right]_{(N\times L)} \begin{bmatrix} \omega_1 & 0 & \cdots & 0 \\ 0 & \omega_2 & \cdots & 0 \\ \vdots & \vdots & \ddots & \vdots \\ 0 & 0 & \cdots & \omega_L \end{bmatrix}_{(L\times L)} +$$

$$[K]'_{(N\times N)} \left[\{\bar{y}_r\}_1 \; \{\bar{y}_r\}_2 \; \cdots \; \{\bar{y}_r\}_L\right]_{(N\times L)} = \left[\{F\}' \; \{F\}' \; \cdots \; \{F\}'\right]_{(N\times L)} \begin{bmatrix} \omega_1^2 & 0 & \cdots & 0 \\ 0 & \omega_2^2 & \cdots & 0 \\ \vdots & \vdots & \ddots & \vdots \\ 0 & 0 & \cdots & \omega_L^2 \end{bmatrix}_{(L\times L)}$$

$$(4.301)$$

or simply as

$$-[Y_r]\left[\,^\diagdown\Omega_\diagdown\right]^2 + i\,[C]'[Y_r]\left[\,^\diagdown\Omega_\diagdown\right] + [K]'[Y_r] = [F]''\left[\,^\diagdown\Omega_\diagdown\right]^2 \qquad (4.302)$$

with

$$\left[\,^\diagdown\Omega_\diagdown\right] = \begin{bmatrix} \omega_1 & 0 & \cdots & 0 \\ 0 & \omega_2 & \cdots & 0 \\ \vdots & \vdots & \ddots & \vdots \\ 0 & 0 & \cdots & \omega_L \end{bmatrix} \qquad (4.303)$$

Equation (4.302) can be separated into its real and imaginary parts:

$$-\mathrm{Re}[Y_r]\left[\,^\diagdown\Omega_\diagdown\right]^2 - [C]'\mathrm{Im}[Y_r]\left[\,^\diagdown\Omega_\diagdown\right] + [K]'\mathrm{Re}[Y_r] = [F]''\left[\,^\diagdown\Omega_\diagdown\right]^2$$

$$-\mathrm{Im}[Y_r]\left[\,^\diagdown\Omega_\diagdown\right]^2 + [C]'\mathrm{Re}[Y_r]\left[\,^\diagdown\Omega_\diagdown\right] + [K]'\mathrm{Im}[Y_r] = [0] \qquad (4.304)$$

where $[C]'$ and $[K]'$ are the unknowns to be calculated. Their calculation gives:

$$[C]' = \left[\text{Im}[Y_r] [\backslash \Omega \backslash]^2 \text{Im}[Y_r]^+ - \left[\text{Re}[Y_r] + [F]'' \right] [\backslash \Omega \backslash]^2 \text{Re}[Y_r]^+ \right]$$

$$\left[\text{Im}[Y_r] [\backslash \Omega \backslash] \text{Re}[Y_r]^+ + \text{Re}[Y_r] [\backslash \Omega \backslash] \text{Im}[Y_r]^+ \right]^{-1} \qquad (4.305)$$

$$[K]' = \left[\text{Im}[Y_r] [\backslash \Omega \backslash] - [C]' \text{Re}[Y_r] \right] [\backslash \Omega \backslash] \text{Im}[Y_r]^+$$

where $\text{Re}[Y_r]^+$ and $\text{Im}[Y_r]^+$ are the pseudo-inverses of $\text{Re}[Y_r]$ and $\text{Im}[Y_r]$, respectively. Knowing $[C]'$ and $[K]'$, the eigenvalues and eigenvectors can be calculated. Although not referred to here, the SVD technique is usually employed in this method to determine the effective number of DOF of the system under study. For more details, see [174] and [175].

4.4.6 Other direct methods in the frequency domain
The Spectral method
The Spectral method, introduced by Klosterman [67] with the objective of identifying close modes, is probably the simplest of all the *MIMO* direct frequency domain methods, where the matrices of the governing differential equation of the system are evaluated from the measured FRF receptances, considering only two frequency points, or sets of pairs of points. Although it works reasonably well for theoretical data, it is not satisfactory in real cases.

The Simultaneous Frequency Domain method (SFD)
The SFD method [176, 177] is a *SIMO* method where the aim is to form the matrices of the system based on responses at several locations on the structure. From a knowledge of those matrices, the eigenvalues and eigenvectors are calculated. Because in the frequency range of interest there are N modes, only N DOF from the p measured ones are effective in the characterisation of the response of the system. These N DOF form what is called the 'independent' set of DOF. Similarly, the p-N measured responses form a set of 'dependent' DOF. Both sets are related through a linear transformation, enabling the calculation of the complex mode shapes, as well as the real ones.

The ability of this method to evaluate the system matrices and therefore the complex and/or real mode shapes is a great advantage. The main problems are related to the correct choice of the 'independent' set of measured coordinates, as a different choice may lead to different answers. In order to determine the correct number of modes to consider in the 'independent' set of coordinates, a first view of the FRFs on the Argand plane can be pursued.

Usually, the number of modes taken are the visible ones plus two, to take into account residual effects. Better results were found when several narrow frequency bands were selected around regions of resonant peaks.

An improved selection method for the number of modes to analyse is given in [177]. This method has been applied very successfully in complex spacecraft structures. Craig and Blair [178] have extended this method, in order to permit multiple exciter testing.

The Multi-Matrix method

This is a *MIMO* direct method [65, 66, 179] where a general matrix input-output polynomial is estimated using frequency responses from several input and output locations. Residual terms rather than additional modes are used to describe the contribution of out-of-range modes. By a principal response analysis, the effective number of modes is determined and the governing system matrices are identified, from which the modal parameters can be evaluated. This method has the ability of handling large amounts of data, being stable and robust, and has proven to be able to deal with closely spaced modes and high modal damping and/or high modal density. In [179] a favourable comparison with the PRCE method is carried out. The main disadvantages of the Multi-matrix method are the considerable computational requirements and complexity for an easy computer implementation. It is, however, one of the most significant contributions in the direction of more automatic processing and analysis of large amounts of data.

4.5 TUNED-SINUSOIDAL METHODS

4.5.1 Introduction

These methods are a special class of modal identification methods in general. They are essentially based upon the experimental 'isolation' or tuning of real modes of vibration, by means of the excitation of the structure at each natural frequency by a set of exciters appropriately distributed in space and time. The *a priori* need of approximately locating the natural frequencies of the structure to be analysed, implies that another identification method must be used in advance. This justifies Ibrahim's comment [75] that this category of methods cannot be considered as genuine identification methods. Nevertheless, these methods constitute one of the oldest approaches to the study of the dynamic properties of structures and are still used nowadays in the aeronautical industry.

Lewis and Wrisley [180] were the first to recognise that real modes of vibration - i.e., those of the undamped structure - could be found if forces applied by several exciters could balance the dissipative forces in the structure and that this could happen if the measured displacements were in quadrature with those applied forces. In such conditions, the dynamic equations of equilibrium are reduced to their homogeneous solution at the undamped natural frequencies of the structure. Moreover, those authors state that the magnitude of the applied forces at the natural frequencies must be proportional to the product of the mass of the structure on which they act with the amplitude of that mass in the mode being excited. Once a mode is tuned, it is a simple matter to determine the modal parameters - the excitation frequency is the natural frequency and the measured motion is the mode shape. Damping can be evaluated by cutting the excitation and measuring the rate of decay of the response. The correctness of the tuning can be assessed by the absence of beating (due to the presence of another mode) in the free decay response.

In 1958, Asher [181] provided a mathematical description to calculate the natural frequencies, by zeroing the determinant of the real part of the FRF matrix, and to calculate the force distribution among the several shakers that must be applied to tune the desired mode. Since then, this procedure has become known as Asher's method, although, also in 1958, Trail-Nash [182] had published a similar procedure, where it was shown that the number of exciters to be used was a function of the effective number of DOF of the structure, and suggested the Kennedy-Pancu method for the localisation of the natural frequencies.

Due to its importance, a description of the basic theory of Asher's method is presented here. Subsequent work related to tuned-sinusoidal methods are - in general - variations and improvements of Asher's method.

4.5.2 Asher's method

This method uses the fact that the application of several shakers that are suitably tuned can excite individual undamped (real) modes of vibration of a structure. Usually, application of the method gives values for the undamped natural frequencies and the relative forces the shakers must apply. An alternative approach is to simulate numerically the application of those forces, instead of actually applying them to the structure. For the basic theory of this method, we can follow [183].

The matrix equilibrium equation of a system can be expressed as

$$[M]\{\ddot{y}(t)\} + [C]\{\dot{y}(t)\} + [K]\{y(t)\} = \{f(t)\} \tag{4.306}$$

If $\{f(t)\} = \{\overline{F}\}\sin\omega t$, the response will be $\{y(t)\} = \{\overline{Y}\}\sin(\omega t - \theta)$. Hence,

$$\left[[K] - \omega^2[M]\right]\{\overline{Y}\}\sin(\omega t - \theta) + [C]\omega\cos(\omega t - \theta)\{\overline{Y}\} = \{\overline{F}\}\sin\omega t \tag{4.307}$$

For all time t,

$$\cos\theta\left[[K] - \omega^2[M]\right]\{\overline{Y}\} + \omega\sin\theta[C]\{\overline{Y}\} = \{\overline{F}\}$$
$$-\sin\theta\left[[K] - \omega^2[M]\right]\{\overline{Y}\} + \omega\cos\theta[C]\{\overline{Y}\} = \{0\} \tag{4.308}$$

For $\theta = 90°$, the displacements are in quadrature with the excitation forces. Thus,

$$\omega[C]\{\overline{Y}\} = \{\overline{F}\} \tag{4.309 a}$$

$$\left[[K] - \omega^2[M]\right]\{\overline{Y}\} = \{0\} \tag{4.309 b}$$

It is clear from (4.309 b) that the condition $\theta = 90°$ corresponds to the undamped solution of the system and the force distribution is given by $\{\overline{F}\}$. This means that if the system is excited with the force distribution $\{\overline{F}\}$, the real modes of vibration can be obtained. Writing the amplitude response of the system in terms of the complex frequency response function $[\alpha(\omega)]$, it follows that

$$\{\overline{Y}\} = \left[[K] - \omega^2[M] + i\omega[C]\right]^{-1}\{\overline{F}\}$$
$$= \left[\text{Re}\left[\alpha(\omega)\right] + i\,\text{Im}[\alpha(\omega)]\right]\{\overline{F}\} \tag{4.310}$$

The condition of the displacements being in quadrature with the forces means that

$$\text{Re}\left[\alpha(\omega)\right]\{\overline{F}\} = \{0\} \tag{4.311}$$

Solving

$$\det\left(\text{Re}\left[\alpha(\omega)\right]\right) = 0 \tag{4.312}$$

enables the determination of the undamped natural frequencies which, when substituted in (4.311), give the force distribution $\{\overline{F}\}$. Moreover, from (4.310), we can find the real modes of vibration, given by

$$\{\phi_u\} = \text{Im}\left[\alpha(\omega)\right]\{\overline{F}\} \tag{4.313}$$

Reviewing this method, we identify the following steps:

1 - From the complex identified modes and frequencies, $[\Phi]$ and $[`\lambda_r^2`]$, calculate $[\alpha(\omega)]$ (given by $[\Phi][`\lambda_r^2 - \omega^2`]^{-1}[\Phi]^T$);
2 - The undamped natural frequencies are calculated from (4.312);
3 - The force distribution is calculated from (4.311);
4 - The real modes are calculated using (4.313).

4.5.3 Other contributions

The necessity for considerable data acquisition capacity for practical applications meant that only in the seventies did these methods start to be more widely used. In 1974, Craig and Su [54] carried out a study on the number and localisation of exciters on a structure, in order to properly tune the real modes of vibration. Considering that the number of measured responses, p, is always greater than the number of exciters, q, it is not usually possible to have all the p responses in quadrature with the forces. These authors introduced the concept of Modal Purity as a criterion to decide whether or not the measurements obtained at the p locations represented a true mode of vibration. This criterion establishes a tolerance for the phase angle of the response with respect to the force and if the phase deviates from 90 or 270 degrees by the amount of the given tolerance, then the test is rejected, meaning that either more exciters are necessary or that their position on the structure must be altered.

As already mentioned, the natural frequencies are calculated in Asher's method by putting the determinant of the real part of the measured FRF matrix to zero, where each element is obtained by exciting the structure at one point and measuring at another point. Craig and Su, in the same work, proposed a method for the localisation of the exciters, by taking several sets of FRFs for different exciter locations and so forming different FRF matrices. Each of those matrices gave results for the natural frequencies, and an analysis of their repeatability provided the most desirable locations for the shakers. The influence of varying the number of shakers was also investigated and numerical cases with close modes analysed with success. It was also found that the force distribution necessary to tune a mode does not agree with the conditions given by Lewis and Wrisley [180].

In 1975, Smith *et al* [184] presented a computerised system for the automatic acquisition, processing and analysis of structural dynamic data.

In 1976, Ibáñez [185] proposed an extension of Asher's method that consisted basically of considering more responses than exciters, with the advantage of reducing the occurrence of spurious resonant frequencies that sometimes happens in the standard method.

In 1978, Hallauer and Stafford [186] presented a detailed revision and discussion of Asher's method, showing its strengths and weaknesses. Systems with non-proportional damping and close modes are discussed, and numerical techniques are presented. Problems associated with the incompleteness of the FRF matrix are also discussed. Numerical simulations are given, where the normal modes calculated by theory are compared with the normal modes regenerated after calculating the force distribution.

Following this work, Gold and Hallauer, in 1979 [187], proposed an analytical application of Asher's method, calculating the mode shapes after evaluating the force distribution vector. However, the calculation of the FRF matrix was not given directly from the measurements, but was generated after having identified it using a least-squares curve-fitting method. This procedure sought to avoid the storage of large amounts of acquired data, storing instead the identified modal parameters. Numerical simulations worked well, but experimental cases failed due to the poor quality of the measured data.

Also in 1979, Ensminger and Turner [188] proposed a variation of Asher's method, again with prior curve-fitting of the measured FRFs, called the Minimum Coincident Response method. The real part of the FRF matrix could be rectangular and, by minimising (by a least-squares technique) the sum of the squares of the in-phase (real) displacements subjected to a normalisation constraint on the quadrature (imaginary) response, it was possible to calculate the force vector as a function of the frequency and also the error (again, a function of the frequency).

Plotting the error function for each measured frequency, the natural frequencies are obtained from the minima on the graph. For these natural frequencies, the force distributions can be calculated and the real mode shapes too. This plot showed itself to be more efficient in the detection of close natural frequencies than the plot of the determinant of the real part of the FRF matrix, because it is visually clearer. This approach also replaces the multi-shaker test by an analytical simulation. However, in 1981, Craig and Chung [189] concluded essentially that Asher's method has greater capability for identifying close modes than the Minimum Coincident Response method.

In 1983, Rades [190] proposed the use of the SVD technique to localise the natural frequencies of the structure when the FRF matrix is rectangular, i.e. when there are more responses than exciters. A matrix is formed by the product of the transpose of the real part of the FRF matrix by itself, and the plots of the singular values of such a matrix with frequency show minima at the natural frequencies. It was shown that these plots give clearer indications of the localisation of the natural frequencies than the usual plots of the determinant of the real part of the FRF matrix.

An automatic procedure, the force appropriation for modal evaluation (FAME) was proposed by Ibáñez and Blakely in 1984 [191], being essentially the automatic implementation of the extended Asher's method developed in [185].

In 1984, Hunt *et al* [192] presented an automatic method based on the perturbations of the exciting frequencies and force ratios to minimise the ratio of coincident response to quadrature response.

An overall review on this type of method can be found in [183]. The practical implementation of these types of procedure to actually test a structure is very expensive and time consuming, as it requires a lot of equipment and because of the difficulty of the tuning process. As Ibrahim notes [75], it can be very dangerous to have a structure in resonance for several minutes, as damage or failure may occur.

4.6   FINAL REMARKS

From all the methods described in this chapter, we can note that in many cases the similarities among them are very clear, even between time and frequency domains. This is not surprising, as all of them start from the same basic dynamic equations of equilibrium. In [193], Allemang *et al* presented the Unified Matrix Polynomial Approach (UMPA), showing that a considerable number of methods both in time and frequency domains are just particular cases of a more general polynomial formulation. That work helps us to understand better the similarities amongst most of the methods presented in this chapter, giving a more general panorama on the subject of identification techniques, and it contributes to a deeper knowledge of the philosophy of modal analysis methods.

Nowadays, the existing commercial software for industrial applications tends to provide very automatic and quick methods of analysis rather than slow interactive techniques, more suited for the development and research environments. On the other hand, those packages usually offer means for controlling and assessing the quality of the analyses, using indicators, charts, various graphical possibilities associated to the re-synthesised curves, etc. One of the most popular means to help the user in taking decisions when doing an analysis is the provision of so-called stabilisation charts. These are often active during global curve fitting of various FRFs. The analysis is repeated several times, taking different frequency points or time intervals, and the quality of the analysis is assessed from the percentage variation of the results obtained. For instance, the user may impose that the estimates on the natural frequencies and damping ratios do not vary more than 1% and 3%, respectively, and then observes the repeatability or stabilisation of those properties in some modes, while a scatter is seen in others. In this way, the user can obtain a picture of which are the reliable answers and which ones are just the result of errors due to the mathematical process used in the analysis or due to some other origin, like noisy data or nonlinear behaviour.

Another conclusion that may be drawn is that there is no such thing as the 'best' method. Some methods work better than others for some applications. The choice of the method to use depends on the available resources, on the available time, on the objectives of the study and on the personal experience of the user.

CHAPTER 5

Coupling Techniques

5.1 INTRODUCTION

It is generally acknowledged that the advent of the Finite Element (FE) method constituted a major step towards the analysis of more complicated static and dynamic structural mechanics problems. Rather than trying to formulate the equilibrium equations of a given problem, treating the structure as a whole continuously defined entity, a new philosophy was then initiated which acted as a catalyst for several researchers. A basic principle inherent to this philosophy was the assumption of the whole structure being composed of individual analytical elements such as beams, plates or shells which, at a final stage, could be assembled to provide a model of the complete system. Nevertheless, the complexity inherent to an increasing number of engineering problems still resulted in limitations of this type of 'discretised' structure, mainly due to the large order of the matrices involved in the process.

A more general approach was then necessary whereby a complex structure could be regarded as being formed of different substructures or components, each of which could first be analysed individually and independently from the others. In this way, before being assembled to form the overall structure, each analysis could be done by whichever method was most convenient and eventually the substructure models could be assembled together to obtain the equations related to the complete structure. This is the idea underlying the now well-known 'substructuring', 'building block' or 'coupling' approaches for solving static and dynamic problems.

Originally, the idea was restricted to the use of purely theoretical models but, as it was often found that certain substructure models could not be properly formulated due to their complexity, the method was developed so as to incorporate models derived by an experimental route too. This different route can not only complement a subsystem theoretical model - by verifying and updating - but can also provide a suitable and reliable subsystem model for direct assembly in the coupling process. This route is now firmly established and an experimental counterpart to the FE modelling is becoming widely used - the techniques being referred to as 'Modal Testing', 'Experimental Modal Analysis' or 'System Identification'. Recent advances in Modal Testing methods and in digital processing have reduced the time required and increased the accuracy and confidence associated with the experimental determination of modal parameters, which are the essential ingredients to construct an experimental model.

Accordingly, an ideal substructure approach should be versatile in terms of being able to incorporate data from either of the sources - the FE method or Modal Testing. Such a method should yet provide other important benefits which may be outlined as:

- each component can be treated by a more accurate and refined model. In certain cases, it may happen that components are still too large to be analysed by conventional experimental means, especially if they have to be suspended to simulate free-free support conditions. The substructure approach allows a further subdivision into other subsystems which are easier to measure;

- any structural modification which has to be applied at any time only involves a re-analysis of the affected part. A design change in one part only implies new data for that modified part which can then be coupled with the remaining unmodified components, without requiring a re-analysis of the rest of the structure;

- each dynamic model may be obtained by theoretical analysis or by testing of the individual subsystems - these are easier to handle than the total system. A mixture of theoretical and experimental subsystem modelling is one of the main requirements for using coupling techniques;

- there is the possibility of creating a library of standard subsystems for which a high level of modelling has already been achieved. Such components can be input, as often as needed, into several assembly processes;

- the location and time for each component analysis may be selected during the design stage, since different organisations on different sites can perform the analysis for each part.

The possibility of attaining a more precise description of the component dynamic properties at the subsystem level leads to a better exploitation of the computational and experimental means available from an individual organisation or team. However, not all the information so obtained is necessarily incorporated into the system model, otherwise no gain would be obtained in terms of efficiently handling the size of the matrices required for the formulation of the final equilibrium equation of the coupled structure. Thus, there is the necessity of reducing, in an optimum way, the size of the matrices at the subsystem level, while retaining a precise description of its dynamic properties. It should be borne in mind that any imprecise formulation of the reduced or condensed model will affect the predicted dynamic behaviour for the whole system.

Generally, the main steps involved in a substructuring technique may be described as:

Step 1 - Partitioning of the whole physical system model into a number of substructures with a proper choice of connection and interior coordinates. At this stage, it must have already been decided which components are going to be subjected to a modal test and which ones are amenable to modelling by the finite element (or other analytical) method. In some cases, it may happen that both experimental and theoretical routes are chosen for the modelling phase of a given component.

Step 2 - Derivation of the respective subsystem models, either by a theoretical or an experimental approach. A valid selection of coordinates, mainly those involved in the connection to other components, and/or modes to be retained should be undertaken and, whenever possible, with an evaluation of the effects of neglecting certain coordinates or modes.

Step 3 - Formulation of the subsystem equations of motion which are generally made by using physical or modal coordinates and, if possible, without requiring the knowledge of the dynamic properties of the remaining components forming the global structure. Such independency is an important requirement, although the interconnecting conditions must be commonly defined by some of the organisations involved in the design.

Step 4 - Construction of reduced-order equations for the global structure by invoking interface displacement compatibility and force equilibrium conditions previously established for the different component models.

The reduction or condensation process - performed at the subsystem stage - leads to a distinction between the different coupling techniques available, which we shall divide into two major categories. On the one hand, there are techniques in which the order of the matrices involved in the final equation of motion is dictated by the number of kept (primary) coordinates pertaining to each reduced subsystem model. On the other hand, there is another category of techniques which also benefit from a reduction performed at the component level, but this time based on the number of modes included. The former approach is referred to as 'Impedance coupling' and the latter as 'Modal coupling'. Before we proceed further on the classification of both categories of methods, some terminology is necessary to characterise the subsystem models.

5.1.1 Subsystem models

Essential to the communication among different organisations and research groups are the terminology and format used to describe the dynamic properties of a system or subsystem model, be it theoretically or experimentally derived. To start with, we shall present the different model formats using the same terminology as Ewins [3] and simultaneously, when pertinent, other similar designations will be referred to.

Spatial models

By discretising a given system, it is essential to assign to each of the N coordinates or degrees-of-freedom (DOFs), the values of the spatially distributed properties i.e., the mass, stiffness and damping. The way to do so is by presenting each of those properties in a matrix form as follows:

[M]   mass matrix (NxN) which provides a means to define the inertia forces assigned to each DOF when they experience an acceleration (the off-diagonal terms contain the inertia coupling information);

[K]   stiffness matrix (NxN) providing a means to define the inherent restoring forces due to the relative displacements at each DOF (as before, the off-diagonal terms express the way the DOF are statically coupled);

[C], [D] viscous and hysteretic damping matrices (NxN), respectively. They are not always used since damping is often neglected in theoretical modelling, which means that the dissipative forces are considered negligible when compared to the previously mentioned ones.

It is important to note that the above-mentioned definitions have a clear physical meaning in the case of lumped parameter systems whereby one can associate a DOF to a lumped region (such as in spring-mass models). This is not the case when matrices result from an FE modelling, since they contain coefficients necessary to satisfy energy balances between the various finite elements connected at the so-called nodal points as described by Zienkiewicz [194].

A complete Spatial model has inherent to it N coordinates and N modes. As will be shown later, Spatial models which are reduced or condensed to n_p primary or master coordinates will have order $n_p < N$, i.e., they will possess information on n_p coordinates and n_p modes only.

Sometimes, Spatial models are referred to as Time Domain models, since equations of motion formulated by using Spatial properties contain the response motions of a system as functions of time.

Modal models

There are situations where the dynamic properties are most conveniently described in terms of natural frequencies and associated mode shapes (see Section 1.4.1). Such convenience may arise from the need to compare data from different sources which use different routes to attain the modal description, as mentioned by Ewins [195], or simply if it is intended to show an animated display of the structure at each corresponding natural frequency.

Mathematically, the mode shapes are represented as vectors, in which each element represents a deflection of one DOF relative to the other (N-1) DOFs in the model. The modal vectors (or eigenvectors) can be grouped together in the so-called modal matrix which is represented by $[\Psi]$ - a square or rectangular matrix containing information on N coordinates and m modes. The eigenvalues, which are intimately related to the m system natural frequencies, can be grouped together forming the terms of a diagonal matrix represented as $[\,{}^{\backslash}\lambda_r^2{}_{\backslash}\,]$. Generally, both are complex matrices whose real and imaginary parts have distinct physical meanings: the kth eigenvalue $\lambda_k^2 = \omega_k^2(1+i\eta_k)$ contains the information related simultaneously to the kth natural frequency ω_k and modal damping η_k of the kth complex mode $\{\psi\}_k = \{a+ib\}_k$, which in turn is represented by a magnitude - the relative amplitude of motion at each DOF - and by a phase angle which expresses a relative delay in the motion of each DOF.

Since the mode shapes represent relative amplitudes, rather than absolute deflections of the structure, the elements of each modal vector are scaled in some manner. Generally, they are scaled in such a way that the largest element is made equal to 1 (for instance, for graphic visualisation purposes).

On other occasions, when Modal models are obtained from different sources, it is convenient to have a consistent scaling factor. This can be achieved by making use of the concept of modal masses and modal stiffnesses.

Due to the orthogonality of the mode shapes relative to the mass and stiffness matrices (see Section 1.4.1), the following relationships, for an undamped system, hold (if m<N):

$$\left[\,{}^{\backslash}m_r{}_{\backslash}\right]_{(mxm)} = [\Psi]^T_{(m \times N)} [M]_{(N \times N)} [\Psi]_{(N \times m)} \tag{5.1}$$

$$\left[\,{}^{\backslash}k_r{}_{\backslash}\right]_{(mxm)} = [\Psi]^T_{(m \times N)} [K]_{(N \times N)} [\Psi]_{(N \times m)} \tag{5.2}$$

m_r and k_r being the generalised modal masses and stiffnesses interrelated as $\lambda_r^2 = \omega_r^2 = k_r/m_r$. If the modal masses are used to scale the mode shapes, a new orthonormal set is obtained:

$$[\Phi]^T_{(m \times N)} [M]_{(N \times N)} [\Phi]_{(N \times m)} = [I]_{(mxm)} \tag{5.3}$$

$$[\Phi]^T_{(m \times N)} [K]_{(N \times N)} [\Phi]_{(N \times m)} = \left[\,{}^{\backslash}\omega_r^2{}_{\backslash}\right]_{(mxm)} \tag{5.4}$$

which is here assumed to be the normalised format for the presentation of mode shapes. In fact, the mode shapes obtained from modal constants which are extracted from measured data by performing a modal analysis process (see Section 1.4.2 and Chapter 4 in general), are mass-normalised.

Response models
According to the description in common use in the analysis of control systems and electrical circuits, we shall assume a structure possessing inputs and outputs which can be interrelated through a kind of 'black box'. The (input) forces $\{f(t)\}$ applied to the system can be related to the (output) responses $\{x(t)\}$ whenever the dynamic characteristics are known. Let us consider a linear system excited with harmonic forces for which the input/output relationship can be written in the frequency domain as

$$\{X(\omega)\} = [H(\omega)]\{F(\omega)\} \quad \text{or} \quad \{F(\omega)\} = [Z(\omega)]\{X(\omega)\} \tag{5.5}$$

where $[H(\omega)]$ and $[Z(\omega)]$ are frequency response transfer functions of the system related as $[Z(\omega)] = [H(\omega)]^{-1}$. From an experimental standpoint, only $H(\omega)$ can be measured directly on the structure, as explained in Section 1.5. Due to the wide use of accelerometers to sense responses on structures, the FRFs are generally measured in terms of accelerance.

The Response models are described by an FRF matrix whose elements (FRFs) can be either all measured or all analytically calculated or, sometimes, a mixture of both. As seen in Section 1.6, it is possible to derive a Response model theoretically by invoking the relationship between Response and Spatial or Response and Modal outlined here for the receptance matrix and for an undamped system:

$$\underset{(N \times N)}{[\alpha(\omega)]} = \left[\underset{(N \times N)}{[K] - \omega^2 [M]} \right]^{-1} \tag{5.6}$$

$$\underset{(N \times N)}{[\alpha(\omega)]} = \underset{(N \times m)}{[\Phi]} \left[\underset{(m \times m)}{\diagdown (\omega_r^2 - \omega^2) \diagdown} \right]^{-1} \underset{(m \times N)}{[\Phi]^T} \tag{5.7}$$

where N and m are the numbers of coordinates and modes, respectively.

For simple components, however, a closed-form solution based on differential equations of equilibrium can be used to relate 'exactly' the responses and excitations, as presented in textbooks by Bishop and Johnson [196] and Timoshenko *et al* [197].

5.1.2 Classification of techniques
The main objective of a coupling process is to obtain a model for the assembled structure. The order of the matrices used to formulate the equations of motion of the assembled structure depends on the order of the matrices used to describe each subsystem model, which in turn should be as condensed as possible. How to achieve the required reduction, which is undertaken independently on each component, is dependent on the selected format to describe their dynamic characteristics.

In fact, the three possible types of a subsystem model (Spatial, Modal and Response) are interrelated, but it is important to note that exact relations hold strictly for models which are considered complete in terms of both coordinates and modes and so approximations are incurred when incomplete models are used.

More specifically, the order of the matrices of the assembled structure depends *either* on the number of coordinates (connection and interior) *or* on the number of kept modes pertaining to each component model. One can say that two groups of coupling techniques emerge from the large variety of methods which have been used in different fields of research and industry, and are classified as:

- Impedance coupling techniques which benefit from the reduction performed on the subsystem models in terms of coordinates;

- Modal coupling techniques which are suitable for the use of reduced models in terms of modes.

The former group deals primarily with the coupling of subsystems whose models are described either by their Spatial or by their Response properties. The first of these types of model is used extensively in the Finite Element method but is rarely used in cases which involve experimental modelling. Although Response models can be obtained by theoretical analysis, they mostly constitute the raw data available from modal tests.

The techniques forming the latter group are applied in those situations when the component models are described by their modal properties - Modal models. This type of model is easily generated from an eigensolution, if a theoretical tool such as the Finite Element method is used, or they can be derived from an identification process carried out on measured FRF data.

Figure 5.1 elucidates the different possibilities of performing the coupling using both experimentally and theoretically derived subsystem models.

5.1.3   Overview of coupling techniques

There have been numerous researchers addressing the substructuring method, especially since the early 1960's. A brief review is presented here, although a more thorough discussion is carried out in [196] and [198] where specific techniques are described or referred to in more detail.

To start with, we shall address the coupling techniques which exploit, in terms of computational efficiency, a reduction performed on the coordinates at the subsystem level. Early works were presented making an analogy between electrical circuits and vibrating systems, one of them from Duncan [199]. The concepts of linear operators and the principle of superposition were employed by Sykes [200] to develop linear multi-terminal network theory for solving periodic steady-state and transient vibration problems of mechanical systems synthesised from a number of small substructures. One of the most significant analytical works in the development of the Mobility (Impedance) concept is that by Bishop and Johnson [196]; from an 'exact' formulation of the Response model of a beam component, the properties of multi-beam assemblies could be formulated. The application of the conceptually simple Impedance Coupling technique is straightforward when the components are amenable to theoretical modelling, but practical complex systems have demanded subsystem Impedances to be derived from measured data rather than a purely theoretical formulation.

The experimental approach to the Impedance Coupling problem was one of the main reasons which motivated a breakthrough to the development of suitable techniques and equipment to measure, assess and analyse data. A comprehensive work was presented by Ewins [201] concerning ground rules, measurement techniques and interpretation and application of measured data with an extensive selected bibliography. The Impedance coupling technique was applied to many

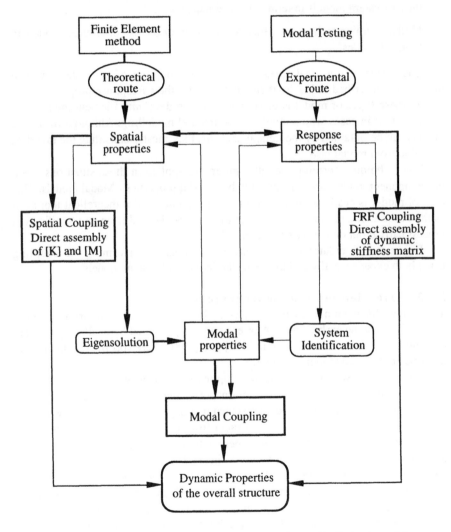

Fig. 5.1 Coupling theoretically and experimentally derived models.

engineering problems, such as those presented by Klosterman [202], Sainsbury and Ewins [203], Ewins *et al* [204], and Heer and Lutes [205], just to quote some of the earliest cases. The main difficulties encountered in those applications were mainly related to the mathematical inconsistency of the measured models and to the inadequacy of experimental means to measure some terms in the FRF matrices of certain components. Mostly, those FRFs were related to rotational response measurements.

Rotational responses are necessary to formulate proper constraints between connected components. Success in the prediction of results for a coupled structure is dependent on how the connection coordinates are measured and included in a coupling process, as demonstrated in some practical applications undertaken by Ewins and Sainsbury [206], and Ewins and Gleeson [207]. For instance, by simply assuming components connected at one single point, if the components respond to excitations in all three planes, it is vital to include the three rotations in addition to the three translations in order to properly formulate the constraint conditions. In terms of Response models, the FRFs related to rotational response/excitations represent 75% (or 60% if symmetry properties are assumed) of the total elements in the corresponding FRF matrix. Although simple to calculate those FRFs analytically, it is not an easy task to perform their measurement.

The FRF matrix of a measured component tends to be ill-conditioned near each resonance frequency, especially when lightly damped structures are dealt with. If any error in the FRFs is present in the vicinity of those regions, which is most likely in measured models causing the undesirable mathematical inconsistency, numerical failures will arise during the coupling process and, as a consequence, meaningless predicted results will be obtained. Lutes and Heer [208] addressed this problem by numerically 'filtering out' elements in the FRF matrix. Another approach was implemented in works by Ewins [209], Gleeson [210] and Imregun *et al* [211], where the inconsistency was removed by subjecting raw data to modal analysis and then, from the modal data base thus obtained, smoothed FRFs were regenerated to improve the predictions.

Another coupling method that we shall discuss is the Modal Coupling technique, also known as Time Domain or Component Mode Synthesis method. There are two modal coupling approaches: the fixed-interface and the free-interface methods.

The fixed-interface method is based on the static substructuring technique proposed by Przemieniecki in 1963 [212]. It was mainly directed towards the use of the finite element technique with the total displacement for each component coordinate being calculated by a superposition of the displacements obtained with fixed and relaxed interface conditions. In 1965, Hurty [213] proposed the Normal mode or Component Mode Synthesis method, at that time focusing his work on structural systems and taking into account both the elastic and mass properties of the component. Later, Craig and Bampton [214] reformulated Hurty's method by simplifying the choice of the groups of modes for the transformation matrix construction.

In the free-interface method the natural modes are obtained from a subsystem vibrating either in its free-interface or completely free support condition. Some of the early works reporting the use of free-interface modes were presented by Gladwell [215] and Goldman [216]. Hou's work [217], which presents some similarities with Goldman's method, uses a less complicated procedure for generating the system transformation matrix.

In some survey papers by Craig [218], Nelson [219], and Hurty *et al* [220], it has been commonly stated that in the free-interface methods like Hou's and Goldman's procedures, very poor accuracy may be obtained for the overall system natural frequencies and mode shapes compared with the accuracy produced by the fixed-interface methods. However, it was then recognised that the free-interface methods could take advantage of using the directly available data from substructure tests as an input into the coupling process, which has led to an improvement of the existing methods by developing a great variety of approaches. Some of these presented by MacNeal [221], Hintz [222], and Rubin [223] were primarily based on a purely analytical description.

Other methods tried to explore the use of experimentally derived modal properties as a basis for the formulation of the equations of motion of each subsystem. In this latter approach, one important early work is Klosterman's thesis [67] which provides a comprehensive study of the experimental determination of modal representations of components including the use of those models in the coupling of substructures.

In the case where the components are rigidly connected, the use of a set of truncated modes to establish the compatibility equations leads sometimes to unacceptable errors in the system response predictions. In such a case, a more accurate definition is necessary either by using more modes or, if these represent an unreasonable number, by providing some information about the effects of the neglected modes. The lack of definition of the component properties may be overcome in two ways; on the one hand, seeking to compensate for a lack of flexibility due to truncation of the set of natural modes by using additional and important information concerning the flexibility effect of the out-of-range modes. On the other hand, by using additional masses attached to the connection points in an attempt to generate a more realistic condition for the component when it is vibrating together with the remaining parts, the localised flexibility properties near the connection area are better represented, since more modes are brought to the frequency range of interest.

The first alternative was presented by MacNeal [221] and Rubin [223] in order to improve the truncated free-free modal representation of a component by including estimates of the residual effects due to the modes higher in frequency than the frequency range of interest. In general, the residual effects are obtained by calculating the component flexibility due to those modes to be retained and then subtracting this from the total flexibility of the respective component. Craig and Chang [224] discussed the coupling of substructures represented by Rubin's component-modes model.

All these works presented a significant improvement to the classical free-interface method, though still using purely analytical representation of component properties.

Since the free-interface method constitutes the most straightforward approach when the required data must be obtained from testing the components, some authors such as Martinez *et al* [225] and Coppolino [226] have directed their work in this direction.

The mass-loading technique may be classified as an intermediate technique between the fixed-interface method and the free-interface method. As mentioned before, the classical free-interface method is more sensitive to the truncation of the elastic modes used to describe the displacement in the connection region, rather than the fixed-interface method. In an attempt to bring the classical free-interface method (which is the most suitable for experimental purposes but does not offer good accuracy) closer to the fixed-interface method (which leads to better accuracy in the results but is not appropriate in experimental cases), the mass-loading technique represents a suitable compromise. It has been used in some specific fields of research, namely in the spacecraft industry.

5.2   IMPEDANCE COUPLING TECHNIQUES
5.2.1   Introduction
The generalised Impedance-based methods for the vibration analysis of complex structural assemblies are examined in the present section in order to evaluate their applicability when incomplete subsystem models are used. The subsystems or component models are described by using a Spatial or Response formulation, and special attention is given to the coordinate incompleteness of these models. However, it is assumed that whatever the performed reduction on the number of coordinates in each subsystem model, it will not be extended to the originally defined set of connecting coordinates which are explicitly required for the formulation of the constraint equations between subsystem models. This has a completely different effect in the coupling analysis, since by ignoring a connection coordinate we are making the subsystem unable to 'pass' some of its dynamic information to another one, when they are acting together.

For example, if a rotation coordinate is eliminated (or ignored) at the junction between two components, these are presumed to have relative motion with respect to that coordinate and would behave as pin-jointed. Thus, the set of the retained coordinates - the primary or master coordinates - always contains the originally defined interface set and in the most extreme situation the reduction process is assumed to condense all the dynamic properties only to the connecting coordinates.

5.2.2   Spatial Coupling method
This is the method mostly used in the Finite Element software packages, since the properties of each component are themselves derived from a suitable assembling of the analytically derived stiffness and mass matrices of the elements. This leads to a description of the component in terms of its spatially distributed properties. The dynamic characteristics of the overall structure are subsequently obtained using the

same assembling technique as used at the component level, but this time with the spatial matrices of the component playing the role of the matrices to be 'added'. The straightforward results of this assembling technique are the spatial matrices of the complete structure, i.e., mass, stiffness and sometimes damping matrices which, since they are known, can be input to an eigensolver to obtain the modal properties or can be used to generate the Response model.

Let us consider two undamped components A and B described by their spatial properties, the corresponding mass and stiffness matrices being of orders N_A and N_B respectively, each being partitioned according to the selected interior and connection coordinates. The equations of equilibrium for each subsystem, acted on only by interconnecting forces, are

$$
\begin{bmatrix} [_A M_{ii}] & [_A M_{ic}] \\ [_A M_{ci}] & [_A M_{cc}] \end{bmatrix} \begin{Bmatrix} \{_A \ddot{u}_i\} \\ \{_A \ddot{u}_c\} \end{Bmatrix} + \begin{bmatrix} [_A K_{ii}] & [_A K_{ic}] \\ [_A K_{ci}] & [_A K_{cc}] \end{bmatrix} \begin{Bmatrix} \{_A u_i\} \\ \{_A u_c\} \end{Bmatrix} = \begin{Bmatrix} \{_A 0_i\} \\ \{_A f_c\} \end{Bmatrix} \tag{5.8}
$$

$$
\begin{bmatrix} [_B M_{ii}] & [_B M_{ic}] \\ [_B M_{ci}] & [_B M_{cc}] \end{bmatrix} \begin{Bmatrix} \{_B \ddot{u}_i\} \\ \{_B \ddot{u}_c\} \end{Bmatrix} + \begin{bmatrix} [_B K_{ii}] & [_B K_{ic}] \\ [_B K_{ci}] & [_B K_{cc}] \end{bmatrix} \begin{Bmatrix} \{_B u_i\} \\ \{_B u_c\} \end{Bmatrix} = \begin{Bmatrix} \{_B 0_i\} \\ \{_B f_c\} \end{Bmatrix} \tag{5.9}
$$

where the subscript c denotes the n_c coordinates involved in the physical connection and i the remaining coordinates for components A and B.

The compatibility of displacements and equilibrium of forces between the subsystems undergoing free vibrations are expressed by the following equations:

$$
\{_A f_c\} = -\{_B f_c\} = \{f_c\} \quad \text{and} \quad \{_A u_c\} - \{_B u_c\} = \{0\} \tag{5.10}
$$

By invoking these equations, the overall system mass and stiffness matrices will be of order $N_C = N_A + N_B - n_c$ and are given by

$$
\underset{(N_C \times N_C)}{[_C M]} = \underset{(N_A \times N_A)}{[_A M]} \oplus \underset{(N_B \times N_B)}{[_B M]} \tag{5.11}
$$

$$
\underset{(N_C \times N_C)}{[_C K]} = \underset{(N_A \times N_A)}{[_A K]} \oplus \underset{(N_B \times N_B)}{[_B K]} \tag{5.12}
$$

the operation sign $\oplus$ meaning the following assembly of the above-mentioned submatrices:

$$
\underset{(N_C \times N_C)}{[_C M]} = \begin{bmatrix} [_A M_{ii}] & [_A M_{ic}] & [0] \\ [_A M_{ci}] & [_A M_{cc}] + [_B M_{cc}] & [_B M_{ci}] \\ [0] & [_B M_{ic}] & [_B M_{ii}] \end{bmatrix} \tag{5.13}
$$

$$\underset{(N_C \times N_C)}{[_C K]} = \begin{bmatrix} [_A K_{ii}] & [_A K_{ic}] & [0] \\ [_A K_{ci}] & [_A K_{cc}] + [_B K_{cc}] & [_B K_{ci}] \\ [0] & [_B K_{ic}] & [_B K_{ii}] \end{bmatrix} \tag{5.14}$$

Since the mass and stiffness matrices of the global system are known, a classical eigensolution will give the natural frequencies and mode shape vectors. This analysis is suitable for the use of theoretically derived subsystems, but it is not generally used in cases where the data available are obtained from modal testing.

5.2.3   FRF Coupling method

The FRF Coupling analysis method makes use of subsystem models derived directly from FRF data (commonly available from experimental studies but seldom from theoretical modelling). The dynamic properties of those models are synthesised in terms of the FRF matrix and generally denoted as [H(ω)] (such as Receptance, Mobility or Accelerance matrices).

As in the preceding analysis, the coordinates involved in the connection between components A and B should be identified and represented by subscript c (and similarly i for the remaining ones) leading to the following partitioned FRF matrices:

$$\underset{(N_A \times N_A)}{[_A H(\omega)]} = \begin{bmatrix} [_A H_{ii}] & [_A H_{ic}] \\ [_A H_{ci}] & [_A H_{cc}] \end{bmatrix} \tag{5.15}$$

$$\underset{(N_B \times N_B)}{[_B H(\omega)]} = \begin{bmatrix} [_B H_{ii}] & [_B H_{ic}] \\ [_B H_{ci}] & [_B H_{cc}] \end{bmatrix} \tag{5.16}$$

By invoking the constraint equations (5.10) used in the previously presented Spatial coupling method, the FRF matrix of the coupled structure is provided by a similar 'addition', as

$$\underset{(N_C \times N_C)}{[_C H(\omega)]} = \left(\underset{(N_A \times N_A)}{[_A H(\omega)]^{-1}} \oplus \underset{(N_B \times N_B)}{[_B H(\omega)]^{-1}} \right)^{-1}_{(N_C \times N_C)} \tag{5.17}$$

In the generalised impedance matrix $[Z(\omega)] = [H(\omega)]^{-1}$ the operation sign $\oplus$ means the following assembly of the corresponding partitioned impedance matrices:

$$
\underset{(N_C \times N_C)}{\left[_C H(\omega) \right]} = \left[\begin{array}{ccc} \left[_A Z_{ii} \right] & \left[_A Z_{ic} \right] & [0] \\ \left[_A Z_{ci} \right] & \left[_A Z_{cc} \right] + \left[_B Z_{cc} \right] & \left[_B Z_{ci} \right] \\ [0] & \left[_B Z_{ic} \right] & \left[_B Z_{ii} \right] \end{array} \right]^{-1}
\qquad (5.18)
$$

The FRF matrices of each substructure (available over a frequency range of interest, which is the same for both structures) are 'added' together frequency by frequency until the whole FRF matrix is completely calculated. Equation (5.18) can be generalised to include more than two subsystems without having to perform once again all the frequency by frequency 'addition', provided the generalised impedance matrices of each component are suitably assembled as shown in figure 5.2 for the case of three components.

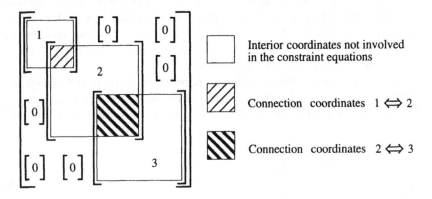

Fig. 5.2 Assembly of Impedance matrices.

As shown in (5.17), the whole system FRF matrix is obtained after three matrix inversions, two of them carried out on the subsystem's FRF matrices and another on the assembled impedance matrix. It is worth noting that those three inversions can be reduced to just one, as it will be mentioned in Section 5.4.2.

5.2.4 Incomplete or reduced subsystem models
Spatial model incompleteness
In many practical situations such as those when using the Finite Element method, the computational limitations often require the order of the final coupled system to be reduced as much as possible. This is achieved by reducing the order (N) of each subsystem model - or in other words, by confining our interest only to a restricted set of coordinates on each subsystem - say, $n_p < N$ - and/or to only some of the modes ($m_k < N$).

These considerations lead to the formulation of condensed, reduced or incomplete models.

The number of DOFs to be retained in the dynamic analysis (at least the DOFs involved in the connection with other components) is specified by the user. In the case of Spatial models, a transformation matrix relating the remaining DOFs (called secondary or slave coordinates) to those retained (primary or master coordinates) is used to reduce the order of the subsystem. The reduction process [227] is often performed upon a transformation which neglects inertia or static contributions for the eliminated DOFs, and is then used to derive the $n_p \times n_p$ spatial matrices of the condensed system, $[M^R]$ and $[K^R]$.

The equation of equilibrium for an undamped subsystem acted on only by forces on the primary (master) coordinates can be written in the following partitioned form:

$$\begin{bmatrix} [M_{ss}] & \vdots & [M_{sp}] \\ \cdots\cdots & \vdots & \cdots\cdots \\ [M_{ps}] & \vdots & [M_{pp}] \end{bmatrix} \begin{Bmatrix} \{\ddot{u}_s\} \\ \cdots \\ \{\ddot{u}_p\} \end{Bmatrix} + \begin{bmatrix} [K_{ss}] & \vdots & [K_{sp}] \\ \cdots\cdots & \vdots & \cdots\cdots \\ [K_{ps}] & \vdots & [K_{pp}] \end{bmatrix} \begin{Bmatrix} \{u_s\} \\ \cdots \\ \{u_p\} \end{Bmatrix} = \begin{Bmatrix} \{0_s\} \\ \cdots \\ \{f_p\} \end{Bmatrix} \qquad (5.19)$$

secondary DOF: $\{u_s\}$ *primary* DOF: $\{u_p\}$
$\qquad\qquad (n_s x1)$ $\qquad\qquad (n_p x1)$

All the coordinates can be related to the primary coordinates by using the following transformation:

$$\begin{Bmatrix} \{u_s\} \\ \cdots \\ \{u_p\} \end{Bmatrix} = \begin{bmatrix} [T] \\ {\scriptstyle(n_s xn_p)} \\ \cdots\cdots \\ [I] \\ {\scriptstyle(n_p xn_p)} \end{bmatrix} \{u_p\} \qquad (5.20)$$

where the matrix [T] is given by

$$\underset{(n_s xn_p)}{[T]} = (1-\beta)\left[-[K_{ss}]^{-1}[K_{sp}]\right] + \beta\left[-[M_{ss}]^{-1}[M_{sp}]\right] \qquad (5.21)$$

β being a reduction coefficient whose limits are $\beta=0$ for static reduction (Guyan reduction [228]) and $\beta=1$ for dynamic reduction.

The matrices describing the reduced Spatial model are given as

$$\underset{(n_p xn_p)}{[M^R]} = \underset{(n_p xN)}{\left[[T]^T \vdots [I]\right]} \begin{bmatrix} [M_{ss}] & \vdots & [M_{sp}] \\ \cdots\cdots & \vdots & \cdots\cdots \\ [M_{ps}] & \vdots & [M_{pp}] \end{bmatrix} \begin{bmatrix} [T] \\ \cdots \\ [I] \end{bmatrix} \qquad (5.22)$$

$\qquad\qquad\qquad\qquad\qquad\qquad\qquad\qquad {\scriptstyle(NxN)} \qquad {\scriptstyle(Nxn_p)}$

$$\begin{bmatrix} K^R \end{bmatrix}_{(n_p x n_p)} = \begin{bmatrix} [T]^T & \vdots & [I] \end{bmatrix}_{(n_p x N)} \begin{bmatrix} [K_{ss}] & \vdots & [K_{sp}] \\ \cdots & & \cdots \\ [K_{ps}] & \vdots & [K_{pp}] \end{bmatrix}_{(NxN)} \begin{bmatrix} [T] \\ \cdots \\ [I] \end{bmatrix}_{(Nxn_p)} \tag{5.23}$$

Whatever value is assumed for the coefficient β, the reduction in the DOFs implies a reduction in the available existing modes as well. The validity of the condensed model generally depends on the mass and stiffness values assigned to the secondary coordinates. Generally, the use of frequency dependent mass matrices improves the accuracy of the reduced model properties, as shown by Kuhar and Stahle [229].

Modal model incompleteness
The incompleteness in the Modal model is generally due to the inherent difficulty in attempting to use experimental data to define a finite model for a continuous subsystem. It is not realistic to undertake measurements either in all the coordinates and/or over a frequency range encompassing all natural frequencies. In the present analysis, however, we shall assume that all the N natural frequencies are known in the frequency range of interest, and the limitation is only concerned with the number of measured coordinates, say $n_p < N$, selected for the FRF measurements.

Let us assume that a subsystem Modal model is described by its (incomplete) modal properties:

$$\begin{bmatrix} \Phi^R \end{bmatrix}_{(n_p x N)}$$
    - rectangular modal matrix (mass-normalised eigenvectors $n_p < N$)

$$\begin{bmatrix} \ddots & \omega_r^2 & \ddots \end{bmatrix}_{(NxN)}$$
    - eigenvalues (diagonal matrix)

It is important to note that a Guyan-reduced Spatial model leads to an incomplete Modal model in terms of coordinates and modes, even though the modal matrix is square.

Depending on the format assumed for the other subsystems' dynamic properties description, the Modal model may be converted as described next.

Modal model ⇨ Response model
In this case, it is assumed that the Modal model is to be converted into the Response model, which is required by the FRF Coupling method. Let us take the receptance FRF matrix, which is given by (see equation (5.7) and also (1.264)):

$$\begin{bmatrix} \alpha(\omega)^R \end{bmatrix}_{(n_p x n_p)} = \begin{bmatrix} \Phi^R \end{bmatrix}_{(n_p x N)} \begin{bmatrix} \ddots & (\omega_r^2 - \omega^2) & \ddots \end{bmatrix}_{(NxN)}^{-1} \begin{bmatrix} \Phi^R \end{bmatrix}_{(Nxn_p)}^T \tag{5.24}$$

This FRF matrix, although limited to only n_p points of interest on the substructure, contains information on all the N modes and is thus accurate for those coordinates retained. In order to study the effects of this reduction, while the models are described by their modal properties, we shall consider an ideal Modal model which possesses the complete information; i.e., $[\Phi]$ and $[\ \omega_r^2 \ ]$ have the dimensions $N \times N$.

The deletion of one coordinate - say, the kth - corresponds to the elimination of the kth row in the modal matrix so that we are ignoring the relative amplitude of that DOF in all the modes. This does not mean that the real system no longer has that DOF; only that we are not including information about its motion. In this way we are condensing the information in previously selected coordinates.

The reduced $n_p \times n_p$ dynamic stiffness matrix will be calculated as $[Z(\omega)^R] = [\alpha(\omega)^R]^{-1}$. In the limiting case when a zero frequency value ω is assumed, that $n_p \times n_p$ reduced matrix $[Z(0)^R]$ corresponds to the well-known Guyan reduced stiffness matrix [228].

Modal model $\Rightarrow$ Spatial model
In this case, the reduced $n_p \times n_p$ modal matrix $[\Phi^R]$ is obtained by eliminating both $(N - n_p)$ coordinates and modes. The reduced dynamic stiffness matrix of the subsystem can be obtained as:

$$\underset{(n_p \times n_p)}{\left[\alpha(\omega)^{-1}\right]^R} = \underset{(n_p \times n_p)}{\left[\Phi^T\right]^{-1}} \underset{(n_p \times n_p)}{\left[\ (\omega_r^2 - \omega^2)_{\searrow}\ \right]} \underset{(n_p \times n_p)}{[\Phi]^{-1}} \tag{5.25}$$

or in terms of its spatial properties, as

$$\underset{(n_p \times n_p)}{\left[Z(\omega)^R\right]} = \underset{(n_p \times n_p)}{[G]} - \omega^2 \underset{(n_p \times n_p)}{[E]} \tag{5.26}$$

with

$$\underset{(n_p \times n_p)}{[G]} = \underset{(n_p \times n_p)}{\left[\Phi^T\right]^{-1}} \underset{(n_p \times n_p)}{\left[\ \omega_r^2 \ {}_{\searrow}\ \right]} \underset{(n_p \times n_p)}{[\Phi]^{-1}} \tag{5.27}$$

$$\underset{(n_p \times n_p)}{[E]} = \underset{(n_p \times n_p)}{\left[\Phi^T\right]^{-1}} \underset{(n_p \times n_p)}{[\Phi]^{-1}} \tag{5.28}$$

Matrices [G] and [E] may be treated as possessing stiffness and mass properties, respectively, for the subsystem, taking for analogy equations (1.261).

It would be expected that the results obtained for the response of a coupled structure using those spatial properties, i.e., by calculating $[\alpha(\omega)^{-1}]^R$, should give the same results as when we use the inverse of the reduced Response model matrix $[\alpha(\omega)^R]^{-1}$. However, it should be noted that matrices [G] and [E] do not represent the true stiffness and mass matrices for the component. In the particular (static) case of $\omega = 0$, those matrices are the same as obtained via a Guyan reduction (see expressions (5.23) and (5.24)).

5.2.5 Procedures to reduce models
The different possibilities of performing a reduction on the original models are now grouped into three procedures which lead to the final format required, either the Spatial or FRF coupling technique.

Although a brief examination is carried out on the ways and effects of reducing the theoretically derived models - Spatial models - attention is mainly focused on the ways and effects of using experimentally derived models - Response models - which by nature are themselves incomplete in terms of coordinates and prone to errors even though they contain information related to both the in- and out-of-range frequency of interest modes.

The attractive and conceptually simple technique that offers the possibility of dealing directly with measured data is the FRF coupling technique. The numerical difficulties associated with the corresponding algorithm in some particular situations, especially in the presence of measured data, has motivated the investigation of new approaches to tackle this problem.

<u>Procedure A</u> using the Spatial Coupling method:

Spatial model   Spatial model

$$[K], [M] \quad \Rightarrow \quad \text{Guyan Reduction} \quad \Rightarrow \quad \left[K^R\right], \left[M^R\right]$$
$$\text{(NxN)} \quad \text{(NxN)} \qquad\qquad\qquad\qquad\qquad (n_p x n_p) \quad (n_p x n_p)$$

(Reduced Stiffness and Mass matrices)

<u>Procedure B</u> using the Spatial Coupling method:

Modal model   Spatial model

$$[\Phi] \quad \Rightarrow \quad \left[\Phi^R\right] \quad \Rightarrow \quad \left[\alpha(\omega)^{-1}\right]^R \quad \Rightarrow \quad [G] \quad\quad [E]$$
$$\text{(NxN)} \qquad (n_p x n_p) \qquad\quad (n_p x n_p) \qquad\quad (n_p x n_p) \quad (n_p x n_p)$$

(Reduced Stiffness and Mass matrices, for $\omega = 0$)

In both previously mentioned procedures, all the subsystems are converted into a final reduced Spatial model. The result for the overall structure is then obtained in terms of its mass and stiffness properties.

Procedure C using the FRF Coupling method:

Modal model   Response model

$$[\Phi] \quad \Rightarrow \quad \left[\Phi^R\right] \quad \Rightarrow \quad [\alpha(\omega)]^R \quad \Rightarrow \quad \left[\alpha(\omega)^R\right]^{-1}$$
$$(\text{NxN}) \qquad\qquad (n_p \text{xN}) \qquad\qquad (n_p \text{x} n_p) \qquad\qquad (n_p \text{x} n_p)$$

Procedure D using the FRF Coupling Method:

Modal model   Response model

$$[\Phi] \quad \Rightarrow \quad \left[\Phi^R\right] \quad \Rightarrow \quad [\alpha(\omega)]^R \quad \Rightarrow \quad \left[\alpha(\omega)^R\right]^{-1}$$
$$(\text{NxN}) \qquad\qquad (n_p \text{x} n_p) \qquad\qquad (n_p \text{x} n_p) \qquad\qquad (n_p \text{x} n_p)$$

In these two last procedures the final format for each component is a reduced Response model, herein assumed as the Receptance model.

5.2.6   The need to use alternative FRF coupling methods
The results achieved with the previously mentioned coupling procedures applied to several examples [198] showed that the incompleteness in terms of coordinates in the component models does not play an important role in the prediction of the global structure results whenever the FRF coupling method is used. It should be noted, however, that the incompleteness referred to is only related to the interior coordinates. At this stage it is therefore assumed that the formulation of the constraints among the subsystems is well undertaken, i.e., one is using the necessary and sufficient number of connecting coordinates which, unfortunately, is not always obvious.

Occurrence of ill-conditioned FRF matrices
The results achieved with the use of the FRF coupling method showed that when experimentally simulated data are used to represent each subsystem Response model, the predicted final response possesses some undesired false peaks [198]. The explanation of this phenomenon can be found by looking at the interrelationship between the Response and the Modal models. In the vicinity of each subsystem resonance frequency the corresponding FRF matrix is largely dominated by one single term, or in other words, the matrix will have order n (number of primary coordinates) but also tends to have rank 1 (the single dominating mode). This means that every FRF matrix tends to be rank-deficient or nearly singular in the vicinity of each subsystem natural frequency, especially in the case of lightly damped structures where the local dominance of a single mode is strongest.

So long as a purely theoretical FRF matrix is of concern near a resonance frequency, the inversion - although applied to a matrix having a high condition number - leads to a correct dynamic stiffness matrix subsequently used in the assembling process. However, when measured FRF matrices are dealt with, their elements are prone to errors which will cause a pronounced deviation of the inverse matrix from the existing but unknown ideal dynamic stiffness matrix. The result of these perturbed matrices on each single component will show up in the predicted response mainly at the frequencies close to each subsystem's natural frequencies, as discussed by Larsson [230].

A remedy suggested by Ewins [209] to avoid the existence of these extra peaks is the smoothing of each FRF before it is used to create the FRF matrix. In order to do so, the Response model is converted to a Modal one which is subsequently used to regenerate the smoothed FRF matrix which will behave better in numerical terms, therefore giving a more accurate prediction of the global Response model. This approach, however, withdraws the main virtue of the FRF coupling technique which is the direct use of what is measured on the actual components without the need of a complementary identification stage.

Besides the occurrence of the ill-conditioned matrices near a resonance frequency, presented above, there are other situations when the required coupling algorithms can behave erratically, even for other frequencies in the range of interest. It suffices to have any linear dependency or near-dependency among some of the rows (or columns) in a given FRF matrix and the inversion may well fail. Unfortunately, this is a common practical problem when dealing with the subsystem models, mainly due to the following two reasons:

- the FRF matrix has been generated from a set of inadequate modal data. The procedure C presented in Section 5.2.5 shows that in the case of a modal matrix containing more coordinates (n_p) than modes (N) or, in practice m_k kept modes, the generated FRF matrix will be of order n_p, although having rank m_k; in that case the matrix is singular, since there exist ($n_p - m_k$) linearly dependent rows (or columns) and it is impossible to calculate the corresponding inverse, even though the subsystem dynamic stiffness matrix generally exists. To prevent such a situation, more modes should be included (up to the number of coordinates) and very likely the undesired singularity will be removed. Should it be impracticable to apply this remedy, other alternatives can be found such as those making use of more sophisticated algorithms enabling the calculation of the closest inverse of a rank-deficient matrix (see Appendix A);

- the FRF matrix has been measured on a structure in which some of the coordinates are situated on locally rigid regions - either interior or interface ones - causing the responses over a certain frequency range to be linearly dependent on each other. In this case the FRF matrix tends to be, or is, rank deficient. Should at least one of the components behave in this manner, the coupling will fail numerically. A way of circumventing this problem is to detect the redundant coordinates prior to the coupling of the components and, if possible, neglect them or, once again to make use of proper algorithms [231].

Model inconsistency

The Response model can be fully created over a certain frequency range by measuring either: (i) all the elements in the FRF matrix (or only the upper or lower triangular matrix assuming a symmetric matrix) or (ii) one single row or column of the corresponding matrix enabling the identification of the corresponding Modal model. In the latter case the full set of FRFs can subsequently be generated over the selected frequency range provided that the effect of the out-of-range modes can be neglected. In practical terms the quality of the measured set of FRFs can be checked by comparing the various estimates of the modal parameters with each other. It often happens that a unique set of modal parameters cannot be extracted directly from the same set of FRFs and then the model is said to suffer from self-inconsistency.

5.2.7 Summary

The main conclusions concerning the application of different procedures to several coupling exercises [232] can be summarised as:

- The use of reduced Spatial models in both procedures A and B leads to results whose accuracy depends on the coordinates eliminated - it is a question of mass value dependency, so the choice of the secondary coordinates should be made according to the relative mass values (when a coordinate with a high relative mass value is eliminated the dynamic information of the subsystem is altered and may affect the overall system behaviour).

- The only procedure that is 'insensitive' to the degree of incompleteness in terms of coordinates for the subsystem model is procedure C, generally giving quite accurate results.

- In procedure D, which is a particular case of procedure C, the results are not so accurate, since the residual effects of the neglected modes are not included.

5.3   MODAL COUPLING TECHNIQUES

5.3.1 Introduction

In this section we shall be concerned with Modal Coupling techniques, also referred to in the related literature as Time Domain or Component Mode Synthesis methods. The basic philosophy is the same as for the previous group, i.e., they permit the use of reduced component models in order to achieve a reduced order in the final equation of motion matrices of the assembled structure. However, unlike the Impedance based methods, which take advantage of the reduction of the number of coordinates, these methods use a reduction performed on the number of modes used to describe each component model while still accounting for all the physical DOFs. By using a Ritz-type transformation, the reduced number of principal coordinates is related to the number of modes that are taken into account for the modal estimation; generally, the information relating to the higher natural frequency modes is discarded. This feature of the method reduces the computational effort required in the system analysis and parallels the modal information in a real test, since it is only possible to measure some of the existing modes in a structure. The objective is to maintain a specified level of accuracy in the dynamic analysis, while using only the lower order modes for computational efficiency.

Amongst the different methods available, two groups emerge and may be classified as fixed-interface and free-interface methods (see figure 5.3). In both these methods, the solution of the original model is approximated by a summation of assumed displacements or Ritz vectors, their amplitudes being the generalised coordinates used to reduce the order of the original model. The Ritz vectors are selected in a predetermined fashion based either on the dynamic characteristics only or on both the dynamic and static characteristics of the individual subsystems.

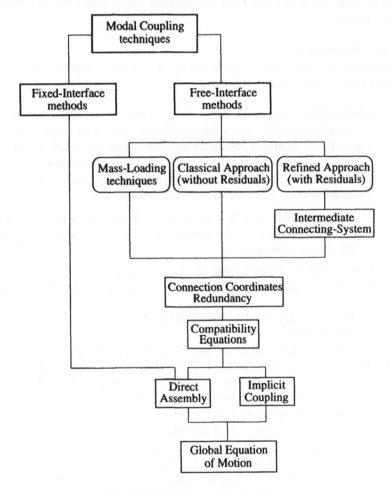

Fig. 5.3 Modal coupling techniques.

Essentially, the methods differ from each other according to the dynamic displacement shapes used to form the truncated set of natural modes. In the first method, the elastic modes pertaining to a fixed-interface component are retained, whereas in the second method, the modes are obtained by assuming the component to be vibrating in a freely-supported condition at its attachment points.

5.3.2 Fixed-Interface methods

The basic idea for the fixed-interface methods relies on the philosophy of the static substructuring technique and is mainly directed towards the use of the finite element method. The total displacements for each component are calculated by a superposition of the displacements obtained with fixed and relaxed interface conditions. The methodology is summarised next, always assuming that the spatial properties of each component are known.

The corresponding equation of motion, according to the selected connecting and interior degrees-of-freedom (DOF), can be written in the following partitioned form:

$$\begin{bmatrix} [M_{ii}] & \vdots & [M_{ic}] \\ \cdots & \vdots & \cdots \\ [M_{ci}] & \vdots & [M_{cc}] \end{bmatrix} \begin{Bmatrix} \{\ddot{u}_i\} \\ \cdots \\ \{\ddot{u}_c\} \end{Bmatrix} + \begin{bmatrix} [K_{ii}] & \vdots & [K_{ic}] \\ \cdots & \vdots & \cdots \\ [K_{ci}] & \vdots & [K_{cc}] \end{bmatrix} \begin{Bmatrix} \{u_i\} \\ \cdots \\ \{u_c\} \end{Bmatrix} = \begin{Bmatrix} \{f_i\} \\ \cdots \\ \{f_c\} \end{Bmatrix} \qquad (5.29)$$

where:

- $\{u_c\}$ and $\{u_i\}$ are the displacements at the n_c connection (primary) and at the n_i interior (secondary) coordinates respectively:

- $\{f_c\}$ and $\{f_i\}$ are the forces at the n_c connection (primary) and at the n_i interior (secondary) coordinates respectively.

Assuming now that the n_c connection coordinates are fixed, $\{u_c\} = \{0\}$, and that no external forces are acting at the interior DOFs, $\{f_i\} = \{0\}$, the corresponding equation of motion becomes:

$$[M_{ii}]\{\ddot{u}_i\} + [K_{ii}]\{u_i\} = \{0\} \qquad (5.30)$$

The associated eigensolution consists of $m = n_i$ mass-normalised eigenvectors $[\Phi_{im}]$ and the respective eigenvalues $[\ \omega^2_{rm}\ \diagdown\ ]$. Each interior DOF displacement can now be approximated by a linear combination of the known fixed-interface modes, as

$$\{u_i\} = [\Phi_{im}]\{p_m\} \qquad (5.31)$$

The second kind of modes required to approximate the displacements at the interior coordinates are the constraint or static modes, which are calculated by relaxing each connection coordinate, but now neglecting the mass properties of the interior DOF. This is in fact a Guyan static reduction [228], as described in Section 5.2.4.

Let us assume that displacements at the interior coordinates $\{u_i\}$ are related to the connection ones $\{u_c\}$ as follows:

$$\underset{(n_i x 1)}{\{u_i\}} = \underset{(n_i x n_c)}{\left[\Phi_{ic}^*\right]} \underset{(n_c x 1)}{\{u_c\}} \tag{5.32}$$

where

$$\left[\Phi_{ic}^*\right] = -\left[K_{ii}\right]^{-1}\left[K_{ic}\right] \tag{5.33}$$

The static or constraint modes are thus the columns of the Guyan transformation matrix $\left[\Phi_{ic}^*\right]$. A Ritz-type transformation matrix can now be constructed in terms of both elastic (fixed-interface) and static (constraint) modes, as presented next:

$$\begin{Bmatrix} \{u_i\} \\ \cdots \\ \{u_c\} \end{Bmatrix} = \begin{bmatrix} \left[\Phi_{im}\right] & \vdots & \left[\Phi_{ic}^*\right] \\ \cdots & & \cdots \\ [0] & \vdots & [I] \end{bmatrix} \begin{Bmatrix} \{p_m\} \\ \cdots \\ \{u_c\} \end{Bmatrix} = [T] \begin{Bmatrix} \{p_m\} \\ \cdots \\ \{u_c\} \end{Bmatrix} \tag{5.34}$$

The main advantage of this subsystem description is that it is possible to truncate the number of modes or modal coordinates (generally the higher natural frequency modes), while still accounting for all the physical DOFs. Assuming that only the first m_k of the total m modes are known, (5.34) can now be written as:

$$\begin{Bmatrix} \{u_i\} \\ \cdots \\ \{u_c\} \end{Bmatrix} = \begin{bmatrix} \left[\Phi_{ik}\right] & \vdots & \left[\Phi_{ic}^*\right] \\ \cdots & & \cdots \\ [0] & \vdots & [I] \end{bmatrix} \begin{Bmatrix} \{p_k\} \\ \cdots \\ \{u_c\} \end{Bmatrix} = [T_k] \begin{Bmatrix} \{p_k\} \\ \cdots \\ \{u_c\} \end{Bmatrix} \tag{5.35}$$

Substituting equation (5.35) into (5.29) (only considering the interconnecting forces) and pre-multiplying by $[T_k]^T$ leads to:

$$[T_k]^T[M][T_k] \begin{Bmatrix} \{\ddot{p}_k\} \\ \cdots \\ \{\ddot{u}_c\} \end{Bmatrix} + [T_k]^T[K][T_k] \begin{Bmatrix} \{p_k\} \\ \cdots \\ \{u_c\} \end{Bmatrix} = [T_k]^T \begin{Bmatrix} \{0\} \\ \cdots \\ \{f_c\} \end{Bmatrix} \tag{5.36}$$

or in a partitioned form,

$$\begin{bmatrix} \left[\tilde{M}_{kk}\right] & \vdots & \left[\tilde{M}_{kc}\right] \\ \cdots & & \cdots \\ \left[\tilde{M}_{ck}\right] & \vdots & \left[\tilde{M}_{cc}\right] \end{bmatrix} \begin{Bmatrix} \{\ddot{p}_k\} \\ \cdots \\ \{\ddot{u}_c\} \end{Bmatrix} + \begin{bmatrix} \left[\tilde{K}_{kk}\right] & \vdots & \left[\tilde{K}_{kc}\right] \\ \cdots & & \cdots \\ \left[\tilde{K}_{ck}\right] & \vdots & \left[\tilde{K}_{cc}\right] \end{bmatrix} \begin{Bmatrix} \{p_k\} \\ \cdots \\ \{u_c\} \end{Bmatrix} = [T_k]^T \begin{Bmatrix} \{0\} \\ \cdots \\ \{f_c\} \end{Bmatrix} \tag{5.37}$$

The next operation is at the system level when all the component matrices are assembled together according to the compatibility and equilibrium equations necessary to describe the physical connections.

Assuming for simplification that $\{_A u_c\}$ and $\{_B u_c\}$ already represent displacements referred to the original global coordinate system u, the final equation for the coupled structure is now given in a new coordinate system by:

$$
\begin{bmatrix}
\begin{bmatrix} _A I_{kk} \end{bmatrix} & [0] & \begin{bmatrix} _A \tilde{M}_{kc} \end{bmatrix} \\
[0] & \begin{bmatrix} _B I_{kk} \end{bmatrix} & \begin{bmatrix} _B \tilde{M}_{kc} \end{bmatrix} \\
\begin{bmatrix} _A \tilde{M}_{ck} \end{bmatrix} & \begin{bmatrix} _B \tilde{M}_{ck} \end{bmatrix} & \begin{bmatrix} _A \tilde{M}_{cc} \end{bmatrix} + \begin{bmatrix} _B \tilde{M}_{cc} \end{bmatrix}
\end{bmatrix}
\begin{Bmatrix}
\{_A \ddot{p}_k\} \\
\{_B \ddot{p}_k\} \\
\{\ddot{u}_c\}
\end{Bmatrix} +
$$

$$
\begin{bmatrix}
\begin{bmatrix} _A \omega_r^2 \backslash \end{bmatrix} & [0] & [0] \\
[0] & \begin{bmatrix} _B \omega_r^2 \backslash \end{bmatrix} & [0] \\
[0] & [0] & \begin{bmatrix} _A \tilde{K}_{cc} \end{bmatrix} + \begin{bmatrix} _B \tilde{K}_{cc} \end{bmatrix}
\end{bmatrix}
\begin{Bmatrix}
\{_A p_k\} \\
\{_B p_k\} \\
\{u_c\}
\end{Bmatrix} =
\begin{Bmatrix}
\{0\} \\
\{0\} \\
\{0\}
\end{Bmatrix}
\tag{5.38}
$$

The natural frequencies and mode shapes $[\Psi]$ for the overall system are obtained by solving the eigenproblem associated with (5.38). The displacements in the original coordinates may be represented as a transformation of a new coordinate system $\{\zeta\}$:

$$
\begin{Bmatrix}
\{_A u_i\} \\
\{_B u_i\} \\
\{u_c\}
\end{Bmatrix} =
\begin{bmatrix}
\begin{bmatrix} _A \Phi_{ik} \end{bmatrix} & [0] & \begin{bmatrix} _A \Phi_{ic}^* \end{bmatrix} \\
\begin{bmatrix} _B \Phi_{ik} \end{bmatrix} & [0] & \begin{bmatrix} _B \Phi_{ic}^* \end{bmatrix} \\
[0] & [0] & [I]
\end{bmatrix}
[\Psi] \{\zeta\}
\tag{5.39}
$$

Fixed-interface methods are widely applied in the cases when the component dynamic properties are described by their mass and stiffness matrices. In general, these methods are expected to be accurate when the final system allows the component of interest to have little motion near the attachment points, or if a flexible component is rigidly linked to a relatively stiff component. The mass associated with the connection DOFs is often neglected but, in turn, the local stiffness is accurately included.

The accuracy of the results predicted using a fixed-interface method can be improved further by expanding the transformation matrix given by (5.35) to include other type of 'modes', such as the attachment modes as presented by Craig and Chang [224]. This group of methods are easy to handle with theoretically defined mass and stiffness component matrices. However, from an experimental point of view they are not recommended, mainly due to two reasons:

- the imposition of a fixed-interface may be easy to implement when we are dealing with theoretical subsystem matrices, but it is very difficult or even impossible to simulate a perfectly fixed support condition during a modal test;

- some of the matrices necessary to calculate the constraint modes are very difficult and tedious to obtain by static or dynamic testing.

5.3.3 Free-Interface methods

This is a group of methods developed with the same basic idea as that outlined in the previous section, but this time using another type of transformation matrix for reduction purposes. In this case, the necessary natural modes incorporated in that matrix are those obtained from a subsystem vibrating either in its free-interface or completely free support condition. This being the most readily simulated condition during an experimental test, it constitutes an attractive technique for the use of combined experimental/theoretical dynamic analysis.

One of the main works using free-interface modes is that of Hou [217], where the components are assumed to be rigidly connected. For convenience of presentation, the free-interface methods will be presented as follows:

- first of all, the free-interface methods will be described according to the type of linkage between subsystems i.e., it may be rigid or flexible. At this stage, interest is confined to the type of existing physical connection, excluding the effects of the mode set truncation;

- in Section 5.3.4, a brief description of the mass-loading technique is given. This is an alternative technique which in fact tries to bring the classical free-interface method closer to the fixed-interface method.

Free-Interface method with rigid connection

Let us consider two undamped subsystems, A and B, which are described by their spatial properties, the corresponding matrices being partitioned according to the selected interior and connection DOFs as represented in figure 5.4.

The equations of equilibrium for both subsystems can be written similarly to (5.29), but this time acted only by the interconnecting forces $\{f_c\}$. For each component individually, the free-interface modes are obtained by setting $\{f_c\} = \{0\}$ and solving the resulting eigenvalue problem, leading to m natural frequencies and mode shapes. For each component the physical displacements $\{u\}$ can now be written as a series expansion of the orthogonal mode shape matrix $[\Phi_m]$ which contains up to six rigid body modes (if the structure is completely unrestrained) plus the elastic free-interface modes.

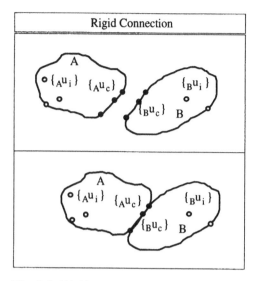

Fig. 5.4 Rigid connection of two subsystems.

$$\left\{\begin{array}{c}\{u_i\}\\ \cdots\cdots\\ \{u_c\}\end{array}\right\}=\left[\begin{array}{c}[\Phi_{im}]\\ \cdots\cdots\\ [\Phi_{cm}]\end{array}\right]\{p_m\}=[\Phi_m]\{p_m\} \qquad (5.40)$$

The vector $\{p_m\}$ contains the generalised or modal coordinates as weighting factors of the series expansion or, in other words, the amplitudes of each independent selected pattern (or mode shape). This equation is exact only for the cases when all the modes are represented.

However, in practical cases only a truncated set of m_k modes is considered or measured, leading to an approximate description of the displacements for each subsystem:

$$\left\{\begin{array}{c}\{_A u\}\\ \cdots\cdots\\ \{_B u\}\end{array}\right\}\approx\left[\begin{array}{c:c}[_A\Phi_k] & [0]\\ \hdashline [0] & [_B\Phi_k]\end{array}\right]\left\{\begin{array}{c}\{_A p_k\}\\ \cdots\cdots\\ \{_B p_k\}\end{array}\right\} \qquad (5.41)$$

Substituting these displacements into the equilibrium equations (5.29) results in a set of uncoupled equations for the disconnected components,

$$[I]\left\{\begin{array}{c}\{_A\ddot{p}_k\}\\ \cdots\cdots\\ \{_B\ddot{p}_k\}\end{array}\right\}+\left[\begin{array}{c:c}[\,_A\omega_{rk}^2\,\cdot\,] & [0]\\ \hdashline [0] & [\,_B\omega_{rk}^2\,\cdot\,]\end{array}\right]\left\{\begin{array}{c}\{_A p_k\}\\ \cdots\cdots\\ \{_B p_k\}\end{array}\right\}=\left[\begin{array}{c:c}[_A\Phi_{ck}^T] & [0]\\ \hdashline [0] & [_B\Phi_{ck}^T]\end{array}\right]\left\{\begin{array}{c}\{_A f_c\}\\ \cdots\cdots\\ \{_B f_c\}\end{array}\right\}$$

$$(5.42)$$

These are the equations which can be established either using the theoretical derivation hitherto presented or using the modal data base available from experimental modal tests. When both subsystems are connected, undergoing free vibrations together, the only forces acting on them are the equal and opposite forces at the interfaces. This equilibrium is expressed as

$$\{_A f_c\} = -\{_B f_c\} \tag{5.43}$$

while the corresponding compatibility equation for the interface displacements is

$$\{_A u_c\} = \{_B u_c\} \tag{5.44}$$

This constraint condition may be reformulated taking into account the approximation assumed in (5.41) and written as

$$\left[[_A \Phi_{ck}] \vdots -[_B \Phi_{ck}] \right] \left\{ \begin{array}{c} \{_A P_k\} \\ \cdots\cdots \\ \{_B P_k\} \end{array} \right\} = [S]\{p\} = \{0\} \tag{5.45}$$

To reduce the order of the equilibrium equations for the global system, matrix [S] and vector {p} may be partitioned as

$$\left[[S_d] \vdots [S_i] \right] \left\{ \begin{array}{c} \{p_d\} \\ \cdots\cdots \\ \{p_i\} \end{array} \right\} = \{0\} \tag{5.46}$$

where $[S_d]$ is a non-singular square matrix and $[S_i]$ is the remaining part of $[S]$. This requires that the total number of modes for both components ($m_{kt} = m_{kA} + m_{kB}$) be greater than the number of connection coordinates n_c.

Vector {p} should then be related to {p_i} through the following transformation:

$$\{p\} = \left\{ \begin{array}{c} \{_A P_k\} \\ \cdots\cdots \\ \{_B P_k\} \end{array} \right\} = [T]\{p_i\} \tag{5.47}$$

where

$$[T] = \left[\begin{array}{c} [S_d]^{-1}[S_i] \\ \cdots\cdots\cdots \\ [I] \end{array} \right] \tag{5.48}$$

To generate matrix [T], a set of m_i vectors must be obtained from matrix [S], while a set of m_d vectors is retained. This requirement may be difficult to satisfy, especially if some of the connection coordinate responses reveal a near dependency which will cause matrix $[S_d]$ to be ill-conditioned or even singular. A suitable process can be devised by applying the Singular Value Decomposition (SVD) technique (see Appendix A) to the matrix $[S]$, keeping only the independent vectors that will constitute a matrix of a defined rank, which may differ from its order - the number n_c of the attachment DOFs. Renaming $\{p_i\}$ as $\{q\}$, just for simplicity, the equilibrium equation referred to the $\{q\}$ coordinates is given as

$$\left[M_q\right]\{\ddot{q}\}+\left[K_q\right]\{q\}=\left\{f_q\right\} \tag{5.49}$$

where

$$\left[M_q\right]=[T]^T[T]$$

$$\left[K_q\right]=[T]^T\begin{bmatrix}\left[\,_A\omega_{rk}^2\,\right] & [0] \\ \hdotsfor{2} \\ [0] & \left[\,_B\omega_{rk}^2\,\right]\end{bmatrix}[T]$$

$$\left\{f_q\right\}=[T]^T\begin{bmatrix}\left[\,_A\Phi_{ck}\right]^T \\ \hdotsfor{1} \\ -\left[\,_B\Phi_{ck}\right]^T\end{bmatrix}\{f_c\} \tag{5.50}$$

The second term of (5.49) vanishes, since no external forces are acting on the coupled system. Thus, the solution of that equation gives the $(m_{kt} - n_c)$ natural frequencies $[\,\omega_r^2\,]$ and mode shapes $[\Psi]$ for the overall system, but referred to the q coordinates. The mode shapes are then transformed to the original coordinates u, according to:

$$[\Phi]=\begin{bmatrix}\left[\,_A\Phi_k\right] & [0] \\ \hdotsfor{2} \\ [0] & \left[\,_B\Phi_k\right]\end{bmatrix}[T][\Psi] \tag{5.51}$$

Free-Interface method with elastic connection
Let us assume now that the previously mentioned subsystems A and B are to be coupled through an intermediate flexible system C as shown in figure 5.5. In this case we assume that the Modal models for each component are already known, whatever method is used for their derivation. The equations of motion for both disconnected components have already been derived.

If the two components are now connected via an elastic system, the only constraint equation which expresses that condition is

$$\left\{_A f_c\right\}=-\left\{_B f_c\right\} \tag{5.52}$$

since, in this case, simple compatibility of displacements is not applicable, i.e.,

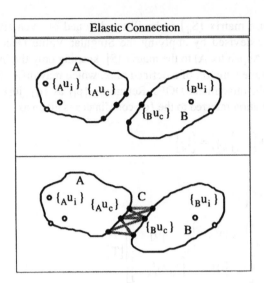

Fig. 5.5 Elastic connection of two subsystems.

$$\left\{ _A u_c \right\} \neq \left\{ _B u_c \right\} \tag{5.53}$$

However, the elastic properties of the connecting system can be represented by means of its stiffness matrix $[K_{cpl}]$, if the mass properties are neglected (although in the case of a more refined approach these can be assigned to the interface coordinates of each component).

The forces applied to component A due to the relative motion of both components are given as

$$\left\{ \begin{matrix} \{ _A f_c \} \\ \cdots\cdots \\ \{ _B f_c \} \end{matrix} \right\} = \left[K_{cpl} \right] \left\{ \begin{matrix} \{ _A u_c \} \\ \cdots\cdots \\ \{ _B u_c \} \end{matrix} \right\} \quad \text{with} \quad \left[K_{cpl} \right] = \left[\begin{matrix} [K_{cc}] & \vdots & -[K_{cc}] \\ \cdots\cdots & \vdots & \cdots\cdots \\ -[K_{cc}] & \vdots & [K_{cc}] \end{matrix} \right] \tag{5.54}$$

Since we have assumed the transformation of coordinates (5.41), (5.42) can now be written as

$$[I] \left\{ \begin{matrix} \{ _A \ddot{p}_k \} \\ \cdots\cdots \\ \{ _B \ddot{p}_k \} \end{matrix} \right\} +$$

$$\left[\left[\begin{matrix} \left[\diagdown \, _A \omega_{rk}^2 \diagdown \right] & \vdots & [0] \\ \cdots\cdots & \vdots & \cdots\cdots \\ [0] & \vdots & \left[\diagdown \, _B \omega_{rk}^2 \diagdown \right] \end{matrix} \right] - \left[\begin{matrix} \left[_A \Phi_{ck}^T \right] & \vdots & [0] \\ \cdots\cdots & \vdots & \cdots\cdots \\ [0] & \vdots & \left[_B \Phi_{ck}^T \right] \end{matrix} \right] \left[K_{cpl} \right] \left[\begin{matrix} \left[_A \Phi_{ck} \right] & \vdots & [0] \\ \cdots\cdots & \vdots & \cdots\cdots \\ [0] & \vdots & \left[_B \Phi_{ck} \right] \end{matrix} \right] \right] \left\{ \begin{matrix} \{ _A p_k \} \\ \cdots\cdots \\ \{ _B p_k \} \end{matrix} \right\} = \{ 0 \}$$

$$\tag{5.55}$$

which is the equation of motion for the final coupled system.

5.3.4 Mass-Loading technique

As stated in Section 5.1.3, this is as an intermediate technique between the fixed-interface and the free-interface methods. By adding some discrete masses at each of the component connection coordinates, the deformation near the interface is increased and a better estimation for the local flexibility is obtained, provided that the auxiliary masses are connected in such a way that the local stiffness properties of the component are not affected. The appropriate size of the mass to be added is dependent on the characteristics of the component to be tested, rather than on those pertaining to the adjoining subsystem. This fulfils the independency requirement for a substructuring procedure.

From an experimental standpoint this technique offers some advantages as stated by Sekimoto [233] and Gwinn [234]:

- it is less time consuming to collect data as compared with the free-interface method including residual effects;

- more modes are brought into the frequency range of interest, thus providing better information about the component model;

- if adequate mass blocks are used, a convenient means of measuring rotational responses is provided (although this might be offset by a reduction in the amplitude of such rotations).

The procedure in coupling the components is similar to that presented for the case of the free-interface method with a rigid connection. The main steps involved are the same, but in this case there are three additional steps which are carried out in order to obtain a better description of the dynamic properties of each subsystem. All the major steps are presented briefly below:

Step 1 - Definition of the original disconnected substructures
The equations of motion are the same as those presented in Section 5.3.3.

Step 2 - Modification of the components by adding some extra masses
The auxiliary masses are added to the connection coordinates. The equation of motion for each substructure is now

$$[M_{mod}]\{\ddot{u}\} + [K]\{u\} = \{f\} \tag{5.56}$$

where the modified mass matrix will be:

$$[M_{mod}] = \begin{bmatrix} [M_{ii}] & [M_{ic}] \\ \hline [M_{ci}] & [M_{cc}] \end{bmatrix} + \begin{bmatrix} [0] & [0] \\ \hline [0] & [\Delta M_{cc}] \end{bmatrix} \tag{5.57}$$

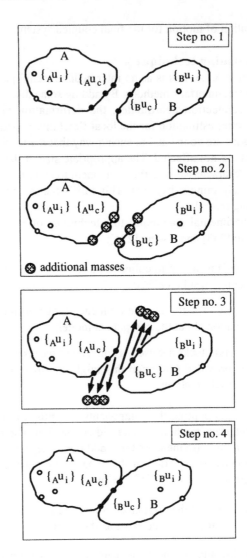

Fig. 5.6 Steps involved in the mass-loading technique.

The eigensolution will lead to the mass-loaded system eigenvalues $[\;\diagdown\omega^2_{r_{mod}}\diagdown\;]$ and associated mass-normalised eigenvectors $[\Phi_{mod}]$, constituting the necessary values to define the modal model for each modified subsystem. The values directly available from a test conducted on the mass-loaded component are the kept or measured modes $[\Phi_{k_{mod}}]$ and the respective natural frequencies $[\;\diagdown\omega^2_{r\,k_{mod}}\diagdown\;]$.

The equation of motion referred to the modal coordinates is then,

$$\left[\Phi_{k_{mod}}\right]^T\left[M_{mod}\right]\left[\Phi_{k_{mod}}\right]\{\ddot{p}_k\}+\left[\Phi_{k_{mod}}\right]^T\left[K_{mod}\right]\left[\Phi_{k_{mod}}\right]\{p_k\}=\{0\} \qquad (5.58)$$

or

$$[I]\{\ddot{p}_k\} + \left[\,\raisebox{0pt}{$\scriptstyle\backslash$}\, \omega^2_{r k_{mod}} \raisebox{0pt}{$\scriptstyle\backslash$} \right]\{p_k\} = \{0\} \tag{5.59}$$

Step 3 - Mass cancellation

The effects of the additional masses now need to be removed before the coupling process is performed. This is achieved by the following analytical process:

$$\left[\Phi_{k_{mod}} \right]^T [M] \left[\Phi_{k_{mod}} \right] = [I] - \left[\Phi_{k_{mod}} \right]^T [\Delta M] \left[\Phi_{k_{mod}} \right] \tag{5.60}$$

and the equation for each component is given as

$$\left[[I] - \left[\Phi_{k_{mod}} \right]^T [\Delta M] \left[\Phi_{k_{mod}} \right] \right]\{\ddot{p}_k\} + \left[\,\raisebox{0pt}{$\scriptstyle\backslash$}\, \omega^2_{r k_{mod}} \raisebox{0pt}{$\scriptstyle\backslash$} \right]\{p_k\} = \{0\} \tag{5.61}$$

Step 4 - Coupling

The coupling process follows now the same procedure presented for the case of the free-interface method with a rigid connection (see Section 5.3.3). All the steps are schematically presented in figure 5.6.

5.4   ALTERNATIVE TECHNIQUES

5.4.1 Limitations of the standard techniques

The theoretical basis of the standard coupling techniques has been presented. Each has its own limitations in certain circumstances and these constitute the reasons for the following discussion.

Impedance Coupling Techniques

As stated before, Spatial coupling is extensively used in applications of the Finite Element method but is rarely used in cases which involve experimental modelling. Reduction methods have been developed to condense Spatial models to primary (master) coordinates - a process of coordinate reduction which inevitably will cause a mode reduction - in order to decrease the computational needs to solve the global problem. However, the selection of those coordinates must be properly made based on a certain criterion.

In contrast to Spatial coupling, there is the FRF coupling technique which is particularly suitable for use with data measured on the components. It makes use of Response models derived directly from experimental data (but seldom from theoretical modelling). The collected data in terms of FRFs defined over a frequency range of interest are used to assemble the FRF matrix which, for every frequency value, expresses the contribution of the in- and out-of-range modes pertaining to each component.

One can say that in physical terms the FRF coupling technique is very attractive since it makes use of models whose dynamic characteristics are fully quantified and thus they do not suffer from modal incompleteness. However, there is a numerical aspect associated with this technique which may cause the coupling procedure to fail. The required FRF matrix of the coupled structure is obtained after three matrix inversions - two of these carried out before and one after the FRF matrices are assembled. Should one of these matrices be near singular, the results will reflect the numerical errors caused by the inversion and will predict the dynamic behaviour of the overall structure erratically.

Unfortunately, when dealing with experimentally derived FRF matrices, one is mostly restricted to the use of models that are inaccurate due to the experimental or systematic errors at the measurement stage, one of these being a slight variation in the resonance frequencies when the model is said to suffer from inconsistency. The FRF matrix then tends to be ill-conditioned, and the inverse is very sensitive to a slight change in one of the FRF matrix elements in the vicinity of every resonance frequency since it will tend to have an order equal to the measured coordinates and to have rank one, due to the dominating effect of one single mode. This local dominance is even stronger in lightly damped structures, and additional peaks tend to appear on the assembled structure FRFs at frequencies which may be misinterpreted as true resonances of the coupled structure.

Another situation which may lead to near singular matrices is caused by local rigidities at the measured coordinates. Over certain frequency ranges, the response in some coordinates may tend to be nearly dependent, and here again the FRF matrix will tend to be rank-deficient. The ways of tackling these numerical difficulties are addressed in [231].

Modal Coupling techniques
Unlike the Impedance-based methods which take advantage of the reduction of the number of coordinates, the Modal Coupling methods use a reduction performed on the number of modes used to describe each component model while still accounting for all the physical coordinates.

In spite of the fact that fixed-interface methods give better predictions than the free-interface methods, they are generally not suitable for handling data which are available from modal tests - the Modal model whose mode shapes are obtained from a component tested in its simulated free-free support condition. This is the main reason for the development of better procedures to improve the accuracy of results predicted using free-interface methods. One cause of the failure of these methods is the poor description of each subsystem's displacements in the interface region due to the reduction performed on the number of modes.

Two feasible alternatives to improve the representation of the actual dynamic properties can be found either by using the previously presented mass-loading technique or by compensating for the lack of flexibility due to the truncation on the number of modes. This latter approach is described; it includes into the coupling process the information on the residual flexibility associated with the neglected or unmeasured modes of each subsystem model, without having to carry all the steps required by the mass-loading technique.

5.4.2 Alternative FRF Coupling technique

The FRF coupling technique discussed so far is suitable for cases where the reduced Response models do not lead to numerical failures in the coupling process. Additionally, it was shown that alternative algorithms are necessary to deal with rank-deficient FRF matrices, at least over certain frequency ranges where for instance the local rigidities cause the rows to be dependent. This is an important aspect to be taken into account in the previously presented FRF Coupling technique, since the final FRF matrix is obtained after carrying out three inversion processes.

A development of the FRF coupling technique by Jetmundsen *et al* [235] has reduced the number of required inversions at each frequency from three to one and, additionally, the size of the matrix for inversion is dictated only by the number of connection coordinates. This refined method, apart from speeding up the calculations, may also behave better, in numerical terms, than the conventional one, since it minimises the crucial inversion operations on matrices which have a smaller order. This may play an important role if subsystems possess dependent coordinates.

The algorithm can be derived taking into account the relationship between the elements of the global FRF matrix and the elements in each subsystem FRF matrix. The coordinates in the FRF matrix of the coupled structure can be partitioned according to three regions, corresponding to:

- the interior coordinates of component A ($_A n_i$) denoted as a;
- the interior coordinates of component B ($_B n_i$) denoted as b;
- the common connection coordinates of component A and B ($_A n_c =_B n_c = n_c$) denoted as c.

The whole FRF matrix can be partitioned, and each partition can now be interrelated with the submatrices composing the FRF matrices of the subsystems, as presented next:

$$
\begin{bmatrix}
\left[H_{aa}\right] & \left[H_{ac}\right] & \left[H_{ab}\right] \\
\left[H_{ca}\right] & \left[H_{cc}\right] & \left[H_{cb}\right] \\
\left[H_{ba}\right] & \left[H_{bc}\right] & \left[H_{bb}\right]
\end{bmatrix} =
$$

$$
\begin{bmatrix}
\left[_A H_{ii}\right] & \left[_A H_{ic}\right] & [0] \\
\left[_A H_{ci}\right] & \left[_A H_{cc}\right] & [0] \\
[0] & [0] & \left[_B H_{ii}\right]
\end{bmatrix}
-
\begin{bmatrix}
\left[_A H_{ic}\right] \\
\left[_A H_{cc}\right] \\
-\left[_B H_{ic}\right]
\end{bmatrix}
\left[\left[_A H_{cc}\right]+\left[_B H_{cc}\right]\right]^{-1}
\begin{bmatrix}
\left[_A H_{ic}\right] \\
\left[_A H_{cc}\right] \\
-\left[_B H_{ic}\right]
\end{bmatrix}^{T}
$$

$$(5.62)$$

As mentioned before, the main advantage of using this formulation over the previously presented method is related to the crucial operation of inversion. Herein only one inversion is required and, additionally, it is applied only to the sum of the submatrices whose order depends only on the number of connection coordinates. More interior coordinates can then be included in the analysis without affecting significantly the required computational time. The result of this will be a quicker calculation of the required FRF matrix and, as in the latter approach, it will be able to deal with redundancies on the interior coordinates whenever they are present in each component.

5.4.3 Alternative Modal Coupling technique

The free-interface method constitutes the most suitable approach for incorporating experimentally derived modal models. Consequently, a refined approach, based on a free-interface methodology with elastic connection, which can include the residual flexibility effects of the unmeasured modes was developed [236] and is summarised next.

The same formulation (equation (5.55)) given for the elastic coupling is used, provided that the stiffness matrix for the intermediate spring-system is constructed as

$$\left[K_{cpl}\right] = \left[\begin{array}{c:c} \left[R_{cc}^{*}\right]^{-1} & -\left[R_{cc}^{*}\right]^{-1} \\ \hdashline -\left[R_{cc}^{*}\right]^{-1} & \left[R_{cc}^{*}\right]^{-1} \end{array} \right] \tag{5.63}$$

where

$$\left[R_{cc}^{*}\right] = \left[\left[_{A}R_{cc}\right] + \left[_{B}R_{cc}\right]\right] \tag{5.64}$$

This equation expresses the series elastic connection between the two components, each of them possessing the residual flexibilities as:

$$\left[_{A}R_{cc}\right] = \left[_{A}\Phi_{ce}\right]\left[\diagdown_{A}\omega_{re}^{2}\diagdown\right]\left[_{A}\Phi_{ce}\right]^{T} \tag{5.65}$$

$$\left[_{B}R_{cc}\right] = \left[_{B}\Phi_{ce}\right]\left[\diagdown_{B}\omega_{re}^{2}\diagdown\right]\left[_{B}\Phi_{ce}\right]^{T} \tag{5.66}$$

where $\left[_{A}\Phi_{ce}\right]$ and $\left[_{B}\Phi_{ce}\right]$ are constituted by the eliminated or unmeasured modes and ω_{re} are the out-of-range natural frequencies.

These equalities are established based on the assumption that $\omega_{re}^{2} >> \omega_{r}^{2}$, and an approximation can be made for the residual flexibility of the eliminated modes in components A and B, i.e., we can say that they respond in a quasi-static manner so that the inertial term can be ignored.

We have, therefore, 'additional' springs located in the boundary region to compensate for the lack of flexibility (associated with the out-of-range modes). The connection between the two subsystems through an intermediate connecting flexible system can be seen in figures 5.7 and 5.8.

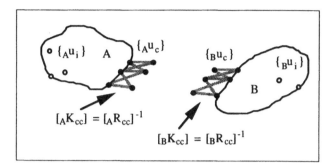

Fig. 5.7 Auxiliary flexible systems used to represent the flexibility contribution of the out-of-range eliminated modes in each separated component.

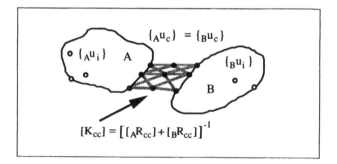

Fig. 5.8 Two auxiliary flexible systems are connected in series to form the 'dummy' interconnecting system.

5.4.4 Final remarks

The main achievement due to the previously presented refinement is the inclusion of the residual flexibility effects which compensate for the truncation in the number of kept or measured modes in each component. The lack of flexibility associated with the description of each component displacement in the connection region causes the classical free-interface methods to predict results with poor accuracy when compared to the fixed-interface methods - in fact, the components are assumed to be stiffer than they actually are supposed to be.

With the inclusion of the residual flexibility information - in fact an approximation when experimental derived models are dealt with - by using a 'dummy' interconnecting flexible system, the two main components are mathematically coupled using the best available information provided by data measured over the frequency range in each component. The results obtained in a theoretical case study [236] permit the additional conclusions:

- the refined approach improves the prediction of the dynamic response of the coupled structure when compared to the classical free-interface method; the valid predicted modes may be taken as the number of kept modes in the incomplete subsystem plus one;

- the selected route for the calculation of the residual flexibility matrix does not affect the predicted dynamic response of the overall structure.

CHAPTER 6

Local Structural Modification

6.1   INTRODUCTION

Structural modification is an area of study that deals with the effects of physical parameter changes on the dynamic properties of a structural system. These physical parameters are related to the mass, stiffness and damping properties of the system, or to a combination of them. It is customary to refer the dynamic properties to the natural frequencies and mode shapes of a system. Structural modification usually comprises two different approaches:

(1) Given prescribed dynamic characteristics such as a new resonance, seek what, where and how much structural modifications will best accomplish it; and

(2) For suggested structural modifications, determine which dynamic characteristic changes will occur.

The first is an inverse problem that is analytically challenging while the second, also known as re-analysis, is a direct problem which requires mathematical and numerical effort. Often, due to practical constraints, structural modifications are only allowed at limited locations on a structure. This leads to what is referred to as 'local structural modification'.

The reason to pursue local structural modification is to improve or optimise the dynamic behaviour of a structure. Designed on specifications, a structure may not always have satisfactory dynamic properties. For instance, its natural frequencies may coincide with some ambient vibration frequencies. Its mode shapes may be hazardous when it is assembled with other structures. The result may be an excessive vibration and short fatigue life.

In most low frequency vibration problems, the fundamental frequency and mode shape of a structure are primarily responsible for its vibration response to an excitation. Thus the ability to shift the fundamental frequency can significantly reduce that response. For an aircraft wing, the lower bending and torsional modes determine its aeroelastic characteristics. Space structures require certain restrictions on their lower resonance frequencies in order not to interfere with the control systems.

For local structural modification to be realistically successful, the problem has to lie within a narrow frequency band. This means either the excitation frequency falls into a narrow band so that modification may help to remove structural resonances from this detrimental band, or there are only a very limited number of locations in the structure whose vibration amplitudes are cause for major concern. Structural modification therefore needs only to target the marked reduction of vibration at those locations. In summary, it is vital that the objective of structural modification be clearly defined and attainable.

The dynamic behaviour of a structural system is determined by the distribution of its mass, stiffness and damping properties. Therefore it is only through the modification of these properties that improved dynamic characteristics of the system can be achieved. The majority of structural modification cases arise because of concerns with detrimental resonant vibrations. As a result, the task is often to alter one or more of the natural frequencies of a structure by means of local structural modification, in order to avoid severe vibration associated with resonance. This leads to the assertion that most structural modifications involve changes of the mass and stiffness properties, since damping has little influence on the natural frequencies of the system.

In theoretical investigation it is convenient, and sometimes necessary, to study structural modification using a mass-spring system. Such a system is physically explicit and technically easy to deal with. For structural modification, it permits separate mass and stiffness changes which in turn allow greater flexibility in achieving a specified structural modification task. In reality, this advantage often diminishes or completely disappears, due to the fact that any modification of a real structure is likely to incur simultaneous mass and stiffness changes. For example, in a simple truss structure, physical parameter changes of some of its elements (such as cross-section modifications), result in changes to both mass and stiffness properties. When used in problem solving, a successful structural modification study provides an efficient approach to predict the effects of structural changes, thus eliminating the need for expensive prototype construction and further testing. In engineering design, structural modification enables a process of optimal design to ensure satisfactory behaviour when a structure is manufactured and is in service.

For a prescribed objective, such as the shift of a natural frequency, there are numerous mathematically possible solutions, all of them leading to that objective. However, some of these solutions can involve drastic material changes and impractical designs, while others may simply be unrealistic modifications. The derivation of the 'best' structural modification for a prescribed objective is therefore an integrated part of the endeavour.

Instead of modifying the mass and stiffness properties of a structure, a different category of structural modification is to adjoin additional degrees-of-freedom to the structure. A typical example is to add a dynamic absorber to a SDOF system in order to eliminate its resonance. This modification can be extended so that a substructure rather than just another DOF can be added to a rationally selected location or locations of the structure in order to improve its dynamic behaviour. The analysis of this type of structural modification closely touches the domain of substructural coupling or substructural analysis discussed in the previous chapter.

Structural modification evolved with the development of structural dynamics. The first meaningful formulation was given by Rayleigh [237] who used perturbation approach to derive an approximate solution in terms of modal coordinates. This approach was later expanded by many researchers to broaden that formulation and to include second order approximations. The practical application of perturbation approach was treated by Stetson and Palma [238] and Sandstrom and Anderson [239]. Their work allows for specified constraints on frequencies and mode shapes, and links physical parameter changes of a structure, such as cross-sectional area of a beam element, to its modal properties. The approximation nature of the perturbation approach hinders its application to large modifications.

The historical development of local structural modification can also be traced back decades, if adding a dynamic absorber is regarded as a type of local structural modification. The works by Den Hartog and Timoshenko were pioneering research at the beginning of this century. A notable study was carried out by Weissenburger [240], who formulated the relationship between the simple lumped mass and stiffness alteration of an undamped linear dynamic system and its dynamic characteristic changes. That work was later expanded by Pomazal and Snyder [241], who analysed the effects of adding springs and dampers to a viscously damped linear system. Their analysis included a transformation of second order equations into a set of first order equations. The complete solution of the original system is a prerequisite in that analysis. Their work was later generalised and improved by Hallquist [242]. These works treated structural modification from the viewpoint of a direct problem. Therefore they constitute re-analyses of a modified system, without having to go through a complete re-eigensolution.

Notable research was also reported by Ram and Blech [243], who theorised the effects of modification at one degree-of-freedom of different mass and stiffness attachments.

Developing in parallel to the re-analysis endeavour was the sensitivity approach. The sensitivity of an eigenvalue problem has been developed by many researchers. Amongst them are Wilkinson [244], Rosenbrock [245], Rogers [246] and Vanhonacker [247]. The focus of their studies was to determine the derivatives of the eigenvalues and eigenvectors of a dynamic system with respect to system changes. This predetermines the limitation of such an approach when used in structural modification. Nevertheless, a great deal of effort has been reported in literature using sensitivity analysis for structural modification, including such examples as Wang *et al* [248], To and Ewins [249], and Skingle and Ewins [250].

Structural modification as an inverse problem is an approach different from that of re-analysis and sensitivity. Significant works were reported by Tusei and Yee

[251, 252], where the authors proposed a method to determine required mass and stiffness changes that would relocate a natural frequency. The advantages of that method are the following ones: (i) it relies on FRF data only at some locations, thus bypassing the problem of not having sufficient FRF data, and (ii) it does not need the mass and stiffness matrices of the original system as pre-requisites for the analysis. That work was later extended by Li *et al* [253, 254] and He and Li [255] to study the relocation of an antiresonance and ways of optimising the properties of a system.

A separate attempt was made by Bucher and Braun [256]. The authors formulated solutions to reallocate eigenvalues and specify eigenvectors by computing necessary mass and stiffness modifications. The solutions require only a partial set of eigensolutions which can be derived from modal testing data. The applied modifications are constrained in such a way as to force the selected number of eigensolutions to reside in the known subspace spanned by the original modal vectors. This circumvents the problem arising from truncation of the modal set. If, however, the measured FRF data are used directly instead of a partial set of derived eigensolutions, then it is possible to avoid this question of subspace spanning.

Table 6.1 summarises the characteristics of the two approaches to structural modification.

Table 6.1 Characteristics of the direct and inverse approaches to structural modification.

Structural Modification	
Direct Methods	Inverse Methods
Specify structural modification and predict new dynamic characteristics	Specify new dynamic characteristics and determine structural modification to accomplish them
Unique solution	Non-unique or no solution
Requires complete spatial model or incomplete modal model	Requires incomplete response model or modal model

Before the discussion and analysis of structural modification, it is important to clarify (see also Chapter 1) certain terminologies used and sometimes confused by modal analysts. A 'Natural Frequency' is a frequency of a dynamic system at which free vibration takes place. A 'Resonance' can refer to the physical phenomenon of maximum vibration or the frequency at which maximum vibration occurs. For the latter, it is more precise to say that resonance is a frequency of excitation for which the same force amplitude will produce maximum vibration response. It is customary to use either resonance or natural frequency in structural modification by ignoring the slight quantitative difference between them (as mentioned in Section 1.2.2 for the case of a SDOF system).

An 'Antiresonance' can also refer to either a physical phenomenon or a frequency. For the latter, it is defined as the frequency at which the ratio of the response at one point to the force at another becomes zero (for an undamped system) or approaches zero (for a damped system). 'Degrees-of-Freedom' are the minimum number of independent coordinates required to completely define the motion of all parts of a system. Although it is often regarded as a synonym of 'point', the latter is ambiguous. A physical point may have several degrees of freedom. Although it is possible to assume modifications at a single DOF, it does not necessarily imply that other DOFs will not be affected.

It is inevitable that the terms 'structure' and 'system' will be found in texts on modal analysis or structural modification. Often, these two terms (or sometimes 'vibratory structure' as well) are intended to mean the same. A system is the idealisation of a structure where mathematical models can be readily established and explored.

When referred to a structure, it is often implicative that its idealisation exists. As a result, theories based on discretisation and linear algebra apply. Alternatively, 'structural system' is also used.

6.2   DYNAMIC CHARACTERISTICS OF A VIBRATORY STRUCTURE

6.2.1 Different mathematical models

The dynamic characteristics of a vibratory structure are manifested mainly by its natural frequencies and mode shapes. These characteristics can be derived from several mathematical models of the structure: (1) the spatial model consisting of its spatial mass, stiffness and damping properties; (2) the response model that comprises a sufficient number of frequency response functions; and (3) the modal model that includes natural frequencies and mode shapes.

Damping properties are important in certain applications; however, their presence does not yield significant changes in natural frequencies. The mode shape changes due to damping usually do not translate critically to the main concerns of a vibratory structure, such as excessive vibration at a given resonance.

The spatial model of an undamped vibratory structure consists of its mass and stiffness matrices, respectively [M] and [K]. They are usually the products of finite element modelling of the structure from its design data. Though mass and stiffness matrices do not explicitly reveal dynamic characteristics, they allow for a simple eigenvalue solution from which the natural frequencies and mode shapes are derived (see Section 1.4.1):

$$\left[[K] - \omega^2[M]\right]\{\psi\} = \{0\} \tag{6.1}$$

Matrices [M] and [K] are only useful for eigenvalue solution if they are complete.

The response model of a vibratory structure comprises its frequency response functions (FRFs). If these functions are derived analytically, then the whole FRF matrix of the structure, denoted $[\alpha(\omega)]$ for receptance FRF, is available. Modal testing is another source of FRF data, where conventionally a column (or a row) of the FRF matrix $[\alpha(\omega)]$ is made available and is denoted as $\{\alpha(\omega)\}$. These FRFs are limited within the frequency range of measurement and, therefore, are incomplete in that regard. By theoretical definition, the receptance FRF matrix of a structure with N DOFs is related to its spatial data by:

$$\left[[K] - \omega^2[M]\right]^{-1} = [\alpha(\omega)]$$ (6.2)

An individual receptance FRF $\alpha_{ij}(\omega)$ is physically defined as the ratio of displacement response at coordinate i to a sole force applied at coordinate j:

$$\alpha_{ij}(\omega) = \left.\frac{X_i(\omega)}{F_j(\omega)}\right|_{F_r = 0, \quad r=1,2,...,N \quad r \neq j}$$ (6.3)

This receptance FRF can be derived directly from the spatial matrices:

$$\alpha_{ij}(\omega) = (-1)^{i+j} \frac{\det\left[[K]_{ij} - \omega^2[M]_{ij}\right]}{\det\left[[K] - \omega^2[M]\right]}$$ (6.4)

where, matrix $[K]_{ij}$ is obtained by deleting the ith row and j^{th} column of matrix [K], and likewise for matrix $[M]_{ij}$. Both matrices are of order (N-1) and are non-symmetric unless i=j.

For convenience in discussion hereafter, an imaginary system called a virtual system exists which comprises mass matrix $[M]_{ij}$ and stiffness matrix $[K]_{ij}$. Therefore, each individual receptance FRF is associated with such a virtual system.

The modal model of a structure includes the natural frequency matrix $[\,{}^{\backprime}\omega_r^2{}_{\backprime}]$ and the corresponding mode shape matrix $[\Phi]$. Here, the mode shape matrix is mass-normalised. The modal analysis theory establishes the relationship between the modal model and the response model:

$$[\alpha(\omega)] = [\Phi]\left[\,{}^{\backprime}\omega_r^2 - \omega^2{}_{\backprime}\,\right]^{-1}[\Phi]^T$$ (6.5)

where

$$\alpha_{ij}(\omega) = \sum_{r=1}^{N} \frac{\phi_{ir}\,\phi_{jr}}{\omega_r^2 - \omega^2}$$ (6.6)

An FRF is characterised by three factors: the resonances where natural frequencies ω_r reside; antiresonances Ω_s or minima between two resonances; and the prominence (or absence) of each resonance which is dictated by the mode shapes $\{\phi_r\}$.

The inter-relationship among the three mathematical models of a dynamic system is demonstrated in table 6.2.

Table 6.2 Relationship among the three models of a dynamic system.

To Derive from	Spatial model	Modal model	Response model
Spatial model		Eigensolution $\left[[K]-\omega^2[M]\right]\{\psi\}=\{0\}$	$[\alpha(\omega)]=\left[[K]-\omega^2[M]\right]^{-1}$
Modal model	$[M]=\left([\Phi][\Phi]^T\right)^{-1}$ $[K]=$ $[M][\Phi]\left[\begin{smallmatrix}\ddots\\ &\omega_r^2\\ &&\ddots\end{smallmatrix}\right][\Phi]^T[M]$		$\alpha_{ij}(\omega)=\sum_{r=1}^{N}\dfrac{\phi_{ir}\,\phi_{jr}}{\omega_r^2-\omega^2}$ $[\alpha(\omega)]=$ $[\Phi]\left[\begin{smallmatrix}\ddots\\ &\omega_r^2-\omega^2\\ &&\ddots\end{smallmatrix}\right]^{-1}[\Phi]^T$
Response model	$[K]=[\alpha(\omega)]^{-1}\big|_{\omega=0}$ $[M]=\dfrac{[\alpha(\omega_1)]^{-1}-[\alpha(\omega_2)]^{-1}}{\omega_2^2-\omega_1^2}$	Modal analysis on measured FRF data $\alpha_{ij}(\omega)=\sum_{r=1}^{N}\dfrac{\phi_{ir}\,\phi_{jr}}{\omega_r^2-\omega^2}$	

6.2.2 Antiresonances of a dynamic structure

Like the resonance, the antiresonance is also an important manifestation of the dynamic characteristics of a structure. Though this subject has already been presented (Section 1.5), it is important to mention it again here, as it is an essential issue in structural modification analysis.

While resonances of a structure are 'global' properties, i.e., the same resonances will appear on any FRFs of the system, antiresonances of a system are 'local' properties. Different FRFs have different antiresonances. Some have no antiresonances at all. Therefore, the reference of antiresonances must be based upon a given FRF.

The antiresonances of a receptance FRF $\alpha_{ij}(\omega)$, which defines the frequency characteristics of a structure between coordinates i and j, are determined by (from equation (6.4)):

$$\det\left[[K]_{ij}-\Omega^2[M]_{ij}\right]=0 \tag{6.7}$$

Mathematically, this is effectively the same as finding the eigenvalues of the following problem:

$$\left[[K]_{ij}-\Omega^2[M]_{ij}\right]\{X\}=\{0\} \tag{6.8}$$

Only those physically 'meaningful' eigenvalues of this equation, typically real positive eigenvalues, are the antiresonances. Other eigenvalues are pure numerical answers which do not have physical interpretation. For an FRF with no antiresonance, none of the eigenvalues of (6.8) will be real and positive.

Equation (6.7) suggests that any structural modification which occurs at coordinate i only or coordinate j only would not alter the antiresonances of FRF $\alpha_{ij}(\omega)$. Physically, since an antiresonance implies that the structure is 'grounded' at that frequency, any structural changes at this location will not make an impact on the response at this frequency.

Although the antiresonances of a system can be determined from its spatial data, an understanding of their formation will be best reached from the viewpoint of modal data. For the sake of simplicity, the receptance $\alpha_{11}(\omega)$ of the 2 DOF system shown in figure 6.1 is used as an example. This receptance can be expressed using equation (6.6) as:

$$\alpha_{11}(\omega) = \frac{\phi_{11}^2}{\omega_1^2 - \omega^2} + \frac{\phi_{12}^2}{\omega_2^2 - \omega^2} \qquad (6.9)$$

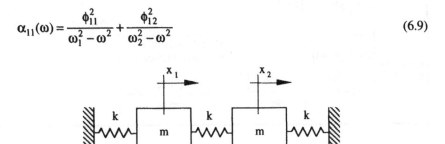

Fig. 6.1 Mass-spring system with 2 DOFs.

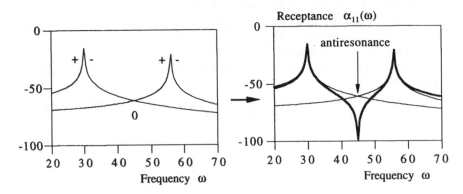

Fig. 6.2 Antiresonance of an FRF of a 2 DOF mass-spring system.

Between two resonances ω_1 and ω_2 there is either an antiresonance or a minimum. Since both modal constants in (6.9) are positive, for a frequency between resonances ω_1 and ω_2 the first term is negative while the second one is positive. Therefore, there will be a frequency between ω_1 and ω_2 for which the two terms cancel out, giving a zero receptance $\alpha_{11}(\omega)$. When the FRF is plotted on a dB graph, this zero receptance will signify the antiresonance. This is illustrated in figure 6.2 (and was also explained in Section 1.5, pages 74 and 75).

The positive and negative signs in figure 6.2 are only explanatory. They do not actually apply to the dB plot of an FRF. Nevertheless, an antiresonance is displayed only on a plot where the response amplitude is logarithmically scaled.

Therefore, from the viewpoint of modal data, the ultimate reason for an antiresonance are the signs of the mode shape elements (which constitute the modal constants). Different combinations of signs result in varying numbers of antiresonances.

6.2.3   Structural modification

The majority of structural modification cases focus on the problem of excessive vibration caused by unwanted resonant vibration when the structure is excited by a narrow-banded force. As a result, the objective of structural modification is primarily to disassociate structural resonances from excitation frequencies. This can often be achieved by relocating (or shifting) a resonance of the structure away from the excitation frequency range. Occasionally, mode shapes may become the concern when, for instance, prescribed mode shape characteristics are needed for two substructures in anticipation of the characteristics of the assembled structure.

Structural modification generally implies changes in mass and stiffness matrices. These changes can be denoted as $[\Delta M]$ and $[\Delta K]$ for mass and stiffness modification matrices, respectively. These matrices are of the same order as $[M]$ and $[K]$. The structural connectivity depicted in matrices $[M]$ and $[K]$ is usually honoured in modification matrices $[\Delta M]$ and $[\Delta K]$. This means that modification only varies the mass or stiffness quantities of existing physical components of a system. When the connectivity is disregarded, it means that unconnected coordinates are now linked together or *vice versa*. This usually has a significant effect on the antiresonances of FRFs of the affected coordinates. If structural modification occurs by adding a subsystem to the original system (such as a dynamic absorber), then analysis based on the eigenvalue equation of the original system is no longer valid.

The counterparts of modification matrices $[\Delta M]$ and $[\Delta K]$ for the virtual system are matrices $[\Delta M]_{ij}$ and $[\Delta K]_{ij}$. Here, matrix $[\Delta M]_{ij}$ is derived by deleting the ith row and j^{th} column of matrix $[\Delta M]$, and likewise for matrix $[\Delta K]_{ij}$.

For structural modification to be local, $[\Delta M]$ and $[\Delta K]$ have to be sparse matrices. Usually, without losing generality, it is possible to denote:

$$[\Delta M] = \begin{bmatrix} [0] & [0] & [0] \\ [0] & [\Delta M^R] & [0] \\ [0] & [0] & [0] \end{bmatrix} \quad \text{and} \quad [\Delta K] = \begin{bmatrix} [0] & [0] & [0] \\ [0] & [\Delta K^R] & [0] \\ [0] & [0] & [0] \end{bmatrix} \qquad (6.10)$$

Here, matrix $[\Delta M^R]$ is a square submatrix reduced from $[\Delta M]$ by retaining coordinates for modification. Likewise for matrix $[\Delta K^R]$. For a mass-stiffness system, $[\Delta M^R]$ is usually a full rank matrix, but $[\Delta K^R]$ can be a singular one. For instance, if structural modification is confined to a single spring connecting coordinates i and j, then matrix $[\Delta K^R]$ will be a 2 by 2 matrix of rank 1. Take as an example the 4 DOF system shown in figure 6.3.

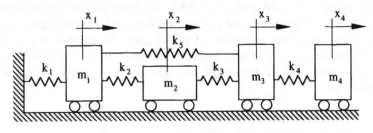

Fig. 6.3 Mass-spring system with 4 DOFs.

If both masses 1 and 4 are modified, then, after rearranging the coordinates, matrix $[\Delta M^R]$ will conform with the appearance given in (6.10) as:

$$[\Delta M^R] = \begin{bmatrix} \Delta m_1 & 0 \\ 0 & \Delta m_4 \end{bmatrix} \tag{6.11}$$

This is an invertible matrix. In a similar way, if the spring between coordinates 2 and 3 is modified, then:

$$[\Delta K^R] = \begin{bmatrix} \Delta k_3 & -\Delta k_3 \\ -\Delta k_3 & \Delta k_3 \end{bmatrix} \tag{6.12}$$

which is a singular matrix.

In structural modification, a neat way to describe physical elements (mass or stiffness elements) in a system is to use a submatrix approach. This approach fractionates the system mass and stiffness matrices into a combination of a series of submatrices, each being of unit rank and representing a physical element of the system. Thus, the mass and stiffness matrices of the 4 DOF system are:

$$[M] = \sum_{r=1}^{4} [M]_r = \sum_{r=1}^{4} m_r \{e_r\} \{e_r\}^T \tag{6.13}$$

$$[K] = \sum_{r=1}^{5} [K]_r = \sum_{r=1}^{5} k_r \{e_{pq}\} \{e_{pq}\}^T \tag{6.14}$$

where matrices $[M]_r$ and $[K]_r$ are submatrices of the rth mass and r^{th} spring respectively. Vectors $\{e_r\}$ and $\{e_{pq}\}$ are Kronecker vectors reflecting only the connectivity of a physical element. This presentation of system matrices is akin to finite element analysis. It sometimes lends considerable convenience to structural modification analysis. For instance, the stiffness matrix of the 4 DOF system can be expressed using submatrices as:

$$
[K] = k_1 \begin{Bmatrix} 1 \\ 0 \\ 0 \\ 0 \end{Bmatrix} \begin{Bmatrix} 1 \\ 0 \\ 0 \\ 0 \end{Bmatrix}^T + k_2 \begin{Bmatrix} 1 \\ -1 \\ 0 \\ 0 \end{Bmatrix} \begin{Bmatrix} 1 \\ -1 \\ 0 \\ 0 \end{Bmatrix}^T + k_3 \begin{Bmatrix} 0 \\ 1 \\ -1 \\ 0 \end{Bmatrix} \begin{Bmatrix} 0 \\ 1 \\ -1 \\ 0 \end{Bmatrix}^T + k_4 \begin{Bmatrix} 0 \\ 0 \\ 1 \\ -1 \end{Bmatrix} \begin{Bmatrix} 0 \\ 0 \\ 1 \\ -1 \end{Bmatrix}^T
$$

$$
+ k_5 \begin{Bmatrix} 1 \\ 0 \\ -1 \\ 0 \end{Bmatrix} \begin{Bmatrix} 1 \\ 0 \\ -1 \\ 0 \end{Bmatrix}^T = \begin{bmatrix} k_1+k_2+k_5 & -k_2 & -k_5 & 0 \\ -k_2 & k_2+k_3 & -k_3 & 0 \\ -k_5 & -k_3 & k_3+k_4+k_5 & -k_4 \\ 0 & 0 & -k_4 & k_4 \end{bmatrix}
$$

$$
(6.15)
$$

The matrices of the virtual system can also be easily presented in the same way. If we denote $\{e_r\}_{(i)}$ as the vector derived from $\{e_r\}$ by deleting its ith element, then (6.13) and (6.14) can be applied to matrices for the virtual system:

$$
[M]_{ij} = \sum_{r=1}^{4} m_r \{e_r\}_{(i)} \{e_r\}_{(j)}^T \qquad (6.16)
$$

$$
[K]_{ij} = \sum_{r=1}^{5} k_r \{e_{pq}\}_{(i)} \{e_{pq}\}_{(j)}^T \qquad (6.17)
$$

For instance, for the receptance FRF $\alpha_{23}(\omega)$, matrix $[K]_{23}$, which is formed by eliminating the second row and the third column of matrix $[K]$, can be expressed as:

$$
[K]_{23} = k_1 \begin{Bmatrix} 1 \\ 0 \\ 0 \end{Bmatrix} \begin{Bmatrix} 1 \\ 0 \\ 0 \end{Bmatrix}^T + k_2 \begin{Bmatrix} 1 \\ 0 \\ 0 \end{Bmatrix} \begin{Bmatrix} 1 \\ -1 \\ 0 \end{Bmatrix}^T + k_3 \begin{Bmatrix} 0 \\ -1 \\ 0 \end{Bmatrix} \begin{Bmatrix} 0 \\ 1 \\ 0 \end{Bmatrix}^T + k_4 \begin{Bmatrix} 0 \\ 1 \\ -1 \end{Bmatrix} \begin{Bmatrix} 0 \\ 0 \\ -1 \end{Bmatrix}^T
$$

$$
+ k_5 \begin{Bmatrix} 1 \\ -1 \\ 0 \end{Bmatrix} \begin{Bmatrix} 1 \\ 0 \\ 0 \end{Bmatrix}^T = \begin{bmatrix} k_1+k_2+k_5 & -k_2 & 0 \\ -k_5 & -k_3 & -k_4 \\ 0 & 0 & k_4 \end{bmatrix}
$$

$$
(6.18)
$$

This concept of submatrix can also be applied to modification matrices. As a result:

$$[\Delta M] = \sum_{r=1}^{u} [\Delta M]_r = \sum_{r=1}^{u} \Delta m_r \{e_r\}\{e_r\}^T \qquad (6.19)$$

$$[\Delta K] = \sum_{r=1}^{v} [\Delta K]_r = \sum_{r=1}^{v} \Delta k_r \{e_{pq}\}\{e_{pq}\}^T \qquad (6.20)$$

and

$$[\Delta M]_{ij} = \sum_{r=1}^{u} (\Delta m_r)_{ij} \{e_r\}_i \{e_r\}_j^T \qquad (6.21)$$

$$[\Delta K]_{ij} = \sum_{r=1}^{v} (\Delta k_r)_{ij} \{e_{pq}\}_i \{e_{pq}\}_j^T \qquad (6.22)$$

where u and v are the number of mass and stiffness elements involved in the modification, respectively.

6.2.4 Structural modification involving only one DOF

Before embarking on a comprehensive analysis of structural modification, it is imperative and essential to understand the physics behind structural modification at a single degree-of-freedom only.

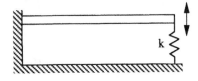

The point FRF (receptance) at the transverse direction of the free end of the cantilevered beam varies as the modification stiffness k increases. The FRF will finally become a null FRF as k becomes excessively large, as the bottom curve shows.

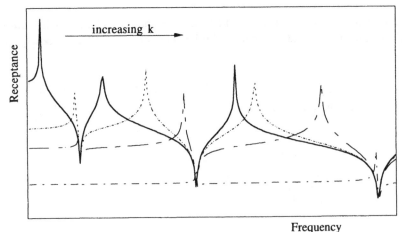

Fig. 6.4 Change of point FRF of a cantilevered beam at its free end due to stiffness modification.

The mass and stiffness modifications at one DOF can bring about significant changes in the resonances and antiresonances of a dynamic system, as illustrated in figure 6.4 where a cantilevered beam is used. The vertical receptance FRF at the free end of the cantilevered beam is shown as the solid line in figure 6.4. Assume a stiffness modification is applied to the vertical coordinate of the free end by linking it to the ground via a spring. Then, according to equations (6.6) and (6.7), this spring will not change the antiresonances of the point FRF at the free end (the stiffness change does not appear in matrix $[K]_{ij}$).

However, adding stiffness to the coordinate will inevitably increase the values of all the natural frequencies. As a result, we can anticipate that, as the stiffness of the spring increases, the FRF will vary as figure 6.4 suggests. Continuous increasing of the stiffness will eventually push the first resonance to meet with the first antiresonance, the second resonance with the second antiresonance, and so on. When the stiffness becomes infinite, these resonance and antiresonance pairs will cancel each other, leaving a null FRF - an expected outcome from a grounded location. An interesting phenomenon observed on the point FRF of this modified cantilevered beam is that the resonances 'incline' towards antiresonances as stiffness k becomes significantly large.

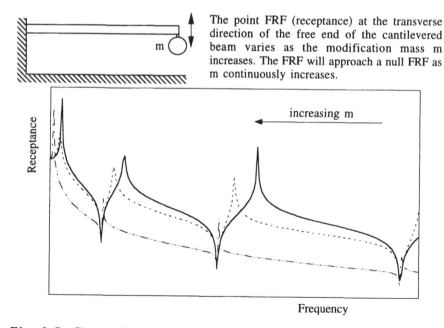

The point FRF (receptance) at the transverse direction of the free end of the cantilevered beam varies as the modification mass m increases. The FRF will approach a null FRF as m continuously increases.

Fig. 6.5　Change of point FRF of a cantilevered beam at its free end due to mass modification.

Still using the same beam as an example, let this time the modification be a concentrated mass added at the free end. The point FRF at the vertical direction before mass modification is shown as the solid line in figure 6.5.

As the mass m increases, all the resonances have to move to lower frequencies. However, this point modification means that the antiresonances of the FRF do not suffer any alteration. As a result, resonances and antiresonances will approach to cancellation, leaving a null FRF - a result expected from an object with infinite mass.

The convergence of each resonance towards the antiresonance below it is an evidence of a large local mass. Physical interpretation of this behaviour of antiresonances can be found in [257]. A slightly more complicated structural modification at one DOF involves adding a SDOF system with a mass and a spring (figure 6.6).

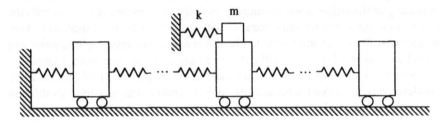

Fig. 6.6 A MDOF system with a SDOF modification at one coordinate.

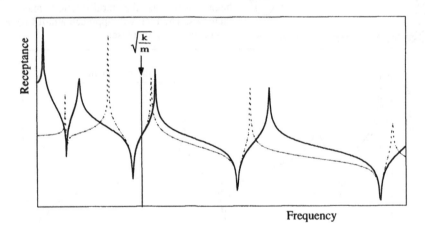

Fig. 6.7 An FRF of the MDOF system before and after the mass and stiffness modification shown in figure 6.6.

In this case, it has been found by Ram and Blech [243] that the natural frequencies of the MDOF system which are lower than the natural frequency ($\sqrt{k/m}$) of the SDOF modification system will increase, while those which are higher will decrease. This is shown in figure 6.7 where the solid line is an FRF of the MDOF system and the dotted line is the same FRF with the SDOF modification.

6.3 SENSITIVITY ANALYSIS AND STRUCTURAL MODIFICATION

The importance of obtaining sensitivities of the dynamic characteristics of a system with respect to the changes of its parameters lies in the fact that they are critical for achieving efficient design modification, and for predicting the 'optimal' structural modifications within a prescribed dynamic characteristics change. In such a case, there will be numerous possible modifications that can accomplish the change. Sensitivity analysis can provide the most effective answer.

From a mathematical point of view, sensitivity analysis implies differentiation. It can be carried out on the modal data or FRF data of a dynamic system. The variables in this analysis are usually system parameters such as mass or stiffness elements. The functions of the analysis can be natural frequencies and mode shapes of the system (modal data), or its FRFs. The assumption of infinitesimal changes demanded by differentiation applies to sensitivity analysis.

6.3.1 Modal model sensitivity analysis

The eigenvalue problem governing an N DOF linear dynamic system is:

$$\left[[K] - \omega_r^2 [M] \right] \{\phi_r\} = \{0\} \qquad r = 1, 2, ..., N \tag{6.23}$$

For the sake of simplicity, assume all eigenvalues of the system are distinct. The partial differentiation of this equation w.r.t. a mass element m_{ij} yields:

$$-\frac{\partial \omega_r^2}{\partial m_{ij}} [M]\{\phi_r\} - \omega_r^2 \frac{\partial [M]}{\partial m_{ij}} \{\phi_r\} - \omega_r^2 [M] \frac{\partial \{\phi_r\}}{\partial m_{ij}} + [K] \frac{\partial \{\phi_r\}}{\partial m_{ij}} = \{0\} \tag{6.24}$$

Pre-multiplying (6.24) by $\{\phi_r\}^T$, considering (6.23) and the fact that mode shapes $\{\phi_r\}$ are mass-normalised, yields:

$$\frac{\partial \omega_r^2}{\partial m_{ij}} + \omega_r^2 \{\phi_r\}^T \frac{\partial [M]}{\partial m_{ij}} \{\phi_r\} = 0 \tag{6.25}$$

Due to the symmetry of mass matrix [M], elements m_{ij} and m_{ji} will change simultaneously. Equation (6.25) can then be recast as:

$$\frac{\partial \omega_r^2}{\partial m_{ij}} + 2 \omega_r^2 \phi_{ir} \, \phi_{jr} = 0 \qquad \text{for } i \neq j \tag{6.26}$$

and

$$\frac{\partial \omega_r^2}{\partial m_{ij}} + \omega_r^2 \phi_{ir}^2 = 0 \qquad \text{for } i = j \tag{6.27}$$

This leads to the sensitivity of the rth squared natural frequency, ω_r^2, w.r.t. the change of the element m_{ij} in the mass matrix:

$$\frac{\partial \omega_r^2}{\partial m_{ij}} = \begin{cases} -2\omega_r^2 \phi_{ir} \phi_{jr} & i \neq j \\ -\omega_r^2 \phi_{ir}^2 & i = j \end{cases} \qquad (6.28)$$

Likewise, the sensitivity of ω_r^2 w.r.t. the change of the element k_{ij} in the stiffness matrix is given by:

$$\frac{\partial \omega_r^2}{\partial k_{ij}} = \begin{cases} 2\phi_{ir} \phi_{jr} & i \neq j \\ \phi_{ir}^2 & i = j \end{cases} \qquad (6.29)$$

Equations (6.28) and (6.29) outline the sensitivity of a natural frequency w.r.t. a spatial parameter change. They provide a guideline for selecting the 'optimal' parameters to obtain a prescribed natural frequency change. However, due to the assumption of infinitesimal changes on which sensitivity analysis is based, these equations become invalid if natural frequency changes become large.

The sensitivity of the rth mode shape $\{\phi_r\}$ w.r.t. the change of the element m_{ij} in the mass matrix can be expressed as a linear combination of all mode shapes, since they are independent of one another:

$$\frac{\partial \{\phi_r\}}{\partial m_{ij}} = \sum_{s=1}^{N} \tau_s \{\phi_s\} \qquad (6.30)$$

where τ_s is the participation factor for the sth mode to the sensitivity of the rth mode due to changes on mass element m_{ij}.

Pre-multiplying (6.24) by $\{\phi_s\}^T$, whith $s \neq r$, and substituting (6.30) into (6.24), gives:

$$\tau_s = \frac{\omega_r^2 (\phi_{js} \phi_{ir} + \phi_{is} \phi_{jr})}{\omega_s^2 - \omega_r^2} \qquad r \neq s, \qquad i \neq j \qquad (6.31)$$

If $r = s$, then the participation factor can be derived by differentiating the orthogonality identity $\{\phi_r\}^T [M] \{\phi_r\} = 1$ w.r.t. the change of the element m_{ij}, leading to:

$$\tau_s = -\phi_{jr} \phi_{ir} \qquad r = s, \qquad i \neq j \qquad (6.32)$$

For $i = j$, the participation factor can be found as:

$$\tau_s = \begin{cases} \dfrac{\omega_r^2 \phi_{is} \phi_{ir}}{\omega_s^2 - \omega_r^2} & \text{for} \quad r \neq s \\ -\dfrac{\phi_{ir}^2}{2} & \text{for} \quad r = s \end{cases} \qquad (6.33)$$

These solutions can be summarised as:

$$\tau_s = \begin{cases} \dfrac{\omega_r^2 \phi_{is}\,\phi_{ir}}{\omega_s^2 - \omega_r^2} & \text{for} \quad r \neq s \quad i = j \\[4mm] -\dfrac{\phi_{ir}^2}{2} & \text{for} \quad r = s \quad i = j \\[4mm] \dfrac{\omega_r^2(\phi_{js}\,\phi_{ir} + \phi_{is}\,\phi_{jr})}{\omega_s^2 - \omega_r^2} & \text{for} \quad r \neq s \quad i \neq j \\[4mm] -\phi_{ir}\,\phi_{jr} & \text{for} \quad r = s \quad i \neq j \end{cases}$$

(6.34)

For an element k_{ij} of the stiffness matrix, the sensitivity of the rth mode shape can also be expressed as:

$$\frac{\partial\{\phi_r\}}{\partial k_{ij}} = \sum_{s=1}^{n} \gamma_s \{\phi_s\}$$

(6.35)

where

$$\gamma_s = \begin{cases} -\dfrac{\phi_{is}\,\phi_{ir}}{\omega_s^2 - \omega_r^2} & \text{for} \quad r \neq s \quad i = j \\[4mm] 0 & \text{for} \quad r = s \quad i = j \\[4mm] -\dfrac{\phi_{js}\,\phi_{ir} + \phi_{is}\,\phi_{jr}}{\omega_s^2 - \omega_r^2} & \text{for} \quad r \neq s \quad i \neq j \\[4mm] 0 & \text{for} \quad r = s \quad i \neq j \end{cases}$$

(6.36)

These sensitivity results shed light on the best locations for a proposed structural modification to fulfil a prescribed objective. However, the assumption of small modifications implied by sensitivity analysis limits the usefulness of these results. An effective way to utilise sensitivity analysis results in structural modification is by iteration: each time, a 'small' modification is applied and sensitivity results are used as a guideline for the next modification.

Sensitivity analysis results can find resemblance in Rayleigh's quotient, proposed earlier in this century. For a small change in the natural frequency, Rayleigh's quotient can be given by:

$$\left(\omega_r + \Delta\omega_r\right)^2 = \frac{\{\phi_r\}^T[[K] + [\Delta K]]\{\phi_r\}}{\{\phi_r\}^T[[M] + [\Delta M]]\{\phi_r\}}$$

(6.37)

as for a small structural modification mode shapes do not change appreciably. Consequently, Rayleigh's quotient provides a good approximation of the natural frequency changes. To and Ewins have shown that Rayleigh's quotient is effectively the first order sensitivity of a dynamic system due to small modifications and this gave rise to a Rayleigh quotient iteration for structural modification [249].

6.3.2 Response model sensitivity analysis

Sensitivity analysis can also be applied to the FRFs of a dynamic system. Suppose that a spatial parameter p_r of a system has changed, then the change rate of the dynamic stiffness matrix of the system will be:

$$\frac{\partial[Z(\omega)]}{\partial p_r} = \frac{\partial\left[[K] - \omega^2[M]\right]}{\partial p_r} \tag{6.38}$$

However, from the theory of matrix derivatives, it is known that

$$\frac{\partial[\alpha(\omega)]}{\partial p_r} = \frac{\partial\left([Z(\omega)]^{-1}[Z(\omega)][Z(\omega)]^{-1}\right)}{\partial p_r} = \frac{\partial[Z(\omega)]^{-1}}{\partial p_r}[Z(\omega)][Z(\omega)]^{-1}$$
$$+ [Z(\omega)]^{-1}\frac{\partial[Z(\omega)]}{\partial p_r}[Z(\omega)]^{-1} + [Z(\omega)]^{-1}[Z(\omega)]\frac{\partial[Z(\omega)]^{-1}}{\partial p_r} \tag{6.39}$$

Hence,

$$\frac{\partial[\alpha(\omega)]}{\partial p_r} = -[\alpha(\omega)]\frac{\partial[Z(\omega)]}{\partial p_r}[\alpha(\omega)] = -[\alpha(\omega)]\left[\frac{\partial[K]}{\partial p_r} - \omega^2\frac{\partial[M]}{\partial p_r}\right][\alpha(\omega)] \tag{6.40}$$

Sensitivity analysis indicates the sensitive locations on a dynamic system for structural modification. It is not, however, capable of predicting the magnitude of physical parameter changes which will accomplish a specified characteristic change. If structural modification is a direct problem, then re-analysis is needed.

6.4 STRUCTURAL MODIFICATION AND RE-ANALYSIS

6.4.1 Modal condensation and structural modification

For a dynamic system with specified structural modifications, it is possible to predict the changes in dynamic characteristics without having to re-work the whole eigenvalue problem. This is the benefit of the re-analysis type of structural modification. The eigenvalue problem which governs the modified system is given as:

$$\left[[K] + [\Delta K] - \omega_*^2[M] - \omega_*^2[\Delta M]\right]\{z\} = \{0\} \tag{6.41}$$

where ω_* is the prescribed natural frequency. The response vector $\{z\}$ can be transformed using the mode shape matrix of the original system, as in:

$$\{z\} = [\phi] \{p\} \tag{6.42}$$

Here, vector $\{p\}$ contains the principal coordinates. The eigenvalue problem (6.41) can then be recast as:

$$\left[\left[\ddots \omega_r^2 \ddots \right] + [\phi]^T [\Delta K][\phi] \right] \{p\} = \omega_*^2 \left[[I] + [\phi]^T [\Delta M][\phi] \right] \{p\} \tag{6.43}$$

This is still a full-sized eigenvalue problem. However, if the mode shape matrix $[\phi]$ comes from experiment, then it usually contains a truncated set of mode shapes. This means the eigenvalue problem presented by equation (6.43) is short of a full-sized one. This approach of re-analysis offers a distinct advantage, i.e., it does not require the mass and stiffness matrices of the original system. For a practical problem-solving case this is particularly useful, since the data available would represent only a limited number of the natural frequencies and mode shapes of a structure and proposed spatial parameter changes. Using this approach, one can analytically study different structural modification proposals for prescribed dynamic characteristic changes without the costly fabrication of prototypes.

It should not be overlooked that (6.43) only provides an approximate solution for structural modification when not all the mode shapes are present. This happens to be the reality if the mode shapes are derived from experiment, since their number is usually insufficient. Should all the mode shapes be used in the equation, it would no longer render a more efficient solution than re-solving the complete eigenvalue problem for the modified system. This feature of approximation caused by modal truncation is, in fact, a common drawback for any methods which utilise modal data rather than FRF data.

6.4.2   Receptance FRFs before and after modification

A valid question in structural modification with response models is what alteration will the proposed structural changes bring to the FRFs of the modified structure. The FRFs of the modified structure $[\alpha(\omega)]_{new}$ and that of the original structure $[\alpha(\omega)]$ relate inherently to each other. The former can be derived from the latter. Assume that for an N DOF system the local structural modification is given by $[\Delta K]$ and is due to local stiffness changes,

$$[\Delta K] = \begin{bmatrix} [\Delta] & [0] \\ [0] & [0] \end{bmatrix} \tag{6.44}$$

Here, singular matrix $[\Delta K]$ is of order N but its rank is r (r < N). The order of the submatrix $[\Delta]$ is m. It is obvious that for local modification $m \ll N$. Matrix $[\Delta K]$ is symmetric and positive semi-definite. Mathematically it can be decomposed as:

$$[\Delta K] = \sum_{k=1}^{r} \rho_k \{\varepsilon_k\}\{\varepsilon_k\}^T \tag{6.45}$$

where ρ_k are real and positive quantities and $\{\varepsilon_k\}$ are linearly independent real vectors (Nx1). This decomposition can be accomplished by an eigenvalue solution or by Singular Value Decomposition (see Appendix A). The original receptance FRF matrix and the new one are linked via the modification matrix $[\Delta K]$ as:

$$[\alpha(\omega)]_{new}^{-1} = [\alpha(\omega)]^{-1} + [\Delta K] \tag{6.46}$$

and by substituting (6.45) into (6.46),

$$[\alpha(\omega)]_{new}^{-1} = [\alpha(\omega)]^{-1} + \sum_{k=1}^{r} \rho_k \{\varepsilon_k\} \{\varepsilon_k\}^T \tag{6.47}$$

Consider a sequence of intermediate receptance matrices:

$$[\alpha(\omega)]_q^{-1} = [\alpha(\omega)]^{-1} + \sum_{k=1}^{q} \rho_k \{\varepsilon_k\} \{\varepsilon_k\}^T \qquad q = 1, 2, ..., r \tag{6.48}$$

Obviously, $[\alpha(\omega)]_{q=r} = [\alpha(\omega)]_{new}$. Then, two consecutive matrices $[\alpha(\omega)]_q$ will associate with each other by:

$$[\alpha(\omega)]_q^{-1} = [\alpha(\omega)]_{q-1}^{-1} + \rho_q \{\varepsilon_q\} \{\varepsilon_q\}^T \tag{6.49}$$

or

$$[\alpha(\omega)]_q = \left[[I] + \rho_q [\alpha(\omega)]_{q-1} \{\varepsilon_q\} \{\varepsilon_q\}^T \right]^{-1} [\alpha(\omega)]_{q-1} \tag{6.50}$$

Using the following mathematical identity:

$$\left[[I] + \{\theta\} \{\theta\}^T \right]^{-1} = [I] - \frac{\{\theta\} \{\theta\}^T}{1 + \{\theta\}^T \{\theta\}} \tag{6.51}$$

equation (6.50) can be simplified as:

$$[\alpha(\omega)]_q = \left[[I] - \frac{\rho_q [\alpha(\omega)]_{q-1} \{\varepsilon_q\} \{\varepsilon_q\}^T}{1 + \rho_q \{\varepsilon_q\}^T [\alpha(\omega)]_{q-1} \{\varepsilon_q\}} \right] [\alpha(\omega)]_{q-1} \tag{6.52}$$

Although appearing complicated, equation (6.52) provides an iterative procedure to derive the receptance FRF matrix of the modified system from its original FRF matrix and modification matrix with great computational economy. This is because no matrix inversion is involved. In summary, the procedure to derive a new FRF matrix is as follows:

(1) decompose a prescribed modification matrix $[\Delta K]$ using eigenvalue solution or Singular Value Decomposition;

(2) for a given frequency ω, estimate matrix $[\alpha(\omega)]_q$ when $q = 2, 3, ..., N$, using (6.52).

When structural modification involves both mass and stiffness changes, it is also possible to derive the FRF matrix of the modified system. This begins with:

$$[\alpha(\omega)]_{new}[Z(\omega) + \Delta Z(\omega)] = [I] \qquad (6.53)$$

leading to the following conclusion:

$$[\alpha(\omega)]_{new} = [\alpha(\omega)]\left[[I] + [\Delta Z(\omega)][\alpha(\omega)]\right]^{-1} \qquad (6.54)$$

Alternatively, the FRF matrix of the modified system can be derived using modal data. Equation (6.53) can be rewritten as:

$$[\alpha(\omega)]_{new}\left[[\phi]^{-T}\left[\,\ \omega_r^2 - \omega^2 \,_\diagdown\,\right][\phi]^{-1} + [\Delta Z(\omega)]\right] = [I] \qquad (6.55)$$

This will lead to the following solution for the modified FRF matrix:

$$[\alpha(\omega)]_{new} = [\phi]\left[\left[\,\ \omega_r^2 - \omega^2 \,_\diagdown\,\right] + [\phi]^T[\Delta Z(\omega)][\phi]\right]^{-1}[\phi]^T \qquad (6.56)$$

6.5   STRUCTURAL MODIFICATION - LUMPED PARAMETER SYSTEMS

The lumped parameter system is very useful in exploring new ideas and algorithms of structural modification and optimisation. This is not because of the usually small size of the system, as may be perceived, but because such systems permit us to separate mass and stiffness modifications. As a result, it is possible to assume some mass or stiffness modification only in the pursuit of structural modification analysis. In addition, the mass matrix for a lumped system is often a diagonal one. This makes mass modification analytically convenient.

Assume structural modifications $[\Delta M]$ and $[\Delta K]$ respectively. The equation of motion of the modified system represents a new eigenvalue question:

$$[[K] + [\Delta K]]\{\theta\} - \omega_*^2[[M] + [\Delta M]]\{\theta\} = \{0\} \qquad (6.57)$$

This is equivalent to:

$$\{\theta\} = [\alpha(\omega_*)]\left[\omega_*^2[\Delta M] - [\Delta K]\right]\{\theta\} \qquad (6.58)$$

Equation (6.58) is the fundamental equation structural modification has to satisfy. It establishes the link between the prescribed natural frequency ω_* and the modifications $[\Delta M]$ and $[\Delta K]$.

6.5.1 Single mass or stiffness modification

Single mass or stiffness modification entails structural modification on one physical mass element or stiffness element alone. This is also called unit rank modification. Such a modification renders a simple yet revealing analytical solution.

If only mass modification is made while stiffness properties are kept unchanged, then equation (6.58) will be recast into:

$$\{\theta\} = \omega_*^2 [\alpha(\omega_*)][\Delta M]\{\theta\} \tag{6.59}$$

Equation (6.59) dictates the relationship between a prescribed natural frequency and mass modification $[\Delta M]$. If mass change only occurs at coordinate k, then (6.59) can be simplified into:

$$\Delta m_k = \left(\omega_*^2 \, \alpha_{kk}(\omega_*)\right)^{-1} \tag{6.60}$$

or

$$\Delta m_k = \frac{1}{\omega_*^2}\left(\sum_{r=1}^{N} \frac{\phi_{kr}^2}{\omega_r^2 - \omega_*^2}\right)^{-1} \tag{6.61}$$

If only stiffness modification is made, then equation (6.58) will become:

$$\{\theta\} = -[\alpha(\omega_*)][\Delta K]\{\theta\} \tag{6.62}$$

If the stiffness change occurs only at coordinate k (an elastic component connecting coordinate k and ground), then (6.62) will become:

$$\Delta k_k = -\alpha_{kk}(\omega_*)^{-1} \tag{6.63}$$

or

$$\Delta k_k = -\left(\sum_{r=1}^{N} \frac{\phi_{kr}^2}{\omega_r^2 - \omega_*^2}\right)^{-1} \tag{6.64}$$

However, a more realistic situation is to modify the stiffness element between coordinates i and j. In this case, (6.62) becomes:

$$\{\theta\} = -[\alpha(\omega_*)]\begin{bmatrix} 0 & & & & 0 \\ & \ddots & & & \\ & & \Delta_{ii} & \cdots & \Delta_{ij} \\ & & \vdots & & \vdots \\ & & \Delta_{ji} & \cdots & \Delta_{jj} & \\ & & & & \ddots \\ 0 & & & & 0 \end{bmatrix}\{\theta\} \tag{6.65}$$

where $\Delta_{ii} = \Delta_{jj} = -\Delta_{ij} = -\Delta_{ji} = \Delta k$. Then, it can be found from (6.65) that:

$$\Delta k = -\frac{1}{\alpha_{ii}(\omega_*) + \alpha_{jj}(\omega_*) - \alpha_{ij}(\omega_*) - \alpha_{ji}(\omega_*)} \tag{6.66}$$

A unit rank modification can be expanded to a more sophisticated one if this can be expressed as a linear combination of unit rank modifications. However, this endeavour may be plausible for re-analysis but would be difficult for the inverse problem. Modification with more than a unit rank is theorised in the following. The real hindrance in reality for some structures is the impossibility of modifying mass and stiffness properties separately. For instance, a mass modification of a truss element in a truss structure may be destined to cause stiffness change. As a result, the analysis of both unit rank and multi-rank modification needs to be extended.

6.5.2 Assigning a resonance frequency to a MDOF system

Theoretically, expression (6.58) allows for the determination of the necessary modifications in order to assign a new natural frequency. Nevertheless, the implementation of that formula still needs some stringent pursuit.

Let only mass modification be proposed at certain coordinates from p to q. The mass modification matrix can be written as:

$$[\Delta M] = \begin{bmatrix} 0 & & & & 0 \\ & \ddots & & & \\ & & \Delta m_p & & \\ & & & \ddots & \\ & & & & \Delta m_q \\ & & & & & \ddots \\ 0 & & & & 0 \end{bmatrix} = \zeta_m \begin{bmatrix} 0 & & & & 0 \\ & \ddots & & & \\ & & \varepsilon_p & & \\ & & & \ddots & \\ & & & & \varepsilon_q \\ & & & & & \ddots \\ 0 & & & & 0 \end{bmatrix} = \zeta_m [\varepsilon] \tag{6.67}$$

Assume the modification ratio matrix $[\varepsilon]$ to be user-defined, dictating the relative mass changes between coordinates with mass modifications. Then, equation (6.59) becomes:

$$\frac{1}{\zeta_m}\{\theta\} = \omega_*^2 [\alpha(\omega_*)][\varepsilon]\{\theta\} \tag{6.68}$$

The size of the matrices and vectors in this equation can be greatly reduced by eliminating rows and columns corresponding to coordinates with zero mass changes. This will result in a reduced eigenvalue problem:

$$\left[\omega_*^2 [\alpha(\omega_*)^R][\varepsilon^R] - \frac{1}{\zeta_m}[I] \right]\{\theta^R\} = \{0\} \tag{6.69}$$

The eigenvalues of this equation are $1/\zeta_m$. For each one, the mass modification matrix can be found from (6.67), as the ratio matrix $[\varepsilon]$ is user-defined.

The feasibility of this mass modification approach hinges on the eigenvalue roots of (6.69). Real eigenvalue roots suggest physically feasible modifications. Since both the number of modifications and the ratios are arbitrary, the solution of real eigenvalues is not analytically guaranteed. However, there may also be more than one real eigenvalue, signifying that more than one modification will secure the assigned resonance.

The stiffness modification approach can be similarly derived. For the ease of illustration, let stiffness modifications be proposed among coordinates p, q and r only. The modification matrix can be written as:

$$
[\Delta K] = \begin{bmatrix} \ddots & & & & & \\ & \Delta k_{pq} + \Delta k_{rp} & \cdots & -\Delta k_{pq} & \cdots & -\Delta k_{rp} & \\ & \vdots & & \vdots & & \vdots & \\ & -\Delta k_{pq} & \cdots & \Delta k_{pq} + \Delta k_{qr} & \cdots & -\Delta k_{qr} & \\ & \vdots & & \vdots & & \vdots & \\ & -\Delta k_{rp} & \cdots & -\Delta k_{qr} & \cdots & \Delta k_{rp} + \Delta k_{qr} & \\ & & & & & & \ddots \end{bmatrix} = \gamma_k [\kappa]
$$

(6.70)

where

$$
\left\{ \Delta k_{pq} \quad \Delta k_{qr} \quad \Delta k_{rp} \right\} = \gamma_k \left\{ \kappa_{pq} \quad \kappa_{qr} \quad \kappa_{rp} \right\}
$$

(6.71)

Assume that the modification ratio matrix $[\kappa]$ is user-defined, representing the relative stiffness changes between coordinates with stiffness modifications. Then, equation (6.62) becomes:

$$
\frac{1}{\gamma_k} \{\theta\} = [\alpha(\omega_*)][\kappa]\{\theta\}
$$

(6.72)

Once again, the size of the matrices and vectors in this equation can be greatly reduced by eliminating rows and columns corresponding to coordinates with zero stiffness changes. The reduced eigenvalue problem is:

$$
\left[[\alpha(\omega_*)^R][\kappa^R] - \frac{1}{\gamma_k}[I] \right] \{\theta^R\} = \{0\}
$$

(6.73)

Now, the eigenvalues of this equation are $1/\gamma_k$, from which the stiffness modification matrix can be found, as $[\Delta K] = \gamma_k[\kappa]$.

Like in mass modification, the real eigenvalues of (6.73) are not analytically guaranteed. If no real eigenvalues exist in the solution, it means that the stiffness changes on the proposed coordinates cannot bring about the proposed new resonance frequency ω_*.

6.5.3 Assigning an antiresonance to an FRF

The technique used to assign a new natural frequency can be employed for assigning an antiresonance. Since antiresonances are local properties, each time this is done it is meant only for a particular FRF. As equation (6.4) suggests, the antiresonances of a given FRF $\alpha_{ij}(\omega)$ are the same as the 'resonances' of the corresponding virtual system comprising mass matrix $[M]_{ij}$ and stiffness matrix $[K]_{ij}$. Therefore, the assignment of an antiresonance to FRF $\alpha_{ij}(\omega)$ will be the same as the assignment of a new natural frequency for the virtual system. The equation for the virtual system modification can be written as:

$$\left[[K]_{ij} + [\Delta K]_{ij}\right]\{X\} - \Omega_*^2 \left[[M]_{ij} + [\Delta M]_{ij}\right]\{X\} = \{0\} \tag{6.74}$$

where modification matrices $[\Delta M]_{ij}$ and $[\Delta K]_{ij}$ are obtained from $[\Delta M]$ and $[\Delta K]$ respectively in the same way matrices $[M]_{ij}$ and $[K]_{ij}$ are derived from $[M]$ and $[K]$ respectively. For instance, if

$$\underset{(N \times N)}{[\Delta M]} = \begin{bmatrix} \ddots & & & & \\ & \Delta m_p & & & \\ & & \ddots & & \\ & & & \Delta m_q & \\ & & & & \ddots & \\ & & & & & \Delta m_r \\ & & & & & & \ddots \end{bmatrix} \tag{6.75}$$

then

$$\underset{((N-1) \times (N-1))}{[\Delta M]_{ij}} = \begin{bmatrix} \ddots & & & & \\ & 0 & \Delta m_p & & \\ & & \ddots & & \\ & & & 0 & \Delta m_q & \\ & & & & \ddots & \\ & & & & & 0 & \Delta m_r \\ & & & & & & \ddots \end{bmatrix} \tag{6.76}$$

Therefore, using the approach for the assignment of new resonance frequencies in expressions (6.67) to (6.73), modification matrices $[\Delta M]_{ij}$ and $[\Delta K]_{ij}$ for the virtual system can be determined separately. Once done, the modifications in the original system $[\Delta M]$ and $[\Delta K]$ can be ascertained from their relationships to matrices $[\Delta M]_{ij}$ and $[\Delta K]_{ij}$ respectively. Vector $\{X\}$ in (6.74) bears no apparent physical meaning. It can be called the 'mode shape' for the 'resonance' Ω_* of the virtual system.

6.5.4 Structural modification without eigenvalue solution

Mass modification

The methods described in the preceding sections to obtain an assigned resonance or antiresonance led to reduced eigenvalue problems. However, when dealing with a mass-spring system, it is possible to formulate structural modifications with the same objectives without leading to an eigenvalue equation. For the sake of simplicity but without losing generality, let us assume that mass modifications are applied at coordinates i, j and k of a mass-spring system. Equation (6.58) becomes:

$$\begin{Bmatrix} \theta_i \\ \theta_j \\ \theta_k \end{Bmatrix} = \omega_*^2 \begin{bmatrix} \alpha_{ii}(\omega_*) & \alpha_{ij}(\omega_*) & \alpha_{ik}(\omega_*) \\ \alpha_{ji}(\omega_*) & \alpha_{jj}(\omega_*) & \alpha_{jk}(\omega_*) \\ \alpha_{ki}(\omega_*) & \alpha_{kj}(\omega_*) & \alpha_{kk}(\omega_*) \end{bmatrix} \begin{bmatrix} \Delta m_i & 0 & 0 \\ 0 & \Delta m_j & 0 \\ 0 & 0 & \Delta m_k \end{bmatrix} \begin{Bmatrix} \theta_i \\ \theta_j \\ \theta_k \end{Bmatrix} \tag{6.77}$$

This equation is the same as:

$$\omega_*^2 \begin{bmatrix} \alpha_{ii}(\omega_*) & \alpha_{ij}(\omega_*) & \alpha_{ik}(\omega_*) \\ \alpha_{ji}(\omega_*) & \alpha_{jj}(\omega_*) & \alpha_{jk}(\omega_*) \\ \alpha_{ki}(\omega_*) & \alpha_{kj}(\omega_*) & \alpha_{kk}(\omega_*) \end{bmatrix} \begin{bmatrix} \theta_i & 0 & 0 \\ 0 & \theta_j & 0 \\ 0 & 0 & \theta_k \end{bmatrix} \begin{Bmatrix} \Delta m_i \\ \Delta m_j \\ \Delta m_k \end{Bmatrix} = \begin{Bmatrix} \theta_i \\ \theta_j \\ \theta_k \end{Bmatrix} \tag{6.78}$$

This is a set of simultaneous equations for mass changes. Vector $\{\theta\}$ is the subset of the mode shape of the modified system corresponding to the prescribed natural frequency ω_*. One sensible selection for $\{\theta\}$ is the mode shape of the original system [253]. Alternatively, it can be arbitrarily determined to dictate the mode shape subset of the modified system.

Equation (6.78) suggests theoretically that we can 'set' the mode shape, as well as a natural frequency, of the modified system by using a series of point mass modifications. However, (6.78) does not guarantee physically meaningful solutions, let alone the practicality of such extensive mass modification.

The FRF data in (6.78) can be computed from an existing FE model or can be measured. Since structural modification is local, the required FRF data will only involve the modified coordinates. This circumvents the problem of modal truncation or incompleteness of measured data.

Alternatively, using a linear transformation:

$$\begin{Bmatrix} X_i \\ X_j \\ X_k \end{Bmatrix} = \begin{bmatrix} \Delta m_i & 0 & 0 \\ 0 & \Delta m_j & 0 \\ 0 & 0 & \Delta m_k \end{bmatrix} \begin{Bmatrix} \theta_i \\ \theta_j \\ \theta_k \end{Bmatrix} \tag{6.79}$$

equation (6.77) becomes:

$$\begin{bmatrix} \Delta m_i & 0 & 0 \\ 0 & \Delta m_j & 0 \\ 0 & 0 & \Delta m_k \end{bmatrix} \begin{bmatrix} \alpha_{ii}(\omega_*) & \alpha_{ij}(\omega_*) & \alpha_{ik}(\omega_*) \\ \alpha_{ji}(\omega_*) & \alpha_{jj}(\omega_*) & \alpha_{jk}(\omega_*) \\ \alpha_{ki}(\omega_*) & \alpha_{kj}(\omega_*) & \alpha_{kk}(\omega_*) \end{bmatrix} \begin{Bmatrix} X_i \\ X_j \\ X_k \end{Bmatrix} = \frac{1}{\omega_*^2} \begin{Bmatrix} X_i \\ X_j \\ X_k \end{Bmatrix} \tag{6.80}$$

This leads to a set of succinct analytical solutions for mass modifications:

$$\Delta m_r = \frac{1}{\omega_*^2} \cdot \frac{X_r}{X_i \alpha_{ri}(\omega_*) + X_j \alpha_{rj}(\omega_*) + X_k \alpha_{rk}(\omega_*)} \qquad r = i, j, k \tag{6.81}$$

This conclusion can be extended to other numbers of point mass modifications. For a single mass modification at coordinate i, it predicts:

$$\Delta m_i = \frac{1}{\omega_*^2 \, \alpha_{ii}(\omega_*)} \tag{6.82}$$

This is how a point mass modification can shift a resonance of a dynamic system.

Stiffness modification
Similarly to mass modification, stiffness modification can be analysed. Again, assume that modification occurs among coordinates i, j and k. Equation (6.58) becomes

$$\begin{Bmatrix} \theta_i \\ \theta_j \\ \theta_k \end{Bmatrix} = -\begin{bmatrix} \alpha_{ii} & \alpha_{ij} & \alpha_{ik} \\ \alpha_{ji} & \alpha_{jj} & \alpha_{jk} \\ \alpha_{ki} & \alpha_{kj} & \alpha_{kk} \end{bmatrix} \begin{bmatrix} \Delta k_{ij} + \Delta k_{ik} & -\Delta k_{ij} & -\Delta k_{ik} \\ -\Delta k_{ij} & \Delta k_{ij} + \Delta k_{jk} & -\Delta k_{jk} \\ -\Delta k_{ik} & -\Delta k_{jk} & \Delta k_{jk} + \Delta k_{ik} \end{bmatrix} \begin{Bmatrix} \theta_i \\ \theta_j \\ \theta_k \end{Bmatrix} \tag{6.83}$$

Using a submatrix expression of modification matrix [ΔK], (6.83) yields:

$$\begin{Bmatrix} \theta_i \\ \theta_j \\ \theta_k \end{Bmatrix} = -\begin{bmatrix} \alpha_{ii} & \alpha_{ij} & \alpha_{ik} \\ \alpha_{ji} & \alpha_{jj} & \alpha_{jk} \\ \alpha_{ki} & \alpha_{kj} & \alpha_{kk} \end{bmatrix} \left[\Delta k_{ij} \begin{bmatrix} 1 & -1 & 0 \\ -1 & 1 & 0 \\ 0 & 0 & 0 \end{bmatrix} + \Delta k_{jk} \begin{bmatrix} 0 & 0 & 0 \\ 0 & 1 & -1 \\ 0 & -1 & 1 \end{bmatrix} + \right.$$

$$\left. \Delta k_{ik} \begin{bmatrix} 1 & 0 & -1 \\ 0 & 0 & 0 \\ -1 & 0 & 1 \end{bmatrix} \right] \begin{Bmatrix} \theta_i \\ \theta_j \\ \theta_k \end{Bmatrix} \tag{6.84}$$

or

$$\begin{Bmatrix} \theta_i \\ \theta_j \\ \theta_k \end{Bmatrix} = - \begin{bmatrix} \alpha_{ii} & \alpha_{ij} & \alpha_{ik} \\ \alpha_{ji} & \alpha_{jj} & \alpha_{jk} \\ \alpha_{ki} & \alpha_{kj} & \alpha_{kk} \end{bmatrix} \left(\Delta k_{ij} \begin{Bmatrix} \theta_i - \theta_j \\ \theta_j - \theta_i \\ 0 \end{Bmatrix} + \Delta k_{jk} \begin{Bmatrix} 0 \\ \theta_j - \theta_k \\ \theta_k - \theta_j \end{Bmatrix} + \Delta k_{ik} \begin{Bmatrix} \theta_i - \theta_k \\ 0 \\ \theta_k - \theta_i \end{Bmatrix} \right)$$

(6.85)

To simplify the analysis, let us use the following variable transformation:

$$\theta_{ij} = \theta_i - \theta_j \qquad \theta_{jk} = \theta_j - \theta_k \qquad \theta_{ik} = \theta_i - \theta_k \tag{6.86}$$

and define a new variable $\alpha_{ij,k\ell}(\omega_*)$ as:

$$\alpha_{ij,k\ell}(\omega_*) = \alpha_{ik}(\omega_*) + \alpha_{j\ell}(\omega_*) - \left[\alpha_{jk}(\omega_*) + \alpha_{i\ell}(\omega_*) \right] \tag{6.87}$$

Equation (6.85) can then be simplified into:

$$- \begin{bmatrix} \alpha_{ij,ij}(\omega_*) & \alpha_{ij,jk}(\omega_*) & \alpha_{ij,ik}(\omega_*) \\ \alpha_{jk,ij}(\omega_*) & \alpha_{jk,jk}(\omega_*) & \alpha_{jk,ik}(\omega_*) \\ \alpha_{ik,ij}(\omega_*) & \alpha_{ik,jk}(\omega_*) & \alpha_{ik,ik}(\omega_*) \end{bmatrix} \begin{bmatrix} \theta_{ij} & 0 & 0 \\ 0 & \theta_{jk} & 0 \\ 0 & 0 & \theta_{ik} \end{bmatrix} \begin{Bmatrix} \Delta k_{ij} \\ \Delta k_{jk} \\ \Delta k_{ik} \end{Bmatrix} = \begin{Bmatrix} \theta_{ij} \\ \theta_{jk} \\ \theta_{ik} \end{Bmatrix} \tag{6.88}$$

As a set of linear equations, this resembles (6.78). Its solution yields the stiffness modifications required at coordinates i, j and k to generate the new resonance ω_*. Unlike mass modification, the physical interpretation of the vector $\{\theta\}$ in this equation is not explicit. This is because of the various variable transformations involved in its derivation. The conclusion given in (6.88) can be extended to more (or less) than three modification coordinates.

Alternatively, (6.88) can be written as:

$$- \begin{bmatrix} \alpha_{ij,ij}(\omega_*) & \alpha_{ij,jk}(\omega_*) & \alpha_{ij,ik}(\omega_*) \\ \alpha_{jk,ij}(\omega_*) & \alpha_{jk,jk}(\omega_*) & \alpha_{jk,ik}(\omega_*) \\ \alpha_{ik,ij}(\omega_*) & \alpha_{ik,jk}(\omega_*) & \alpha_{ik,ik}(\omega_*) \end{bmatrix} \begin{bmatrix} \Delta k_{ij} & 0 & 0 \\ 0 & \Delta k_{jk} & 0 \\ 0 & 0 & \Delta k_{ik} \end{bmatrix} \begin{Bmatrix} \theta_{ij} \\ \theta_{jk} \\ \theta_{ik} \end{Bmatrix} = \begin{Bmatrix} \theta_{ij} \\ \theta_{jk} \\ \theta_{ik} \end{Bmatrix} \tag{6.89}$$

Using the linear transformation:

$$\begin{Bmatrix} X_{ij} \\ X_{jk} \\ X_{ik} \end{Bmatrix} = \begin{bmatrix} \Delta k_{ij} & 0 & 0 \\ 0 & \Delta k_{jk} & 0 \\ 0 & 0 & \Delta k_{ik} \end{bmatrix} \begin{Bmatrix} \theta_{ij} \\ \theta_{jk} \\ \theta_{ik} \end{Bmatrix} \tag{6.90}$$

equation (6.89) can be written as:

$$\begin{bmatrix} \Delta k_{ij} & 0 & 0 \\ 0 & \Delta k_{jk} & 0 \\ 0 & 0 & \Delta k_{ik} \end{bmatrix} \begin{bmatrix} \alpha_{ij,ij}(\omega_*) & \alpha_{ij,jk}(\omega_*) & \alpha_{ij,ik}(\omega_*) \\ \alpha_{jk,ij}(\omega_*) & \alpha_{jk,jk}(\omega_*) & \alpha_{jk,ik}(\omega_*) \\ \alpha_{ik,ij}(\omega_*) & \alpha_{ik,jk}(\omega_*) & \alpha_{ik,ik}(\omega_*) \end{bmatrix} \begin{bmatrix} X_{ij} \\ X_{jk} \\ X_{ik} \end{bmatrix} = -\begin{Bmatrix} X_{ij} \\ X_{jk} \\ X_{ik} \end{Bmatrix}$$

(6.91)

This leads to a set of simple *formulæ* for estimating stiffness modifications:

$$\Delta k_{ij} = -\frac{X_{ij}}{X_{ij}\alpha_{ij,ij}(\omega_*) + X_{jk}\alpha_{ij,jk}(\omega_*) + X_{ik}\alpha_{ij,ik}(\omega_*)}$$

(6.92 a)

$$\Delta k_{jk} = -\frac{X_{jk}}{X_{ij}\alpha_{jk,ij}(\omega_*) + X_{jk}\alpha_{jk,jk}(\omega_*) + X_{ik}\alpha_{jk,ik}(\omega_*)}$$

(6.92 b)

$$\Delta k_{ik} = -\frac{X_{ik}}{X_{ij}\alpha_{ik,ij}(\omega_*) + X_{jk}\alpha_{ik,jk}(\omega_*) + X_{ik}\alpha_{ik,ik}(\omega_*)}$$

(6.92 c)

This conclusion can be extended to cover more (or less) than three stiffness modifications. However, it is intuitively useful to derive from this conclusion a single stiffness modification between the ith and the jth coordinates:

$$\Delta k_{ij} = -\frac{1}{\alpha_{ij,ij}(\omega_*)} = -\frac{1}{\alpha_{ii}(\omega_*) + \alpha_{jj}(\omega_*) - \alpha_{ji}(\omega_*) - \alpha_{ij}(\omega_*)}$$

(6.93)

which involves all point and transfer receptances associated with the modified coordinates i and j. This is the same as the prediction given in equation (6.66).

For a single stiffness modification between the ith co-ordinate and the ground, $\alpha_{jj}(\omega_*) = \alpha_{ji}(\omega_*) = \alpha_{ij}(\omega_*) = 0$, and we have:

$$\Delta k_i = -\frac{1}{\alpha_{ii}(\omega_*)}$$

(6.94)

6.6   SIMULTANEOUS MASS AND STIFFNESS MODIFICATION

The structural modification theory for a lumped system deals with separate mass and stiffness changes. For a structure with a finite element model, certain physical parameters such as mass density or elasticity only affect mass or stiffness properties. That theory still applies. However, if the geometrical dimensions have to change, both the mass and stiffness matrices will change with different orders.

332

The theory for lumped systems needs to be broadened to accommodate such structural modifications.

If for a dynamic system, the change of a structural parameter γ will result in linear change of the mass and stiffness of some elements, then from equations (6.67) and (6.70), the modification matrices will become:

$$[\Delta M] = \Delta\gamma[\epsilon] \quad \text{and} \quad [\Delta K] = \Delta\gamma[\kappa] \tag{6.95}$$

The eigenvalue equation for a prescribed resonance ω_* and that for a prescribed antiresonance Ω_* of the ij^{th} receptance FRF will be, respectively:

$$\frac{1}{\Delta\gamma}\{\theta\} = [\alpha(\omega_*)]\left[\omega_*^2[\epsilon] - [\kappa]\right]\{\theta\} \tag{6.96}$$

$$\frac{1}{\Delta\gamma}\{X\} = [\alpha(\Omega_*)]_{ij}\left[\Omega_*^2[\epsilon]_{ij} - [\kappa]_{ij}\right]\{X\} \tag{6.97}$$

Equations (6.96) and (6.97) characterise an idealised case for structural modification. An example of this is the cantilevered beam in figure 6.8. If vertical transverse vibration is of interest and the horizontal dimension of some parts of the beam change, then it will result in proportional mass and stiffness changes.

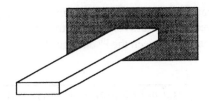

Fig. 6.8 A cantilevered beam.

However, for some finite element types, a physical parameter change will result in changes of element matrices non-proportional to it. Take as an example a three-dimensional brick element shown in figure 6.9 a). The second moment of area I_y and the mass of the element are linear functions of dimension b. However, this is not true if dimension h is to be changed. In figure 6.9 b), neither the mass nor the second moment of area of the element are linear functions of the diameter D.

In fact, a change Δ in the value of the diameter D will be associated with quadratic mass change and quadruple bending stiffness change. The modification matrices will become:

$$[\Delta M] = (D + \Delta)^2[\epsilon] - D^2[\epsilon] = \Delta^2[\epsilon] + 2D\Delta[\epsilon] \tag{6.98}$$

$$[\Delta K] = (D + \Delta)^4[\kappa] - D^4[\kappa] = \sum_{r=1}^{4} C_4^r D^{4-r}\Delta^r[\kappa] \tag{6.99}$$

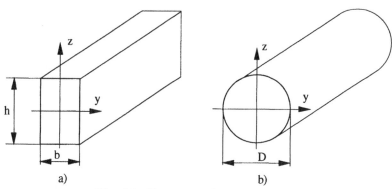

Fig. 6.9 Two types of finite elements.

where C_4^r is the binomial function[†]. Substituting (6.98) and (6.99) into (6.57) leads to an eigenvalue problem for structural modification. In general, (6.57) will have the following form:

$$\Delta^p[A]_p\{\theta\} + \Delta^{p-1}[A]_{p-1}\{\theta\} + ... + \Delta[A]_1\{\theta\} + [I]\{\theta\} = \{0\}$$ (6.100)

This is not a conventional eigenvalue problem for variable D and its solution has to be found using a state-space transformation to convert the equation into a conventional eigenvalue equation [254]. Assuming that:

$$\{\theta\} = \{\Theta\}e^{\Delta\tau}$$ (6.101)

then

$$\frac{\partial^r}{\partial\tau^r}\{\theta\} = \{\theta\}^{(r)} = \Delta^r\{\theta\}$$ (6.102)

Therefore, defining a state-space vector:

$$\{V\} = \left\{\{\theta\}^{(p-1)} \quad \{\theta\}^{(p-2)} \quad ... \quad \{\theta\}^{(0)}\right\}^T$$ (6.103)

equation (6.100) will become:

$$\left[[A]_p \quad [A]_{p-1} \quad ... \quad [A]_1\right]\{V\}^{(1)} + \left[[0] \quad [0] \quad ... \quad [I]\right]\{V\} = \{0\}$$ (6.104)

This equation can be appended by a number of identities to form the following eigenvalue equation:

[†] $C_4^r = \dfrac{4!}{r!(4-r)!}$

$$\Delta \begin{bmatrix} [A]_p & [A]_{p-1} & [A]_{p-2} & \cdots & [A]_1 \\ [0] & [I] & [0] & \cdots & [0] \\ [0] & [0] & [I] & \cdots & [0] \\ \vdots & \vdots & \vdots & \cdots & \vdots \\ [0] & [0] & [0] & \cdots & [0] \end{bmatrix} \{V\} + \begin{bmatrix} [0] & [0] & [0] & \cdots & [I] \\ -[I] & [0] & [0] & \cdots & [0] \\ [0] & -[I] & [0] & \cdots & [0] \\ \vdots & \vdots & \vdots & \vdots & \vdots \\ [0] & [0] & \cdots & -[I] & [0] \end{bmatrix} \{V\} = \{0\}$$

$$(6.105)$$

This is a conventional eigenvalue problem. The eigenvalues obtained from this equation will be the possible changes of a physical parameter which will result in a prescribed resonance frequency for the modified system.

The assignment of an antiresonance for a receptance FRF renders the same theoretical treatment, except that the numerical work involved increases because of the non-symmetrical matrices of the virtual system.

6.7 OPTIMISATION OF DYNAMIC PROPERTIES BY LOCAL STRUCTURAL MODIFICATION

The objective of optimising the dynamic properties of a system aims at avoiding resonance vibration. This is possible when the structure operates within a limited frequency range. It is then desirable to modify the structure so that resonances can be moved outside the frequency range of major concern. This may be achieved by either relocating resonances, or by creating a cancellation of a resonance and an antiresonance within the frequency range. Relocation of resonances affects the whole structure because of their 'global' nature. Cancellation of a resonance and an antiresonance physically means creating a nodal point at a given coordinate. This is useful when vibration control is sought at a given location, rather than for the whole structure. A good example is the need for vibration reduction at the tip of a cutting tool on a lathe.

Let us revisit the modification equation (6.57),

$$[[K] + [\Delta K]] \{\theta\} - \omega_*^2 [[M] + [\Delta M]] \{\theta\} = \{0\} \qquad (6.106)$$

Assuming that the modification matrices $[\Delta M]$ and $[\Delta K]$ are such that they satisfy:

$$\left[[\Delta K] - \omega_*^2 [\Delta M] \right] \{\psi\} = \{0\} \qquad (6.107)$$

and that ω_* and $\{\psi\}$ are a natural frequency and the corresponding mode shape of the original system, (6.106) will become (6.1) which is the eigenvalue equation of the original system. This means that the modifications $[\Delta M]$ and $[\Delta K]$ on the original system do not alter one of its natural frequencies and corresponding mode shape. Such modifications effectively fix that natural frequency and its mode shape while altering other natural frequencies and mode shapes. This opens a useful avenue for manipulating the dynamic properties of a system by local structural modification. For example, it becomes possible to fix one natural frequency and shift another, creating a frequency range with no resonances.

If the structural connectivity of the original system is to be honoured, modifications by $[\Delta M^R]$ and $[\Delta K^R]$, which are the submatrices of modifications $[\Delta M]$ and $[\Delta K]$ that satisfy (6.107), would form an imaginary dynamic system. This system, referred to as the 'modification system', usually contains a full rank mass matrix $[\Delta M^R]$ and a positive semi-definite stiffness matrix $[\Delta K^R]$. Therefore, it can also be called a 'semi-definite system'. No apparent physical meaning is needed for this modification system since it is not a genuine dynamic system. It may even have negative mass or stiffness which means the reduction of mass and stiffness values of certain elements in the system to be modified. The eigenvalue equation of this modification system is:

$$\left[[\Delta K^R] - \omega^2 [\Delta M^R] \right] \{\psi^R\} = \{0\} \tag{6.108}$$

If one of its eigenvalues coincides with an eigenvalue of the original system, then the modification system, when added to the original one, will not alter that particular eigenvalue. This makes it possible to fix one resonance of a system while shifting others. For instance, let us consider a simple 5 DOF mass-spring system as shown in figure 6.10 and a 2 DOF modification system, illustrated in figure 6.11, which consists of two masses connected by a uni-directional spring.

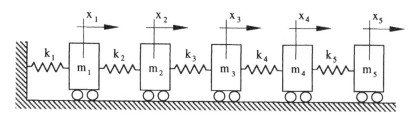

Fig. 6.10   A five degree-of-freedom mass-spring system.

Fig. 6.11   A two degree-of-freedom modification system.

The natural frequencies and mode shapes of this system are:

$$\left[`\omega_r^2 \diagdown \right] = \begin{bmatrix} 0 \\ \dfrac{\Delta k_3(\Delta m_2 + \Delta m_3)}{\Delta m_2 \Delta m_3} \end{bmatrix} \quad \text{and} \quad [\psi] = \begin{bmatrix} 1 & -\Delta m_3 \\ 1 & \Delta m_2 \end{bmatrix} \tag{6.109}$$

If this 2 DOF system is to be 'added' to coordinates 2 and 3 of the 5 DOF system for structural modification, then it is possible to select the stiffness Δk_3

and the masses Δm_2 and Δm_3, so that the non-zero natural frequency of the 2 DOF system matches a natural frequency of the 5 DOF system, and the mode shape of the 2 DOF system becomes the subset of that mode shape of the 5 DOF system. As a result, when the 2 DOF system is coupled to the 5 DOF system, as illustrated in figure 6.12, it will not change that particular resonance of the system while all other resonances will be altered to different extents. This shows that while modifying a structure, it is possible to 'fix' one of its resonances. Similarly, it is possible to fix an antiresonance of a system by carefully selecting the modifications to be introduced.

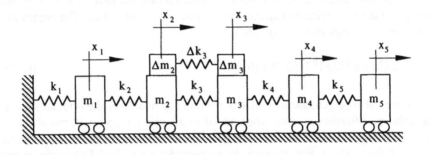

Fig. 6.12 The modified five degree-of-freedom mass-spring system.

The question now is, given the framework of a modification system (i.e. its DOFs and connectivity), how to determine its mass and stiffness quantities so that, when added to a dynamic system, it will not alter a natural frequency of the system. From (6.108), it is clear that equal multiples of both matrices $[\Delta M^R]$ and $[\Delta K^R]$ will also satisfy the equation. This suggests that what can be determined is merely the ratios among the mass and stiffness quantities, not their absolute values.

Using the submatrix approach, (6.108) can be recast as:

$$\left[\sum_s \kappa_s \{e_{pq}^R\}\{e_{pq}^R\}^T - \omega_*^2 \sum_r \varepsilon_r \{e_r^R\}\{e_r^R\}^T\right]\{\psi^R\} = \{0\} \tag{6.110}$$

where s is the number of stiffness elements and r the number of mass elements in the modification system, whereas ω_* and $\{\psi^R\}$ are the 'fixed' natural frequency and the subset of its mode shape, respectively. Equation (6.110) can be mathematically manipulated to form a set of standard linear simultaneous equations of the form:

$$[A]\{\varepsilon_1 \ \varepsilon_2 \ ... \ \varepsilon_r \ \kappa_1 \ \kappa_2 \ ... \ \kappa_s\}^T = \{0\} \tag{6.111}$$

Matrix $[A]$ is a coefficient matrix derived from the known quantities ω_*^2, $\{e_{pq}^R\}$, $\{e_r^R\}$ and $\{\psi^R\}$. The non-trivial solution of (6.111) provides the mass and stiffness ratios of the modification system which will satisfy the requirement of 'fixing' the natural frequency ω_* of the system to be modified.

Once the mass and stiffness ratios are found, they can be used in (6.96) to determine the coefficient $\Delta\gamma$ which will ascertain the absolute mass and stiffness quantities of the modification system. The frequency ω_* in (6.96) will now be the prescribed new natural frequency which the addition of the modification system to the original system should produce. Furthermore, the mass and stiffness ratios can be brought to (6.97) to determine coefficient $\Delta\gamma$ which will locate a new antiresonance of the original system, while keeping the resonance ω_* unchanged.

The ability to fix a natural frequency of a system during structural modification opens up interesting possibilities for modifying and optimising the dynamic characteristics of a structure.

The first is to create a frequency range of no resonance. This can be accomplished by fixing a resonance while shifting or relocating an adjacent resonance away from it. In doing so, a modification system needs to be conceived first. The mass and stiffness modification ratios of such a system can then be derived from the natural frequency and the subset of its mode shape of the original system using (6.110). Once these ratios are determined, equation (6.96) can be used to determine factor $\Delta\gamma$ which, together with the ratios, will dictate the mass and stiffness modification quantities.

A dividend from the ability to fix a natural frequency is the possibility of fixing an antiresonance of a system. As described before, the antiresonances of a given FRF $\alpha_{ij}(\omega)$ are the square roots of the eigenvalues from:

$$\left[[K]_{ij} + [\Delta K]_{ij}\right]\{X\} - \Omega_*^2\left[[M]_{ij} + [\Delta M]_{ij}\right]\{X\} = \{0\} \qquad (6.112)$$

Fixing an antiresonance of an FRF is analytically the same as fixing a resonance of the corresponding virtual system. Thus, structural modifications should satisfy the following condition in order that antiresonance Ω_* be fixed:

$$\left[[\Delta K]_{ij} - \Omega_*^2[\Delta M]_{ij}\right]\{X\} = \{0\} \qquad (6.113)$$

Like (6.110), this equation can also be transformed into:

$$\left[\sum_u \kappa_u \left\{e_{pq}^R\right\}_{(i)} \left\{e_{pq}^R\right\}_{(j)}^T - \Omega_*^2 \sum_v \varepsilon_v \left\{e_r^R\right\}_{(i)} \left\{e_r^R\right\}_{(j)}^T\right]\{X^R\} = \{0\} \qquad (6.114)$$

where u is the number of stiffness elements and v the number of mass elements in the modification system whereas Ω_* and $\{X^R\}$ are the 'fixed' antiresonance and the subset of its eigenvector of the virtual system, respectively. Expression (6.114) can be mathematically transformed to form a set of standard linear simultaneous equations:

$$[B]\{\varepsilon_1 \quad \varepsilon_2 \quad \dots \quad \varepsilon_v \quad \kappa_1 \quad \kappa_2 \quad \dots \quad \kappa_u\}^T = \{0\} \qquad (6.115)$$

Matrix [B] is a coefficient matrix derived from the known quantities Ω_*^2, $\{e_{pq}^R\}_{(i)}$, $\{e_{pq}^R\}_{(j)}$, $\{e_r^R\}_{(i)}$, $\{e_r^R\}_{(j)}$ and $\{X^R\}$. The non-trivial solution of (6.115) provides the mass and stiffness ratios of the modification system which, when added to the original system, will fix its antiresonance Ω_*.

Again, the absolute values of the mass and stiffness elements in the modification system will be determined by either equation (6.96) or (6.97). If successful, the former will relocate a resonance while keeping the antiresonance Ω_* unchanged, or the latter will relocate an antiresonance while keeping the antiresonance Ω_* unchanged.

The second possibility derived from fixing a natural frequency is the cancellation of a resonance and an antiresonance of a dynamic system. The ability to fix an antiresonance while relocating a resonance has paved the way for this application. Theoretically, if a resonance and an antiresonance of a system occur at the same frequency, they will cancel each other, leaving a nodal point at a mode shape. For the FRF involved, that means that the vibration mode will become unobservable. If this cancellation is to be 'man-made', then a question arises as to whether this should be accomplished by fixing a resonance and moving an antiresonance to 'meet' it, or should it be done the opposite way? The answer can only be to fix an antiresonance and shift a resonance over to cancel it. The opposite approach would change the mode shape of the fixed resonance, thus contradicting the assumption that adding the modification system does not change the original mode shape.

6.8 STRUCTURAL MODIFICATION BY ADDITIONAL DOFs

In addition to changing the mass and stiffness elements of a dynamic system, structural modification can also be carried out by appending to the system additional DOFs. This analysis, in effect, becomes a substructural coupling analysis which has been discussed in detail in Chapter 5.

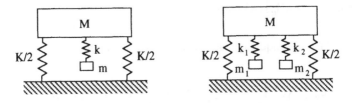

Fig. 6.13 A SDOF system with one or two dynamic absorbers.

The simplest structural modification in this category is the traditional SDOF dynamic absorber. For a SDOF system with one dynamic absorber as shown in figure 6.13, the antiresonance of the point FRF at mass M will reside at $\sqrt{k/m}$. A tuned absorber would assign this frequency to where the resonance of the SDOF system is, this achieving the best effect for the absorber. If, however, two absorbers are used, then it is easy to show that the two antiresonances of the 3 DOF system formed will be the roots of:

$$\begin{vmatrix} k_1 - \omega^2 m_1 & 0 \\ 0 & k_2 - \omega^2 m_2 \end{vmatrix} = 0 \qquad\qquad (6.116)$$

Therefore, the natural frequencies of the two absorbers will become the antiresonances of the point FRF. This conclusion applies to a SDOF system having a higher number of distributed dynamic absorbers.

For a MDOF system with a SDOF modification, it is possible to predict the effect of modifying the natural frequencies of the system [243]. As an example, let us consider the system of figure 6.6 except that the SDOF modification now becomes an additional DOF of the system, as shown in figure 6.14.

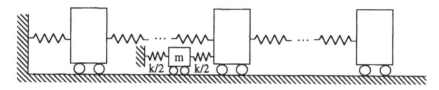

Fig. 6.14 A MDOF system with a SDOF modification.

Then, the natural frequencies of the MDOF system which are less than the natural frequency of the SDOF modification system $\sqrt{k/m}$ will increase, while those of the system above the natural frequency of the SDOF modification system will decrease.

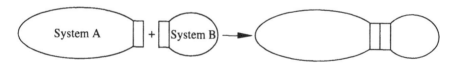

Fig. 6.15 System A modified by system B.

A more general case of structural modification by additional DOFs is illustrated in Figure 6.15 where a dynamic system is appended by a subsystem. The FRF matrix of the modified system can be derived using either impedance coupling methods or modal coupling methods, as described in Chapter 5.

6.9   SUMMARY

The objectives of structural modification are often linked to requirements of specific dynamic characteristics of a structure. A practical objective would be to relocate a resonance to circumvent the excessive vibration it generates. Sophisticated objectives may also arise, such as creating a pole-zero cancellation at a particular coordinate or achieving a specified mode shape.

It is theoretically formulated and practically feasible to change the dynamic characteristics of a system by changing its local spatial properties. Concretely, we can specify a resonance and determine necessary modifications at designated coordinates. In addition, we can specify the subset of a mode shape to be obtained after the new resonance. There is a certain degree of complication involved in formulating the solution for physical parameter changes which result in simultaneous mass and stiffness changes.

Sensitivity analysis is often used in structural modification. It is particularly useful to predict the 'best' locations for modification. If modifications are small, the analysis can also be used to quantify necessary changes to realise a specified dynamic characteristic objective.

When do we need a complete set of modal data? If modification is done on non-rotational coordinates such as by mass and stiffness elements only, then there is no need to involve the FRF data or modal data from those coordinates. This will be a huge relief for modal testing since rotational coordinates are far from easy to measure (see Section 3.4.8). It is also possible to use FRF or modal data from only a limited number of coordinates on which structural modification will take place.

To derive the re-analysis solution, accurate results will emerge if more coordinates are used. For local structural modification as an inverse problem to satisfy a specific objective, only FRF or modal data from the coordinates where modification will occur will be needed. As a result, incompleteness of modal data is circumvented.

CHAPTER 7

Finite Element Updating Using Measured Data

7.1   INTRODUCTION

In the previous chapters, we have established that the testing, acquisition and analysis of structures for the purposes of modal analysis are neither trivial nor straightforward. However, adopting rigorous testing procedures and attention to detail we can confidently assume that our measured data - whether in the time, modal or frequency domain - provide a truer representation of the dynamics of the structure. It is on this basis that we are able to use these data for engineering applications such as product development and life-cycle costing.

An application of modal analysis which has attracted particular interest is Finite Element (FE) model updating. To date, several hundred papers have been published in the field originating from both research and industry. The enthusiasm for this topic is easily understood: analytical models require verification if they are to be used in anger. The aim of these papers is to provide a measure of confidence in the dynamic behaviour of the structure. The work in this field has various levels of ambition; these have been summarised [258] as:

"(i)   a model which reproduces exactly all the measured modal properties (m natural frequencies, n mode shape amplitudes for each mode);

(ii)   a model which reproduces all of the measured FRF properties (thereby correctly representing information from the unmeasured modes, as well as those identified explicitly by the modal analysis;

(iii)   a model which is capable of reproducing all of the measured FRFs and/or modal properties (i.e. at points which were not included in the modal test);

(iv)   a model which, in addition to the above requirements, exhibits the correct connectivity; and,

(v)   a model which possesses all of the correct mass, stiffness and damping elements (and which is capable of reproducing all of the dynamic properties, including those which are as yet unmeasured)."

It can be seen that each of these criteria becomes increasingly more difficult and elusive to achieve and, as yet, no single strategy - which may include several techniques - has been shown to be systematic in its degree of success. This being the case, it is difficult to justify the use of the word 'updating' for this area of work, as many techniques simply seek to reconcile the inevitable differences between the analytical and experimental models. As the analytical model is rarely the only source of differences between the FE and experimental data, it is perhaps more appropriate to use the term 'reconciliation' rather than validation or updating. Indeed, several workers [259, 260] have identified possible difficulties induced by both analytical and test data.

The problematic areas that are associated with finite element models are summarised by the following:

(i) approximation of boundary conditions;
(ii) discretisation of distributed parameter systems;
(iii) estimation of the physical properties of structural materials;
(iv) approximation/omission of damping representation, or assumption of proportional damping;
(v) inadequate modelling of joints; and
(vi) condensation of FE models to make them compatible with the test degrees of freedom.

Although difficulties arising from the test and analysis procedure have been discussed in Chapter 3, it is now perhaps appropriate to mention potential problems that are relevant to model reconciliation:

(i) the number of measured degrees of freedom is limited and may be different from the analytical model;
(ii) difficulty of measuring rotational degrees of freedom;
(iii) limited number of identified mode shapes;
(iv) complex mode shapes obtained;
(v) unmodelled errors in measurement (noise, nonlinearity etc.);
(vi) poor modal analysis of the experimental results; and
(vii) some modes not excited or, if excited, then not identified; conversely, overanalysis leads to false modes occurring.

Although it is wise to consider the practical limitations of the experimental data, they are frequently ignored as the ultimate application of a reconciliation process is often only to replicate the measured or unmeasured characteristics of the experimental model. Accordingly, such procedures have tended to be referred to as correlation, validation or updating methods by assuming that the measured data are correct, which is justifiable given the primary goals of the techniques, but in practice may not be entirely accurate. However, given that this is the normal application of reconciliation techniques, one still has to judge the success of the updated model relative to its original state and the experimental model.

7.2   COMPARISON, LOCATION AND CORRELATION METHODS

Given the abundance of reconciliation methods, as discussed in the previous section, it is impossible to describe them all in detail. It should be pointed out that Friswell and Mottershead's book on model updating [261] provides a rigorous and comprehensive guide to the field, which is essential reading for anyone whose main research is in this area. The aim of the overview presented here is to lay special emphasis on the methods that either have stood the test of time, or are of current interest and relevance. Before describing these methods in detail, it is appropriate to define the categories into which they fall:

(i)   *Comparison* methods - these are used for a preliminary assessment of the compatibility of the FE and experimental models. They are limited to giving an indication of which modes correspond to each other and do not attempt to explain why, in a spatial sense, they might differ;

(ii) *Location* methods - as the name suggests, these aim to provide information as to where differences exist between the two models without describing whether they are caused by mass or stiffness irregularities; and finally,

(iii) *Correlation* methods - are the most sophisticated of the three approaches in that they attempt to apply localised perturbations to the mass and/or stiffness properties or to the elemental parameters of the FE model, the goal being to achieve a modal and spatial model which accurately represents the physical characteristics of the real structure.

The logical approach which is normally adopted is to apply comparison, location and correlation methods sequentially. This is based on the rationale that in order to use location techniques a prior assessment has to be made as to whether they are 'comparable' systems on consideration of their modal properties. Similarly, several correlation methods rely on accurate location of the regions of discrepancy between the FE and experimental models; others, however, incorporate their own idealised locations in order to achieve accurate model updating.

Thus, the purpose of the next section is to outline the various strategies which are currently adopted for model reconciliation and to show which fall into the categories of comparison, location or correlation. It should be noted that for all of the methods described, the problem of incompatibility between the FE and measured data must be resolved.

7.3   INCOMPATIBILITY BETWEEN MEASURED AND FE DATA

Before discussing methods to achieve reconciliation between the models it is essential to draw the reader's attention to the problem of coordinate incompleteness[†]. Inevitably, there will be an incompatibility in the size of the FE and test models to be coupled/compared. This difference manifests itself in two

---

[†] See also Chapter 5.

forms, firstly as an insufficiency in the number of modes and secondly as a lack of measured DOFs. The first of these difficulties arises from the limited bandwidth of data acquisition and the fact that increased modal density at higher frequencies makes modal extraction problematic beyond moderately low order modes. The luxury of a second modal test to acquire more data if the first modal analysis reveals quirks or limitations is sufficiently rare not to be relied upon. There is little that can be done to resolve this after the completion of the test; we simply have to make the best of the available data by meticulous modal identification (see Chapter 4). Fortunately, more options are available when considering the lack of measured DOFs. It is inevitable that an experimental test will omit some DOFs. However, it is advisable to ensure that experimental effort is used effectively and that the measured data give a full modal description - including local modes - across the measured frequency range. Lim [262] suggests an 'effective impedance distribution vector' method which defines optimum actuator and response locations. This selects DOFs on the basis of linear modal independence from the other coordinates. Despite this approach to the problem, coordinate incompleteness will exist but can be mitigated by adopting the following strategies, either singly or in combination:

(i) the larger model could have the non-measured (slave) DOFs condensed out in such a way as not to perturb either the modal or spatial properties of the system. In practice this is difficult to achieve as methods generally allow retention of kinetic or potential energy, or compromise both, but system connectivity is unavoidably perturbed; or

(ii) the smaller of the two models - the experimental - could be expanded to match its larger FE counterpart.

Given that the goal of updating is to provide a 'correct' FE model, it is obviously advantageous that the final model remains in this coordinate system. Although a complete coordinate description of a reduced set of modes can be obtained from the FE model by using Lanczos [263] and subspace iteration [264] techniques, there will always be some DOFs that will be unmeasurable. These occur through physical restrictions in the measurement of the structure and the difficulties encountered in measuring rotational DOFs. Consequently, it is desirable to expand the experimental data onto the corresponding FE coordinate set, the proviso being that the expansion ratio - the number of FE DOFs in relation to the measured DOFs - is not so large that the experimental information is too diluted to provide localised information for updating. If this is the case then reduction - despite its shortcomings - is necessary.

7.3.1 Finite Element model reduction

When FE model reduction is unavoidable, the user should be aware that spatial information will either be lost or dispersed. These characteristics can significantly compromise the effectiveness of the location or updating method. This being said, sufficient information is retained to allow *comparison* methods to be used with impunity.

In Section 5.2.4., we have discussed both static/Guyan [228] and dynamic [265] reduction techniques. For the purposes of model reconciliation, Guyan's technique is most favoured as it preserves the low order modal properties. Two further methods that are frequently mentioned in the literature are:

System Equivalent Reduction Expansion Process (SEREP)
This technique [266] maps the complete FE eigenvectors onto the experimental slave DOFs by means of the transformation matrix

$$[T] = \begin{bmatrix} [\phi_{Ap}] \\ [\phi_{As}] \end{bmatrix} \left[[\phi_{Ap}]^T [\phi_{Ap}] \right]^{-1} [\phi_{Ap}]^T \tag{7.1}$$

where $[[\phi_{Ap}]]^T [\phi_{Ap}]]^{-1} [\phi_{Ap}]^T$ is the Moore-Penrose pseudo-inverse $[\phi_{Ap}]^+$ and the subscripts A, p and s refer to the analytical model, master (primary) and slave (secondary) DOFs respectively.

The reduced system matrices are then obtained by substituting the transformation matrix into equations (5.22) and (5.23) (with the addition of the identity matrix).

The main advantages of the SEREP process are the following ones: (i) the reduced model has exactly the same frequencies and mode shapes as the full system for the selected modes of interest and (ii) the quality of the results is insensitive to the selection of the full system DOFs that are kept in the reduced model.

Improved Reduction System (IRS)
This method [267] develops Guyan's technique to include a contribution from the slave inertias. Thus, a new transformation matrix is generated by

$$[T] = \begin{bmatrix} [I] \\ [-K_{ss}]^{-1} [K_{sp}] \end{bmatrix} + \begin{bmatrix} [0] & [0] \\ [0] & [K_{ss}]^{-1} \end{bmatrix} [M] \begin{bmatrix} [I] \\ [-K_{ss}]^{-1} [K_{sp}] \end{bmatrix} [M^R]^{-1} [K^R] \tag{7.2}$$

The inclusion of the slave inertial effects enhances the reduced system characteristics. It does not require the solution of the full system's eigenvalue problem.

For a thorough discussion of reduction techniques, readers are advised to consider Avitable's paper on the subject [268].

7.3.2   Expansion of measured modal data

A greater variety of expansion techniques compared to reduction methods has been developed in recent years. This perhaps reflects the relative value of having modal completeness for model reconciliation. In addition, Gysin [269] demonstrated that the effectiveness of the expansion methods has a significant influence on the overall location and updating procedure.

Four schools of expansion method exist [31] which have a varying dependency on the FE model:

(i) *analytical eigenvector substitution.* This method directly substitutes the analytical slave DOFs for the unmeasured data, making no use of the available measurements;

(ii) *direct data generation using the FE model and experimental data.* The SEREP method - discussed above - and the modal projection techniques create a transformation matrix from the master DOFs to the complete FE coordinate system. Kidder's method [270] creates a spatial transformation formulated using the analytical mass and stiffness matrices and partitioning the generalised eigenvalue problem;

(iii) *indirect data generation using the FE model.* A geometric fit of the experimental data is enhanced by knowledge acquired from the FE model [271]; and

(iv) *expansion of experimental data in isolation.* These methods interpolate the experimental data by fitting a continuous function though the data by means of cubic splines [272], surface splines [273] or polynomial fits [274].

With such a variety of techniques, all formulated using different starting points, it is easy to visualise that the accuracy of the expansion can undermine the subsequent reconciliation process. Accordingly, we will discuss the attributes of some of these methods in more detail.

Kidder's method

This is the most commonly used expansion method, which is effectively derived from an inverse Guyan reduction. Starting from the partitioned eigenvalue problem

$$\left[\begin{bmatrix} [K_{pp}] & [K_{ps}] \\ [K_{sp}] & [K_{ss}] \end{bmatrix} - \omega_r^2 \begin{bmatrix} [M_{pp}] & [M_{ps}] \\ [M_{sp}] & [M_{ss}] \end{bmatrix}\right] \left\{ \begin{matrix} \{\phi_p\} \\ \{\phi_s\} \end{matrix} \right\}_r = \left\{ \begin{matrix} \{0\} \\ \{0\} \end{matrix} \right\} \tag{7.3}$$

where the subscript r corresponds to the rth mode. From (7.3) it can be shown[†] that by substituting the measured eigenvalues and mode shapes, the expanded slave degrees of freedom are given by

$$\left\{ {}_x\phi_s^E \right\}_r = -\left[[K_{ss}] - \omega_r^2 [M_{ss}] \right]^{-1} \left[[K_{sp}] - \omega_r^2 [M_{sp}] \right] \left\{ {}_x\phi_p \right\}_r \tag{7.4}$$

This method has the advantage that by using the partitioned unreduced analytical system matrices, the physical connectivity properties are consistent with

[†] See Section 8.4.2.

the structure of the final model. The implication of this is that the expansion depends on the basic *form* of the original FE model.

Expansion using analytical modes

Although missing DOFs in the measured data can be directly replaced by substituting corresponding analytical data, this may well cause discontinuities in the resulting mode shape vector. A reliable approach suggested by Lipkins and Vandeurzen [275] uses the eigensolution from the FE model to expand the measured modes, so that the complete coordinates in the experimental model are generated by assuming that each mode is constructed from a linear combination of the analytical models.

The approach is developed in the following way. The full analytical modal matrix can be partitioned:

(i) by rows into n master and N-n slave DOFs; and
(ii) by columns into m chosen and N-m discarded modes.

$$[\phi_A]_{(N \times N)} = \begin{bmatrix} [_A\phi_{11}] & \vdots & [_A\phi_{12}] \\ \cdots\cdots & \cdots & \cdots\cdots \\ [_A\phi_{21}] & \vdots & [_A\phi_{22}] \end{bmatrix} \begin{matrix} \}n \\ \\ \}N-n \end{matrix} \tag{7.5}$$

$$\underbrace{}_{m} \quad \underbrace{}_{N-m}$$

Then the expanded experimental modes (superscript E, subscript X) can be constructed from a linear combination of the chosen analytical modes by setting

$$\begin{Bmatrix} \{_X\phi_1^E\} \\ \cdots\cdots \\ \{_X\phi_2^E\} \end{Bmatrix} = \begin{bmatrix} [_A\phi_{11}] \\ \cdots\cdots \\ [_A\phi_{21}] \end{bmatrix} \{\upsilon\} \tag{7.6}$$

for some unknown set of coefficients $\{\upsilon\}$. Then

$$\{_X\phi_1^E\} = [_A\phi_{11}]\{\upsilon\} \tag{7.7}$$

$$\{_X\phi_2^E\} = [_A\phi_{21}]\{\upsilon\} \tag{7.8}$$

where

$$\{\upsilon\} = [_A\phi_{11}]^+\{_X\phi_1\} \tag{7.9}$$

Note that:

(i) if n = m, the pseudo-inverse in (7.9) becomes the standard inverse and $\{_X\phi_1^E\} = \{_X\phi_1\}$. This means that the expanded mode fits the experimental data exactly;

(ii)   if n>m, a linear combination of analytical modes is found that fits the experimental data in a least-squares sense. Consequently, $\{_X \phi_I^E\}$ is a smoothed counterpart of $\{_X \phi_I\}$; and

(iii) if N< m, then the pseudo-inverse in (7.9) is rank deficient and the solution is likely to be physically meaningless.

Waters [31] points out that the success of this modal projection approach is critically dependent on the selected set of analytical modes, which must include a reasonable counterpart to each experimental mode. This can be achieved by means of using comparison methods discussed in the next section. Gysin's work [269] concludes that it is this form of expansion which is most effective for updating. It should also be noted that modal projection using analytical modes is a specific case of the SEREP approach to expansion [266].

Spline techniques for expansion

Spline and polynomial techniques for fitting experimental data simply interpolate and extrapolate unmeasured data from the existing measured information. Therefore, they are the only methods to be able to map the measured data onto the FE geometry if the node points are not coincident. Spline functions in particular offer the advantage that controlled smoothing can be applied to absorb the vagaries of measurement noise.

The surface spline method [273] is based on small static deflections of an infinite plate subjected to point loads. Using a closed form solution, a load distribution can be found that yields a static deflection representative of the mode shape to be interpolated.

The deflection, W, of an infinite plate subjected to a point load P is given by the radial distribution

$$W(r) = A + Br^2 + \frac{P}{16\pi D} r^2 \ln r^2 \qquad (7.10)$$

where: r is the distance from the point load; and
         D is the plate rigidity.

It is shown in [273] that by applying the constraint $W \to 0$ as $r \to \infty$, the deflection of the plate at point (x, y) due to point loads at each of n measured DOFs is given by

$$W(x,y) = a_0 + a_1 x + a_2 y + \sum_{i=1}^{n} \frac{P_i}{16\pi D} r_i^2 \ln r_i^2 \qquad (7.11)$$

where

$$r_i^2 = (x - x_i)^2 + (y - y_i)^2 \qquad (7.12)$$

and P_i is the load at the ith DOF at (x_i, y_i). The modification to (7.11) required to include smoothing in the spline is

$$W(x, y) = a_0 + a_1 x + a_2 y + \sum_{i=1}^{n} \frac{P_i}{16\pi D} \left(r_i^2 \ln r_i^2 + C_i \right) \tag{7.13}$$

where C_i is proportional to the spring flexibility at the ith master DOF. Note that the interpolation is unsmoothed if all of the spring flexibilities are set to zero whilst the shape function tends to a least-squares plane fit as all spring flexibilities tend to infinity.

A further development of spline methods is to fit the mode shape disparity [271]. In addition, by including 'intuitive' smoothing which detects the amount of noise, expanded mode shapes can be obtained without dispersing the true differences between the experimental and FE models [31]. Note also that this method is able to generate missing FRFs (as are Kidder's and modal projection techniques) - essential for frequency domain updating procedures - by applying splines separately to the real and imaginary components. A word of caution however: splines should only be used for continuous mode shapes/FRFs as discontinuities will create mathematically feasible, but physically unjustifiable, shape functions.

7.4   COMPARISON TECHNIQUES
The first stage of any reconciliation exercise is to determine how closely the experimental and analytical models correspond. This is usually achieved in the modal domain, which presents a happy medium between the raw state of the experimental data - time domain information - and that of the analytical model - the system matrices. Note that before comparison of the data sets can be achieved, the issue of coordinate incompleteness has to be resolved.

Comparison techniques are to a greater or lesser extent based on the orthogonality conditions

$$[\phi]^T [M][\phi] = [I] \quad \text{and} \quad [\phi]^T [K][\phi] = \left[\ddots \, \omega_r^2 \, \ddots \right] \tag{7.14}$$

where $[\phi]$ represents the mass normalised eigenvectors, either from test or analysis, and $[I]$ and $\left[\ddots \, \omega_r^2 \, \ddots \right]$ are the diagonal identity and eigenvalue matrices respectively. Note that using the experimental eigenvector matrix - which inevitably will differ from that produced by the analysis - will cause the conditions of orthogonality to be contravened. This will result in off-diagonal terms appearing in the identity or eigenvalue matrix and perturbations to the leading diagonal terms away from their predicted values. Taking the mass orthogonality condition, as this is the one most commonly used, one can identify correlated modes by values of near unity and their degree of correlation by their deviation from this value.

An extension of equation (7.14) is the Normalised Cross Orthogonality (NCO) [276], where the modes are not necessarily mass normalised and which gives a clearer indication of modal correlation:

$$NCO\left(\{\psi_X\}_i, \{\psi_A\}_j\right) = \frac{\left(\{\psi_X\}_i^T [M_A] \{\psi_A\}_j\right)^2}{\left(\{\psi_X\}_i^T [M_A] \{\psi_X\}_i\right)\left(\{\psi_A\}_j^T [M_A] \{\psi_A\}_j\right)} \qquad (7.15)$$

and is real and bounded between 0 and 1. Arbitrarily scaled experimental and analytical modes, $\{\psi_X\}_i$ and $\{\psi_A\}_j$ are similar, or correlated, if their NCO value is close to 1 and uncorrelated if the NCO value is small. Given a set of m experimental modes and n analytical modes the NCO yields an $m \times n$ matrix from which the Correlated Mode Pairs (CMPs) are conspicuous by their high values.

Perhaps the most common means of comparing experimental and analytical models is by means of the Modal Assurance Criterion (MAC) [277] which is defined as follows:

$$MAC\left(\{\phi_X\}_i, \{\phi_A\}_j\right) = \frac{\left|\{\phi_X\}_i^T \{\phi_A^*\}_j\right|^2}{\left(\{\phi_X\}_i^T \{\phi_X^*\}_i\right)\left(\{\phi_A\}_j^T \{\phi_A^*\}_j\right)} \qquad (7.16)$$

The MAC is used extensively in the modal analysis community for measuring model similarity and establishing CMPs. Here, the modes may be complex. Its greatest advantage over the orthogonality based formulations is that it does not require coordinate complete experimental data, or reduced structural matrices - partitioning of the analytical mode shapes is sufficient. Although highly effective for many structures, the MAC has been unreliable in the correlation of localised modes that are described by only a very few DOFs.

The NCO differs from the MAC in its weighting by the analytical mass matrix and in taking only real modes. By virtue of the orthogonality condition, the NCO yields the identity matrix when presented with perfectly correlated modes. It is worth noting that the MAC will not provide a true identity matrix even if the analytical and experimental models are the same, as the orthogonal conditions (7.14) are not maintained due to the omission of the structural matrices.

A more discriminating comparison technique has been proposed by Waters [31]. The Normalised Modal Difference (NMD), is proposed in quantifying the accuracy of modal data without the use of FE system matrices.

$$Error\left(\{\phi\}\right) = NMD\left(\{\phi_X\}, \{\phi_A\}\right) \qquad (7.17)$$

where the NMD between two modes $\{\phi_X\}$ and $\{\phi_A\}$ is defined as

$$NMD\left(\{\phi_X\}, \{\phi_A\}\right) = \frac{\left\|\{\phi_X\} - \gamma\{\phi_A\}\right\|_2}{\left\|\gamma\{\phi_A\}\right\|_2} \qquad (7.18)$$

where γ is the Modal Scale Factor (MSF), given by

$$\gamma = \text{MSF}\left(\{\phi_X\}, \{\phi_A\}\right) = \frac{\left|\{\phi_X\}^T\{\phi_A^*\}\right|}{\{\phi_A\}^T\{\phi_A^*\}} \tag{7.19}$$

Physically, the NMD represents the fraction, on average, by which each DOF differs between the two modes. For example, each DOF is in error by 10% on average given an NMD value of 0.1. The NMD is closely related to the Modal Assurance Criterion by the following formula:

$$\text{NMD} = \sqrt{\frac{1 - \text{MAC}}{\text{MAC}}} \tag{7.20}$$

To demonstrate the NMD in relation to the MAC, it is appropriate to introduce a simple case study. Figure 7.1 shows an FE model consisting of 22 shell elements and 36 nodes. The structure is a cantilevered plate of uniform thickness with a planform typifying a civil aircraft wing. Each node has a translational freedom in the out of plane (z) direction and rotational freedoms about the x and y axes (θ_x, θ_y). This uniform model represents the simulated analytical structure. A second model for which the Young's modulus has been increased by 100% at the mid-wing region (indicated by the two shaded elements in figure 7.1), is the basis for the 'experimental' structure.

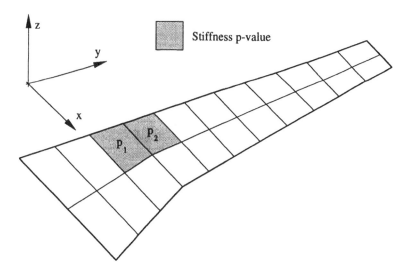

Fig. 7.1 FE model of cantilevered wing showing localised error.

Table 7.1 shows the MAC values of correlated mode pairs between the experimental and analytical mode sets, and the associated natural frequencies. The localised FE error can be seen to have only marginally affected the modal parameters. The NMD is clearly more sensitive to mode shape differences than the

MAC. However, unlike the MAC the NMD is not bounded by unity and so makes comparison of weakly correlated modes more difficult to distinguish.

Table 7.1 Comparison of modal parameters between 'experimental' model and analytical model with localised error.

Mode number	Analytical frequency (Hz)	'Experimental' frequency (Hz)	MAC	NMD (%)
1	13.53	14.29	0.999	2.4
2	64.88	62.73	0.999	2.7
3	154.0	154.0	0.962	20.0
4	175.1	176.3	0.979	14.5
5	313.1	307.4	0.941	25.1
6	371.9	356.1	0.987	11.6
7	512.0	512.3	0.992	9.1
8	596.1	601.9	0.995	7.0
9	754.7	754.2	0.997	5.1
10	852.4	851.7	0.996	6.1

7.5   LOCATION TECHNIQUES

Detecting the locations of differences between the test structure and the FE model is the most exacting task in the correlation procedure. This activity normally does not attribute errors to either of the models, it simply indicates the region(s) of difference. It is a relatively easy task to rectify anomalies if we know where they exist. Research in the area of fault/damage detection [278-280] has shown that detecting these regions presents a considerable challenge.

Perhaps the most directly informative technique to identify errors is to compare mode shape animations. All too frequently this trivial activity is bypassed in order to explore more glamorous approaches. It is far better to eradicate the mundane sources of difference by observation. Errors such as faulty cabling, or signal conditioning in the experimental test, or an oversight in grounding the correct nodes in the FE model are often culprits which can undermine the whole process. The test errors in particular must be identified and corrected early in the reconciliation process, as a second modal test is not a luxury which can be relied upon in a tight design cycle. It is surprising therefore that relatively few algorithms have been developed specifically for this task. However, the correlation procedures outlined in the next section implicitly attempt to locate as well as rectify errors.

Probably the simplest formal error location technique is the Coordinate Modal Assurance Criterion (COMAC) [281]. The COMAC is the reverse of the MAC in that it measures the correlation at each DOF averaged over the set of CMPs:

$$COMAC(i) = \frac{\left(\sum_{j=1}^{n_{CMP}} |\phi_A(i,j)\, \phi_X^*(i,j)|\right)^2}{\sum_{j=1}^{n_{CMP}} |\phi_A(i,j)|^2 \sum_{j=1}^{n_{CMP}} |\phi_X(i,j)|^2}$$
(7.21)

The COMAC can be useful as a supplementary guide to error location, but has no physical basis. It is advisable to use this method in cases where the structure is tested and modelled in a free-free configuration. Grounding of the structure will amplify the effect of anti-nodes away from the region of constraint and dominate the error prediction.

The Dynamic Force Balance method [282] is an error location technique that does have a physical basis. The experimental modal parameters fail to satisfy the analytical eigendynamic equation due to errors in the FE model causing a residual force term

$$\left[[K_A] - (\omega_X)_r^2 [M_A]\right]\{\phi_X\}_r = \{f\}$$
(7.22)

Large residual components are taken to indicate regions of error in either mass or stiffness. The method is not widely used due to its bias towards predicting errors at anti-nodes. This method, like all methods based on the equations of motion, requires coordinate complete measurements.

7.6　CORRELATION TECHNIQUES

The literature in this area far outweighs all of the other activities related to model reconciliation put together. This is understandable given that the holy grail of 'the right model' can only be achieved through correlation. Given the diverse nature of the field, it is difficult to address all of the available techniques and categorise the methods neatly into schools. In certain cases omission and brevity have been called for and readers should not consider that their favoured method has been summarily judged.

7.6.1　Least-Squares Updating
Theoretical development in the frequency domain
Suppose that a structure is modelled by FE analysis and that by applying some unknown parametric changes an updated FE model can be derived that mimics the measured response of the structure.

Then applying a single-input *unit* force to both the updated and original FE models gives

$$[Z_U(\omega)]\{\alpha_X(\omega)\} = \{f\} = [Z_A(\omega)]\{\alpha_A(\omega)\} \qquad (7.23)$$

Equation (7.23) can be rearranged in the form

$$[\Delta Z(\omega)]\{\alpha_X(\omega)\} = \{I\}_j - [Z_A(\omega)]\{\alpha_X(\omega)\} \qquad (7.24)$$

where $[\Delta Z(\omega)]$ is the error in the dynamic stiffness matrix, $\{I\}_j$ denotes the jth column of the identity matrix and j indicates the location of the input force. Whilst the relatively linear variation of the dynamic stiffness matrix with respect to the updating parameters makes for a stable updating problem [283], the right-hand side of (7.24) - often referred to as the input error- is inherently ill-conditioned [284].

An established remedy is to pre-multiply (7.24) by the analytical FRF matrix:

$$[\alpha_A(\omega)][\Delta Z]\{\alpha_X(\omega)\} = \{\alpha_A(\omega)\}_j - \{\alpha_X(\omega)\} = \{\Delta\alpha(\omega)\} \qquad (7.25)$$

The right-hand side of (7.25) is the disparity between analytical and experimental FRFs, or output error.

Equation (7.25) is exact irrespective of the size or nature of the errors. To proceed towards a solution for the system error matrices we must assume a form for the errors. This is achieved by selecting a set of N_p design parameters to vary, $\{P\}$, to account for the discrepancy in response between experiment and analysis. It is convenient to non-dimensionalise the updating parameters as follows

$$p_i = \left(P_i - P_i^o\right)/P_i^o \qquad (7.26)$$

where $\{P^o\}$ are the parameter values for the original FE model. The non-dimensionalised updating parameters - or p-values - represent the fractional changes in the design variables.

The dynamic stiffness matrix for the updated FE model, $[Z_U]$, is a function of $\{p\}$ and can be expressed as a Taylor expansion about the dynamic stiffness matrix for the original FE model, $[Z_A]$, as follows:

$$[Z_U] = [Z_A] + [\Delta Z] = [Z_A] + \sum_{i=1}^{N_p} \frac{\partial[Z]}{\partial p_i} p_i + O\left(p_i^2\right) \qquad (7.27)$$

Retaining only first order terms,

$$[\Delta Z] = \sum_{i=1}^{N_p} \frac{\partial[Z]}{\partial p_i} p_i \qquad (7.28)$$

Substituting for $[\Delta Z]$ in (7.25) and rearranging gives

$$[S(\omega)]\{p\} = \{\Delta\alpha(\omega)\} \qquad (7.29)$$

where

$$[S(\omega)] = [\alpha_A(\omega)]\left[\frac{\partial[Z]}{\partial p_1}\{\alpha_X(\omega)\} \quad \cdots \quad \frac{\partial[Z]}{\partial p_{N_p}}\{\alpha_X(\omega)\}\right] \qquad (7.30)$$

Each row of the sensitivity matrix, [S], defines the sensitivities of the response at a particular DOF to the set of p-values.

Note that (7.29) is a set of N linear equations for N_p unknowns at a single excitation frequency, ω. Given N_f measured frequency points, N_f sets of equations can be combined to form an over-determined problem with $N_f \times N$ equations for N_p unknowns. Such a set of equations can be solved simultaneously in a least-squares sense by application of Singular Value Decomposition (see Appendix A).

Note also that the rows of equation (7.29) can be partitioned in the event of coordinate incompleteness so as to consider the response disparity at only master DOFs. However, the sensitivity matrix on the left-hand side still requires measured data at every FE DOF if model reduction is to be avoided.

Updating using modal sensitivity

The modal [285] and FRF sensitivity methods are analogous approaches set in different response domains. Both methods treat the experimental model as a perturbation in design parameters about the original FE model. From estimates of the sensitivity of the dynamic response or modes of vibration to parameter variations, a combination of changes is found that accounts for the disparity between analytical and experimental values.

Theory of modal sensitivity

In Section 6.3.1 the sensitivities of the eigenvalues and eigenvectors with respect to changes in the mass and stiffness matrix elements were presented. Here, the sensitivities of the eigenvalues and eigenvectors with respect to changes in specified design parameters will be discussed.

Let $(\omega_A)_r^2$ and $\{\phi_A\}_r$ be an eigensolution to the eigendynamic equation for the undamped analytical model, so

$$[K_A]\{\phi_A\}_r = (\omega_A)_r^2[M_A]\{\phi_A\}_r \qquad (7.31)$$

Suppose that {p} is a set of N_p updating parameters that are assigned to design variables of the FE model. Then $(\omega_A)_r^2$ and $\{\phi_A\}_r$ are functions of {p}.

Now suppose that by means of a MAC correlation the experimental eigenpair $(\omega_X)_r^2, \{\phi_X\}_r$ has been found to correspond to the analytical eigensolution, $(\omega_A)_r^2$ and $\{\phi_A\}_r$. Then the experimental eigensolution can be expressed as a Taylor expansion - in terms of the updating parameters - about the analytical eigensolution:

$$\left(\omega_X\right)_r^2 = \left(\omega_A\right)_r^2 + \sum_{i=1}^{N_p} \frac{\partial \left(\omega_A\right)_r^2}{\partial p_i} p_i + O\left(p_i^2\right)$$

$$\left\{\phi_X\right\}_r = \left\{\phi_A\right\}_r + \sum_{i=1}^{N_p} \frac{\partial \left\{\phi_A\right\}_r}{\partial p_i} p_i + O\left(p_i^2\right)$$

(7.32)

Assuming that the required changes in the updating parameters are small, (7.32) can be linearised and rearranged to give

$$\Delta(\omega)_r^2 \approx \sum_{i=1}^{N_p} \frac{\partial \left(\omega_A\right)_r^2}{\partial p_i} p_i$$

(7.33)

$$\left\{\Delta\phi\right\}_r \approx \sum_{i=1}^{N_p} \frac{\partial \left\{\phi_A\right\}_r}{\partial p_i} p_i$$

(7.34)

The problem is to calculate the derivatives of the eigensolution, or eigensensitivities. The solution was given by Fox and Kapoor [286], back in 1968, although similar results for the sensitivity of the eigenvalues alone had already been developed earlier by other authors, as referred to in their article.

Sensitivity of the eigenvalues
Starting from equation (7.31) and pre-multiplying it by $\left\{\phi_A\right\}_r^T$,

$$\left\{\phi_A\right\}_r^T \left[\left[K_A\right] - \left(\omega_A\right)_r^2 \left[M_A\right]\right] \left\{\phi_A\right\}_r = 0$$

(7.35)

Differentiating (7.35) w. r. t. p_i,

$$\frac{\partial \left\{\phi_A\right\}_r^T}{\partial p_i} \left[\left[K_A\right] - \left(\omega_A\right)_r^2 \left[M_A\right]\right] \left\{\phi_A\right\}_r + \left\{\phi_A\right\}_r^T \frac{\partial \left[\left[K_A\right] - \left(\omega_A\right)_r^2 \left[M_A\right]\right]}{\partial p_i} \left\{\phi_A\right\}_r$$

$$+\left\{\phi_A\right\}_r^T \left[\left[K_A\right] - \left(\omega_A\right)_r^2 \left[M_A\right]\right] \frac{\partial \left\{\phi_A\right\}_r}{\partial p_i} = 0$$

(7.36)

Due to (7.31), the first and third terms of (7.36) are zero and thus,

$$\left\{\phi_A\right\}_r^T \frac{\partial \left[\left[K_A\right] - \left(\omega_A\right)_r^2 \left[M_A\right]\right]}{\partial p_i} \left\{\phi_A\right\}_r = 0$$

(7.37)

The term in the middle gives:

$$\frac{\partial\left[[K_A]-(\omega_A)_r^2[M_A]\right]}{\partial p_i} = \frac{\partial[K_A]}{\partial p_i} - \frac{\partial(\omega_A)_r^2}{\partial p_i}[M_A] - (\omega_A)_r^2\frac{\partial[M_A]}{\partial p_i} \qquad (7.38)$$

Because of the orthogonality conditions (7.14), expression (7.37) becomes:

$$\{\phi_A\}_r^T\frac{\partial[K_A]}{\partial p_i}\{\phi_A\}_r - \frac{\partial(\omega_A)_r^2}{\partial p_i} - (\omega_A)_r^2\{\phi_A\}_r^T\frac{\partial[M_A]}{\partial p_i}\{\phi_A\}_r = 0 \qquad (7.39)$$

and finally,

$$\frac{\partial(\omega_A)_r^2}{\partial p_i} = \{\phi_A\}_r^T\left[\frac{\partial[K_A]}{\partial p_i} - (\omega_A)_r^2\frac{\partial[M_A]}{\partial p_i}\right]\{\phi_A\}_r \qquad (7.40)$$

Sensitivity of the eigenvectors
As the eigenvectors are linearly independent, it is possible to state that their sensitivity w. r. t. p_i can be expressed as a linear combination of the eigenvectors themselves:

$$\frac{\partial\{\phi_A\}_r}{\partial p_i} = \sum_{j=1}^{N} a_{rj}^{(i)}\{\phi_A\}_j \qquad (7.41)$$

Going back to (7.31) and differentiating it, it follows that

$$\frac{\partial\left[[K_A]-(\omega_A)_r^2[M_A]\right]}{\partial p_i}\{\phi_A\}_r + \left[[K_A]-(\omega_A)_r^2[M_A]\right]\frac{\partial\{\phi_A\}_r}{\partial p_i} = \{0\} \qquad (7.42)$$

Substituting (7.41) in (7.42),

$$\frac{\partial\left[[K_A]-(\omega_A)_r^2[M_A]\right]}{\partial p_i}\{\phi_A\}_r + \left[[K_A]-(\omega_A)_r^2[M_A]\right]\sum_{j=1}^{N} a_{rj}^{(i)}\{\phi_A\}_j = \{0\}$$

$$\qquad (7.43)$$

or

$$\sum_{j=1}^{N} a_{rj}^{(i)}\left[[K_A]-(\omega_A)_r^2[M_A]\right]\{\phi_A\}_j = -\frac{\partial\left[[K_A]-(\omega_A)_r^2[M_A]\right]}{\partial p_i}\{\phi_A\}_r \qquad (7.44)$$

Pre-multiplying (7.44) by $\{\phi_A\}_s^T$, with $s \neq r$, leads to

$$\sum_{j=1}^{N} a_{rj}^{(i)} \{\phi_A\}_s^T \left[[K_A] - (\omega_A)_r^2 [M_A] \right] \{\phi_A\}_j =$$

$$-\{\phi_A\}_s^T \frac{\partial \left[[K_A] - (\omega_A)_r^2 [M_A] \right]}{\partial p_i} \{\phi_A\}_r \tag{7.45}$$

Due to the orthogonality properties, the l. h. s. of (7.45) is equal to zero, except for j = s:

$$a_{rj}^{(i)} \left((\omega_A)_j^2 - (\omega_A)_r^2 \right) = -\{\phi_A\}_j^T \frac{\partial \left[[K_A] - (\omega_A)_r^2 [M_A] \right]}{\partial p_i} \{\phi_A\}_r \qquad j \neq r \tag{7.46}$$

Developing the r. h. s. of (7.46), we obtain

$$a_{rj}^{(i)} \left((\omega_A)_j^2 - (\omega_A)_r^2 \right) = -\{\phi_A\}_j^T \frac{\partial [K_A]}{\partial p_i} \{\phi_A\}_r + \frac{\partial (\omega_A)_r^2}{\partial p_i} \{\phi_A\}_j^T [M_A] \{\phi_A\}_r$$

$$+ (\omega_A)_r^2 \{\phi_A\}_j^T \frac{\partial [M_A]}{\partial p_i} \{\phi_A\}_r \tag{7.47}$$

As j ≠ r, the second term on the r. h. s. of (7.47) is zero and so,

$$a_{rj}^{(i)} = \frac{1}{(\omega_A)_r^2 - (\omega_A)_j^2} \{\phi_A\}_j^T \left[\frac{\partial [K_A]}{\partial p_i} - (\omega_A)_r^2 \frac{\partial [M_A]}{\partial p_i} \right] \{\phi_A\}_r \qquad j \neq r \tag{7.48}$$

It is clear from (7.37) that for j = r coefficients $a_{rj}^{(i)}$ have to be calculated separately. Differentiating $\{\phi_A\}_r^T [M_A] \{\phi_A\}_r = 1$, it follows that

$$2\{\phi_A\}_r^T [M_A] \frac{\partial \{\phi_A\}_r}{\partial p_i} = -\{\phi_A\}_r^T \frac{\partial [M_A]}{\partial p_i} \{\phi_A\}_r \tag{7.49}$$

Substituting (7.41) in (7.49),

$$2\{\phi_A\}_r^T [M_A] \sum_{j=1}^{N} a_{rj}^{(i)} \{\phi_A\}_j = -\{\phi_A\}_r^T \frac{\partial [M_A]}{\partial p_i} \{\phi_A\}_r \tag{7.50}$$

or

$$2 \sum_{j=1}^{N} a_{rj}^{(i)} \{\phi_A\}_r^T [M_A] \{\phi_A\}_j = -\{\phi_A\}_r^T \frac{\partial [M_A]}{\partial p_i} \{\phi_A\}_r \tag{7.51}$$

and due to the orthogonality properties,

$$a_{rr}^{(i)} = -\frac{1}{2}\{\phi_A\}_r^T \frac{\partial[M_A]}{\partial p_i}\{\phi_A\}_r \qquad (7.52)$$

Combining (7.48) and (7.52) and substituting into (7.41), the following result is obtained:

$$\frac{\partial\{\phi_A\}_r}{\partial p_i} = \sum_{j=1;j\neq r}^{N} \frac{\{\phi_A\}_j\{\phi_A\}_j^T}{(\omega_A)_r^2 - (\omega_A)_j^2}\left[\frac{\partial[K_A]}{\partial p_i} - (\omega_A)_r^2\frac{\partial[M_A]}{\partial p_i}\right]\{\phi_A\}_r$$

$$-\frac{1}{2}\{\phi_A\}_r\{\phi_A\}_r^T\frac{\partial[M_A]}{\partial p_i}\{\phi_A\}_r \qquad (7.53)$$

In [287] the authors explore the case where a truncated modal basis is used, i.e., when the summation in (7.41) extends to a number of modes smaller than N.

Now that we know how to evaluate the eigensensitivities, we can return to expressions (7.33) and (7.34) and arrange them so to constitute $(N+1)$ equations for the N_p unknown updating parameters. By using m correlated mode pairs, $m(N+1)$ equations can be obtained, leading to a system of linear equations:

$$[A]\{p\} = \{\Delta\xi\} \qquad (7.54)$$

where

$$\underset{(m(N+1)\times N_p)}{[A]} = \begin{bmatrix} \dfrac{\partial(\omega_A)_1^2}{\partial p_1} & \cdots & \dfrac{\partial(\omega_A)_1^2}{\partial p_{N_p}} \\[2mm] \dfrac{\partial\{\phi_A\}_1}{\partial p_1} & \cdots & \dfrac{\partial\{\phi_A\}_1}{\partial p_{N_p}} \\[2mm] \vdots & & \vdots \\[2mm] \dfrac{\partial(\omega_A)_m^2}{\partial p_1} & \cdots & \dfrac{\partial(\omega_A)_m^2}{\partial p_{N_p}} \\[2mm] \dfrac{\partial\{\phi_A\}_m}{\partial p_1} & \cdots & \dfrac{\partial\{\phi_A\}_m}{\partial p_{N_p}} \end{bmatrix} \qquad \underset{(m(N+1)\times1)}{\{\Delta\xi\}} = \begin{Bmatrix} \Delta(\omega)_1^2 \\[2mm] \Delta\{\phi\}_1 \\[2mm] \vdots \\[2mm] \Delta(\omega)_m^2 \\[2mm] \Delta\{\phi\}_m \end{Bmatrix}. \qquad (7.55)$$

This is essentially of the same form as the corresponding updating problem using the FRF sensitivity method discussed previously, and certain parallels can be drawn:

(i) p-values can be assigned to elemental or macro-elemental matrices which then become the system matrix derivatives in (7.40) and (7.53);

(ii)   the problem is over-determined provided that $m(N+1) > N_p$ and a least-squares solution for the p-values is obtainable by SVD:

$$\{p\} = [A]^+ \{\Delta \xi\} \tag{7.56}$$

(iii) the rows of the sensitivity matrix can be of different orders of magnitude, forcing some equations to dominate in the determination of the solution. The sensitivity matrix can be balanced, such that all equations carry equal weight, or a weight in accordance with the confidence in the measurements;

(iv)   mode shapes are insensitive to parameter changes at some DOFs causing noise amplification effects. A minimum sensitivity threshold can be employed to eliminate equations due to the least sensitive DOFs; and

(v)   an iterative process is required because the modal parameters are nonlinear with respect to the updating parameters.

Experimental sensitivities
When large discrepancies exist between the experimental and analytical models, the validity of the Taylor series truncations in equation (7.32) are undermined and the iterative process is prone to divergence. Lin *et al* [288] propose a modification to the modal sensitivity method that enables convergence for larger magnitudes of FE modelling errors.

This is achieved by using both analytical and experimental modal data in the formulation of the eigensensitivities. Equations (7.40) and (7.53) become:

$$\frac{\partial(\omega_A)_r^2}{\partial p_i} = \{\phi_A\}_r^T \frac{\partial[K_A]}{\partial p_i} \{\phi_x\}_r - (\omega_x)_r^2 \{\phi_A\}_r^T \frac{\partial[M_A]}{\partial p_i} \{\phi_x\}_r \tag{7.57}$$

$$\frac{\partial\{\phi_A\}_r}{\partial p_i} = \sum_{j=1; j \neq r}^{N} \frac{\{\phi_A\}_j \{\phi_A\}_j^T}{(\omega_x)_r^2 - (\omega_A)_j^2} \left[\frac{\partial[K_A]}{\partial p_i} - (\omega_x)_r^2 \frac{\partial[M_A]}{\partial p_i} \right] \{\phi_x\}_r$$

$$- \frac{1}{2} \{\phi_A\}_r \{\phi_A\}_r^T \frac{\partial[M_A]}{\partial p_i} \{\phi_x\}_r \tag{7.58}$$

Rigorous derivations for these *formulæ* can be found in [288]. As a result of this modification:

(i)   when there is a large disparity between experimental and analytical modes, the sensitivities are not calculated exclusively on the basis of the incorrect analytical modes. This enables convergence for a wider domain of updating problems; and

(ii)   calculation of the experimental eigensensitivities requires measured modal displacements at every DOF in the FE model, causing coordinate incompleteness in all practical cases.

Modes or FRFs for correlation?
Unfortunately it is impossible to give hard and fast rules as to whether modal or frequency domain data are more appropriate for correlation. Some of the issues which need to be considered when adopting a particular approach are:

(i)   modal extraction is subject to errors and omission, a problem exacerbated by increased modal density at high frequencies;

(ii)   although FRFs give more equations, the only new information occurs due to the residual terms outside the measurement bandwidth;

(iii)   modal parameters need weighting between the natural frequencies and the mode shapes. Natural frequencies are more reliable and sensitive to parametric changes. However, excessive weighting towards natural frequencies will cause mode shape information to be lost, thereby encouraging the updating problem to become ill-conditioned;

(iv)   for FRFs the weighting is implicit and depends on the proximity of the chosen frequency points to resonance. Weighting is therefore less controllable; and

(v)   FRFs are more appropriate for the updating of damping due to the variation of phase over the measured bandwidth [289]. However, this asset can be a problem for updating undamped models.

The arguments for the use of FRF data become stronger as updating applications become more ambitious. Increasing the number of updating parameters including damping means that over-determination of the problem to combat ill-conditioning and noise effects will dictate an increased data requirement, currently only available in the frequency domain. This inevitably will require measurements over a higher frequency bandwidth, making modal extraction problematic in regions of high modal density.

Choice of updating parameters and generic elements
It is often assumed that the inadequacy of the FE model can be corrected by error matrices of the form

$$[\Delta M] = \sum_{i=1}^{N_e} m_i [M_e]_i \qquad [\Delta K] = \sum_{i=1}^{N_e} k_i [K_e]_i \qquad (7.59)$$

(for an undamped system) where N_e is the number of elemental matrices suspected to be in error and the p-value set is given by

$$\{p\} = \left\{m_1, \cdots, m_{N_e}, k_1, \cdots, k_{N_e}\right\} \qquad (7.60)$$

The physical consequences of this assumption are that:

(i) the connectivity of the FE model is retained; and

(ii) elemental inertias (for mass p-values) and elemental Young's moduli (for stiffness p-values) are scaled uniformly in all directions. This ensures that properties such as isotropic construction and beam element cross-section shapes are invariant under model updating.

There is an additional advantage to this choice of p-values. The derivatives in equation (7.30) need not be calculated numerically but are simply given by:

$$\frac{\partial[Z]}{\partial m_i} = -\omega^2 [M_e]_i \quad \text{and} \quad \frac{\partial[Z]}{\partial k_i} = [K_e]_i \qquad (7.61)$$

Gladwell and Ahmadian [290] have made a significant contribution to the topic of parameter selection by defining families of 'generic' finite elements, each element having more parameters than standard finite elements. They achieve this by considering what combinations of the modes of the elements maintain the required properties of the element. These required properties include the mass matrix being positive definite, the stiffness matrix being positive semi-definite, there being an appropriate number of rigid body, symmetric and anti-symmetric modes and the position of the centre of mass and principal axes of inertia remaining unchanged.

The parameters of these generic elements may then be updated. This guarantees that the updated model will be a valid FE model, whilst allowing the update to include general effects that may have been ignored in the original model such as flexure, shear, Poisson's ratio and joint flexibility.

These generic elements encompass many of the standard elements as special cases of the extra parameter values. As an example, the generic rod (bar) element has three parameters: mass, stiffness and an extra parameter which is called S_2. If $S_2 = 1$ then the element has a lumped mass distribution, and if $S_2 = \sqrt{3}/3$ then the element has a consistent mass distribution. Other values of S_2 give different distributions, some of which have lower discretisation errors than the two distributions mentioned. Other generic elements are also given in the paper, such as a beam element, an in-plane frame element and a triangular plate element.

The more complex generic elements have several extra parameters, and their use in model updating has to be carefully considered. There are many possible approaches, such as only using generic elements at unknown modelling positions like joints, or allowing every element to be a generic element but using the same values of the extra parameters for all the elements, or limiting the choice of extra parameters used thus ignoring some possible effects.

Ill-conditioning

A set of linear equations of the form

$$[A]\{x\} = \{b\} \tag{7.62}$$

is determined if $[A]$ has full rank, i.e., there are as many linearly independent rows as there are columns. However, the rank of $[A]$ is unclear if the rows are nearly linear combinations of each other. In this case, the solution is determined by almost parallel lines in multi-dimensional space and is therefore highly sensitive to noise on $\{b\}$. Therefore, the correct question to ask is not whether $[A]$ has full rank, but whether it is well conditioned. All least-squares methods yield a set of linear equations of the form of equation (7.62), and since $\{b\}$ is based on measured response data which are subject to noise contamination, the conditioning of $[A]$ is of crucial importance [291, 292].

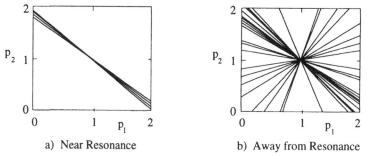

a) Near Resonance   b) Away from Resonance

Fig. 7. 2 Updating equations for two p-values from simulated noise-free FRFs.

To demonstrate the conditioning of the updating problem it is instructive to consider briefly the simple simulated structure which was depicted in figure 7.1. Here, just two updating parameters are considered and the true solution is known to be $\{p_1, p_2\} = \{1,1\}$. In this - somewhat contrived - case, each DOF from each frequency point contributes a scalar equation of the form

$$ap_1 + bp_2 = c \tag{7.63}$$

which can be plotted in $p_1 - p_2$ space. Figure 7.2 illustrates the updating equations (7.25) - the output error in the frequency domain - when two different frequency points are chosen, one close to resonance and the other away from resonance. Note that:

(i)   all equations pass through the known solution $\{1,1\}$ because in this case the simulated experimental FRFs are noise-free. Any pair of linearly independent equations is sufficient to solve the problem;

(ii) the equations due to the frequency point near resonance are nearly parallel and in the interest of good conditioning cannot be relied upon to provide more than one independent relation between p_1 and p_2; and

(iii) the equations due to the frequency point away from resonance are more diverse.

Duplicity of equations is in general desirable to average out noise effects. However, it is essential that there are sufficient (in this case two) equations that are not nearly dependent. It is therefore of concern that frequencies near resonance can essentially contribute a single relation. The response sensitivities at frequency points close to and away from resonance will be examined in turn.

Equations close to resonance

From equation (7.30), the sensitivity of the ith DOF with respect to the jth p-value (for an undamped system) is given by

$$(S(\omega))_{i,j} = \{\alpha_A(\omega)\}_i^T \left[\frac{\partial[K]}{\partial p_j} - \omega^2 \frac{\partial[M]}{\partial p_j} \right] \{\alpha_X(\omega)\} \tag{7.64}$$

If ω is close to both an experimental and an analytical natural frequency then a single mode will dominate each response, so

$$\{\alpha_A(\omega)\} \approx c_A(\omega)\{\phi_A\} \qquad \{\alpha_X(\omega)\} \approx c_X(\omega)\{\phi_X\} \tag{7.65}$$

for some frequency dependent values, $c_A(\omega)$ and $c_X(\omega)$. Substituting these expressions in (7.64),

$$(S(\omega))_{i,j} \approx c_A(\omega)\,c_X(\omega)\{\phi_A\}_i^T \left[\frac{\partial[K]}{\partial p_j} - \omega^2 \frac{\partial[M]}{\partial p_j} \right] \{\phi_X\} \tag{7.66}$$

This can be recognised as being closely related to the modified eigenvalue sensitivity proposed by Lin *et al* [288] (see (7.57)):

$$\frac{\partial \omega^2}{\partial p_j} \approx \{\phi_A\}_i^T \left[\frac{\partial[K]}{\partial p_j} - \omega_X^2 \frac{\partial[M]}{\partial p_j} \right] \{\phi_X\} \tag{7.67}$$

Combining (7.66) and (7.67) gives

$$(S(\omega))_{i,j} \approx \begin{cases} \dfrac{\omega^2}{\omega_X^2} c_A(\omega) c_X(\omega) \dfrac{\partial \omega^2}{\partial p_j} & \text{for mass updating parameters} \\[4mm] c_A(\omega) c_X(\omega) \dfrac{\partial \omega^2}{\partial p_j} & \text{for stiffness updating parameters} \end{cases} \tag{7.68}$$

The receptance sensitivity matrix near a natural frequency is then approximated by

$$[S(\omega)] \approx$$

$$c_A(\omega)c_X(\omega) \begin{bmatrix} \dfrac{\partial \omega^2}{\partial p_1} & \cdots & \dfrac{\partial \omega^2}{\partial p_{N_e}} & \dfrac{\partial \omega^2}{\partial p_{N_e+1}} & \cdots & \dfrac{\partial \omega^2}{\partial p_{2N_e}} \\ \vdots & & & & & \vdots \\ \dfrac{\partial \omega^2}{\partial p_1} & \cdots & \dfrac{\partial \omega^2}{\partial p_{N_e}} & \dfrac{\partial \omega^2}{\partial p_{N_e+1}} & \cdots & \dfrac{\partial \omega^2}{\partial p_{2N_e}} \end{bmatrix} \left[\mathrm{diag} \left\{ \dfrac{\omega^2}{\omega_X^2}, \cdots, \dfrac{\omega^2}{\omega_X^2}, 1, \cdots, 1 \right\} \right]$$

$$(7.69)$$

which is of rank 1. Hence the sensitivity matrix from a single frequency close to analytical and experimental resonance results in a set of almost identical equations that are highly ill-conditioned. This implies that if only frequencies close to resonance are used and experimental and analytical natural frequencies are not too disparate, then the frequency range must cover as many modes as there are updating parameters to avoid ill-conditioning. Although selecting resonant frequencies alone is an unnatural choice for FRF updating, there is a real issue concerning the recommended balance between resonance and off-resonance frequency points. Should only off-resonance frequencies be used? To answer this question, the nature of the updating equations from off-resonance frequency points must be examined.

Equations away from resonance

At off-resonance frequencies, the response is composed of a summation of all the modes of the structure, the adjacent modes usually being more dominant than others. The contribution from each mode is dependent on the excitation and measurement locations and so different DOFs react differently to the same change in updating parameters. For this reason, the equations at each off-resonance frequency point are diverse and therefore contain more information about the inter-dependencies of the unknown variables.

From figure 7.2 b) it is apparent that in this case most equations have a negative gradient indicating that changes in p_1 and p_2 have similar effects on the response. For a well conditioned problem, measurements are needed that can distinguish between changes in p_1 and changes in p_2. The equations with positive gradient are therefore of special interest since they represent responses for which p_1 and p_2 have opposing effects.

Now consider the response sensitivities obtained from the same off-resonance frequency point plotted as a vector field as shown in figure 7.3. In this way, the components of each vector represent the receptance sensitivities of a single DOF due to unit changes in p_1 and p_2, and the length of the vector indicates the magnitude of the sensitivity. This shows that those DOFs where p_1 and p_2 have an opposite effect on the response (indicated by the hollow arrows) are actually insensitive to both! This is a consistent observation over the whole frequency range and has undesirable consequences that will be discussed later.

Inclusion of frequency points away from resonance potentially improves the conditioning of the updating problem due to the diversity in sensitivity ratios from DOF to DOF. However, those DOFs that are most able to dispel ill-conditioning are relatively insensitive to the updating parameters.

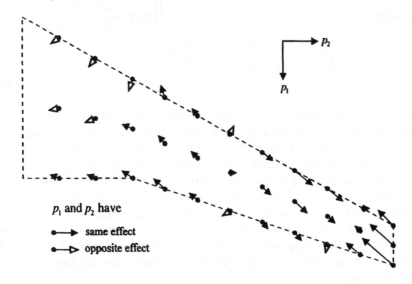

Fig. 7.3 Vector plot of receptance sensitivities.

Sources and effects of ill-conditioning

As seen before - in the case of the FRF output error updating method - the updating problem can become ill-conditioned or even degenerate. However, this characteristic is by no means specific to this or any other method, but is a manifestation of the inverse problem of model updating itself. Physically speaking, ill-conditioning arises from attempting to determine updating parameters which affect a set of response data in very similar ways. Whilst the formulation of the updating problem can be tuned to optimise the information available, it is the choice of updating parameters and responses that is ultimately responsible for ensuring a well conditioned problem.

Figure 7.4 a) shows an updating solution for the wing model in figure 7.1 when the stiffnesses of all eleven leading edge plate elements are chosen as updating parameters. For the frequency range spanning the first 8 modes of vibration, the high level of dependency between parameters has resulted in an ill-conditioned problem which is characterised by the oscillatory pattern in the solution. (The corruption is particularly severe near the wing tip due to the insensitivity of the response to stiffness changes in this region.) Removal of the tip-most stiffness from the updating parameter set results in an improved solution - shown in figure 7.4 b) - due in part to the reduced dependency of the remaining ten parameters. Alternatively, the full set of eleven parameters can be retained and their dependency reduced by altering the set of response data. Figure 7.4 c) shows that by extending the baseband of the response to include a further five modes the parameters are no longer indistinguishable from each other.

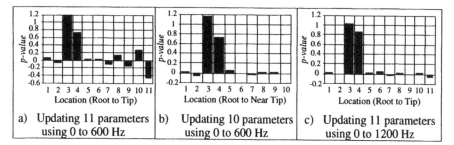

a) Updating 11 parameters using 0 to 600 Hz

b) Updating 10 parameters using 0 to 600 Hz

c) Updating 11 parameters using 0 to 1200 Hz

Fig. 7.4 Effects of parameter and response sets on updating.

The problem of ill-conditioning is a key issue in least-squares updating for which a solution must be found if these methods are to be applied robustly. The previous examples have illustrated the underlying principles by which some of the following strategies are pinned:

(i)   *precursory error location.* Successful implementation of error location techniques such as those described in Section 7.5 are often essential in order to reduce the number of parameters whilst not discarding any which are significantly in error;

(ii) *use of super- or generic elements* to reduce the number of updating parameters. A word of warning, however: selecting large elements which go beyond the boundary of the true error causes this error to be dispersed within the new element. This can cause adjacent elements to act in a compensatory manner, again resulting in the oscillatory behaviour evident in figure 7.4 a);

(iii) *use of higher frequency range (or more modes).* Although desirable, this can be difficult to achieve in practice;

(iv) *selective sensitivity.* Ben-Haim [293, 294] proposes a technique of selective sensitivity to uncouple updating parameters thus improving the conditioning of the updating problem. A multiple input force is determined that renders the response sensitive to some parameters and insensitive to others. A drawback with this method is the large number of shakers that is required to apply the force shape;

(v)   *perturbed boundary condition testing and measurement of static deflections.* Lammens [295] suggests extending the set of response data to include test results for the structure under different boundary conditions, thus augmenting the diversity of parametric sensitivities. Static tests can also be used to supplement stiffness sensitivities;

(vi) *regularisation.* This involves modifying the least-squares norm,

$$\| r \|^2 = \| [A] \{x\} - \{b\} \|^2 \qquad (7.70)$$

so as to penalise large changes to the FE model that could otherwise occur when noise is amplified by an ill-conditioned updating problem. The general form of the modified norm [296] is given by

$$\| r \|^2 = \| [A] \{x\} - \{b\} \|^2 + \lambda^2 \| [C] \{b\} \|^2 \qquad (7.71)$$

where [C] is a user defined transformation matrix and λ is the regularisation parameter. Natke [296] suggests that [C] should reflect the level of noise apparent on the experimental data. This and other regularisations appear to be justifiable on the grounds that they relax the strict least-squares solution criterion in favour of a nearby, but physically more credible, solution. A minimum change type of solution can also be obtained by truncated SVD. Singular values are discarded from the pseudo-inverse culminating in a minimum norm solution to a rank deficient problem. However, it is inevitable that a certain amount of useful information in the truncated terms will be lost, causing the solution to compromise the structure's physical integrity. D'Ambrogio *et al* [297] propose criteria for the retention of singular values based on response and/or natural frequency norms.

Coordinate incompleteness
Having covered least-squares updating in some detail, it is worth including some brief comment on related matters. The first implementation of frequency domain least-squares updating - known as the Response Function Method (RFM) - was proposed by Lin and Ewins [298], which bypassed the modal data altogether using an iterative approach.

Fritzen adopts a sensitivity approach to FRF updating [299] which is akin to modal sensitivity and equivalent to the RFM. Linear equations of the form of (7.29) are derived by considering forced responses of the experimental and analytical models at discrete frequency points. Matrix [S] now represents the response sensitivities to unit changes in the updating parameters.

The FRF sensitivity approach to model updating requires measured response data at every DOF in the FE model, resulting in coordinate incompleteness in all practical cases. Lin and Fritzen complete the experimental response vector by adopting analytical counterparts at slave DOFs which is effectively a very simple form of FRF expansion. Bretl [300] takes this approach one step further and uses analytical counterparts even when measured values are available. Over-dilution of the experimental responses by analytical counterparts often causes divergence of the solution, especially when the difference between the experimental and analytical models is large. Accordingly, D'Ambrogio *et al* [301] and Lammens *et al* [302] advocate reduction of the FE model to the measured DOFs to overcome coordinate incompleteness.

7.6.2   Direct optimisation

It can be seen form the previous section that considerable effort has been expended trying to minimise a norm between two vectors using the least-squares approach. This approach can be interpreted as a form of optimisation. The methods we will discuss in this section are concerned with some of the techniques of optimisation and other ways by which they can be applied to the correlation process.

Direct Updating methods

One of the earliest attempts to update FE models was proposed in 1978 by Baruch [303]. Baruch assumed that the mass matrix was correct and minimised the distance between the original and updated FE models defined by the Euclidean norm,

$$\varepsilon = \left\| [M_A]^{-\frac{1}{2}} [[K_X] - [K_A]] [M_A]^{-\frac{1}{2}} \right\| \tag{7.72}$$

subject to the constraints that the updated stiffness matrix is (i) symmetric, i.e.,

$$[K_X] = [K_X]^T \tag{7.73}$$

and (ii) respecting the orthogonality condition

$$[\phi_X]^T [K_X] [\phi_X] = \left[{}^{\backprime}\omega_X^2 {}_{\backprime} \right] \tag{7.74}$$

The objective function in (7.72) is minimised using Lagrange multipliers to give the following expression for the stiffness error matrix:

$$\begin{aligned}[\Delta K] = &-[K_A][\phi_X][\phi_X]^T [M_A] - [M_A][\phi_X][\phi_X]^T [K_A] \\ &+[M_A][\phi_X][\phi_X]^T [K_A][\phi_X][\phi_X]^T [M_A] \\ &+[M_A][\phi_X][{}^{\backprime}\omega_X^2 {}_{\backprime}][\phi_X]^T [M_A]\end{aligned} \tag{7.75}$$

A similar approach was later adopted by Berman [304] to update the mass matrix, and modified by Caesar [305] to constrain the total mass of the system. These methods have been largely discarded in recent years because the updated spatial properties bear little resemblance to those of the original model. This is to be expected when, in practice, the measured modal matrix, $[\phi_X]$, in (7.75) is highly rectangular due to modal truncation [306].

Method of Steepest Descent

Soma and Gola [307] use an optimisation approach similar to Baruch's modal based method that minimises the distance between the experimental and analytical models. Here, the objective function by which the distance is measured is based on FRF data. The function is chosen as the following summation over N_f frequency points:

$$\varepsilon = \frac{1}{N_f} \sum_{i=1}^{N_f} \left\{ \{\alpha_X(\omega)\}_i - \{\alpha_A(\omega)\}_i \right\}^T [W(\omega)] \left\{ \{\alpha_X(\omega)\}_i - \{\alpha_A(\omega)\}_i \right\} \quad (7.76)$$

where $[W(\omega)]$ is a weighting matrix. This distance is then minimised by the method of 'steepest descent' [308]. The major advantage with this method is that coordinate incompleteness can be easily overcome by partitioning of the response vectors. This is possible because the objective function does not feature the mass and stiffness matrices explicitly and therefore does not directly encompass the equations of motion. However, the uniqueness of the solution can be suspect.

Robust methods for function minimisation

Many approaches to model updating use an optimisation approach, attempting to find the minimum of a function that encapsulates the difference between the analytical and experimental results. Most of the optimisation algorithms used simply find the nearest local minima to the starting position, whereas the actual optimum point may be in a significantly different location.

Recently two new approaches to optimisation that are capable of finding global minima have been developed; these two approaches are the Genetic Algorithm and Simulated Annealing [309]. Both of these algorithms are probabilistic search algorithms derived from analogies with natural phenomena: Simulated Annealing from a thermodynamic cooling process and Genetic Algorithms from natural evolution.

There has been much interest in applying these algorithms to model updating in recent years. The algorithms will be considered in turn, and then the model updating issues.

Simulated Annealing (SA)

Simulated Annealing was derived from an analogy with the annealing process of material physics by Kirkpatrick [310] in 1983. It is a generalisation on the random walk optimisation algorithm, which consists of simply 'wandering' about the search space for a given number of iterations, and selecting the best point visited as the global minimum.

Simulated Annealing introduces the concept of a temperature, which acts as a control parameter for the algorithm. The temperature is set to a high value initially and is slowly lowered as the algorithm proceeds. When the temperature is high, most of the points that are visited are accepted whether they are an improvement on the current position or not. As the temperature falls, fewer and fewer detrimental transitions are accepted, until the algorithm 'freezes', hopefully at the global minimum point. Thus, Simulated Annealing can be seen as a random walk that slowly 'condenses' onto the global minimum point. See figure 7.5 for a flowchart outlining the algorithm.

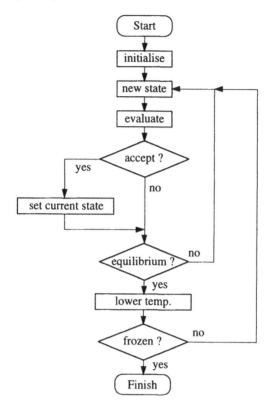

Fig. 7.5 Simulated Annealing flowchart.

For a discussion of both the theory behind Simulated Annealing and the implementation details, refer to [311]. Implementation details include the choice of the initial temperature, and the method of lowering the temperature throughout the simulated annealing run. These details are referred to as an *annealing schedule*. Other implementation details include the method of generating subsequent points in the search space for the algorithm to visit, called the *neighbourhood function*. The appropriate choice of neighbourhood function and annealing schedule can have a significant effect on the performance of the algorithm.

Genetic Algorithms (GAs)
Genetic Algorithms are optimisation algorithms based on an analogy with natural evolution. In natural evolution, members of a population compete with each other to survive and reproduce successfully. If the genetic makeup of an individual member of a population gives that individual an advantage over its rivals, then it is more likely to breed successfully. Consequently the combination of genes that

confer this advantage is likely to spread across the population. This is a natural optimisation method that may also be simulated.

The basic Genetic Algorithm was suggested by Holland in 1975 [312, 313]. The algorithm acts on a population of binary-string chromosomes. Each of these chromosomes is a discretised representation of a point in the search space, and as such has a fitness value given by the objective function. Three genetic operators are applied to the population to generate a new population: *reproduction, crossover* and *mutation*. The nature of these operators is such that each subsequent generation tends to have an average fitness level higher than the previous generation. See figure 7.6 for an outline of the algorithm.

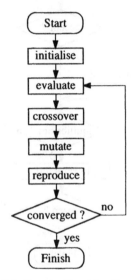

Fig. 7.6 Genetic Algorithm flowchart.

The operators will be briefly discussed. The reproduction operator assigns each chromosome a relative probability of being reproduced according to the fitness of the chromosomes. The fittest chromosomes may typically be reproduced several times; the least fit chromosomes may not be reproduced at all. The crossover operator mixes genetic information amongst the population by snipping pairs of chromosomes at a random point along their length, and swapping over the snipped sections. The crossover operator has the potential to join successful genetic fragments together to form fitter individuals.

The mutation operator has the potential to re-introduce genetic information that has been lost, back into the population. After crossover has occurred, each binary digit of each chromosome has a small probability of mutating. Mutation is defined as an inversion of a binary digit. There are many other genetic operators that are possible, but these three operators produce the basic Genetic Algorithm. The implementation details for the Genetic Algorithm include the size of the initial population, which depends on the length of the chromosomes, which in turn depends on the level of discretisation acceptable. A population that is too large results in much wasted effort, whereas too small a population results in failure to locate the global minimum.

Other important implementation details include the reproduction probabilities; these probabilities control how quickly or slowly the population converges to a result.

Model updating using Genetic Algorithms and Simulated Annealing

The use of Genetic Algorithms for model updating has been considered by several workers [314-319]. Most have tried the standard Genetic Algorithm on small problems, and have reported successful results.

Rajeev and Krishnamoorthy [315] considered using the Genetic Algorithm for structural optimisation, e.g. minimising the weight of a structure that satisfies certain static conditions. A larger model was considered (a 160 bar tower), with successful results. However, a high level of discretisation was used to achieve this result, i.e., each of the twelve parameters considered could take only one of sixteen possible values.

Larson and Zimmerman [316] used a Genetic Algorithm to update three stiffness parameters of a simulated truss system, and briefly considered the effects of noise. The objective function they considered used the first five natural frequencies and mode shapes of the truss. The results they reported were very successful, even with 15% noise applied to the mode shapes.

The approach considered by Friswell *et al* [317] differed slightly in that they restricted the changes made to the model (a cantilevered beam) to be in one or two places, i.e., a damage detection algorithm. They also included a penalty function to weight the Genetic Algorithm away from using two locations. After finding the error locations, an eigensensitivity algorithm was subsequently used to ascertain the appropriate stiffness values of the error locations. This overcame the problem of the Genetic Algorithm discretising the updating parameters. Four separate simulated tests were reported, each of which located the error positions and appropriate updating parameters successfully.

Mares and Surace [318] applied a Genetic Algorithm to an eight DOF mass/spring system, attempting to identify six stiffness parameters even in the presence of limited modal data and moderate noise contamination. Again, successful results were reported. Dunn [319] has developed the ideas of updating FE models using Genetic Algorithms, by suggesting improvements to the standard Genetic Algorithm for updating using Gray scaling and multiple restarts. Gray scaling reforms the standard binary system by allowing a single digit inversion to increment the chromosome, so changes are continuous rather than stepped. Restarting initiates the optimisation from several points to find the most appropriate position in the error space prior to the full optimisation, thus minimising computational effort.

Levin and Lieven [320] have compared Simulated Annealing to Genetic Algorithms for model updating purposes and concluded that Simulated Annealing is slightly superior, mainly because it is not necessary to discretise the updating parameters. Frequency domain data were used for the objective functions, and the more difficult problem of location and updating of structure using many p-values.

It was found that despite the use of a powerful optimisation technique, including more updating parameters can produce disappointing results. Figure 7.7 shows the p-values produced by using Simulated Annealing on a ten-element cantilevered beam using twenty p-values (mass and stiffness for every element).

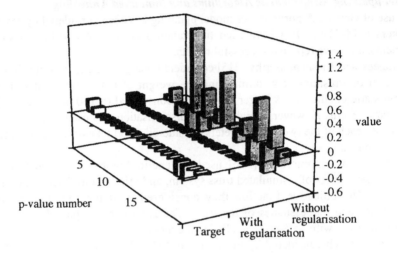

Fig. 7.7 Updating using weighted objective functions.

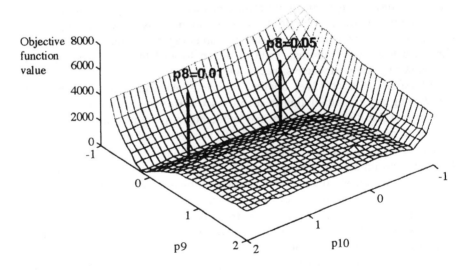

Fig. 7.8 Error surface of objective function, with minimum positions highlighted (all other p-values set to zero).

The results appear disappointing, with the p-values oscillating wildly around the correct values. However, the updated and target FRFs are virtually identical in the frequency range of interest. This is a well-known problem of FE model updating, found in many updating methods (c.f. figure 7.4). It stems from the nature of the dynamic behaviour of structures; the 'oscillating' p-value structure actually does behave very similarly to the correct structure at low and medium order modes. This suggests that higher frequency ranges are desirable to update FE models, which is widely known. Unfortunately it is often impractical to measure the high frequency responses of structures.

The sensitivity of the objective function can be seen by considering the error surface produced by an objective function (figure 7.8) when varying two of the updating parameters (the stiffnesses at the tip).

The error surface contains a nearly flat trough which holds the global minimum point. A small change to another p-value can slightly change the gradient of the trough, drastically changing the position of the global minimum. Figure 7.8 shows the position of the minimum when the eighth p-value is changed from 0.01 to 0.05. This is why the function is so hard to minimise, even with the best optimisation algorithms and least-squares techniques currently available.

Possible ways of reducing these oscillations were considered [320]. One approach is the use of regularisation terms in the objective function to reduce the oscillations, encouraging more physically sensible updating parameters. Equation (7.77) penalises large changes between adjacent updating parameters. Note that the form of this equation is similar to (7.71), in which regularisation is implemented in a least-squares sense.

$$f(\{p\}) = \sum_j \sum_\omega \left| \log\left(\left\| {}_X\alpha_{jk}(\omega) \right\|_2 \right) - \log\left(\left\| {}_A^U\alpha_{jk}(\omega) \right\|_2 \right) \right|$$
$$+ \; W_p \sum_i \sum_h \left| p_i - p_h^{i\_nbr} \right| \tag{7.77}$$

where

W_p	=	weighting factor for p-values
p_i	=	i^{th} p-value
$p_h^{i\_nbr}$	=	h^{th} neighbouring p-value of ith p-value
${}_X\alpha_{jk}$	=	(simulated) experimental receptance point
${}_A^U\alpha_{jk}$	=	updated analytical receptance
$\{p\}$	=	vector of p-values
$f(\{p\})$	=	objective function

The results of using this regularisation function are also shown in figure 7.7. It is clear that regularisation can have a significant effect on the results. Care must be taken when using these functions to find an appropriate balance between the regularisation terms and the model updating terms.

A second approach is to reduce the number of p-values considered, possibly by the use of superelements. This has two desirable consequences: firstly, they reduce the number of p-value oscillations that are possible, and secondly they reduce the ability of the remaining oscillations to mimic the correct behaviour of the structure. Unfortunately, if the superelements are too large then the resulting p-values will not be capable of capturing the correct changes to model, and the update will fail. A possible iterative approach to model updating is to update the model with large superelements initially, whilst maintaining the oscillating p-value regularisation, and then reduce the size of the superelements (and increase the number of p-values) until a successful update is performed.

Note that both SA and GAs tend to give results in the neighbourhood of the global minimum. If more accuracy is desired then the result may be used as the starting point of a further optimisation technique that finds the nearest local minimum, i.e. the nonlinear simplex algorithm [314]. It is also generally noted that these algorithms are slow to run, but the superior performance obtained relative to conventional optimisation, generally outweighs the speed concerns.

7.6.3 Neural Networks

Neural Networks have been applied successfully to many diverse application areas. They are particularly applicable to problems where plenty of examples are available but it is difficult to specify an explicit algorithm, such as character recognition, time series prediction, etc. [321, 322]. They are generally appropriate for difficult classification problems and nonlinear regression problems.

The phrase neural network refers to a computational structure derived from the study of biological neurons. Neural Networks consist of a number of simple processing units, *neurons*, linked to each other. All neurons have multiple inputs and a single output but there are many different types of neuron. The architecture of a Neural Network is the number and type of neurons present in the network, and how they are connected to each other and to the outside world. The two most common types of neural network used are Multi-Layer Perceptron (MLP) and Radial Basis Function (RBF) networks. For a good text on many of the issues surrounding Neural Networks refer to Bishop [323].

Neural Networks have certain properties that make them attractive to many applications. The two main properties that are of use for model updating are the capacity to generalise, i.e., the production of a sensible response from previously unseen data, and robustness in the presence of noise.

Neural Networks need to be trained before they can be used, and this can be a slow process. This requires a *training algorithm* and a set of *training data*. Training data consist of a set of input vectors and a corresponding set of output vectors. Training a network consists of adjusting the network parameters (the neuron weights) until the output of the network when presented with the input vectors matches the target output vectors to within a given tolerance. Often some of the training data are held back as a test set to check that adequate generalisation has occurred. Two common training algorithms used are the Back-Propagation algorithm for MLP [324] networks and the Orthogonal Least-Squares (OLS) algorithm [325] for RBF networks. For more details refer to Bishop [323].

The architecture of a typical RBF network is shown in figure 7.9.

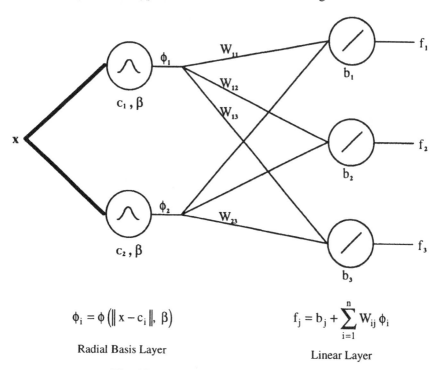

$$\phi_i = \phi\left(\left\|x - c_i\right\|, \beta\right) \qquad\qquad f_j = b_j + \sum_{i=1}^{n} W_{ij}\,\phi_i$$

Radial Basis Layer   Linear Layer

Fig. 7.9 Schematic of a small RBF network.

The overall response characteristic of a RBF network is given by

$$f_j = b_j + \sum_{i=1}^{n} W_{ij}\,\phi\left(\left\|\{x\} - \{c_i\}\right\|, \beta\right) \tag{7.78}$$

where

$\{x\}$ is the value of the input vector
$\{c_i\}$ is the centre of the ith neuron (a network parameter)
β is the spread constant of the network (same for every neuron, a network parameter)
$\|\cdot\|$ represents the Euclidean norm (2-norm)
ϕ is the radial basis function of the network, typically a Gaussian function
f_j is the output of the jth linear neuron (j^{th} network output)

b_j is the bias of the jth linear neuron
W_{ij} is the weight between the ith RBF neuron and the jth linear neuron
    (a network parameter)

The only architectural decision that remains for this network is the number of RBF neurons that are present in the first layer. This decision can be critical as it decides the amount of 'freedom' that the network has. A network with too little freedom is not capable of learning the training data. A network with too much freedom will simply remember the training data and not generalise properly, a condition known as *overtraining*, which is analogous to an over-prescribed polynomial fit.

Model updating using Neural Networks
Various workers are beginning to experiment with neural networks for model updating purposes. Atalla and Inman [326] decided to use RBF networks, and managed to update a 3-DOF problem successfully using experimental frequency domain data. Levin and Lieven [327] again decided to use RBF networks and managed to update a twenty p-value problem successfully, using simulated noisy modal data.

Obviously, to update a dynamic FE model it is first necessary to obtain experimental data from the physical model. However, even for small models the FRFs contain too many data points to use realistically with neural networks. It is necessary to reduce the number of data points to manageable proportions. One possible technique of achieving this goal is sensitivity analysis of the FRFs to select important data points. This is the approach considered by Atalla and Inman. An alternative approach is to work in the modal domain; this approach is considered in [327]. Note that all training data must be appropriately scaled, or the response of the network will be dominated by a few data points.

The model updating procedure consists of generating training data from the initial FE model by varying the updating parameters. Each training vector consists of the selected modal or frequency information obtained from the FE model, and each target vector consists of the updating parameter settings. A neural network is then trained with these data. The aim is to produce a network that is capable of deciding what updating parameters produced the experimental data shown to the network. After an appropriate network has been trained, the experimental data are applied to the network, resulting in suggested updating parameters that may be applied to the FE model. It is advisable to iterate the process [327], i.e., update the updated model again using the same procedure. The iterative procedure terminates when no further changes are made to the model.

Before updating can begin it is necessary to choose the number of p-values and the location of each p-value. This choice is a complex problem. If too few p-values are taken then the networks will be unable to generalise over the experimental data, and if too many are chosen then different settings of the p-values may produce almost exactly the same results, an observation of equal relevance to Simulated Annealing and FE sensitivity problems.

It is also necessary to consider what training data will be used to train the network, as the data are expensive to generate. Since Atalla and Inman are updating three parameters they can show training vectors representing every combination of the parameters to the network at little expense. However, the expense of this becomes prohibitive as the number of p-values increases. A more selective approach can be achieved by showing the network the effect of adjusting each p-value alone, and various combinations of p-values.

The neural network updating approach has several advantages over many other updating methods. The most significant advantage is resistance to noise, which is a key requirement for any successful updating method. A further advantage is that there is no prior specification on what the updating parameters should be. Any set of parameters may be chosen, from very low level to very high level parameters, including generic elements, and an appropriate training data set generated.

With the neural network updating method the problem of coordinate incompatibility is neatly side-stepped. The neural networks implicitly reconcile the analytical and experimental models, instead of requiring explicit expansion or reduction.

The main disadvantage of this method is that it is computationally expensive, because a modal solution is required to generate every training vector.

7.6.4   Other methods

It is useful to include here the methods which do not natural fall into any of the previous sections on Correlation. These are techniques which are frequently mentioned in the literature, or are of current interest.

The Error Matrix Method (EMM), proposed by Sidhu and Ewins [328], uses pseudo flexibility and accelerance matrices generated from the truncated m-dimensional modal space. The stiffness error matrix is given by

$$[\Delta K] = [K_A]\left[\left[K_A^*\right]^{-1} - \left[K_X^*\right]^{-1}\right][K_A] \tag{7.79}$$

where

$$\underset{(N\times N)}{\left[K_A^*\right]^{-1}} = \underset{(N\times m)}{[\phi_A]}\underset{(m\times m)}{\left[\,^{\backprime}\omega_A^2\,_{\backprime}\right]^{-1}}\underset{(m\times N)}{[\phi_A]^T}$$

$$\underset{(N\times N)}{\left[K_X^*\right]^{-1}} = \underset{(N\times m)}{[\phi_X]}\underset{(m\times m)}{\left[\,^{\backprime}\omega_X^2\,_{\backprime}\right]^{-1}}\underset{(m\times N)}{[\phi_X]^T} \tag{7.80}$$

and the asterisk denotes pseudo-matrices.

Similarly, the mass error matrix is given by

$$[\Delta M] = [M_A]\left[\left[M_A^*\right]^{-1} - \left[M_X^*\right]^{-1}\right][M_A] \tag{7.81}$$

where

$$\left[M_A^*\right]^{-1}_{\substack{(N\times N)}} = \left[\phi_A\right]_{\substack{(N\times m)}}\left[\phi_A\right]^T_{\substack{(m\times N)}} \qquad \left[M_X^*\right]^{-1}_{\substack{(N\times N)}} = \left[\phi_X\right]_{\substack{(N\times m)}}\left[\phi_X\right]^T_{\substack{(m\times N)}} \qquad (7.82)$$

The method is iterative due to its linearised formulation and is only convergent for small [ΔK] and [ΔM]. The EMM has also been shown to be very sensitive to measurement noise [329] and disregards the connectivity of the original FE model. As a result, the updated matrices are often physically unrealisable from which no real eigensolution can be found.

Ibrahim's Two Response method [330] is algebraically equivalent to the Response Function Method although the equations are expressed in terms of the separate input and output data rather than transfer functions. The difference lies in the fact that the solution is not chosen on a least-squares basis over a selection of frequency points. Instead, just two frequency points are chosen yielding two independent sets of updating equations. Solutions to each set are sought such that (i) the two solutions differ only by a scaling factor; and (ii) the scaling factor is as close to unity as possible.

Inman [331] proposes a constrained eigenstructure assignment method which in effect is a pole placement technique for eigenvalue adjustment adopted from control theory. In the context of updating, the unstable poles are the analytical eigenvalues that are mapped onto the stable - measured - poles by defining a closed loop system. An inconvenience associated with the method is that the input and output variables are defined in terms of displacement and velocity, as dictated by control law. This means that the associated transformation matrices refer to a first order ordinary differential equation rather than the second order systems generally used in structural dynamics. This in turn means that the selection of the transformation matrices is difficult to associate with physical parameters and does not guarantee the integrity of the final system matrices. However, the method does ensure that the updated eigensolution is a perfect match for the experimental data.

Another school of combined location and correlation techniques is based on Dynamic Reaction methods. Methods of this type [332-334] explore the residues in the variational expression of the equilibrium equations and also the experimental mode shapes, in the resolution of a series of linear static problems. The eigenvalues and eigenvectors obtained from the experiments are considered as exact, the difference between analytical and experimental results being expressed by non-zero reaction forces over the degrees-of-freedom of the structure. The calculation of the correcting parameters represents, at each iteration, a quadratic optimisation problem without constraints.

A different approach, due to the works of Ladevèze and Reynier [335-338], is based on the establishment of a distance between theory and tests through the definition of an error measure on the constitutive relation. For the model of the real structure, the reference solution is constituted by a displacement field verifying the kinematic constraints and a stress field verifying the equilibrium equation in free vibration. In terms of a finite element discretisation, the problem results in the minimisation of the following error:

$$E^2(\{U\},\{V\}) = \{U - V\}^T [K] \{U - V\}$$

$$+ \frac{r}{1-r}\{[\Pi]\{U\} - [\Pi]\{V_X\}\}^T [K_r]\{[\Pi]\{U\} - [\Pi]\{V_X\}\} \quad (7.83)$$

under the constraint

$$[K]\{U\} = \omega_X^2 [M]\{V\} \quad (7.84)$$

where $\{U\}$ is the displacement field associated with the stress field at the nodes of the elements, $\{V\}$ is the displacement field (also at the nodes) verifying the kinematic constraints, $[\Pi]$ is a projection operator allowing the extraction of the measured part of the displacements from the column of the generalised ones, $[K_r]$ is the condensed stiffness matrix (e.g., via a Guyan reduction), the subscript x means experimental and r is a factor (between 0 and 1) that takes into consideration the confidence in the experimental results. From the minimisation of (7.83), $\{U\}$ and $\{V\}$ are known, and an indicator of the localisation of the errors is given by:

$$\Gamma_s = \sum_{k=1}^{q} \left(\frac{\|\{U\} - \{V\}\|_s^2}{\frac{1}{2}\left(\|\{U\}\|^2 + \|\{V\}\|^2\right)} \right)_k \quad (7.85)$$

where Γ_s is the error at each element, $\|.\|$ are energy norms and the summation reflects the contribution of the various mode shapes (q).

After locating the errors, an optimisation process is undertaken to correct those errors, using a conjugate gradient algorithm. Extension of this method to incorporate results from forced vibration tests using FRFs can be found in [339, 340].

7.7   CONCLUDING REMARKS

Finite element updating in its simplest sense is the minimisation of differences between the experimental and the analytical model. This implies that a norm should be created which represents the variable(s) to minimise. Correlation methods have been defined in time, frequency, modal and spatial domains. The process of minimisation takes place by the application of an optimisation procedure to the norm. Many optimisation techniques have been investigated by workers in the field: Lagrange multipliers, least-squares, Simplex, Genetic Algorithms and Simulated Annealing to name but a few. It should be borne in mind that the correct solution using any of these techniques will be unique. Thus, the solution process has to find the global minimum of the norm with respect to the updating parameters. Unfortunately, application of the norm alone usually provides an error surface in parameter space, which either:

(i) undulates aggressively causing the optimisation process to find only a local minimum; or

(ii) is too shallow and leads to a solution which although close in terms of the norm is a long way from the correct selection and quantification of updating parameters.

It appears that the condition of the solution can be improved by adding constraints to the norm. Baruch did this by Lagrange multipliers at the outset of model updating. More recently, penalty functions have been used to reflect the experimental limitations. On this basis, model updating requires two components:

(i) the definition of an objective function which includes constraints so that its error surface clearly depicts a global minimum; and

(ii) an optimisation procedure which can accurately locate the global minimum of the objective function.

In isolation many appropriate techniques already exist, for example regularisation for improved conditioning and Simulated Annealing for optimisation. Perhaps the way forward is to draw on the best parts of individual 'schools' of correlation and amalgamate them into a common strategy. The choice of those components is left to the reader.

CHAPTER 8

Nonlinear Modal Analysis

8.1   INTRODUCTION

Most of the theories upon which structural dynamic analysis is founded rely heavily on the assumption that the dynamic behaviour of the structure is linear. By this it is meant that (i) if a given loading is - for example - doubled, the resulting deflections are doubled and (ii) the deflection due to two (or more) simultaneously applied loads is equal to the sum of the deflections caused when the loads are applied one at a time. This superposition principle of linear systems can be mathematically expressed as

$$x\left(\alpha\,f(t)\right) = \alpha\,x\left(f(t)\right) \tag{8.1}$$

$$x\left(\sum_{k=1}^{n} f_k(t)\right) = \sum_{k=1}^{n} x\left(f_k(t)\right) \tag{8.2}$$

where x is the deflection, f(t) is the loading force and α is a constant. Linear mathematical models of structures based on this superposition principle have been proven very useful in numerous engineering applications. From general theoretical considerations of the superposition principle, successful methods have been developed and applied to the dynamic analysis of linear structures as discussed in detail in the previous chapters.

Failure to obey the superposition principle implies that the structure is nonlinear. In fact, most practical engineering structures have a certain degree of nonlinearity due to nonlinear dynamic characteristics of structural joints, boundary conditions and material properties. For practical purposes, in many cases, they are regarded as linear structures, whenever the degree of nonlinearity is small and therefore insignificant in the response range of interest, whereby linear modal analysis methods can be applied to analyse their dynamic characteristics.

For other structures, however, the effect of nonlinearity may become so significant that it has to be taken into account and nonlinear modal analysis methods have to be used instead.

Similarly to linear modal analysis, the main objective of nonlinear modal analysis is to establish, from measured vibration test data, a mathematical model of a structure. Since such a structure does not obey the superposition principle, its mathematical model becomes non-unique, being dependent on vibration amplitude. Corresponding to different vibration amplitude levels, one requires, in theory, different mathematical models to describe a nonlinear structure. Because of this amplitude dependency, special considerations will have to be made during measurement and analysis. It is often supposed that unless a real measurement is taken, the existence of nonlinearity in a practical structure cannot be foreseen based on analytical prediction, nor can the degree of nonlinearity be analytically quantified. Experimental investigation becomes essential in the identification of dynamic characteristics. In fact, a nonlinear modal analysis process starts at the stage of frequency response function measurement. In order to reveal satisfactorily the hidden structural nonlinearity(ies), the range of excitation forces to be applied in a test will have to be carefully selected. The choice of the excitation method becomes very important and will have to be made in accordance to the required subsequent analysis.

The existence of nonlinearities also adds complications to the procedures involved in modal analysis. Before starting to analyse measured FRF data, one needs to determine whether the nonlinearity in the measurement range is strong enough to warrant a nonlinear modal analysis. Once it is confirmed as necessary, the choice of the method to use depends on the type of measured FRF data. After having identified the modal parameters (which are functions of the vibration amplitude), further procedure is needed in order to possibly establish the physical characteristics of the existing nonlinearity(ies). Though nonlinear modal analysis is far more difficult and complicated, in comparison with conventional linear modal analysis, significant progress has been made over the past years and many practical and useful methods have been developed.

This chapter presents some of the major work done in recent years on detection, identification and location of nonlinearities, based on measurement of the input/output dynamic characteristics of real structures. First, we shall examine existing techniques for the detection and identification of nonlinear behaviour using measured first-order FRFs. Then, discussions on the identification of nonlinearity using higher-order FRFs will be made. Finally, the problem of detecting localised nonlinearity(ies) by correlating analytical finite element models and measured vibration test data will be addressed.

8.2 NONLINEAR MODAL ANALYSIS BASED ON FIRST-ORDER FREQUENCY RESPONSE FUNCTIONS

8.2.1 Measurement of first-order FRFs

Since the main objective of nonlinear modal analysis is to build mathematical models of nonlinear structures based on measured input/output dynamic characteristics, it becomes necessary, prior to the introduction of any identification techniques, to discuss how the dynamic characteristics of these kinds of structures can be measured. First of all, it is necessary to explain what is meant by the so-called first-order frequency response functions. In concept, first-order FRFs are the extension of FRFs of linear systems to nonlinear ones. As with the measurement of FRFs, in the case of sinusoidal excitation, the first-order FRF $H_1(\omega)$ of a nonlinear structure is defined as the spectral ratio of the response $X(\omega)$ to the force $F(\omega)$ at the excitation frequency: $H_1(\omega) = X(\omega)/F(\omega)$. During the estimation of $H_1(\omega)$, all the harmonic components (subharmonics, superharmonics and combinational resonances) are ignored and only the fundamental frequency component of the response is retained. As for the linear case (see Section 2.3.1), for random or transient excitation, the first-order FRF is defined as the ratio of the cross-spectrum of the force and response to the auto-spectrum of the force: $H_1(\omega) = S_{fx}(\omega)/S_{ff}(\omega)$ (or its equivalent form $H_1(\omega) = S_{xx}(\omega)/S_{xf}(\omega)$). The first-order FRFs of a nonlinear structure measured using sinusoidal and random excitations are in general different and their relationship will be discussed later.

For linear structures, the first-order frequency response functions (often referred to simply as frequency response functions) are unique and, therefore, will not vary according to different excitation techniques and conditions. For nonlinear structures, however, the measured first-order FRFs are, in general, non-unique. They depend not only on the excitation conditions (input force levels), but also on the different excitation signals used to measure them. Therefore, the first problem will necessarily be to decide on a proper means of excitation so that nonlinearity can easily be revealed and then identified.

There are three types of excitation methods widely used in vibration study practice: sinusoidal, random and transient. Although this subject has already been addressed and discussed previously (mainly in Chapters 2 and 3), it seems important to recall it here, so that the comparison between the use of those main types of excitation on both linear and nonlinear structures becomes clearer.

Sinusoidal excitation technique
In testing a linear structure, if the input is a sinusoid, the response will also be a sinusoid with the same frequency as that of the excitation, and the FRF at this excitation frequency is simply the ratio of the amplitudes (usually complex) of the response and input signals. This observation of its special characteristics naturally made the sinusoidal excitation the first choice of excitation signal at the very beginning of structural dynamic testing; it still remains one of the most preferred excitation techniques in today's modal testing practice because of its uniqueness and precision, although other techniques, such as those based on random and transient excitations, have also been developed.

The main advantages of sinusoidal excitation are: (i) the input force level can be accurately controlled, hence it becomes possible to excite the structure at specified required response levels and (ii) since all the input energy is concentrated at one frequency each time, and the noise and harmonic components in the response signal are averaged out through an integration process, the signal-to-noise ratio is generally good in comparison with other excitation methods. As in most cases the study of nonlinearity requires either response or force control, the characteristics of (i) become important in the successful identification of structural nonlinearity.

When the response level is set to be constant during a measurement (response amplitude is kept constant at different excitation frequencies), a nonlinear structure is said to be linearised and the measured first-order FRFs can be analysed using standard linear modal analysis methods in exactly the same way as for the FRFs measured from a linear structure. On the other hand, when the input force is kept constant during a measurement (the amplitude of the input force is set to be constant at different excitation frequencies), the measured first-order FRFs are nonlinear (they are characteristically different from FRFs measured from linear structures) and in this case, special nonlinear methods have to be used to analyse them.

The main drawback of the sinusoidal excitation technique is that it is relatively slow when compared with many of the other techniques used in modal testing practice. The reason for this is that the method is based on frequency by frequency measurements and, at each frequency, time is needed for the transient response components to decay and the system to settle to its steady-state vibration. However, it is believed that in many applications, correct measurement of the dynamic characteristics of a structure becomes more important than the measurement time involved, especially in cases where the structure to be tested is nonlinear, and detailed analysis of its nonlinear characteristics is required.

Measurement using random excitation

The term 'random' applies to the amplitudes of the excitation force which, in statistical terms, have a Gaussian or Gaussian-like probability distribution. Wide-band random excitation is widely used in structural dynamic testing because it approximates more closely the statistical characteristics of vibration service environments than does a pure sinusoidal excitation.

With this type of excitation, individual time records in the analyser contain data with random amplitude and phase for each frequency component. On average, however, the spectrum is flat and continuous, containing energy at approximately the same level for every frequency in the range of interest. Spectrum distribution is easy to control in a random test, and it can be limited to cover the same frequency range as that used in the analysis.

Excitation is random and continuous in time, but record length is finite, and so the recorded signals (force and response) are, in general, nonperiodic. However, during the signal analysis, these nonperiodic signals are assumed to be periodic and as a result leakage errors occur in the estimation of FRFs, as already explained in Section 2.2. These errors can be minimised by using window functions, or weighting, which act as a soft entry and exit for the data in each record.

A suitable weighting function to use with random data is the Hanning window. In order to eliminate the leakage problem, a pseudo-random excitation signal can be used instead of a true random signal. A pseudo-random signal is periodic and repeats itself with every record of analysis. A single time record of a pseudo-random signal resembles a true random wave form, with Gaussian-like amplitude distribution. However, the spectral properties are quite different from those of a random signal because of its periodicity. First, the periodic nature of the pseudo-random signal removes the leakage error entirely so that a rectangular window must be used in the analysis and secondly, the spectrum becomes discrete, only containing energy at the frequencies sampled in the analysis. For measurement of linear systems, random and pseudo-random excitations are attractive to researchers and practitioners because of their potential time-saving in obtaining FRFs. In a random and pseudo-random excitation measurement, the structure is excited simultaneously at every frequency within the range of interest. It is this wide-band excitation characteristic that makes the random and pseudo-random excitation faster than sinusoidal excitation.

From the measured first-order FRF point of view, a random test in general linearises nonlinear behaviour due to the randomness of the amplitude and phase of the input force signal and the effect of spectrum averaging. As a result, the measured first-order FRFs are linear. However, linearisation of a nonlinear structure when using random excitation does not mean that it is impossible to identify nonlinearity using this type of excitation. Testing at different input excitation levels (power spectra), the corresponding measured linearised first-order FRFs are different; and if a set of these FRFs are taken, identification of a nonlinearity could, in some cases, become possible. Also, as will be discussed later, the conventional random test technique can be extended to measure higher-order FRFs of a nonlinear structure, providing valuable information concerning the nature of the nonlinearity and that can be analysed to achieve the objective of nonlinearity identification.

Pseudo-random excitation, on the other hand, is in general not suitable for the first-order FRF measurement of nonlinear structures. This is because a pseudo-random signal is periodic and so contains limited discrete frequency components. When such an input signal is applied to a nonlinear structure, modulation and intermodulation distortion will be generated due to nonlinearity and, unfortunately, these distortion products (e.g. $2f_1$, $3f_1$, ... due to modulation of input component at frequency f_1) will fall exactly on the other frequency components of the input signal (e.g. $2f_1$, $3f_1$, ...). Thus, distortion products add to the output and therefore interfere with the measurement of FRFs [341]. Unlike random excitation, in which these distortions can be averaged out, pseudo-random is periodic, and so averaging has no effect on the measured FRF.

Although the first-order FRFs measured using random excitation are different when the input force spectrum levels are different, these differences could be very small when practical tests are considered. One reason for this is the dropout of the input force spectrum around resonance frequencies due to the impedance mismatch between the test structure and the electromagnetic shaker. Since the energy input around structural resonance(s) is mainly responsible for the vibration level of a structure, dropout of the input force spectrum around resonance(s) means that the structure cannot be easily driven into its very nonlinear regime, and the measured

FRFs corresponding to different input force levels will not, in general, be very different from one another. With sinusoidal excitation, this impedance mismatch can be compensated using a feedback control system, but for random excitation, such a compensation seems to be difficult and this is a practical problem for the identification of nonlinearity using a random test.

Measurement using transient excitation

One of the most popular excitation techniques used nowadays in structural dynamic testing is transient excitation, sometimes referred to as 'impact testing'. This popularity is due to transient excitation having certain unique characteristics when compared to shaker-based excitation techniques. The main advantages of using transient excitation can be summarised as:

(i) transient excitation does not require a shaker to generate the input excitation force; this is usually produced using an impactor such as a hand-held hammer, and thus the structure remains unmodified during the test;

(ii) because there is no attachment required in the test, transient excitation provides easier access to the measurement points of the structure; and

(iii) transient excitation requires less equipment (no shaker and its related power amplifier involved) and measurement time; therefore, it is ideal for mobile experiments.

An ideal impulse is the delta function $\delta(t)$ which, after being Fourier transformed, produces a force spectrum with equal amplitude at all frequencies. Unfortunately, this ideal impulse is practically impossible to achieve. The waveform which can be produced by an impact is a transient (short time duration) energy transfer event whose spectrum is continuous, with a maximum amplitude at zero frequency and amplitude decaying with increasing frequency. The spectrum shape of the transient signal is mainly determined by the time duration of the signal. The shorter the time duration of the signal, the broader the range of energy distribution in the frequency domain. On the other hand, the time duration of an impact is determined by the mass and stiffness of both the impactor and the structure. Therefore, by properly choosing the material of the hammer tip and its mass, it is possible to generate the required transient signal with desired spectrum characteristics. However, the spectrum can only be controlled at the upper frequency limit, which means the technique is not suitable for zoom analysis.

Although it has been suggested that the high crest factor of transient excitation makes it possible for the nonlinear behaviour of a structure to be provoked and then possibly identified, there has not been much evidence so far which seems to support the advantage of using transient excitation to identify structural nonlinearity based on the measured first-order FRFs.

Comments on practical nonlinearity measurements

As discussed above, there are three main types of excitation techniques available for the vibration testing of a structure. Each of them has its advantages and disadvantages and a proper choice of the excitation technique depends, in general, on the measurement accuracy required and time available for the test. For linear structures, since the measured FRFs are, in theory, unique and independent of the excitation, all techniques should be equally applicable. For nonlinear structures, however, the choice of excitation becomes important for the hidden nonlinearities to be revealed and then identified because, in this case, the measured dynamic properties are excitation-dependent.

Transient excitation is one of the most often-used techniques in structural dynamic testing because of its simplicity and speed in obtaining FRFs. It requires less equipment and is therefore suitable for mobile experiments. Since there is no shaker involved, the structure remains unmodified during the test. The coherence functions obtained from transient tests, being an indication of the measurement quality, are usually better than those from random tests in the sense that low coherence only occurs at antiresonances due to the low signal-to-noise ratio of the response signal. In contrast, in the case of random excitation, low coherence occurs not only at antiresonances but also at resonances, due to the dropout of input force spectrum around them, caused by impedance mismatch between the test structure and shaker. As for the identification of nonlinearity, although it is believed that it might be possible to use transient excitation because of its high crest factor which provokes the structural nonlinearity, few studies have been carried out to support this argument.

When random excitation is used, the measured first-order FRFs are always linear, whether the structure is linear or not. In the case when the test structure is linear, the measured FRFs are unique and will not vary according to different excitation levels, while if the test structure is nonlinear, a series of linearised first-order FRFs will be obtained corresponding to different excitation levels. These measured first-order FRFs can be used to detect whether a structure is linear or not by comparing their values for different excitation levels. In cases where only an approximate linear model of a nonlinear structure is of interest, regardless of the type of nonlinearity the structure possesses, these linearised FRFs can often provide an accurate linear approximation of the nonlinear structure from a response prediction point of view. Moreover, conventional random excitation technique can be extended to the case of higher-order FRF measurement based on the Wiener theory of nonlinear systems [342]. As will be discussed later, higher-order FRFs can be used in some cases to identify the type of structural nonlinearity and, together with the measured first-order FRFs, to predict responses due to certain inputs more accurately than those obtained using measured first-order FRFs alone.

In the case where accurate quantification of structural nonlinearity is required, e.g., how the modal and/or spatial model of a nonlinear structure will change at different vibration response levels, sinusoidal excitation is generally regarded as the best choice because of its flexibility of input force level control. There are two different types of controlled sinusoidal measurement technique commonly used for testing a nonlinear structure, referred to as the 'constant response' and 'constant force' measurement procedures. In constant response measurements, the response amplitude of the nonlinear test structure at a certain coordinate is kept constant at different excitation frequencies by adjusting the input force levels and, as a result, the measured first-order FRFs are linear. However, corresponding to different response levels, the measured first-order FRFs are different; by analysing them using linear modal analysis methods, a relationship between modal model and response levels can be established. The problem here is that measurement is extremely time-consuming and therefore expensive. Also, the measured range of response amplitude, which is important in nonlinearity analysis, could be limited because of the dramatic changes of receptance amplitude around resonances, especially when the structure is very lightly damped.

In the case of constant force measurements, the amplitude of the input force is kept constant at each of the different excitation frequencies. Due to the varying receptance amplitudes, the response amplitudes are different at different measurement frequencies and, therefore, the measured first-order FRFs are nonlinear and contain information of a series of linearised FRFs measured at constant response amplitudes. Such nonlinear first-order FRFs are used from time to time in nonlinearity investigations and it will be shown later that they can be analysed based on nonlinear modal analysis methods developed to establish the relationship between the modal model and response levels of the structure.

In practical measurements, because of the existence of different types of nonlinearity, care must be taken in determining the necessary excitation range so that nonlinearity(ies) can be exposed to a satisfactory extent. In general, they can be categorised into four different types. For the majority of nonlinearities commonly encountered in practice (either stiffness or damping nonlinearities), increasing the excitation force level will be similar to increasing the degree of nonlinearity. Examples of such situations are cubic stiffness and quadratic damping. In some cases such as backlash, the structure will remain linear until its response exceeds a certain limit. For frictional damping, on the other hand, increasing the excitation level will decrease the degree of nonlinearity and for some nonsymmetric nonlinearities such as bilinear and quadratic stiffness, the nonlinearity will have no effect on the measured first-order FRFs; in order to identify these special types of nonlinearity, the introduction of higher-order FRFs becomes necessary.

From these observations, it follows that it is clearly important to choose the response range and thus the required excitation range properly, so that nonlinearity can be exposed and then identified satisfactorily. For most nonlinear structures in practice, relatively high excitation levels are recommended. For certain others, such as structures with frictional damping, the situation can be the reverse, and in order to identify such nonlinearities, excitation levels should be set as low as possible. Lastly, for structures whose nonlinearities have no effect on the measured first-order FRFs, measurement of higher-order FRFs is recommended.

8.2.2 Nonlinear modal analysis methods

Once a structure is suspected of being nonlinear and its first-order FRFs are measured as discussed above, it becomes necessary to take nonlinearity into account in the subsequent modal analysis of the data. In practical analysis, in order to understand the nature and extent of the nonlinearity, three requirements must be achieved by appropriate application of modal analysis methods. First, the existence of nonlinearity needs to be detected. Second, the extent of the nonlinearity needs to be quantified. Finally, the physical characteristics of the nonlinearity need to be established.

It is believed that the first requirement (detection) is comparatively easy to achieve. In fact by simply comparing the differences between measured FRFs at different excitation levels the existence of nonlinearity can be discovered. The quantification of nonlinearity, that is the establishment of the relationship between modal parameters and response levels, can be achieved by analysing measured first-order FRFs. The last objective and the most difficult task in nonlinear modal analysis is the identification of the physical characteristics of nonlinearity - the relationship between the structural spatial properties (such as stiffness) and response amplitudes. The establishment of such a nonlinear spatial model can only be achieved by correlating between an analytical model and vibration test data.

Trying to achieve these various objectives of nonlinearity analysis, extensive research work has been carried out and a number of papers have been published in recent years. Methods which are commonly used in practical nonlinear modal analysis are reviewed and their advantages and disadvantages are discussed next.

Detection of structural nonlinearities

The first important task in the analysis of nonlinear structures is the detection of the existence of nonlinearity. If the effect of structural nonlinearity is not so obvious and the structural behaviour is totally dominated by linear characteristics, then the introduction of nonlinear analysis should, perhaps, not be recommended. Nonlinear modal analysis becomes necessary only when a structure is found to be quite nonlinear. To execute such a task as nonlinearity detection, many different techniques have been developed over the years. The simplest methods are those which are based on the use of raw measured first-order FRF data without any post-measurement data processing. Detection techniques of this type are the Overlaid Bode Plot method [3] and the Inverse Receptance method [343]. Other detection methods which require more complicated post-measurement data processing are the Isometric Damping Plot technique [3] and the Hilbert Transform method [344].

Overlaid Bode Plot method

The basis of using the Overlaid Bode Plot method is that, due to the existence of nonlinearities, the measured first-order FRFs will be distorted systematically from the corresponding linear FRFs. Since linear FRFs are very well recognised, the existence of a nonlinearity can be revealed by examining the abnormal behaviour of the measured first-order FRFs. In the case of sinusoidal excitation, the measured FRF data are obtained using different constant force levels or different constant response levels. Then, the Bode plots of those measured FRFs are overlaid and the existence of a nonlinearity can be established.

To illustrate this method, some FRFs from a practical nonlinear structure were measured. As shown in figure 8.1, when the excitation force level increases, the distortion in the measured FRF data also increases and the apparent resonance frequency (the frequency of maximum FRF amplitude) drops. The existence of a softening stiffness nonlinearity in the structure is clearly demonstrated.

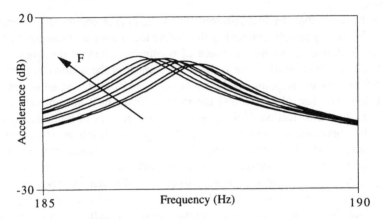

Fig. 8.1 Measured first-order FRFs of a practical nonlinear structure.

Inverse Receptance method

As an alternative but more versatile technique, FRF data can also be displayed in inverse form to detect the existence of nonlinearity. It is nothing more than the Inverse method (see Section 4.4.1), for real modes, applied to nonlinear behaviour detection. The advantage of displaying data in this format is that, in cases where the modal constant of the mode to be analysed is effectively real, this technique can not only detect the existence of nonlinearity, but can also give an indication as to whether the nonlinearity exists in the stiffness or damping. Suppose the residual contribution of other modes has been subtracted or can be neglected. The receptance data around the rth mode of the structure, considering hysteretic damping, can be expressed as (see Section 1.4.2):

$$\alpha_{jk}(\omega) = \frac{{_r}A_{jk}}{\omega_r^2 - \omega^2 + i\eta_r\omega_r^2} \tag{8.3}$$

Rewriting (8.3) in its inverse form and assuming the modal constant ${_r}A_{jk}$ to be real (and writing the receptance as α_r, for simplicity),

$$\frac{1}{\alpha_r} = \frac{\omega_r^2 - \omega^2}{{_r}A_{jk}} + i\frac{\eta_r\omega_r^2}{{_r}A_{jk}} = \text{Re}\left(\frac{1}{\alpha_r}\right) + i\,\text{Im}\left(\frac{1}{\alpha_r}\right) \tag{8.4}$$

From (8.4), it can be seen that if the FRF data are expressed in their inverse form, ω_r (related to stiffness nonlinearity) and η_r (related to damping nonlinearity) can be identified separately from the real and imaginary parts of the inverse receptance.

When FRF data are obtained from linear structures, the relationships $Re(1/\alpha_r)$ *versus* ω^2 and $Im(1/\alpha_r)$ *versus* ω^2 are straight lines for the case of hysteretic damping as shown in (8.4). Any distortion from a straight line gives indication of the existence of nonlinearity.

When FRF data from nonlinear structures are to be analysed, the effect of a stiffness nonlinearity will show up in the real part of the inverse of receptance data while the effect of a damping nonlinearity will appear only in the imaginary part of the data.

To illustrate this point, FRF data with stiffness and damping nonlinearity measured from simulated analogue circuits are analysed and the real and imaginary parts of the inverse FRF *versus* ω^2 are shown in figure 8.2. In the case of stiffness nonlinearity, distortion of the inverse FRF data only appears in the real part. On the other hand, when damping nonlinearity is considered, its effect is clearly observed to be confined to the imaginary part of the inverse receptance data.

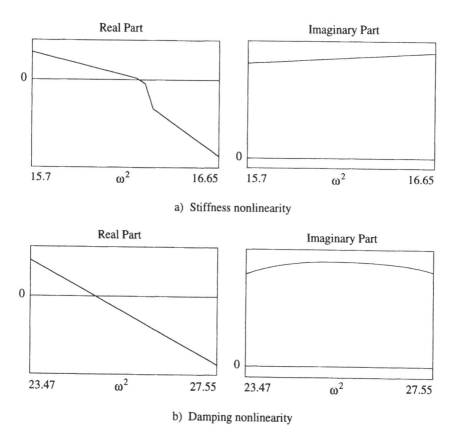

a) Stiffness nonlinearity

b) Damping nonlinearity

Fig. 8.2 Inverse Receptance of nonlinear FRF data.

Isometric Damping Plot technique

As already mentioned in Section 4.4.1, in the presentation of the Circle-fitting method, it has been established [127] that structural nonlinearity can be detected by inspection of the isometric damping plots (see figure 4.7) which can be calculated from measured FRF data. The argument which supports this technique is generally believed to be that structural nonlinearity (usually stiffness nonlinearity) distorts the spacing of frequency response data points around the Nyquist circle from their positions in a linear situation. Since the distortion caused by nonlinearity is systematic, the consequent distortion of the damping estimate plot using different pairs of points around the Nyquist circle will display a specific pattern depending on the type of nonlinearity. These patterns in the damping plot can then be recognised and compared to detect and possibly to identify the type of nonlinearity.

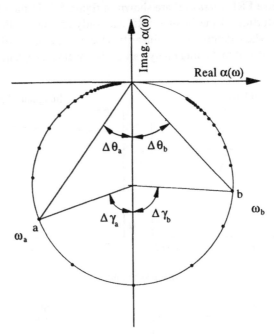

Fig. 8.3 Nyquist circle of receptance data.

Suppose that the residual effect of other modes can be neglected and that the modal constant is effectively real for the mode to be analysed. Then, the receptance around the rth mode can be expressed as in (8.3). When α_r is plotted in the Nyquist plane, a circle as shown in figure 8.3 is obtained. If the data are measured on a linear structure, then the damping loss factor of the rth mode, η_r, can be calculated using the following expression:

$$\eta_r = \frac{\omega_a^2 - \omega_b^2}{\omega_r^2} \frac{1}{tg(\Delta\theta_a) + tg(\Delta\theta_b)} \tag{8.5}$$

which was deducted in Section 4.4.1 (expressions (4.168) to (4.171)).

When different combinations of points (ω_b, ω_a) are used, a flat plane, which is the surface plot of the estimated damping loss factors $\eta_r(\omega_b, \omega_a)$ against its two variables ω_b and ω_a, can be obtained in the case of linear FRF data. However, if the measured FRF data are from a nonlinear structure, distortion of the isometric damping plot (no longer a flat surface) calculated based on (8.5) will, in general, be expected as shown in figure 8.4.

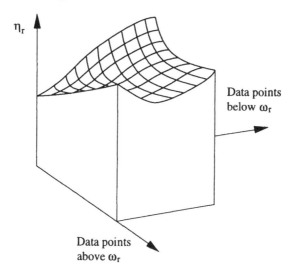

Fig. 8.4 Damping plot obtained from nonlinear FRF data.

Hilbert Transform method

The Hilbert transform (see, for example [344]) is an integral transform, defined as the convolution of a real-valued function f(t) extending from $-\infty \le t \le +\infty$, with $1/\pi t$:

$$\tilde{f}(t) = \mathscr{H}\left[f(t)\right] = f(t) * 1/\pi t \tag{8.6}$$

where $*$ means convolution.

The result, $\tilde{f}(t)$, is also a real-valued function and it should be noted that, unlike the Fourier transform, the Hilbert transform does not change the domain of the function: a function of time is transformed into another function of time, a function of frequency is transformed into another function of frequency.

The definition of convolution between A(t) and B(t) is C(t) such that:

$$C(t) = A(t) * B(t) = \int_{-\infty}^{+\infty} A(\tau) B(t-\tau) d\tau \tag{8.7}$$

where τ is a dummy, real-valued variable. Thus, (8.6) becomes:

$$\tilde{f}(t) = \mathscr{H}[f(t)] = \int_{-\infty}^{+\infty} \frac{f(\tau)}{\pi(t-\tau)} d\tau \tag{8.8}$$

Similarly, in the frequency domain, we have

$$\tilde{f}(\omega) = \mathscr{H}[f(\omega)] = \int_{-\infty}^{+\infty} \frac{f(\Omega)}{\pi(\omega-\Omega)} d\Omega \tag{8.9}$$

If $f(z)$ is a function of complex variable, the expression for the Hilbert transform can be deduced from Cauchy's integral formula, which is

$$f(z) = \frac{1}{i2\pi} \oint_C \frac{f(\alpha)}{\alpha - z} d\alpha \tag{8.10}$$

where C is an appropriate contour in the complex plane, where $f(z)$ must be analytic.

When (8.10) is applied to the frequency response function $H(\omega)$ of a linear and stable system, it can be shown that

$$H(\omega) = -\frac{1}{i\pi} PV \int_{-\infty}^{+\infty} \frac{H(\Omega)}{\Omega - \omega} d\Omega \tag{8.11}$$

where PV means principal value in Cauchy's sense, i.e., avoiding the singularity at $\omega = \Omega$. As $H(\omega) = \text{Re}(H(\omega)) + i\,\text{Im}(H(\omega))$, (8.11) gives

$$\text{Re}(H(\omega)) = -\frac{1}{\pi} PV \int_{-\infty}^{+\infty} \frac{\text{Im}(H(\Omega))}{\Omega - \omega} d\Omega \tag{8.12}$$

$$\text{Im}(H(\omega)) = \frac{1}{\pi} PV \int_{-\infty}^{+\infty} \frac{\text{Re}(H(\Omega))}{\Omega - \omega} d\Omega \tag{8.13}$$

Taking advantage of the even and odd nature of the real and imaginary parts of the FRF, respectively, (8.12) and (8.13) can be expressed using as limits of integration 0 and $+\infty$:

$$\text{Re}\,(H(\omega)) = -\frac{2}{\pi} PV \int_0^{+\infty} \frac{\text{Im}\,(H(\Omega))\Omega}{\Omega^2 - \omega^2} d\Omega \tag{8.14}$$

$$\text{Im}\,(H(\omega)) = \frac{2\omega}{\pi} PV \int_0^{+\infty} \frac{\text{Re}\,(H(\Omega))}{\Omega^2 - \omega^2} d\Omega \tag{8.15}$$

known as Kramer-Kronig's relationships.

It is therefore clear that for a causal, linear and stable system the real and imaginary parts of an FRF are related between themselves and thus, each of them is enough to represent the behaviour of such a system, as one can be evaluated from the other. Therefore, the philosophy of the application of the Hilbert transform in the detection of nonlinear behaviour is that if the system is nonlinear, then the calculation of the real part of the FRF from the imaginary part (or *vice versa*) should not reproduce the true measured real part results.

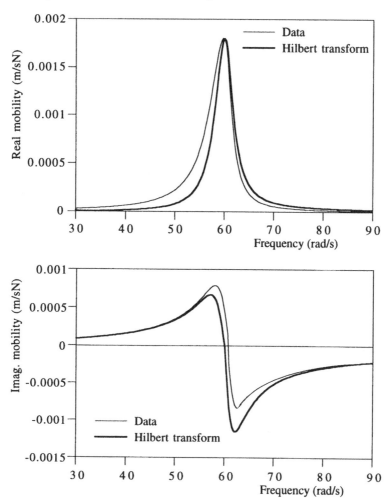

Fig. 8.5 Use of the Hilbert transform to detect nonlinearity.

There is, however, a problem in applying this concept to practical cases. As expressions (8.14) and (8.15) reveal, the Hilbert transform has a kind of non-local nature, as to have (for example) the real part at a particular frequency ω, one has to integrate from zero to infinity. Although these expressions are easy to compute numerically, in a real structure only a limited frequency range can be measured. As a consequence, unless some kind of correction is introduced to take care of the response outside the measured frequency range, we can easily have a misleading result: the calculated real part does not match the measured one due to the out-of-range response influence and not due to a true nonlinear behaviour.

An application example is shown in figure 8.5, for a hardening spring type of nonlinearity, in terms of real and imaginary FRF mobilities, as is usual in this type of application.

8.2.3 Quantitative analysis of nonlinearity

Once the existence of structural nonlinearity is established, the next thing one wants to do is to quantify the extent of such a behaviour. To establish a quantitative measure of nonlinearity, several methods have been developed. For a nonlinear structure, its modal parameters are in general vibration amplitude dependent. Therefore, it will be necessary to establish relationships between nonlinear modal parameters and their corresponding vibration amplitudes. Such relationships are important especially from a design point of view, as one can predict variations in natural frequencies, damping and mode shapes when structural vibration amplitude changes. Two methods of nonlinearity quantification which are often used in practice are discussed here: these are the Inverse Receptance method [345] and the general Nonlinear Complex Mode method [346].

Inverse Receptance method

As discussed in the previous section, nonlinearities in FRF data will cause distortion on the inverse receptance data plot and this has been employed for the detection of the existence of nonlinearity. These inverse receptance data can be further used to quantify nonlinearity. For a nonlinear SDOF system, the natural frequency $\omega_n(X)$ and damping loss factor $\eta(X)$ are, in general, response amplitude (X) dependent. With this in mind, the inverse of receptance can be expressed as:

$$\frac{1}{\alpha(\omega)} = \frac{\omega_n^2(X) - \omega^2}{A} + i\frac{\eta(X)\,\omega_n^2(X)}{A} \tag{8.16}$$

where A is the modal constant, assumed as real. Separating (8.16) into its real and imaginary parts gives

$$Re\left(\frac{1}{\alpha(\omega)}\right) = \frac{\omega_n^2(X) - \omega^2}{A} \tag{8.17}$$

$$Im\left(\frac{1}{\alpha(\omega)}\right) = \frac{\eta(X)\,\omega_n^2(X)}{A} \tag{8.18}$$

Suppose that the input force signal $F(\omega)$ is also recorded during the measurement, so that the response amplitude at each frequency can be easily calculated by:

$$X(\omega) = |F(\omega)| \, |\alpha(\omega)|$$ (8.19)

It becomes clear that if the modal constant A can be estimated by some means, then the relationships of $\omega_n(X)$ *versus* X and $\eta(X)$ *versus* X can be calculated based on (8.17) and (8.18), as:

$$\omega_n^2(X) = \omega^2 + A\,\mathrm{Re}\left(\frac{1}{\alpha(\omega)}\right)$$ (8.20)

$$\eta(X) = \frac{A}{\omega_n^2(X)}\,\mathrm{Im}\left(\frac{1}{\alpha(\omega)}\right)$$ (8.21)

where X is the response amplitude at frequency ω, which can be calculated from (8.19). The calculation of the modal constant A in this method is based on a trial-and-error approach until satisfactory results are obtained.

To illustrate the application of the Inverse Receptance method, FRF data measured from an analogue computer circuit with cubic hardening stiffness nonlinearity have been analysed and the results are shown in figure 8.6. The effect of a hardening stiffness nonlinearity is clearly demonstrated, since the identified natural frequency increases with vibration amplitude.

However, the assumption that the modal constant of the mode to be analysed should be real and constant was made during the development of the Inverse Receptance method. Unfortunately, such assumption is generally incorrect, restricting the range of practical applications of the method, as discussed next.

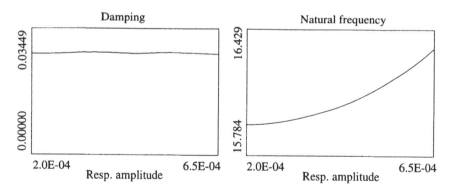

Fig. 8.6 Quantitative identification using the Inverse Receptance method.

Nonlinear Complex Mode method
It is believed that for practical structures, most of the damping comes from joints [347]. Therefore, practical structures possess very non-proportional damping distribution, and genuine complex modes exist. Detailed numerical simulations on the existence of genuine complex modes are given in [3] and experimental evidence

of this fact can indeed be found [348]. Also, for a general nonlinear MDOF structure, since natural frequencies and damping loss factors are assumed to be vibration amplitude dependent, so are modal constants. Thus, in order to develop a method which is applicable to general nonlinear practical structures, it must be assumed that the modal constants of modes to be analysed are in general complex and are vibration amplitude dependent. These observations have led to the development of the general Nonlinear Complex Mode method [347].

In the case of sinusoidal excitation, when a nonlinear structure vibrates at a specific amplitude, there will be an equivalent (linearised) stiffness and damping model as far as the first-order FRF is concerned. Therefore, measured FRF data generally contain information on a series of linear models. What the Nonlinear Complex Mode method seeks to do is to calculate the modal parameters of these linear models together with their corresponding response amplitudes so that the relationship between modal parameters and response amplitude can be established. Due to the nature of resonance, it is always possible to find two frequency points in the measured FRF data - one on either side of the resonance - which have the same (or very similar) response amplitude. These two data points constitute a specific linear model corresponding to that specific response amplitude in the sense that all the modal parameters necessary to determine that linear(ised) model can be calculated by just using these two receptance data points. The thus determined modal parameters are associated with that specific response level. Therefore, if there are many point pairs with different response amplitudes available around that resonance, a relationship between modal parameters of the mode and response amplitudes can be established.

Suppose $\alpha(\omega_1)$ and $\alpha(\omega_2)$ are known to correspond to a certain specific response level X, one on either side of the resonance. Assuming that the residual effect is negligible or has been removed (its influence on analysis accuracy is discussed in detail in [348]), the following two equations can be established:

$$\alpha(\omega_1) = \frac{A_r + i\,B_r}{\omega_r^2 - \omega_1^2 + i\,\eta_r \omega_r^2} \tag{8.22}$$

$$\alpha(\omega_2) = \frac{A_r + i\,B_r}{\omega_r^2 - \omega_2^2 + i\,\eta_r \omega_r^2} \tag{8.23}$$

Since (8.22) and (8.23) are complex algebraic equations, the four modal parameters ω_r, η_r, A_r and B_r can be obtained, giving:

$$\omega_r^2 = \frac{(R_2 - R_1)(R_2\omega_2^2 - R_1\omega_1^2) + (I_2 - I_1)(I_2\omega_2^2 - I_1\omega_1^2)}{(R_2 - R_1)^2 + (I_2 - I_1)^2} \tag{8.24}$$

$$\eta_r = \frac{(I_2 - I_1)(R_2\omega_2^2 - R_1\omega_1^2) + (R_2 - R_1)(I_2\omega_2^2 - I_1\omega_1^2)}{\omega_r^2[(R_2 - R_1)^2 + (I_2 - I_1)^2]} \tag{8.25}$$

$$A_r = \frac{(\omega_2^2 - \omega_1^2)[(R_2 - R_1)(R_2R_1 - I_2I_1) + (I_2 - I_1)(R_1I_2 + R_2I_1)]}{(R_2 - R_1)^2 + (I_2 - I_1)^2} \qquad (8.26)$$

$$B_r = \frac{(\omega_2^2 - \omega_1^2)[(R_2 - R_1)(I_2R_1 - R_2I_1) + (I_2 - I_1)(R_2R_1 - I_2I_1)]}{(R_2 - R_1)^2 + (I_2 - I_1)^2} \qquad (8.27)$$

where R_1, R_2, I_1 and I_2 are defined as:

$$R_1 = Re(\alpha(\omega_1)), \ I_1 = Im(\alpha(\omega_1)), \ R_2 = Re(\alpha(\omega_2)) \text{ and } I_2 = Im(\alpha(\omega_2)) \quad (8.28)$$

These parameters represent the linear model which corresponds to the chosen response amplitude. By examining different point pairs similar to $\alpha(\omega_1)$ and $\alpha(\omega_2)$, the characteristics $\omega_r(X)$, $\eta_r(X)$, $A_r(X)$ and $B_r(X)$ of the original nonlinear structure against vibration amplitude X can be revealed.

The proposed method has been applied to several systems with various types of nonlinear stiffness or damping in order to fully assess its feasibility [348]. To give a practical application, first-order FRF data measured from a practical nonlinear structure as shown in figure 8.1 were analysed using the proposed method. By analysing one of the curves shown in figure 8.1, the relationship between natural frequency and response amplitude can be established as shown in figure 8.7, from which one can conclude that the structure possesses a softening type of stiffness nonlinearity. It is also worth mentioning that in this case, the mode to be analysed is quite complex with mode complexity of about 15°, as shown in the Nyquist plot of the measured FRF data of figure 8.8.

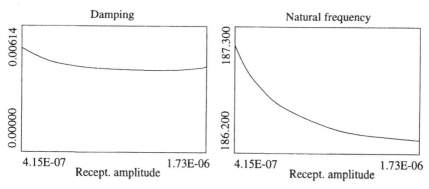

Fig. 8.7 Quantification using the Nonlinear Complex Mode method.

8.3   IDENTIFICATION OF NONLINEARITY USING HIGHER-ORDER FREQUENCY RESPONSE FUNCTIONS

As discussed in Section 8.2, modal analysis techniques have been developed to identify nonlinear structural dynamic characteristics based on the analysis of measured classical first-order FRFs and have been found to be quite successful in cases where the effect of structural nonlinearities shows up in the measured data. However, due to the symmetrisation effect and the approximate nature of the first-order FRF measurements, for some nonsymmetric nonlinear systems, the thus-measured first-order FRFs are those from their equivalent symmetric counterparts, and the harmonic components which are usually present in the response signal of nonlinear systems subject to sinusoidal excitation are filtered out. This symmetrisation of nonsymmetrical nonlinearities and the elimination of harmonic components mean that first-order FRF analysis is not very appropriate for the analysis of structures with such a behaviour. In fact, for some nonlinear systems such as quadratic and bilinear systems (bilinear and quadratic stiffness nonlinearities [348]), first-order FRF analysis is incapable of analysing them at all. Moreover, from the response prediction point of view, calculations made using first-order FRFs can only be accurate in some cases because, mathematically, this means that only the linear term of a nonlinear function at a certain point has been retained. To enable identification in more complicated situations and to improve the capability and accuracy of response prediction, research work on higher-order FRF analysis has been carried out.

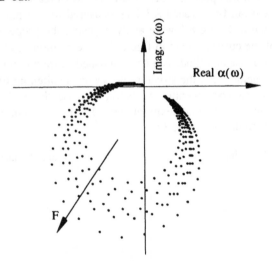

Fig. 8.8 Measured FRF data showing mode complexity.

The mathematical basis of higher-order FRF analysis lies in the Volterra series theory which, being the functional series representation of nonlinear systems and with its rigorous mathematical base, has been found to be quite effective in the characterisation of general nonlinear systems. The theory was first introduced into nonlinear circuit analysis in 1942 by Wiener [342] who later extended and applied it in a general way to a number of problems. Since Wiener's early work, many papers have been published dealing with this subject in systems and communication engineering [349-351]. However, it was not until recently that the theory was applied to the identification of nonlinear mechanical structures [352-354] and found to be quite useful. This section introduces the basic theory of Volterra series and of their relation to the higher-order FRFs and how the higher-order FRFs generalise the linear system theory to cover nonlinear systems. The harmonic probing method for Volterra kernel measurement using multi-tone input [355] and the correlation technique for Wiener kernel measurement using random input [356] will be discussed, and the relationship between the Volterra and Wiener kernels is studied. Possible ways of curve-fitting or surface-fitting measured higher-order FRFs so that parametric or non-parametric models of a nonlinear structure can be established are discussed.

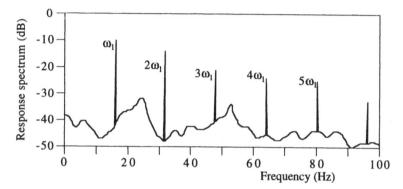

a) Response spectrum due to pure sinusoidal input.

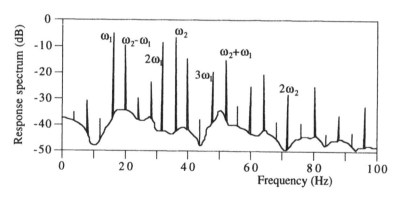

b) Response spectrum due to multi-tone input.

Fig. 8.9 Response spectrum of a nonlinear system.

8.3.1 Basic characteristics of nonlinear systems

Since a linear system must satisfy the principle of superposition, a sinusoid can be regarded as an eigenfunction of the system. That is, for a sinusoid applied to a linear system, the system changes only its amplitude and phase angle, without distorting its wave form. A nonlinear system, however, is characterised by the transfer of energy between frequencies. For a sinusoidal input $f(t)=A\sin\omega t$, the system will - in general - generate harmonic frequency components in addition to the fundamental frequency component, as shown in figure 8.9 a) (the background curve is due to numerical inaccuracy). If a multi-tone input $f(t) = A\sin\omega_1 t + B\sin\omega_2 t$ is applied (the input signal has two or more frequency components where A, B can be complex numbers to accommodate the different phases of these two waveforms), then in addition to the fundamental frequencies (ω_1, ω_2) and their harmonics ($n\omega_1$, $n\omega_2$), there will also be combinational frequency components ($\omega_1 + \omega_2$, $\omega_2 - \omega_1$, etc.) as shown in figure 8.9 b). In order to establish an input/output model of a nonlinear system which can not only predict the fundamental frequency, but also the harmonics and combinational frequencies, the Volterra series theory of nonlinear systems was developed.

8.3.2 Volterra series and higher-order FRFs
A nonlinear function $f(x)$ can, in general, be represented as a Taylor series at a certain point (e.g. $x = x_0$) and this series approaches $f(x)$ when the variable x is not far from that point. Similarly, a nonlinear system can, in general, be characterised by a Volterra series which converges when the nonlinearity of the system satisfies certain general conditions [357]. Volterra series have been described as 'power series with memory' which express the output of a nonlinear system in 'powers' of the input. A wide class of nonlinear systems encountered in engineering can be represented by Volterra series. Given an input $f(t)$, the output $x(t)$ of a time-invariant system can be expressed as follows:

$$x(t) = \int_{-\infty}^{+\infty} h_1(\tau_1) f(t - \tau_1) d\tau_1 + \int_{-\infty}^{+\infty}\int_{-\infty}^{+\infty} h_2(\tau_1, \tau_2) f(t - \tau_1) f(t - \tau_2) d\tau_1 d\tau_2$$

$$+ \int_{-\infty}^{+\infty}\int_{-\infty}^{+\infty}\int_{-\infty}^{+\infty} h_3(\tau_1, \tau_2, \tau_3) f(t - \tau_1) f(t - \tau_2) f(t - \tau_3) d\tau_1 d\tau_2 d\tau_3$$

$$+ \ldots + \int_{-\infty}^{+\infty} \ldots \int_{-\infty}^{+\infty} h_n(\tau_1, \ldots, \tau_n) f(t - \tau_1) \ldots f(t - \tau_n) d\tau_1 \ldots d\tau_n \quad (8.29)$$

or, in a more compact form,

$$x(t) = \sum_{n=1}^{\infty} \int_{-\infty}^{+\infty} \ldots \int_{-\infty}^{+\infty} h_n(\tau_1, \tau_2, \ldots, \tau_n) \prod_{r=1}^{n} f(t - \tau_r) d\tau_1 \ldots d\tau_n \quad (8.30)$$

where $h_n(\tau_1, \ldots, \tau_n)$ are the Volterra kernels which describe the system. In pure mathematical terms, they simply represent weighting functions. Physically, the first-order kernel $h_1(\tau)$ is the impulse response due to the linear part of the nonlinear system and the higher-order kernels can thus be viewed as higher-order impulse responses which serve to characterise the various orders of nonlinearity. In the special case where the system is linear, all the higher-order kernels except $h_1(\tau)$ are zero. Like a Taylor series representation of a nonlinear function, the Volterra series representation of a general nonlinear system is theoretically infinite and, as will be discussed later, the effort for computing the nth-order kernel increases exponentially as n increases so that one has to be satisfied with the first few kernels only (usually, up to the third kernel). Fortunately, good approximations can be obtained for most engineering problems by just considering these first few kernels and this is why this theory has been widely applied to the characterisation of practical nonlinear systems.

The main problem is, therefore, the calculation of the kernels $h_n(\tau_1, ..., \tau_n)$. Without loss of generality, we shall give an example to show how the first two kernels can be estimated. Let the input $f(\tau)$ be sampled at time intervals T, considering several impulses, as illustrated in figure 8.10.

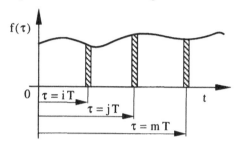

Fig. 8.10 Input $f(\tau)$, considering several impulses.

The response at time $t_k = kT$, from (8.29), will be given by:

$$x(kT) = \sum_{i=0}^{p} h_1(iT)f(kT - iT) + \sum_{i=0}^{p}\sum_{j=0}^{p} h_2(iT, jT)f(kT - iT)f(kT - jT) + e(kT)$$

$$(8.31)$$

for $k \geq p$, where $e(kT)$ represents higher order terms. Expanding (8.31),

$$x(kT) = f(kT)h_1(0) + ... + f(kT - pT)h_1(pT)$$

$$+ f^2(kT)h_2(0,0) + f(kT)f(kT - T)h_2(0,1) + ... \qquad (8.32)$$

$$+ f^2(kT - pT)h_2(p,p) + e(kT)$$

which may be alternatively written as

$$x(kT) = \left\{ f(kT) \ ... \ f(kT - pT) \ \vdots \ f^2(kT) \ ... \ f^2(kT - pT) \right\} \begin{Bmatrix} h_1(0) \\ \vdots \\ h_1(pT) \\ \cdots\cdots\cdots \\ h_2(0,0) \\ \vdots \\ h_2(p,p) \end{Bmatrix} + e(kT)$$

$$(8.33)$$

The dimensions of the vectors in (8.33) are $(p+1)(p+2)$. Collecting $N \geq (p+1)(p+2)$ responses, we have the following problem to solve:

$$\{x\} = [U]\{h\} + \{e\} \tag{8.34}$$

Minimising the squared error $\{e\}^T\{e\}$ leads to a solution in a least-squares sense, for kernels h_1 and h_2:

$$\{h\} = \left([U]^T[U]\right)^{-1}[U]^T\{x\} \tag{8.35}$$

The nth-order Volterra kernel transform or n^{th}-order Volterra transfer function is defined as the corresponding n-dimensional Fourier transform:

$$H_n(\omega_1, \omega_2, ..., \omega_n) = \int_{-\infty}^{+\infty} ... \int_{-\infty}^{+\infty} h_n(\tau_1, \tau_2, ..., \tau_n) e^{-i(\omega_1\tau_1 + \omega_2\tau_2 + ... + \omega_n\tau_n)} d\tau_1 ... d\tau_n \tag{8.36}$$

For a linear system, if its frequency response functions (only the first-order) have been determined, the output $x(t)$ can be calculated for any form of inputs. The same argument holds if all the Volterra transfer functions $H_n(\omega_1, \omega_2, ..., \omega_n)$ have been determined; since $H_n(\omega_1, \omega_2, ..., \omega_n)$ are unique (independent of input and output of the system), the Volterra series representation is mathematically very attractive, because under this representation the identification of a nonlinear system is reduced to the determination of these unique Volterra kernel transforms. However, due to the interactions between kernels, these uniquely defined Volterra transfer functions cannot be uniquely measured in practice and all that can be measured are their experimental counterparts, the so-called higher-order frequency response functions which are only approximations of Volterra transfer functions and which are, in general, input/output dependent.

The nth-order FRF $H_n(\omega_1, \omega_2, ..., \omega_n)$, which can be measured in practice, is defined as the output component $X(\omega_1, \omega_2, ..., \omega_n)$ of $x(t)$ at frequency $\omega = \omega_1 + \omega_2 + ... + \omega_n$ due to the input $f(t) = A_1\cos\omega_1 t + A_2\cos\omega_2 t + ... + A_n\cos\omega_n t$ (here A_i can be complex to accommodate the different phases and the frequencies ω_i must be incommensurable[†]) divided by the input spectra, that is [355],

$$H_n(\omega_1, \omega_2, \cdots, \omega_n) = \frac{X(\omega_1 + \omega_2 + ... + \omega_n)}{\prod_{r=1}^{n} A_r} \tag{8.37}$$

The relationship between the measured n^{th}-order FRF $H_n(\omega_1, \omega_2, ..., \omega_n)$ and its theoretical counterpart, the nth-order Volterra transfer function $H_n(\omega_1, \omega_2, ..., \omega_n)$ can be written as

---

[†] This means that ω_1 and ω_2 (for example) cannot be expressed as the ratio of two integers n_1/n_2.

$$H_n\left(\omega_1,\omega_2,...,\omega_n\right) = \mathbf{H}_n\left(\omega_1,\omega_2,...,\omega_n\right) + \text{higher order terms} \qquad (8.38)$$

In general, there will be some contribution from the higher-order Volterra kernels, and the measured FRF $\mathbf{H}_n(\omega_1, \omega_2, ..., \omega_n)$ is only an approximation of the uniquely defined Volterra kernel transform $H_n(\omega_1, \omega_2, ..., \omega_n)$. Based on this observation, the Volterra kernel $h_n(\tau_1, \tau_2, ..., \tau_n)$ and its transform $H_n(\omega_1, \omega_2, ..., \omega_n)$ have direct physical meaning and interpretation.

It is worth pointing out here that the Volterra kernel transforms $H_n(\omega_1, \omega_2, ..., \omega_n)$ are mathematically unique. However, the nth-order FRFs $\mathbf{H}_n(\omega_1, \omega_2, ..., \omega_n)$ are usually input-output dependent like the classical first-order FRF $\mathbf{H}_1(\omega)$ measured using a sine wave excitation. Since we are only able to deal with truncated series, these measured FRFs will, in some cases, give more accurate representation than their equivalent Volterra kernel transforms which are by no means measurable. For a given nonlinear system whose equations of motion are known, the higher-order Volterra transfer functions can be analytically determined [355]. To illustrate some of the features of Volterra transfer functions and higher-order FRFs, consider a nonlinear SDOF system given as

$$m\ddot{x} + c\dot{x} + kx + k_1 x^2 + k_2 x^3 = f(t) \qquad (8.39)$$

The theoretically unique second-order Volterra transfer function of the system is shown in figure 8.11, with two independent variables ω_1 and ω_2 which are the input frequencies. The peaks are associated with the resonance of the system [353].

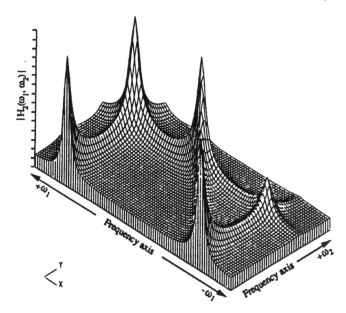

Fig. 8.11 Typical second-order Volterra transfer function.

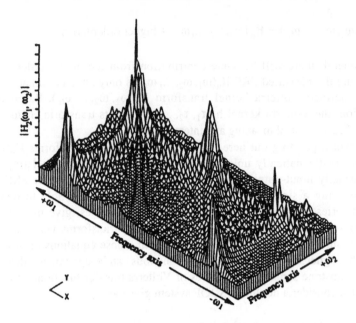

Fig. 8.12 Simulated measured second-order FRF using multi-tone input.

On the other hand, by setting $f(t) = A_1 \sin \omega_1 t + A_2 \sin \omega_2 t$ and numerically integrating the response $x(t)$ based on (8.39), the spectral component of $X(\omega)$ at frequency $\omega_1 + \omega_2$ can be computed. Then, based on (8.37), a simulated measured second-order FRF $H_2(\omega_1, \omega_2)$ can be obtained as shown in figure 8.12.

In comparison with its analytical counterpart, the simulated measured second-order FRF $H_2(\omega_1, \omega_2)$ shows quite similar characteristics but there exist some spurious spikes which are due to the fact that during the simulation, ω_1 and ω_2 were not made incommensurable [355] - a condition which is very difficult to achieve in practice. Nevertheless, as compared with the analytical second-order transfer function, the simulated measured second-order FRF is quite accurate and this demonstrates the feasibility of practical measurement of the higher-order FRFs of nonlinear structures.

8.3.3 Wiener series and determination of Wiener kernels
In the Wiener theory of nonlinear systems, if the input $f(t)$ is a white Gaussian time series with auto-correlation function $R_{ff}(\tau) = A\delta(\tau)$, then the output $x(t)$ of a nonlinear system can be expressed by the orthogonal expression:

$$x(t) = \sum_{n=1}^{\infty} G_n[k_n; f(t)]^{\dagger} \qquad (8.40)$$

$\dagger$ The semicolon is conventionally used here, although it simply means that G_n is a function of both k_n and $f(t)$.

in which $\{k_n(\tau_1, \tau_2, ..., \tau_n)\}$ is the set of Wiener kernels of the nonlinear system which, like the set of Volterra kernels $\{h_n(\tau_1, \tau_2, ..., \tau_n)\}$, serves to describe the system, and $\{G_n\}$ is a complete set of orthogonal functionals [357]. For a linear system, all the higher-order kernels except k_0 and $k_1(\tau)$ are zero.

The Wiener G-functionals are a set of non-homogeneous Volterra functionals defined as:

$$G_n[k_n; f(t)] = k_{0(n)} + \sum_{r=1}^{n} \int_{-\infty}^{+\infty} ... \int_{-\infty}^{+\infty} k_{r(n)}(\tau_1, \tau_2, ..., \tau_n) \prod_{s=1}^{r} f(t - \tau_r) d\tau_1 ... d\tau_n$$

(8.41)

where $k_{n(n)} \equiv k_n$ is known as the nth-order Wiener kernel and $k_{n-1(n)}, ..., k_{0(n)}$ are known as the derived Wiener kernels of the Wiener G-functional. The G-functionals $G_n[k_n; f(t)]$ satisfy

$$\overline{T_m\big[h_m; f(t)\big] G_n\big[k_n; f(t)\big]} = 0 \qquad \text{for} \qquad m < n \qquad (8.42)$$

where $T_m[h_m; f(t)]$ is the mth-order Volterra functional and the over bar means taking the average of the process. Theoretically, all the derived Wiener kernels of nth-order ($k_{n-1(n)}, ..., k_{0(n)}$) can be determined uniquely by the leading n^{th}-order Wiener kernel k_n when (8.42) is satisfied for all integer values of $m < n$ and, therefore, in the notation $G_n[k_n; f(t)]$, only the leading term k_n is specified. The first few $G_n[k_n; f(t)]$ of a general nonlinear system are given as [357]:

$$G_0\big[k_0; f(t)\big] = k_0 \qquad (k_0 \text{ is a constant}) \qquad (8.43)$$

$$G_1\big[k_1; f(t)\big] = \int_{-\infty}^{+\infty} k_1(\tau_1) f(t - \tau_1) d\tau_1 \qquad (8.44)$$

$$G_2\big[k_2; f(t)\big] = \int_{-\infty}^{+\infty}\int_{-\infty}^{+\infty} k_2(\tau_1, \tau_2) f(t - \tau_1) f(t - \tau_2) d\tau_1 d\tau_2$$
$$- A \int_{-\infty}^{+\infty} k_2(\tau_1, \tau_1) d\tau_1 \qquad (8.45)$$

$$G_3\big[k_3; f(t)\big] = \int_{-\infty}^{+\infty}\int_{-\infty}^{+\infty}\int_{-\infty}^{+\infty} k_3(\tau_1, \tau_2, \tau_3) f(t - \tau_1) f(t - \tau_2) f(t - \tau_3) d\tau_1 d\tau_2 d\tau_3$$
$$- 3A \int_{-\infty}^{+\infty}\int_{-\infty}^{+\infty} k_3(\tau_1, \tau_2, \tau_2) f(t - \tau_1) d\tau_1 d\tau_2 \qquad (8.46)$$

The relationship between the Volterra kernel $h_n(\tau_1, \tau_2, ..., \tau_n)$ and the Wiener kernel $k_n(\tau_1, \tau_2, ..., \tau_n)$ is that the system's n^{th}-order Volterra kernel is equal to the system's n^{th}-order Wiener kernel plus the sum of all the (even or odd order) derived Wiener kernels that are of the nth-order, that is,

$$h_n\left(\tau_1, \tau_2, \cdots, \tau_n\right) = k_n\left(\tau_1, \tau_2, \cdots, \tau_n\right) + k_{n-2(n)}\left(\tau_1, \tau_2, \cdots, \tau_{n-2}\right) + \cdots \tag{8.47}$$

From (8.47), it can be seen that since the derived Wiener kernels are determined uniquely by their leading Wiener kernel, a given system's Volterra kernels can be obtained uniquely from the system's Wiener kernels (leading Wiener kernels). Also, it should be noted from (8.45) and (8.46) that as the input level A→0, the derived kernels approach zero and the leading Wiener kernels approach the Volterra kernels. On the other hand, it should be pointed out that unlike the Volterra kernels, which are mathematically unique, the Wiener kernels are input-output dependent and since the Volterra kernels $h_n(\tau_1, \tau_2, ..., \tau_n)$ are uniquely determined by the Wiener kernels $k_n(\tau_1, \tau_2, ..., \tau_n)$, the measured Wiener kernels $k_n(\tau_1, \tau_2, ..., \tau_n)$ also uniquely determine the system.

As in the case of Volterra series, under Wiener series representation the identification problem becomes one of determining all the Wiener kernels which describe the system. The orthogonality property of G_n and the statistical properties of Gaussian noise enable the Wiener kernels to be determined using a cross-correlation technique. The first four kernels are given [356] as

$$k_0 = \overline{x(t)} \tag{8.48}$$

$$k_1(\tau) = \frac{1}{A}\overline{x(t)f(t-\tau)} \tag{8.49}$$

$$k_2(\tau_1, \tau_2) = \frac{1}{2A^2}\overline{[x(t) - k_0]f(t-\tau_1)f(t-\tau_2)} \tag{8.50}$$

$$k_3(\tau_1, \tau_2, \tau_3) = \frac{1}{3!A^3}\overline{\left[x(t) - G_1[k_1; f(t)]\right]f(t-\tau_1)f(t-\tau_2)f(t-\tau_3)} \tag{8.51}$$

Equations (8.48) to (8.51) serve as a basis for the measurement of Wiener kernels. If the nth-order Wiener kernel $k_n(\tau_1, \tau_2, ..., \tau_n)$ has been measured, then its corresponding n^{th}-order FRF $\mathbf{H}_n(\omega_1, \omega_2, ..., \omega_n)$ is defined as

$$\mathbf{H}_n\left(\omega_1, \omega_2, \cdots, \omega_n\right) \equiv K_n\left(\omega_1, \omega_2, \cdots, \omega_n\right) \tag{8.52}$$

where $K_n(\omega_1, \omega_2, ..., \omega_n)$ is the nth-order Wiener kernel transform of $k_n(\tau_1, \tau_2, ..., \tau_n)$. As discussed, when the input is low (the power spectrum of the input A → 0), the measured Wiener kernels approach their corresponding Volterra kernels and, therefore, the measured $\mathbf{H}_n(\omega_1, \omega_2, ..., \omega_n)$ based on (8.52) approaches the Volterra kernel transform $H_n(\omega_1, \omega_2, ..., \omega_n)$.

To see how higher-order FRFs can be measured based on Wiener series using random input, numerical simulation on the experimental determination of second-order FRF of the same system described by (8.39) has been carried out. The input signal f(t) is random noise with flat spectrum up to the cut-off frequency, which is much higher than the frequency range of interest. Next the response x(t) of the system due to f(t) is computed and the second-order Wiener kernel $k_2(\tau_1, \tau_2)$ can then be determined based on expression (8.50), which is shown in figure 8.13. The corresponding second-order FRF $H_2(\omega_1, \omega_2)$ which is the two-dimensional Fourier transform of $k_2(\tau_1, \tau_2)$, is shown in figure 8.14. In comparison with its analytical Volterra transfer function, the simulated measured second-order FRF is quite accurate, demonstrating the practical feasibilities of higher-order FRF measurement using random input based on Wiener series.

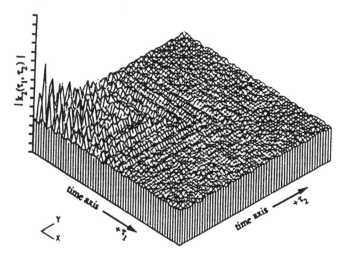

Fig. 8.13 Typical second-order Wiener kernel.

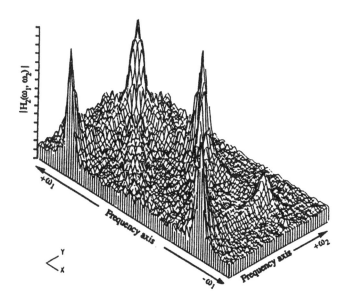

Fig. 8.14 Simulated measured second-order FRF using random input.

8.3.4 Identification of nonlinearity using higher-order FRFs

The theoretical basis and measurement techniques for higher-order FRFs have been discussed. The remaining question which needs to be answered is: "what information about the nature of nonlinearity of a system can be derived from those measured higher-order FRFs?" First, the existence of second-order FRFs indicates the existence of nonsymmetric nonlinearity of a system - a result that, for some systems such as quadratic and bilinear systems as mentioned before, cannot be achieved on the basis of analysis of classical first-order FRFs. Secondly, the parameters of a nonlinear system can be identified based on the analysis of higher-order FRFs together with first-order ones. Depending on whether physical parameters or modal parameters are of interest, a 'state-space analysis' method [353] or a 'modal-space analysis' method [354] can be employed.

In the 'state-space analysis', *a priori* information about the total number of degrees-of-freedom and the physical connectivity of the system to be analysed should be given. Mathematically, measured first- and second-order FRF data are functions of all the physical parameters (mass, nonlinear stiffness and damping elements). Therefore, from given measured first- and second-order FRFs, these parameters can, in theory, be calculated provided that enough data have been measured. The physical parameter identification problem of a nonlinear system can therefore be formulated mathematically as the solution of the following linear algebraic equation:

$$[A] \{p\} = \{b\} \tag{8.53}$$

where $\{p\}$ is the unknown vector of physical parameters to be determined, and $[A]$ and $\{b\}$ are the coefficient matrix and vector formed using the measured first- and second-order (or higher-order) FRFs.

The 'modal-space analysis' is based on the mathematical observation that, for a general nonlinear system, its second-order FRF can be decomposed as [354]

$$H_2(\omega_1,\omega_2) = \sum_{n=1}^{2N} \frac{A_n}{i\omega_1 + \omega_n} \sum_{m=1}^{2N} \frac{A_m}{i\omega_2 + \omega_m} \sum_{s=1}^{2K} \frac{A_s}{i\omega_1 + i\omega_2 + \omega_s} \tag{8.54}$$

where N is the number of degrees-of-freedom of the system and 2K represents the number of poles corresponding to 'nonlinear coupling modes' which are the combinational resonances of the system. When one of the variables (ω_1 or ω_2) is fixed, (8.54) reduces to the following polynomial form:

$$H_2(\omega_1,\omega_2) = \frac{P(\omega_1,\omega_n,\omega_m,\omega_s)}{Q(\omega_1,\omega_n,\omega_m,\omega_s)}; \quad n = 1, 2N; m = 1, 2N; s = 1, 2K \tag{8.55}$$

Curve-fitting of this polynomial function can be made using the well-developed polynomial curve-fitting algorithms used in linear modal analysis as discussed in Chapter 4. Then all the ω_n and ω_m, which are the natural frequencies of the system, can be obtained and the analytical model of $H_2(\omega_1,\omega_2)$ in its polynomial expression can be established. Such analytical models can be used for further applications such as response prediction, physical parameter estimation and prediction of harmonic and combinational resonances.

8.4   LOCATION OF STRUCTURAL NONLINEARITIES

It is usually believed that, if they exist, structural nonlinearities are localised in terms of spatial coordinates, as a result of nonlinear dynamic characteristics of structural joints, boundary conditions and material properties such as plasticity. The ability to locate a structure's localised nonlinearity has, consequently, some important engineering applications. First, the information about where the structural nonlinearity is may offer opportunities to separate the structure into linear and nonlinear subsystems so that these can be analysed separately based on nonlinear substructuring analysis [358]. Second, since nonlinearity is often caused by the improper connection of structural joints, its location may give an indication of a malfunction or of poor assembly of the system. Third, from a materials property point of view, the stress at certain parts of the structure during vibration can become so high that the deformation of that part becomes plastic and the dynamic behaviour becomes nonlinear. In this case, the location process may offer the possibility of failure detection. Finally, location information is essential if a nonlinear mathematical model of the structure is to be established.

In practical measurements, the data measured are usually quite limited (both measured modes and coordinates are incomplete) and this is especially true when a nonlinear structure is considered. It is therefore believed that the task of locating a structure's localised nonlinearity can only become possible by correlating an analytical model, which may contain modelling errors but can represent the structure to some degree of accuracy, with the results from dynamic tests of the structure. Fortunately, due to the development of analytical modelling techniques, though an analytical model of a structure may contain modelling errors, these errors are usually of second order when compared with the analytical model itself in the sense of the Euclidean norm. With such an analytical model available, it will be shown in this section that by correlating the analytical model with the measured dynamic test data, location of nonlinearity can be achieved.

8.4.1 Location using measured modal data

The location method [359] described here is developed based on the correlation between an analytical model which contains modelling errors and dynamic test data which are measured at different response levels. Before embarking on detailed discussion, it is necessary to mention that the 'modes' of a nonlinear structure are difficult to define (if indeed they exist at all!) in an exact mathematical sense [360, 361] because of the existence of harmonic response components. So, the term 'modes of a nonlinear structure' is used to mean the natural frequencies and mode shapes which are derived from the analysis of measured first-order FRFs, in which only the fundamental frequency component of the response is of interest. For most nonlinear mechanical structures, the thus-obtained natural frequencies and mode shapes are response level dependent. As far as stiffness nonlinearity is concerned, the stiffness matrix of the structure corresponding to different response levels will be different and, therefore, if this difference in stiffness matrix can be calculated in some way, the problem of nonlinearity location can be resolved.

Suppose that the eigenvalues and eigenvectors of the rth mode (which is sensitive to the localised nonlinearity) corresponding to a lower vibration level, X_1, are $\lambda_1, \{\phi_1\}$ and those corresponding to a higher vibration level, X_2, are $\lambda_2, \{\phi_2\}$ and that these have been obtained from the analysis of measured first-order FRFs based on nonlinear modal analysis methods discussed earlier. Suppose also that the analytical model which contains second-order modelling errors (corresponding to lower vibration level) is available. Then, from the eigendynamic equations, the following relationships can be established:

$$\left[[K_a]+[\Delta K]-\lambda_1[[M_a]+[\Delta M]]\right]\{\phi_1\}=\{0\} \tag{8.56}$$

$$\left[[K_a]+[\Delta K]+[\Delta K_n]-\lambda_2[[M_a]+[\Delta M]]\right]\{\phi_2\}=\{0\} \tag{8.57}$$

where subscripts a and n stand for analytical and nonlinear, respectively. Post-multiplying (8.57) by $\{\phi_1\}^T$, it follows that

$$\left[[\Delta K]+[\Delta K_n]-\lambda_2[\Delta M]\right]\{\phi_2\}\{\phi_1\}^T = -\left[[K_a]-\lambda_2[M_a]\right]\{\phi_2\}\{\phi_1\}^T \tag{8.58}$$

Post-multiplying (8.56) by $\{\phi_2\}^T$ gives

$$\left[[\Delta K]-\lambda_1[\Delta M]\right]\{\phi_1\}\{\phi_2\}^T = -\left[[K_a]-\lambda_1[M_a]\right]\{\phi_1\}\{\phi_2\}^T \tag{8.59}$$

Subtracting (8.59) from (8.58) and rearranging, leads to

$$[\Delta K]\left[\{\phi_2\}\{\phi_1\}^T-\{\phi_1\}\{\phi_2\}^T\right]+[\Delta K_n]\{\phi_2\}\{\phi_1\}^T$$

$$+[\Delta M]\left[-\lambda_2\{\phi_2\}\{\phi_1\}^T+\lambda_1\{\phi_1\}\{\phi_2\}^T\right] \tag{8.60}$$

$$=\left[[K_a]-\lambda_1[M_a]\right]\{\phi_1\}\{\phi_2\}^T-\left[[K_a]-\lambda_2[M_a]\right]\{\phi_2\}\{\phi_1\}^T$$

Since $\{\phi_2\}$ is a perturbed mode shape of $\{\phi_1\}$, due to stiffness change of nonlinearity, $\{\phi_2\}\{\phi_1\}^T-\{\phi_1\}\{\phi_2\}^T$ is of second order compared to $\{\phi_2\}\{\phi_1\}^T$ in the sense of the Euclidean norm (one can notice that all the diagonal elements of $\{\phi_2\}\{\phi_1\}^T-\{\phi_1\}\{\phi_2\}^T$ are zero). As a result, if the modelling errors $[\Delta K]$, $\lambda_1[\Delta M]$ and stiffness change due to nonlinearity $[\Delta K_n]$ are of the same order of magnitude (also in the sense of the Euclidean norm), then to a first order approximation, (8.60) becomes

$$
[\Delta K_n]\{\phi_2\}\{\phi_1\}^T = \big[[K_a]-\lambda_1[M_a]\big]\{\phi_1\}\{\phi_2\}^T
$$
$$
-\big[[K_a]-\lambda_2[M_a]\big]\{\phi_2\}\{\phi_1\}^T
\tag{8.61}
$$

In the special case of $[\Delta K] = [0]$ and $[\Delta M] = [0]$ (no modelling errors), (8.61) becomes an exact statement for $[\Delta K_n]$. The principle of nonlinearity location process based on (8.61) is illustrated in figure 8.15. If the nonlinearity is localised, then $[\Delta K_n]$ will be a very sparse matrix (only those elements where the structural nonlinearity is located are non-zero) and, as shown in the same picture, the dominant non-zero elements of the resultant matrix after the matrix multiplication will indicate the location of the nonlinearity. Also, it should be noticed that during the location process, only one measured mode is required and it is recommended that the mode which is the most sensitive to nonlinearity in the measured frequency range should be used. Extra modes can be used to check the consistency and reliability of the location results.

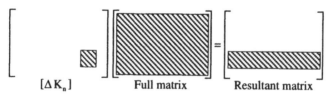

$$[\Delta K_n] \qquad\qquad \text{Full matrix} \qquad\qquad \text{Resultant matrix}$$

Fig. 8.15 Principle of nonlinearity location.

8.4.2 Expansion of measured coordinates

In the theoretical development of the location method, it is assumed that the measured coordinates are complete. In practice, however, this is very difficult to achieve because certain coordinates are physically inaccessible, such as internal DOFs, and the rotational coordinates are very difficult to measure. Consequently, the unmeasured coordinates have to be interpolated before the location process can be carried out. This interpolation can be achieved by using the analytical model itself based on Kidder's expansion method [270][†].

Although the analytical model contains modelling errors, in order to interpolate the unmeasured coordinates it is assumed that between the analytical model and the rth measured and unmeasured sub-modes the following relationship holds:

$$
\left[\begin{bmatrix}[K_{pp}] & [K_{ps}] \\ [K_{sp}] & [K_{ss}]\end{bmatrix} - \omega_r^2 \begin{bmatrix}[M_{pp}] & [M_{ps}] \\ [M_{sp}] & [M_{ss}]\end{bmatrix}\right]\begin{Bmatrix}\{\phi_p\} \\ \{\phi_s\}\end{Bmatrix}_r = \begin{Bmatrix}\{0\} \\ \{0\}\end{Bmatrix}
\tag{8.62}
$$

[†] This method has already been mentioned in Section 7.3.2.

where $\{\phi_p\}_r$ and $\{\phi_s\}_r$ are the rth measured and unmeasured sub-modes, respectively. Upon multiplying out (8.62), the following two equations are obtained:

$$\left[\left[K_{pp}\right]-\omega_r^2\left[M_{pp}\right]\right]\left\{\phi_p\right\}_r+\left[\left[K_{ps}\right]-\omega_r^2\left[M_{ps}\right]\right]\left\{\phi_s\right\}_r=\{0\} \qquad (8.63)$$

$$\left[\left[K_{sp}\right]-\omega_r^2\left[M_{sp}\right]\right]\left\{\phi_p\right\}_r+\left[\left[K_{ss}\right]-\omega_r^2\left[M_{ss}\right]\right]\left\{\phi_s\right\}_r=\{0\} \qquad (8.64)$$

Theoretically, $\{\phi_s\}_r$ can be calculated from either (8.63) or (8.64). However, when the number of measured coordinates is less than that of the unmeasured ones, which is quite usual in practice, (8.63) becomes underdetermined in terms of the solution of $\{\phi_s\}_r$ (the coefficient matrix is rank deficient), and it is therefore recommended that (8.64) should be used to interpolate $\{\phi_s\}_r$ as follows:

$$\left\{\phi_s\right\}_r=-\left[\left[K_{ss}\right]-\omega_r^2\left[M_{ss}\right]\right]^{-1}\left[\left[K_{sp}\right]-\omega_r^2\left[M_{sp}\right]\right]\left\{\phi_p\right\}_r \qquad (8.65)$$

It has been found that the interpolation of unmeasured coordinates based on (8.65) is quite accurate for the lower modes. If some coordinates are measured where the structural nonlinearity exists, the thus-interpolated mode shapes can be used to achieve a successful nonlinearity location. Also, it can be shown mathematically that the located errors will only occur in the measured coordinates if the unmeasured ones are interpolated based on (8.65). This is briefly illustrated below.

Expression $[[K_a]-\lambda_1[M_a]]\{\phi_1\}$ (and similarly $[[K_a]-\lambda_2[M_a]]\{\phi_2\}$) on the right-hand side of (8.61) can be re-written as:

$$[K_a]-\lambda_1[M_a]]\{\phi_1\}=\left[\left[\begin{array}{c:c}[K_{pp}] & [K_{ps}] \\ \hdashline [K_{sp}] & [K_{ss}]\end{array}\right]-\lambda_1\left[\begin{array}{c:c}[M_{pp}] & [M_{ps}] \\ \hdashline [M_{sp}] & [M_{ss}]\end{array}\right]\right]\left\{\begin{array}{c}\{\phi_p\}_1 \\ \{\phi_s\}_1\end{array}\right\}$$

$$=\left\{\begin{array}{c}\left[\left[K_{pp}\right]-\lambda_1\left[M_{pp}\right]\right]\left\{\phi_p\right\}_1+\left[\left[K_{ps}\right]-\lambda_1\left[M_{ps}\right]\right]\left\{\phi_s\right\}_1 \\ \hdashline \left[\left[K_{sp}\right]-\lambda_1\left[M_{sp}\right]\right]\left\{\phi_p\right\}_1+\left[\left[K_{ss}\right]-\lambda_1\left[M_{ss}\right]\right]\left\{\phi_s\right\}_1\end{array}\right\}$$

$$(8.66)$$

When $\{\phi_s\}_r$ is interpolated based on (8.65), it is easy to see that in (8.66) the elements corresponding to the unmeasured coordinates (the lower part) are zero and (8.66) becomes

$$\left\{\begin{array}{c}\left[\left[K_{pp}\right]-\lambda_1\left[M_{pp}\right]\right]\left\{\phi_p\right\}_1+\left[\left[K_{ps}\right]-\lambda_1\left[M_{ps}\right]\right]\left\{\phi_s\right\}_1 \\ \hdashline \left[\left[K_{sp}\right]-\lambda_1\left[M_{sp}\right]\right]\left\{\phi_p\right\}_1+\left[\left[K_{ss}\right]-\lambda_1\left[M_{ss}\right]\right]\left\{\phi_s\right\}_1\end{array}\right\}=\left\{\begin{array}{c}\{R_1\} \\ \hdashline \{0\}\end{array}\right\} \qquad (8.67)$$

Upon substitution of (8.67), (8.61) becomes

$$[\Delta K_n]\{\phi_2\}\{\phi_1\}^T = \left\{\begin{matrix} \{R_1\} \\ \cdots\cdots \\ \{0\} \end{matrix}\right\}\{\phi_2\}^T - \left\{\begin{matrix} \{R_2\} \\ \cdots\cdots \\ \{0\} \end{matrix}\right\}\{\phi_1\}^T = \begin{bmatrix} [R] \\ \cdots \\ [0] \end{bmatrix} \qquad (8.68)$$

8.4.3 Application example

In what follows, the practical applicability of this proposed nonlinearity location method will be demonstrated with an example. The experimental system is an essentially linear frame structure made of mild steel, as shown in figure 8.16,

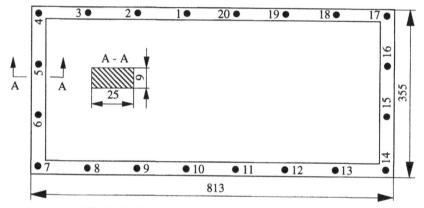

Fig. 8.16 Frame structure used in the experiment.

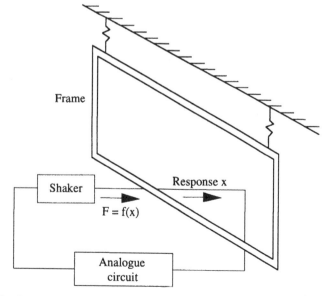

Fig. 8.17 Simulated nonlinear structure with localised nonlinearity.

coupled to a SDOF system with nonlinear stiffness. This SDOF system is simulated using an electromagnetic shaker by feeding the displacement signal of its moving table through a nonlinear analogue circuit and then back to the shaker to produce a force which satisfies a prescribed nonlinear function $F = f(x)$. This is illustrated in figure 8.17. A linear finite element model is established by discretising the frame into 20 elements with 3 DOFs (one translation and two rotations) at each of the 20 nodes. During modal testing, translational coordinates at all 20 nodes are measured at different controlled constant vibration amplitudes and one typical such measurement is presented in figure 8.18. This indicates that for the first mode, a shift of 4 Hz in natural frequency, corresponding to the lowest and the highest vibration amplitudes used in the test, was observed.

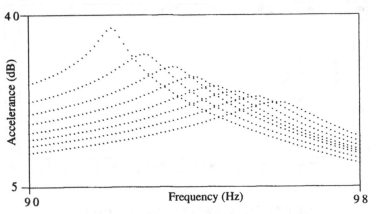

Fig. 8.18 Measured FRF data at different vibration amplitudes.

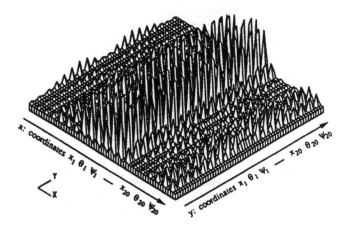

Fig. 8.19 Results of nonlinearity location.

By performing detailed modal analysis, mode shape vectors corresponding to the lowest and the highest vibration amplitudes can be established. These incomplete mode shape vectors are then expanded, based on the known FE model, to include unmeasured rotational coordinates. These expanded mode shape vectors and the FE model are then combined based on the proposed method to locate the nonlinearity.

The results for the location are shown in figure 8.19. Theoretically, the errors should be contained between two translational coordinates right in the middle of the plot, since the localised nonlinearity was introduced between nodes 10 and 11. However, the errors are distributed so that almost a third of the coordinates are contaminated. The reason for this is that, since the exact coordinate where the localised stiffness nonlinearity is introduced has neither been measured nor included in the FE model, this missing coordinate is expected to cause some spatial leakage and is responsible for the scattering of the location results. Nevertheless, the location results should be considered as quite successful when one considers that this is an actual practical application.

8.4.4 Extension of the method to FRF data

The location technique developed above can be generalised when measured FRF data are used. Suppose that the ith column $\{\alpha_i(\omega)\}_1$ of the receptance matrix (corresponding to lower vibration level X_1) and $\{\alpha_i(\omega)\}_2$ (corresponding to higher vibration level X_2) around the rth mode (which is sensitive to nonlinearity) have been measured and, also, the analytical model which contains second order modelling errors (corresponding to lower vibration level X_1) is available. The dynamic stiffness and receptance matrices of a system satisfy

$$[Z(\omega)][\alpha(\omega)] = [I] \tag{8.69}$$

Thus, taking the ith column of both sides of (8.69), the following equations can be written:

$$\left[[K_a] + [\Delta K] - \omega_j^2[[M_a] + [\Delta M]]\right]\{\alpha_i(\omega_j)\}_1 = \{e_i\} \tag{8.70}$$

$$\left[[K_a] + [\Delta K] + [\Delta K_n] - \omega_k^2[[M_a] + [\Delta M]]\right]\{\alpha_i(\omega_k)\}_2 = \{e_i\} \tag{8.71}$$

where $\{e_i\}$ is a vector with its ith element equal to unity and all the others zero, and ω_j and ω_k are the measured frequency data points chosen. Post-multiplying (8.71) by $\{\alpha_i(\omega_j)\}_1^T$ gives

$$\left[[\Delta K] + [\Delta K_n] - \omega_k^2[\Delta M]\right]\{\alpha_i(\omega_k)\}_2\{\alpha_i(\omega_j)\}_1^T =$$

$$-\left[[K_a] - \omega_k^2[M_a]\right]\{\alpha_i(\omega_k)\}_2\{\alpha_i(\omega_j)\}_1^T + \{e_i\}\{\alpha_i(\omega_j)\}_1^T \tag{8.72}$$

Post-multiplying (8.70) by $\{\alpha_i(\omega_k)\}_2^T$, one has

$$\left[[\Delta K]-\omega_j^2[\Delta M]\right]\left\{\alpha_i(\omega_j)\right\}_1\left\{\alpha_i(\omega_k)\right\}_2^T =$$
$$-\left[[K_a]-\omega_j^2[M_a]\right]\left\{\alpha_i(\omega_j)\right\}_1\left\{\alpha_i(\omega_k)\right\}_2^T +\left\{e_i\right\}\left\{\alpha_i(\omega_k)\right\}_2^T \tag{8.73}$$

Subtracting (8.73) from (8.72) and rearranging, it follows that

$$[\Delta K]\left[\left\{\alpha_i(\omega_k)\right\}_2\left\{\alpha_i(\omega_j)\right\}_1^T -\left\{\alpha_i(\omega_j)\right\}_1\left\{\alpha_i(\omega_k)\right\}_2^T\right]+$$

$$[\Delta K_n]\left\{\alpha_i(\omega_k)\right\}_2\left\{\alpha_i(\omega_j)\right\}_1^T +$$

$$[\Delta M]\left[\omega_j^2\left\{\alpha_i(\omega_j)\right\}_1\left\{\alpha_i(\omega_k)\right\}_2^T -\omega_k^2\left\{\alpha_i(\omega_k)\right\}_2\left\{\alpha_i(\omega_j)\right\}_1^T\right]$$

$$=[K_a]\left[\left\{\alpha_i(\omega_j)\right\}_1\left\{\alpha_i(\omega_k)\right\}_2^T -\left\{\alpha_i(\omega_k)\right\}_2\left\{\alpha_i(\omega_j)\right\}_1^T\right]+$$

$$[M_a]\left[\omega_k^2\left\{\alpha_i(\omega_k)\right\}_2\left\{\alpha_i(\omega_j)\right\}_1^T -\omega_j^2\left\{\alpha_i(\omega_j)\right\}_1\left\{\alpha_i(\omega_k)\right\}_2^T\right]+$$

$$\left\{e_i\right\}\left\{\left\{\alpha_i(\omega_j)\right\}_1^T -\left\{\alpha_i(\omega_k)\right\}_2^T\right\} \tag{8.74}$$

Unlike the case of location using modal data, here we have the chance to properly choose ω_j and ω_k in the measured frequency range, so that the following function is minimised:

$$\min \frac{\left\|\left\{\alpha_i(\omega_k)\right\}_2\left\{\alpha_i(\omega_j)\right\}_1^T -\left\{\alpha_i(\omega_j)\right\}_1\left\{\alpha_i(\omega_k)\right\}_2^T\right\|}{\left\|\left\{\alpha_i(\omega_k)\right\}_2\left\{\alpha_i(\omega_j)\right\}_1^T\right\|} \tag{8.75}$$

If the modelling errors $[\Delta K]$, $\omega_j^2[\Delta M]$ and stiffness change due to nonlinearity $[\Delta K_n]$ are of the same order of magnitude in the sense of the Euclidean norm, then to a first order approximation, (8.74) becomes

$$[\Delta K_n]\left\{\alpha_i(\omega_k)\right\}_2\left\{\alpha_i(\omega_j)\right\}_1^T =[K_a]\left[\left\{\alpha_i(\omega_j)\right\}_1\left\{\alpha_i(\omega_k)\right\}_2^T -\left\{\alpha_i(\omega_k)\right\}_2\left\{\alpha_i(\omega_j)\right\}_1^T\right]$$

$$+[M_a]\left[\omega_k^2\left\{\alpha_i(\omega_k)\right\}_2\left\{\alpha_i(\omega_j)\right\}_1^T -\omega_j^2\left\{\alpha_i(\omega_j)\right\}_1\left\{\alpha_i(\omega_k)\right\}_2^T\right]$$

$$+\left\{e_i\right\}\left\{\left\{\alpha_i(\omega_j)\right\}_1^T -\left\{\alpha_i(\omega_k)\right\}_2^T\right\} \tag{8.76}$$

Like (8.61) in the case of location using modal data, (8.76) can be used to locate localised stiffness nonlinearity(ies) using measured FRF data at different vibration levels.

The Singular Value Decomposition Technique (SVD)

A.1   INTRODUCTION

The purpose of this Appendix is to give a simple introduction to the Singular Value Decomposition technique (SVD), which enhances its main characteristics and highlights some common applications, giving to the reader a first contact with this powerful numerical method.

A.2   THE RANK OF A MATRIX

The concept of rank of a matrix is directly related to the linear dependency of the rows (or columns) of that matrix. For example, an NxN matrix whose rows are linearly independent will have rank = N (full rank). If one of the rows is a linear combination of the others, then the rank will be N-1. In other words, the rank of a matrix equals its number of linearly independent rows. An MxN matrix with M ≥ N is said to be of 'full rank' if its rank = N, or 'rank-deficient' if its rank < N. For a square matrix, rank deficiency implies that the matrix is singular, i.e., the determinant equals zero.

The classical way of calculating the rank of a matrix is by means of Gauss elimination. An NxN matrix with rank = r < N will have N-r zero rows after a Gauss elimination. If we think of the rows of a matrix as being vectors, two linearly dependent rows mean two parallel vectors. Thus, the most linearly independent case for two rows is when the corresponding vectors are perpendicular, or in general, orthogonal. In practice, we can have anything in between these two extremes. If two vectors are almost but not exactly parallel, then the two corresponding rows of the matrix will not be linearly dependent, but almost so and, in general, we will not have exact zero rows after a Gauss elimination, but a set of relatively small elements. These rows of small elements must be compared with the other rows of the matrix in order to assess the rank of the matrix. To compare rows, or vectors, may not be an easy task, especially for large matrices. The problem is even more complicated if the elements of the rows are complex quantities. It would be preferable to have a means of comparison in terms of scalars. The SVD of a matrix allows for such a comparison.

A.3   THE SINGULAR VALUE DECOMPOSITION
The SVD of an MxN real matrix [A] is given by:

$$\underset{(MxN)}{[A]} = \underset{(MxM)}{[U]} \ \underset{(MxN)}{[\Sigma]} \ \underset{(NxN)}{[V]}^T \tag{A.1}$$

where [U] and [V] are orthogonal matrices, i.e.,

$$[U]^T [U] = [U][U]^T = [V]^T [V] = [V][V]^T = [I] \tag{A.2}$$

and

$$[U]^T = [U]^{-1} \quad \text{and} \quad [V]^T = [V]^{-1} \tag{A.3}$$

$[\Sigma]$ is a real matrix with elements $\sigma_{ij} = \sigma_i$ for i=j and $\sigma_{ij} = 0$ for i≠j. The values σ_i are called the singular values of matrix [A]. Without loss of generality we shall assume them to be in decreasing order $(\sigma_1 > \sigma_2 > ... > \sigma_N)$.

$$[\Sigma] = \left.\begin{bmatrix} \sigma_1 & 0 & \cdots & 0 \\ 0 & \sigma_2 & \cdots & 0 \\ \vdots & \vdots & \ddots & \vdots \\ 0 & 0 & \cdots & \sigma_N \\ \hdashline 0 & 0 & \cdots & 0 \end{bmatrix}\right\} \begin{array}{l} \\ \\ \text{N} \\ \\ \text{M} - \text{N} \end{array} \tag{A.4}$$

The relationship between this decomposition and the rank of the matrix [A] is that the value of the rank is equal to the number of non-zero singular values (the multiplication by orthogonal matrices does not alter the value of the rank). For a 3x3 matrix with one linearly-dependent row, σ_3 would be zero. The advantage of the SVD to calculate the rank is that if the considered row is not totally linearly dependent, we would obtain a small value for σ_3, instead of zero, but now we only have to compare this small value with the other singular values. Having established a criterion for the rejection or acceptance of small singular values, we would have an answer concerning the value of the rank. This criterion may depend on the accuracy of the expected results. If [A] is a complex matrix, then (A.1) becomes

$$\underset{(MxN)}{[A]} = \underset{(MxM)}{[U]} \ \underset{(MxN)}{[\Sigma]} \ \underset{(NxN)}{[V]}^H \tag{A.5}$$

where the superscript ^{H} denotes complex conjugate (Hermitian) transpose. [U] and [V] are unitary† matrices, i.e.,

$$[U]^H [U] = [U][U]^H = [V]^H [V] = [V][V]^H = [I] \qquad (A.6)$$

and

$$[U]^H = [U]^{-1} \quad \text{and} \quad [V]^H = [V]^{-1} \qquad (A.7)$$

The singular values σ_i are the non-negative square-roots of the eigenvalues of the matrix $[A]^T [A]$, if $[A]$ is real, and of $[A]^H [A]$, if $[A]$ is complex. Because $[A]^T [A]$ is symmetric and $[A]^H [A]$ is Hermitian, their eigenvalues are always real and therefore both equations (A.1) and (A.5) provide real singular values.

We shall consider the case where $M \geq N$ (without loss of generality), because it is the most common case in engineering applications. If $M \leq N$, then we could decompose $[A]^T$ instead. For simplicity, we shall assume $[A]$ to be real from now on. Because there is a close relationship between the singular values of $[A]$ and the eigenvalues of $[A]^T [A]$ (the same as the eigenvalues of $[A][A]^T$) [362], the columns of $[U]$ and $[V]$ are particular choices of, respectively, the eigenvectors of $[A][A]^T$ and $[A]^T [A]$, and are called the left and right singular vectors. For this reason, the algorithms to compute the SVD are very similar to the ones used to compute eigenvalues and eigenvectors. It is possible to use the SVD as an eigensolver, but for this application the SVD is not advantageous, as current eigensolvers are very well developed and are quite reliable. Usually, the SVD computation is performed in two stages: first, a reduction of $[A]$ to a bidiagonal form using Householder transformations and, second, a reduction of the superdiagonal elements to a negligible size, using the QR algorithm. These numerical methods can be studied in [244].

Detailed application to the SVD can be found in [363]. Improved algorithms can be found in [364-366]. In [363], an ALGOL program is given and in [367] a FORTRAN subroutine is presented. In [368], a subroutine for the case when $[A]$ is complex can be found. Another relevant article [369] gives details about the algorithms used. In [364], a list of references is given for the use of the SVD in several applications. The book by Golub [370] is also an excellent reference. A detailed work by Otte [371] develops the application of SVD and related techniques to modal analysis problems.

A.4   APPLICATIONS OF THE SVD
A.4.1   Recalculation of [A]
Besides the calculation of the rank of a matrix, many other applications exist for the SVD. The first one is the recalculation of matrix $[A]$ after having removed small non-zero singular values and calculated the rank. For instance, if $[A]$ is a 5x3

---

† This designation replaces the term 'orthogonal' when the matrices are complex.

matrix, [U] will be 5x5, [Σ] 5x3 and [V]^T 3x3. If the rank is 2, then only the first two columns of [U] and [V] are effectively operational, and [A] (5x3) can be recomputed with [U] as 5x2, [Σ] 2x2 and [V]^T 2x3. This procedure is quite convenient in the majority of cases, as it leads to an improved [A] matrix.

A.4.2 Condition number

Another simple application is the calculation of the condition number of a matrix. After performing the decomposition, the condition number can be expressed as the ratio $\sigma_{max}/\sigma_{min}$ where σ_{min} is the smallest non-zero singular value. This calculation can serve as an indicator of potential problems, as a high value reflects an ill-conditioned matrix.

A.4.3 Linear system of equations

The SVD can be very useful in solving over-determined linear systems of equations of the form

$$\underset{(MxN)(Nx1)}{[A]} \underset{}{\{x\}} = \underset{(Mx1)}{\{b\}} \tag{A.8}$$

where M>N. Applying the SVD on [A], we obtain

$$\underset{(MxM)(MxN)(NxN)}{[U]\ [Σ]\ [V]}^T \underset{(Nx1)}{\{x\}} = \underset{(Mx1)}{\{b\}} \tag{A.9}$$

or

$$\underset{(MxN)(NxN)}{[Σ]\ [V]}^T \underset{(Nx1)}{\{x\}} = \underset{(MxM)}{[U]}^T \underset{(Mx1)}{\{b\}} \tag{A.10}$$

or

$$\underset{(MxN)(Nx1)}{[Σ]\ \{z\}} = \underset{(Mx1)}{\{d\}} \tag{A.11}$$

with

$$\{z\} = [V]^T \{x\} \tag{A.12}$$

$$\{d\} = [U]^T \{b\} \tag{A.13}$$

Equation (A.11) represents a set of uncoupled M equations with N unknowns. From (A.11), we have

$$\sigma_j z_j = d_j \qquad \text{for } j \leq N \text{ and } \sigma_j \neq 0 \tag{A.14 a}$$

$$0 \cdot z_j = d_j \qquad \text{for } j \leq N \text{ and } \sigma_j = 0 \tag{A.14 b}$$

$$0 = d_j \qquad \text{for } j > N \tag{A.14 c}$$

Equations (A.14 b) and (A.14 c) will only be consistent if $d_j = 0$ for $\sigma_j = 0$ or $j > N$. The range of [A], i.e., the set of {b} for which [A] {x} = {b} has a solution {x}, implies that d_j has to be zero for $\sigma_j = 0$ if $j \leq N$ or for $j > N$. If this does not happen, then {b} does not belong to the range of [A] and (A.8) has no exact solutions. In this case, z_j cannot be determined from (A.14 b), although approximate solutions can be obtained by setting z_j to zero whenever $\sigma_j = 0$. This corresponds to the shortest solution (minimisation of $\left\| [A]\{x\} - \{b\} \right\|^2$) in a least-squares sense. Having calculated {z}, the vector {x} can be recovered from (A.12), by doing

$$\{x\} = [V]\{z\} \tag{A.15}$$

A.4.4 Pseudo-inverse

The NxM matrix [A]⁺ is called the Moore-Penrose pseudo-inverse of [A] if the following conditions are satisfied:

i) $[A][A]^+[A] = [A]$

ii) $[A]^+[A][A]^+ = [A]^+$ $\qquad\qquad$ (A.16)

iii) $[A][A]^+$ is symmetric

iv) $[A]^+[A]$ is symmetric

[A]⁺ always exists and it is unique. If [A] is square and non-singular, then $[A]^+ = [A]^{-1}$ and if [A] is rectangular and of full rank, then $[A]^+ = ([A]^T[A])^{-1}[A]^T$. In this last case, if [A] is complex, then $[A]^+ = ([A]^H[A])^{-1}[A]^H$. If [A] is not of full rank, the best way to calculate the pseudo-inverse is via the SVD. The pseudo-inverse is related to the least-squares problem, as the value of {x} that minimises $\left\| [A]\{x\} - \{b\} \right\|^2$ can be given by $\{x\} = [A]^+\{b\}$. Considering (A.1) and calculating the pseudo-inverse, we obtain:

$$\underset{(NxM)}{[A]^+} = \underset{(NxN)}{\left([V]^T\right)^+} \underset{(NxM)}{[\Sigma]^+} \underset{(MxM)}{[U]^+} \tag{A.17}$$

Because [U] and [V] are orthogonal and full rank matrices, the pseudo-inverse coincides with the classical inverse and (A.3) holds. Therefore,

$$\underset{(NxM)}{[A]^+} = \underset{(NxN)}{[V]} \underset{(NxM)}{[\Sigma]^+} \underset{(MxM)}{[U]^T} \tag{A.18}$$

$[\Sigma]^+$ is an NxM real diagonal matrix, constituted by the inverse values of the non-zero singular values σ_i. Each element of [A]⁺ can be computed more efficiently by

$$a_{ij}^+ = \sum_{\sigma_k \neq 0} \frac{v_{ik} u_{jk}}{\sigma_k} \tag{A.19}$$

where v_{ik} and u_{jk} are the corresponding elements of [V] and [U]^T. The summation excludes the values of σ_k that are zero. In practical terms, considering only the singular values that are bigger than a critical value τ, we have

$$a_{ij}^+ = \sum_{\sigma_k > \tau} \frac{v_{ik} u_{jk}}{\sigma_k} \tag{A.20}$$

This practical condition of acceptance for the singular values σ_k implies that the first condition of (A.16) is no longer true[†] and in this case it should be replaced by the condition:

$$\left\| [A][A]^+[A] - [A] \right\| < \tau \tag{A.21}$$

The other three conditions of (A.16) remain valid. Since one of the four conditions of (A.16) is not fulfilled, the pseudo-inverse will not be unique, but amongst the possible answers, condition (A.21) is the one that minimises $\left\| [A]^+ \right\|$, providing the minimum error for the least-squares problem.

A.4.5 Determinants
In some problems, we want to evaluate the values of z that make [A(z)] singular, i.e., the values of z such that the determinant of [A(z)] is zero. Using the SVD, we have:

$$\det[A] = \det[U] \det[\Sigma] \det[V]^T \tag{A.22}$$

Because [U] and [V] are orthogonal matrices, their determinants are ± 1, and so,

$$\det[A] = \pm \prod_{i=1}^{N} \sigma_i \tag{A.23}$$

It is possible to determine exactly the value of det[A]. But if we only want to know when its value is zero, one does not even have to calculate the product of the singular values, but only to investigate the variation of σ_N (the smallest singular value) and evaluate the value of z that makes σ_N a minimum.

[†] As a consequence, $[A]^+[A]$ will not equal the unit matrix.

APPENDIX B

Orthogonal Functions

A set of functions $f_1(x)$, $f_2(x)$, ... $f_n(x)$ is said to be mutually orthogonal over an interval $a \le x \le b$ if

$$\int_a^b f_i(x)f_j(x)\,dx = 0 \qquad i \ne j$$

$$\int_a^b f_i^2(x)\,dx = Q_i > 0 \qquad i = j$$

(B.1)

and if $f_i(x)$ are real, continuous and not identically zero. $f_i(x)$ can be normalised by defining

$$g_i(x) = f_i(x)\big/\sqrt{Q_i}$$

(B.2)

so that

$$\int_a^b g_i(x)g_j(x)\,dx = \begin{cases} 0 & i \ne j \\ 1 & i = j \end{cases}$$

(B.3)

When the data to curve-fit are not of equal reliability, we can introduce a measure of the relative precision of the value to be assigned to $g_i(x)$ by means of a weighting function $q(x) \ge 0$. The functions are then said to be orthonormal relative to the weighting function $q(x)$, over $[a, b]$:

$$\int_a^b q(x)g_i(x)g_j(x)\,dx = \begin{cases} 0 & i \ne j \\ 1 & i = j \end{cases}$$

(B.4)

If the functions $g_i(x)$ are to be calculated at L discrete points, the integral will be replaced by a summation over that number of points:

$$\sum_{k=1}^{L} q(x_k)g_i(x_k)g_j(x_k) = \begin{cases} 0 & i \neq j \\ 1 & i = j \end{cases}$$
(B.5)

Usually, $g_i(x_k)$ will be polynomials. When these polynomials are complex, (B.5) becomes

$$\text{Re}\left(\sum_{k=1}^{L} q(x_k)g_i^*(x_k)g_j(x_k)\right) = \begin{cases} 0 & i \neq j \\ 1 & i = j \end{cases}$$
(B.6)

where * denotes complex conjugate.

References

[1] Cooley, J. W., Tukey, J. W., "An Algorithm For The Machine Calculation Of Complex Fourier Series", *Mathematics of Computation*, Vol. 19, April 1965, pp. 297-301.

[2] Bergland, G. D., "A Guided Tour Of The Fast Fourier Transform", *IEEE Spectrum*, Vol. 6, July 1969, pp. 41-52.

[3] Ewins, D. J., "Modal Testing: Theory And Practice", *Research Studies Press Ltd.*, England, 1984.

[4] Tomlinson, G. R., "Force Distortion In Resonance Testing Of Structures With Electro-Dynamic Vibration Exciters", *Journal of Sound and Vibration*, Vol. 63, No. 3, 1979, pp. 337-350.

[5] Rao, D. K., "Electrodynamic Interaction Between A Resonating Structures", *Proceedings of the 5th International Modal Analysis Conference (IMAC V)*, London, U. K., 1987, pp. 1142-1145.

[6] Shelley, S., Zhang, Q., Luo, X. N., Allemang, R. J., Brown, D. J., "Investigation And Modelling Of Shaker Dynamic Effects", *Proceedings of the 13th International Seminar on Modal Analysis*, C51, Leuven, Belgium, 1988.

[7] Mitchell, L. D., "Improved Method For The Fast Fourier Transform (FFT) Calculation Of The Frequency Response Function", *ASME Journal of Mechanical Design*, Vol. 104, 1982, pp. 277-279.

[8] Elliott, K. B., Mitchell, L. D., "The Improved Frequency Response Function And Its Effect On Modal Circle Fits", *ASME Journal of Applied Mechanics*, Vol. 51, 1984, pp. 657-663.

[9] Mitchell, L. D., Cobb, R. E., Deel, J. C. and Luk, Y. W., "An Unbiased Frequency Response Function Estimator", *International Journal of Analytical and Experimental Modal Analysis*, Jan. 1988, pp. 12-19.

[10] Goyder, H. G. D., "Frequency Response Testing A Non-linear Structure In A Noisy Environment With A Distorting Shaker", *Proceedings of the Spring Conference, Institute of Acoustics*, 1981, pp. 37-40.

[11] Wellstead, P. E., "Reference Signals For Closed-Loop Identification", *International Journal of Control*, Vol. 26, 1977, pp. 945-962.

[12] Wellstead, P. E., "Non-Parametric Methods Of System Identification", *Proceedings of the 5th IFAC Symposium on Identification and System Parameter Estimation*, 1979, pp. 115-129.

430

[13] Cawley, P., "Rapid Measurement Of Modal Properties Using FFT Analysers With Random Excitation", *ASME Journal of Vibration, Acoustics, Stress and Reliability in Design*, Vol. 108, Oct. 1986, pp. 394-398.
[14] Allemang, R. J., Rost, R. W., Brown, D. J., "Multiple Input Estimation Of Frequency Response Functions: Excitation Considerations", *ASME Paper*, No.83-DET-73, 1983.
[15] Bendat, J. S., Piersol, A. G., "Random Data: Analysis And Measurement Procedures", *John Wiley & Sons*, 1986.
[16] Leuridan, J., "The Use Of Principal Inputs In Multiple-Input Multiple-Output Data Analysis ", *International Journal of Analytical and Experimental Modal Analysis*, July 1986, pp. 1-8.
[17] McConnell, K. G., "Vibration Testing: Theory And Practice", *Wiley Interscience*, 1995.
[18] "Primer On Best Practice In Dynamic Testing", *Dynamic Testing Agency*, HMSO, 1993.
[19] Harris, C. M., Crede, C. E., "Shock And Vibration Handbook", *McGraw-Hill*, 1976.
[20] Patrick, G. B., "Practicalities Of Acquiring Valid Data During Modal Tests", *Proceedings of the 3rd International Modal Analysis Conference (IMAC III)*, Orlando, Florida, U. S. A., 1983, pp. 845-849.
[21] Ewins, D. J., "Once Is Not Enough", Editorial, *Sound and Vibration*, Aug. 1993.
[22] Thomas, R. S., Nolan, T. W., "Once Is Not Enough - A Few More Thoughts", Editorial, *Sound and Vibration*, April 1994.
[23] Kientzy, D., Richardson, M., Blakely, K., "Using Finite Element Data To Set-Up Modal Tests", *Sound and Vibration*, June 1989.
[24] "Cada LINK/PRETEST User Manual", *LMS International*, Leuven, Belgium, 1991.
[25] "Methods For Calibration Of Shock And Vibration Pickups", *U.S.A. Standard S2.2*, 1959.
[26] "American National Standard For The Selection Of Calibrations And Tests For Electrical Transducers Used For Measuring Shock And Vibration", *ANSI*, 1970.
[27] Tustin, W., Lin, F., "Accelerometer Calibration Using A Laser Doppler Displacement Meter", *Sound and Vibration*, April 1995.
[28] Mitchell, L. D., Elliot, K. B., "A Method For Designing Stingers For Use In Mobility Testing", *Proceedings of the 2nd International Modal Analysis Conference (IMAC II)*, Orlando, Florida, U. S. A., 1984, pp. 872-876.
[29] Dossing, O., "The Enigma Of Dynamic Mass", *Sound and Vibration*, Nov. 1990.
[30] Mace, B. R., "The Effects Of Transducer Inertia On Beam Vibration Measurements", *Journal of Sound and Vibration*, 145, 1991, pp. 365-379.
[31] Waters, T. P., "Finite Element Model Updating Using Measured Frequency Response Functions", *Ph.D. Thesis*, Department of Aerospace Engineering, University of Bristol, U. K., 1995.
[32] Gleeson, P. T., "Limitations Of Accelerometers In The Measurement Of Rotational Mobilities", *Dynamics Section Report*, Imperial College of Science and Technology, University of London, U.K., Feb. 1973.
[33] O'Callahan, J.C., Lieu, I. W., Chou, C. M., "Determination Of Rotational Degrees Of Freedom For Moment Transfers In Structural Modifications", *Proceedings of the 3rd International Modal Analysis Conference (IMAC III)*, Orlando, Florida, U. S. A., 1985, pp. 465-470.
[34] Haisty, B. S., Springer, W. T., "A Simplified Method For Extracting Rotational Degree-Of-Freedom Information From Modal Test Data", *International Journal of Analytical and Experimental Modal Analysis*, Vol. 1, No. 3, July 1986, pp. 35-39.

[35] Sanderson, M. A., Fredo, C. R., "Direct Measurement Of Moment Mobility, Part I: A Theoretical Study", *Journal of Sound and Vibration*, 179, 1995, pp. 669-684.

[36] Sanderson, M. A., Fredo, C. R., "Direct Measurement Of Moment Mobility, Part II: An Experimental Study", *Journal of Sound and Vibration*, 179, 1995, pp. 685-696.

[37] Sanderson, M. A., "An Improved Method For Moment Excitation In Mechanical Mobility Measurements Using Manipulated Source Signals", *Proceedings of the 21st International Seminar on Modal Analysis*, Leuven, 1996, pp. 927-938.

[38] "The Fundamentals Of Modal Testing", *Hewlett Packard Application Note No. 243-3*.

[39] Silva, J. M. M., Maia, N. M. M., Ribeiro, A. M. R., "Some Applications Of Coupling/Uncoupling Techniques In Structural Dynamics - Part 1: Solving The Mass Cancellation Problem", *Proceedings of the 15th International Modal Analysis Conference (IMAC XV)*, Orlando, Florida, U.S.A, 1997, pp. 1431-1439.

[40] Maia, N. M. M., Silva, J. M. M., Ribeiro, A. M. R., "Some Applications Of Coupling/Uncoupling Techniques In Structural Dynamics - Part 2: Generation Of The Whole FRF Matrix From Measurements On A Single Column - The Mass Uncoupling Method (MUM)", *Proceedings of the 15th International Modal Analysis Conference (IMAC XV)*, Orlando, Florida, U.S.A, 1997, pp. 1440-1452.

[41] Maia, N. M. M., Silva, J. M. M., Ribeiro, A. M. R., "Some Applications Of Coupling/Uncoupling Techniques In Structural Dynamics - Part 3: Estimation Of Rotational Frequency-Response-Functions Using MUM", *Proceedings of the 15th International Modal Analysis Conference (IMAC XV)*, Orlando, Florida, U.S.A, 1997, pp. 1453-1462.

[42] Peterson, E. L., Rusen, W. M., Mouch, T. A., "Modal Excitation: Force Drop-Off At Resonances", *Proceedings of the 8th International Modal Analysis Conference (IMAC VIII)*, Kissimmee, Florida, U. S. A., 1990, pp. 1226-1231.

[43] Olsen, N. L., "Using And Understanding Electrodynamic Shakers In Modal Applications", *Proceedings of the 4th International Modal Analysis Conference (IMAC IV)*, Los Angeles, California, U. S. A., 1986, pp. 1160-1167.

[44] Salter, J. P., "Steady State Vibrations", *Kenneth Mason Press*, 1969.

[45] To, W. M., Ewins, D. J., "A Closed-Loop Model For Single/Multi-Shaker Modal Testing", *Mechanical Systems and Signal Processing*, Vol. 5, No. 4, 1991, pp. 305-316.

[46] Mitchell, L. D., "Modal Test Methods - Quality, Quantity, And Unobtainable", *Sound and Vibration*, Nov. 1994.

[47] Smallwood, D. O., Lauffer, J. P., "Qualification Of Frequency-Response Functions Using The Rigid-Body Response", *International Journal of Analytical and Experimental Modal Analysis*, Vol. 5, No. 2, April 1990, pp. 115-122.

[48] Stein, K. P., "Experimental Error? No! Experimenter's Error!, A Measurement Engineer's View Of Experimental Modal Analysis", *Proceedings of the 3rd International Modal Analysis Conference (IMAC III)*, Orlando, Florida, U. S. A., 1985, pp. 824-830.

[49] Henderson, F. N., "The Influence Of Rotational Coordinates In A Dynamic Coupling Analysis", *M.Sc. Thesis*, Imperial College of Science and Technology, University of London, U.K., Oct.1984.

[50] Thrane, N., "Zoom-FFT", *Brüel & Kjær TR No. 2-1980*, Using Digital Filters and FFT Techniques, Selected reprints from Technical Review, 1985.

432

[51] Olsen, N., "Excitation Functions For Structural Frequency Response Measurements", *Proceedings of the 2nd International Modal Analysis Conference (IMAC II)*, Orlando, Florida, U. S. A., 1984, pp. 894-902.

[52] Friswell, M. I., Penny, J. E. T., "Stepped Multisine Modal Testing Using Phased Components", *Mechanical Systems and Signal Processing*, Vol. 4, No. 2, 1990, pp. 145-156.

[53] Friswell, M. I., Penny, J. E. T., "Stepped Sine Testing Using Recursive Estimation", *Mechanical Systems and Signal Processing*, Vol. 7, No. 6, 1993, pp. 477-491.

[54] Craig Jr., R. R., Su, Y. W. T., "On Multiple-Shaker Resonance Testing", *AIAA Journal*, Vol. 12, No. 7, July 1974, pp. 924-931.

[55] Cooper, J. E., Hamilton, M. J., Wright, J. R., "Experimental Evaluation Of Normal Mode Force Appropriation Methods Using A Rectangular Plate", *Proceedings of the 10th International Modal Analysis Conference (IMAC X)*, San Diego, California, U. S. A., 1992, pp. 1326-1332.

[56] Alexiou, K., Wright. J. R., "Comparison Of Some Direct Multi-Point Force Appropriation Methods", *International Journal of Analytical and Experimental Modal Analysis*, Vol. 8, No. 2, April 1993, pp. 119-136.

[57] Holmes, P., "Advanced Applications Of Normal Mode Testing", *Ph.D. Thesis*, University of Manchester, 1996.

[58] White, R. G., Pinnington, R. J., "Practical Application Of The Rapid Frequency Sweep Technique For Structural Frequency Response Measurement", *Aeronautical Journal*, May 1982.

[59] Williams, R., Vold, H., "The Multiphase-Step-Sine Method For Experimental Modal Analysis", SDRC.

[60] Williams, R., Crowley, J., Vold, H., "The Multivariate Mode Indicator Function In Modal Analysis", *Proceedings of the 3rd International Modal Analysis Conference (IMAC III)*, Vol. I, Orlando, Florida, U. S. A., 1985, pp. 66-70.

[61] Bishop, R. E. D., Gladwell, G. M. L., "An Investigation Into The Theory Of Resonance Testing", *Philosophical Transactions of the Royal Society of London*, Vol. 255 A 1055, 1963, pp. 241-280.

[62] Pendered, J. W., Bishop, R. E. D., "The Determination Of Modal Shapes In Resonance Testing", *Journal of Mechanical Engineering Science*, Vol. 5, No. 4, 1963, pp. 379-385.

[63] Kennedy, C. C., Pancu, C. D. P., "Use Of Vectors In Vibration Measurement And Analysis", *Journal of the Aeronautical Sciences*, Vol. 14, No. 11, Nov. 1947, pp. 603-625.

[64] Walker, P. B., Clegg, A. W., "The Technique Of Wing Resonance Tests", *Report No. A. D. 3105/M. T. 11767*, Royal Aircraft Establishment, South Farnborough, England, Feb. 1939.

[65] Leuridan, J., "Some Direct Parameter Model Identification Methods Applicable For Multiple Modal Analysis", *Ph.D. Dissertation*, Department of Mechanical and Industrial Engineering, University of Cincinnati, 1984.

[66] Vold, H., Leuridan, J., "A Generalized Frequency Domain Matrix Estimation Method For Structural Parameter Identification", *7th International Seminar on Modal Analysis*, Katholieke Universiteit Leuven, Belgium, Sept. 1982.

[67] Klosterman, A., "On the Experimental Determination And Use Of Modal Representation Of Dynamic Characteristics", *Ph.D. Thesis*, University of Cincinnati, 1971.

[68] Mitchell, L. D., "A Perspective View Of Modal Analysis", *Keynote Address, Proceedings of the 6th International Modal Analysis Conference (IMAC VI)*, Kissimmee, Florida, U. S. A., 1988, pp. xvii-xxi.

[69] Stroud, R. C., "Excitation, Measurement And Analysis Methods For Modal Testing", *Combined Experimental Analytical Modeling Of Dynamic Structural Systems, AMD*, Vol. 67, presented at The Joint ASCE/ASME Mechanics Conference, Albuquerque, New Mexico, June 1985, pp. 49-78.

[70] Rades, M., "Frequency Domain Experimental Modal Analysis Techniques", *The Shock and Vibration Digest*, Vol. 17, No. 6, June 1985, pp. 3-15.

[71] Füllekrug, U., "Survey Of Parameter Estimation Methods In Experimental Modal Analysis", *Proceedings of the 5th International Modal Analysis Conference (IMAC V)*, London, U. K., 1987, pp. 460-467.

[72] Allemang, R. J., "Experimental Modal Analysis", *The Winter Annual Meeting of the ASME*, Boston, Massachusetts, U. S. A., 1983, pp. 1-29.

[73] Snoeys, R., Sas, P., Heylen, W., Van Der Auweraer, H., "Trends In Experimental Modal Analysis", *Proceedings of the 10th International Seminar on Modal Analysis*, Part III, Katholieke Universiteit Leuven, Belgium, 1985.

[74] Zhang, L., Yao, Y., "Advances In Modal Identification - From SISO, SIMO To MIMO Methods", *Proceedings of the 5th International Modal Analysis Conference (IMAC V)*, London, U. K., 1987, pp. 1003-1007.

[75] Ibrahim, S. R., "Modal Identification Techniques Assessment And Comparison", *Proceedings of the 10th International Seminar on Modal Analysis*, Part III, Katholieke Universiteit Leuven, Belgium, 1985.

[76] Mergeay, M., "General Review Of Parameter Estimation Methods By Modal Analysis", *7th International Seminar on Modal Analysis*, Katholieke Universiteit Leuven, Belgium, Sept. 15-17, 1982.

[77] Brown, D. L., Allemang, R. J., Zimmerman, R., Mergeay, M., "Parameter Estimation Techniques For Modal Analysis", *SAE Technical Paper Series*, No. 790221, 1979.

[78] Allemang, R. J., "Experimental Modal Analysis Bibliography", *Proceedings of the 2nd International Modal Analysis Conference (IMAC II)*, Orlando, Florida, U. S. A., 1984, pp. 1085-1097.

[79] Mitchell, L. D., Mitchell, L. D., "Modal Analysis Bibliography, An Update, 1980-1983", *Proceedings of the 2nd International Modal Analysis Conference (IMAC II)*, Orlando, Florida, U. S. A., 1984, pp. 1098-1114.

[80] Park, B.-H., Kim, K.-J., "Vector ARMAX Modeling Approach In Multi-Input Modal Analysis", *Mechanical Systems And Signal Processing*, Vol. 3, No. 4, Oct. 1989, pp. 373-387.

[81] Spitznogle, F. R., Quazi, A. H., "Representation And Analysis Of Time-Limited Signals Using A Complex Exponential Algorithm", *The Journal of The Acoustical Society of America*, Vol. 47, No. 5 (Part 1), 1970, pp. 1150-1155.

[82] Spitznogle, F. R., et al, "Representation And Analysis Of Sonar Signals, Vol. 1: Improvements In The Complex Exponential Signal Analysis Computational Algorithm", *Texas Instruments, Inc.*, Report No. U1-829401-5, Office of Naval Research- Contract No. N00014-69-C0315, 1971, 37 pp.

[83] Prony, R., "Essai Expérimental Et Analitique Sur Les Lois De La Dilatabilité Des Fluides Élastiques Et Sur Celles De La Force Expansive De La Vapeur De L'Eau Et De La Vapeur De L'Alkool, À Différentes Températures", *Journal de L'École Polytechnique (Paris)*, Vol. 1, Cahier 2, Floréal et Prairial, An. III, 1795, pp. 24-76.

[84] Vold, H., Kundrat, J., Rocklin, G. T., Russel, R., "A Multi-Input Modal Estimation Algorithm For Mini-Computers", *SAE Technical Paper Series*, No. 820194, 1982.

[85] Vold, H., Rocklin, G. T., "The Numerical Implementation Of A Multi-Input Modal Estimation Method For Mini-Computers", *Proceedings of the 1st*

International Modal Analysis Conference (IMAC I), Orlando, Florida, U. S. A., 1982, pp. 542-548.

[86] Deblauwe, F., Allemang, R. J., "The Polyreference Time-Domain Technique", *Proceedings of the 10th International Seminar on Modal Analysis*, Part IV, Katholieke Universiteit Leuven, Belgium, 1985.

[87] Brillhart, R. D., Hunt, D. L., Crowley, S. M., "Comparison Of Modal Parameter Estimation Methods For Highly Damped Structures", *Proceedings of the 6th International Modal Analysis Conference (IMAC VI)*, Kissimmee, Florida, U. S. A., 1988, pp. 705-711.

[88] Ibrahim, S. R., Mikulcik, E. C., "A Time Domain Modal Vibration Test Technique", *The Shock and Vibration Bulletin*, Vol. 43, No. 4, 1973, pp. 21-37.

[89] Ibrahim, S. R., Mikulcik, E. C., "The Experimental Determination Of Vibration Parameters From Time Responses", *The Shock and Vibration Bulletin*, Vol. 46, No. 5, 1976, pp. 187-196.

[90] Ibrahim, S. R., Mikulcik, E. C., "A Method For The Direct Identification Of Vibration Parameters From The Free Response", *The Shock and Vibration Bulletin*, Vol. 47, No. 4, 1977, pp. 183-198.

[91] Ibrahim, S. R., "A Modal Identification Algorithm For Higher Accuracy Requirements", *AIAA Paper*, No. 84-0928, *Proceedings of the 25th Structures, Structural Dynamics and Materials Conference*, Palm Springs, California, May 1984, pp. 117-122.

[92] Ibrahim, S. R., "Modal Confidence Factor In Vibration Testing", *The Shock and Vibration Bulletin*, Vol. 48, No. 1, 1978, pp. 65-75.

[93] Pappa, R.S., Ibrahim, S.R., "A Parametric Study Of The Ibrahim Time Domain Identification Algorithm", *The Shock and Vibration Bulletin*, Vol. 51, No. 3, 1981, pp. 43-72.

[94] Ibrahim, S. R., "An Upper Hessenberg Sparse Matrix Algorithm For Modal Identification of Mini-computers", *Proceedings of the 10th International Seminar on Modal Analysis*, Part III, Katholieke Universiteit Leuven, Belgium, 1985.

[95] Juang, J.-N., Pappa, R. S., "An Eigensystem Realization Algorithm For Modal Parameter Identification And Model Reduction", *Journal of Guidance, Control, and Dynamics*, Vol. 8, No. 5, Sept.-Oct. 1985, pp. 620-627.

[96] Juang, J.-N., "Mathematical Correlation Of Modal-Parameter-Identification Methods Via System Realization Theory", *International Journal of Analytical and Experimental Modal Analysis*, Vol. 2, No. 1, Jan. 1987.

[97] Juang, J.-N., Pappa, R. S., "Effects Of Noise On Modal Parameters Identified By The Eigensystem Realization Algorithm", *Journal of Guidance, Control, and Dynamics*, Vol. 9, No.3, May-June 1986, pp. 294-303.

[98] Gersch, W., "Estimation Of The Autoregressive Parameters Of A Mixed Autoregressive Moving-Average Time Series", *IEEE Transactions on Automatic Control*, Vol. AC-15, Oct. 1970, pp. 583-588.

[99] Gersch, W., Luo, S., "Discrete Time Series Synthesis Of Randomly Excited Structural System Response", *The Journal of the Acoustical Society of America*, Vol. 51, No. 1 (Part 2), 1972, pp. 402-408.

[100] Gersch, W., Sharpe, D. R., "Estimation Of Power Spectra With Finite-Order Autoregressive Models", *IEEE Transactions on Automatic Control*, Vol. AC-18, 1973, pp. 367-369.

[101] Gersch, W., Nielsen, N. N., Akaike, H., "Maximum Likelihood Estimation Of Structural Parameters From Random Vibration Data", *Journal of Sound and Vibration*, Vol. 31, No. 3, 1973, pp. 295-308.

[102] Gersch, W., "On The Achievable Accuracy Of Structural System Parameter Estimates", *Journal of Sound and Vibration*, Vol. 34, No. 1, 1974, pp. 63-79.

[103] Gersch, W., Foutch, D. A., "Least Squares Estimates Of Structural System Parameters Using Covariance Function Data", *IEEE Transactions on Automatic Control*, Vol. AC-19, Dec. 1974, pp. 898-903.

[104] Gersch, W., "Parameter Identification: Stochastic Process Techniques", *The Shock and Vibration Digest*, Vol. 7, No. 11, Nov. 1975, pp. 71-86.

[105] Gersch, W., Liu, R. S.-Z., "Time Series Methods For The Synthesis Of Random Vibration Systems", *Transactions of the ASME, Journal of Applied Mechanics*, Vol. 43, No. 1, Mar. 1976, pp. 159-165.

[106] Gersch, W., Martinelli, F., "Estimation Of Structural System Parameters From Stationary And Non-Stationary Ambient Vibrations: An Exploratory-Confirmatory Analysis", *Journal of Sound and Vibration*, Vol. 65, No. 3, 1979, pp. 303-318.

[107] Gersch, W., Brotherton, T., "Estimation Of Stationary Structural System Parameters From Non-Stationary Random Vibration Data: A Locally Stationary Model Method", *Journal of Sound and Vibration*, Vol. 81, No. 2, 1982, pp. 215-227.

[108] D'Alessandro, G., Giorcelli, E., Garibaldi, L., De Stefano, A., "ARMAV Model Technique For Multiple Input/Output System Identification", *Proceedings of the 9th International Modal Analysis Conference (IMAC IX)*, Vol. I, Florence, Italy, 1991, pp. 246-249.

[109] Piombo, B., Giorcelli, E., Garibaldi, L., Fasana, A., "Structures Identification Using ARMAV Models", *Proceedings of the 11th International Modal Analysis Conference (IMAC XI)*, Vol. I, Kissimmee, Florida, U. S. A., 1993, pp. 588-592.

[110] Giorcelli, E., Fasana, A., Garibaldi, L., Riva, A., "Modal Analysis And System Identification Using ARMAV Models", *Proceedings of the 12th International Modal Analysis Conference (IMAC XII)*, Vol. I, Honolulu, Hawaii, U. S. A., 1994, pp. 676-680.

[111] Giorcelli, E., Garibaldi, L., Riva, A., Fasana, A., "ARMAV Analysis Of Queensborough Bridge Ambient Data", *Proceedings of the 14th International Modal Analysis Conference (IMAC XIV)*, Vol. I, Dearborn, Michigan, U. S. A., 1996, pp. 466-469.

[112] Leuridan, J., Vold, H., "A Time Domain Linear Model Estimation Technique For Multiple Input Modal Analysis", *The Winter Annual Meeting of the ASME*, Boston, Massachusetts, 1983, pp. 51-62.

[113] Zaghlool, S. A., "Single-Station Time-Domain (SSTD) Vibration Testing Technique: Theory And Application", *Journal of Sound and Vibration*, Vol. 72, No. 2, 1980, pp. 205-234.

[114] Swevers, J., De Moor, B., Van Brussel, H., "Stepped Sine System Identification, Errors-In-Variables And The Quotient Singular Value Decomposition", *Mechanical Systems and Signal Processing*, Vol. 6, No. 2, Mar. 1992, pp. 121-134.

[115] Lee, J.-K., Park, Y. S., "The Complex Envelope Signal And An Application To Structural Modal Parameter Estimation", *Mechanical Systems and Signal Processing*, Vol. 8, No. 2, Mar. 1994, pp. 129-144.

[116] Yang, Q. J., Zhang, P. Q., Li, C. Q., Wu, X. P., "A System Theory Approach To Multi-Input Multi-Output Modal Parameters Identification Methods", *Mechanical Systems and Signal Processing*, Vol. 8, No. 2, Mar. 1994, pp. 159-174.

[117] Desforges, M. J., Cooper, J. E., Wright, J. R., "Spectral And Modal Parameter Estimation From Output-Only Measurements", *Mechanical Systems and Signal Processing*, Vol. 9, No. 2, Mar. 1995, pp. 169-186.

[118] Brincker, R., Demosthenous, M., Manos, G. C., "Estimation Of The Coefficient Of Restitution Of Rocking Systems By The Random Decrement Technique",

Proceedings of the 12th International Modal Analysis Conference (IMAC XII), Vol. I, Honolulu, Hawaii, U. S. A., 1994, pp. 528-534.

[119] Asmussen, J. C., Brincker, R., "Estimation Of Frequency Response Functions By Random Decrement", *Proceedings of the 14th International Modal Analysis Conference (IMAC XIV)*, Vol. I, Dearborn, Michigan, U. S. A., 1996, pp. 246-252.

[120] Ibrahim, S. R., Brincker, R., Asmussen, J. C., "Modal Parameter Identification From Responses Of General Unknown Random Inputs", *Proceedings of the 14th International Modal Analysis Conference (IMAC XIV)*, Vol. I, Dearborn, Michigan, U. S. A., 1996, pp. 446-452.

[121] Asmussen, J. C., Ibrahim, S. R., Brincker, R., "Random Decrement And Regression Analysis Of Traffic Responses Of Bridges", *Proceedings of the 14th International Modal Analysis Conference (IMAC XIV)*, Vol. I, Dearborn, Michigan, U. S. A., 1996, pp. 453-458.

[122] Brincker, R., De Steffano, A., Piombo, B., "Ambient Data To Analyse The Dynamic Behaviour Of Bridges: A First Comparison Between Different Techniques", *Proceedings of the 14th International Modal Analysis Conference (IMAC XIV)*, Vol. I, Dearborn, Michigan, U. S. A., 1996, pp. 477-482.

[123] Pendered, J. W., Bishop, R. E. D., "A Critical Introduction To Some Industrial Resonance Testing Techniques", *Journal of Mechanical Engineering Science*, Vol. 5, No. 4, 1963, pp. 345-367.

[124] Pendered, J. W., "Theoretical Investigation Into The Effects Of Close Natural Frequencies In Resonance Testing", *Journal of Mechanical Engineering Science*, Vol. 7, No. 4, 1965, pp. 372-379.

[125] Woodcock, D. L., "On The Interpretation Of The Vector Plots Of Forced Vibrations Of A Linear System With Viscous Damping", *The Aeronautical Quaterly*, Feb. 1963, pp. 45-62.

[126] Marples, V., "The Derivation Of Modal Damping Ratios From Complex-Plane Response Plots", *Journal of Sound and Vibration*, Vol. 31, No. 1, 1973, pp. 105-117.

[127] Ewins, D. J., Sidhu, J., "Modal Testing And The Linearity Of Structures", *Mécanique Matériaux Électricité*, No. 389-390-391, 1982, pp. 297-302.

[128] Silva, J. M. M., Maia, N. M. M., "Single Mode Identification Techniques For Use With Small Microcomputers", *Journal of Sound and Vibration*, Vol. 124, No. 1, 1988, pp. 13-26.

[129] Robb, D. A., "User's Guide To Program MODENT", Department of Mechanical Engineering, Imperial College of Science and Technology, London, U. K.

[130] Talapatra, D. C., Haughton, J. M., "Modal Testing Of A Large Test Facility For Space Shuttle Payloads", *Proceedings of the 2nd International Modal Analysis Conference (IMAC II)*, Orlando, Florida, U. S. A., 1984, pp. 1-7.

[131] Kirshenboim, J., Ewins, D. J., "A Method For The Derivation Of Optimal Modal Parameters From Several Single-Point Excitation Tests", *Proceedings of the 2nd International Modal Analysis Conference (IMAC II)*, Orlando, Florida, U. S. A., 1984, pp. 991-997.

[132] Dobson, B. J., "Modal Analysis Using Dynamic Stiffness Data", Royal Naval Engineering College (RNEC), TR-84015, June 1984.

[133] Dobson, B. J., "A Straight-Line Technique For Extracting Modal Properties From Frequency Response Data", *Mechanical Systems and Signal Processing*, Vol. 1, No. 1, Jan. 1987, pp. 29-40.

[134] Maia, N. M. M., "A Global Version of Dobson's Method", *Proceedings of the 10th International Modal Analysis Conference (IMAC X)*, Vol. II, San Diego, Califórnia, U. S. A., 1992, pp. 907-911.

[135] Maia, N. M. M., Ribeiro, A. M. R., Silva, J. M. M., "A New Concept in Modal Analysis: The Characteristic Response Function (CRF)", *International Journal of Analytical and Experimental Modal Analysis*, Vol. 9, No. 3, July 1994, pp. 191-202.

[136] Maia, N. M. M., Silva, J. M. M., Ribeiro, A. M. R., "A New Modal Identification Method Using An Advanced Characteristic Response Function", *Proceedings of the 13th International Modal Analysis Conference (IMAC XIII)*, Vol. II, Nashville, Tennessee, U. S. A, 1995, pp. 1069-1075.

[137] Maia, N. M. M., Silva, J. M. M., Ribeiro, A. M. R., "Identification Of Structural Dynamic Properties With Modal Constant Consistency", *Proceedings of Vibration and Noise*, Venice, Italy, 1995, pp. 366-374.

[138] Goyder, H. G. D., "Methods And Applications Of Structural Modelling From Measured Structural Frequency Response Data", *Journal of Sound and Vibration*, Vol. 68, No. 2, 1980, pp. 209-230.

[139] Petrick, L., "Obtaining Global Frequency And Damping Estimates Using Single Degree-Of-Freedom Real Mode Methods", *Proceedings of the 2nd International Modal Analysis Conference (IMAC II)*, Orlando, Florida, U. S. A., 1984, pp. 425-431.

[140] Ewins, D. J., Gleeson, P. T. , "A Method For Modal Identification Of Lightly Damped Structures", *Journal of Sound and Vibration* , Vol. 84, No. 1, 1982, pp. 57-79.

[141] Maia, N. M. M., Ewins, D. J., "A New Approach For The Modal Identification Of Lightly Damped Structures", *Mechanical Systems and Signal Processing*, Vol. 3, No. 2, 1989, pp. 173-193.

[142] Schmerr, L. W., "A New Complex Exponential Frequency Domain Technique For Analysing Dynamic Response Data", *Proceedings of the 1st International Modal Analysis Conference (IMAC I)*, Orlando, Florida, U. S. A., 1982, pp. 183-186.

[143] Richardson, M. H., Formenti, D. L., "Parameter Estimation From Frequency Response Measurements Using Rational Fraction Polynomials", *Proceedings of the 1st International Modal Analysis Conference (IMAC I)*, Orlando, Florida, U. S. A., 1982, pp. 167-181.

[144] National Bureau of Standards, "Handbook Of Mathematical Functions - Chapter 22 - Orthogonal Polynomials", 1964.

[145] Chihara, T. S., "An Introduction To Orthogonal Polynomials", *Gordon and Breach Science Publishers*, London, U.K., 1978.

[146] Ascher, M., Forsythe, G. E., "SWAC Experiments On The Use Of Orthogonal Polynomials For Data Fitting", *J. Assoc. Comput. Mach.*, Vol. 5, Jan. 1958.

[147] Forsythe, G. E., "Generation And Use Of Orthogonal Polynomials For Data-Fitting With A Digital Computer", *Journal of the Society for Industrial and Applied Mathematics*, Vol. 5, 1957.

[148] Kelly, L. G., "Handbook Of Numerical Methods And Applications - Chapter 5 - Curve Fitting And Data Smoothing", *Addison-Wesley Pub. Co. Inc.*, 1967.

[149] Smiley, R. G., Wey, Y. S., Hogg, K. D., "A Simplified Frequency Domain MDOF Curve-Fitting Process", *Proceedings of the 2nd International Modal Analysis Conference (IMAC II)*, Vol. I, Orlando, Florida, U. S. A., 1984, pp. 432-436.

[150] Adcock, J., Potter, R., "A Frequency Domain Curve Fitting Algorithm With Improved Accuracy", *Proceedings of the 3rd International Modal Analysis Conference (IMAC III)*, Orlando, Florida, U. S. A., 1985, pp. 541-545.

[151] Richardson, M. H., Formenti, D. L., "Global Curve-Fitting Of Frequency Response Measurements Using The Rational Fraction Polynomial Method", *Proceedings of the 3rd International Modal Analysis Conference (IMAC III)*, Orlando, Florida, U. S. A., 1985, pp. 390-397.

438

[152] Richardson, M. H., "Global Frequency And Damping Estimates From Frequency Response Measurements", *Proceedings of the 4th International Modal Analysis Conference (IMAC IV)*, Los Angeles, California, U. S. A., 1986, pp. 465-470.

[153] Zhang,L., Kanda, H., "The Algorithm And Application Of A New Multi-Input-Multi-Output Modal Parameter Identification Method", *The 56th Shock and Vibration Symposium*, Oct. 1985.

[154] Zhang,L., Kanda, H., Brown, D., Allemang, R., "A Polyreference Frequency Domain Method For Modal Parameter Identification", *ASME Paper*, No. 85-DET-106, 1985, pp. 1-6.

[155] Zhang,L., Kanda, H., Lembregts, F., "Some Applications Of Frequency Domain Polyreference Modal Parameter Identification Method", *Proceedings of the 4th International Modal Analysis Conference (IMAC IV)*, Vol. II, Los Angeles, California, U. S. A., 1986, pp. 1237-1245.

[156] Lembregts, F., Leuridan, J., Zhang,L., Kanda, H., "Multiple Input Modal Analysis Of Frequency Response Functions Based On Direct Parameter Identification", *Proceedings of the 4th International Modal Analysis Conference (IMAC IV)*, Vol. I, Los Angeles, California, U. S. A., 1986, pp. 589-598.

[157] Gaukroger, D. R., Skingle, C. W., Heron, K. H., "Numerical Analysis Of Vector Response Loci", *Journal of Sound and Vibration* , Vol. 29, No. 3, 1973, pp. 341-353.

[158] Brittingham, J. N., Miller, E. K., Willows, J. L., "Pole Extraction From Real-Frequency Information", *Proceedings of the IEEE*, Vol. 68, No. 2, 1980, pp. 263-273.

[159] Juang, J.-N., Suzuki, H., "An Eigensystem Realization Algorithm In Frequency Domain For Modal Parameter Identification", *Transactions of the ASME, Journal of Vibrations, Acoustics, Stress, and Reliability in Design*, Vol. 110, Jan. 1988, pp. 24-29.

[160] Fillod, R., Lallement, G., Piranda, J., Raynaud, J. L., "Méthode Globale D'Identification Modale", *Mécanique, Matériaux, Électricité*, Mar.-Apr. 1984.

[161] Fillod, R., Lallement, G., Piranda, J., Raynaud, J. L., "Global Method Of Modal Identification", *Proceedings of the 3rd International Modal Analysis Conference (IMAC III)*, Vol. II, Orlando, Florida, U. S. A., 1985, pp. 1145-1151.

[162] Fillod, R., Lallement, G., Piranda, J., Raynaud, J. L., "Identification À Partir D'Une Excitation Multipoints À Phase Variable", *GAMI, Colloque Vibrations et Chocs*, Lyon ECL, France, June 10-12, 1986.

[163] Giménez, J. G., Carrascosa, L. I., "On The Use Of Constraint Equations In The Global Fitting Algorithms Of Transfer Functions", *Proceedings of the 2nd International Modal Analysis Conference (IMAC II)*, Orlando, Florida, U. S. A., 1984, pp. 998-1003.

[164] Zeng, X., Wicks, A. L., "Modal Parameter Identification By Matrix Decomposition", *Proceedings of the 13th International Modal Analysis Conference (IMAC XIII)*, Vol. I, Nashville, Tennessee, U. S. A., 1995, pp. 35-39.

[165] Van Der Auweraer, H., Leuridan,J., "Multiple Input Orthogonal Polynomial Parameter Estimation", *Mechanical Systems and Signal Processing*, Vol. 1, No. 3, 1987, pp. 259-272.

[166] Wu, W.-Z., Lwo, T.-W., "Global Modal Parameter Estimation Combined With A Statistical Criterion", *Proceedings of the 6th International Modal Analysis Conference (IMAC VI)*, Kissimmee, Florida, U. S. A., 1988, pp. 121-128.

[167] Shih, C. Y., Tsuei, Y. G., Allemang, R. J., Brown, D. L., "A Frequency Domain Global Parameter Estimation Method For Multiple Reference Frequency Response Measurements", *Mechanical Systems and Signal Processing*, Vol. 4, No. 2, Oct. 1988, pp. 349-365.

[168] Lin, G. L., Garibaldi, L., "Global MDOF Curve Fitting With Separated Global Parameters Using Frequency Response Function", *Proceedings of the 9th International Modal Analysis Conference (IMAC IX)*, Vol. I, Florence, Italy, 1991, pp. 254-260.

[169] Lin, G. L., Garibaldi, L., "A Global Partial Fraction Curve Fitting Using Frequency Response Functions", *Proceedings of the 8th International Modal Analysis Conference (IMAC VIII)*, Vol. I, Kissimmee, Florida, U. S. A., 1990, pp. 407-412.

[170] Shih, C. Y., Tsuei, Y. G., Allemang, R. J., Brown, D. L., "Complex Mode Indicator Function And Its Application To Spatial Domain Parameter Estimation", *Mechanical Systems and Signal Processing*, Vol. 4, No. 2, Oct. 1988, pp. 367-377.

[171] Rades, M., "A Comparison Of Some Mode Indicator Functions", *Mechanical Systems and Signal Processing*, Vol. 8, No. 4, Jul. 1994, pp. 459-474.

[172] Chatelet, E., Piranda, J., "Modal Identification By Squared Amplitude Fitting Methods", *Proceedings of the 13th International Modal Analysis Conference (IMAC XIII)*, Vol. I, Nashville, Tennessee, U. S. A., 1995, pp. 40-46.

[173] Chalko, T. J., Gershkovich, V., Haritos, N., "Direct Simultaneous Modal Approximation Method", *Proceedings of the 14th International Modal Analysis Conference (IMAC XIV)*, Vol. II, Dearborn, Michigan, U. S. A., 1996, pp. 1130-1136.

[174] Link, M., Vollan, A., "Identification Of Structural System Parameters From Dynamic Response Data", *Z. Flugwiss. Weltraumforsch*, Vol. 2, No. 3, 1978, pp. 165-174.

[175] Link, M., "Identification Of Physical System Matrices Using Incomplete Vibration Test Data", *Proceedings of the 4th International Modal Analysis Conference (IMAC IV)*, Los Angeles, California, U. S. A., 1986, pp. 386-393.

[176] Coppolino, R. N., "A Simultaneous Frequency Domain Technique For Estimation Of Modal Parameters From Measured Data", *SAE Technical Paper Series*, No. 811046, 1981.

[177] Coppolino, R. N., Stroud, R. C., "A Global Technique For Estimation Of Modal Parameters From Measured Data", *Proceedings of the 4th International Modal Analysis Conference (IMAC IV)*, Los Angeles, California, U. S. A., 1986, pp. 674-681.

[178] Craig Jr., R. R., Blair, M. A., "A Generalized Multiple-Input, Multiple-Output Modal Parameter Estimation Algorithm", *AIAA Journal*, Vol. 23, No. 6, June 1985, pp. 931-937.

[179] Leuridan, J. M., Kundrat, J. A., "Advanced Matrix Methods For Experimental Modal Analysis - A Multi-Matrix Method For Direct Parameter Excitation", *Proceedings of the 1st International Modal Analysis Conference (IMAC I)*, Orlando, Florida, U. S. A., 1982, pp. 192-200.

[180] Lewis, R. C., Wrisley, D. L., "A System For The Excitation Of Pure Natural Modes Of Complex Structures", *Journal of the Aeronautical Sciences*, Vol. 17, No. 11, 1950, pp. 705-722.

[181] Asher, G. W., "A Method Of Normal Mode Excitation Utilizing Admittance Measurements", *Proceedings of the National Specialists' Meeting in Dynamics and Aeroelasticity, Institute of the Aeronautical Sciences*, 1958, pp. 69-76.

[182] Trail-Nash, R. W., "On The Excitation Of Pure Natural Modes In Aircraft Resonance Testing", *Journal of Aerospace Sciences*, Vol. 25, Dec. 1958, pp. 775-778.

[183] Zaveri, K., "Modal Analysis Of Large Structures - Multiple Exciter Systems", *Brüel & Kjær publications*, Nov. 1984.

[184] Smith, S., Stroud, R. C., Hanema, G. A., Hallauer, W.L., Yee, R. C., "MODALAB - A Computarized Data Acquisition And Analysis System For Structural Dynamic Testing", *ISA - ASI 75230*, 1975, pp. 183-189.

[185] Ibáñez, P., "Force Appropriation By Extended Asher's Method", *SAE Paper*, No. 760873, 1976.

[186] Hallauer Jr., W. L., Stafford, J. F., "On The Distribution Of Shaker Forces In Multiple-Shaker Modal Testing", *The Shock and Vibration Bulletin*, Sept. 1978.

[187] Gold, R. R., Hallauer, W. L. Jr., "Modal Testing With Asher's Method Using A Fourier Analyser And Curve Fitting", *Instrum. Aerospace Indust.*, 25, Adv. Test Measure., 16, Part I, *Proceedings of the 25th International Symposium*, Anaheimn, California, U. S. A., 1979, pp.185-192.

[188] Ensminger, R., Turner, M. J., "Structural Parameter Identification From Measured Vibration Data", *AIAA/ASME/ASCE/AHS 20th Structural Dynamics Materials Conference*, St. Louis, AIAA Paper, No. 79-0829, 1979, pp. 410-416.

[189] Craig Jr., R. R., Chung, Y. T., "Modal Analysis Using A Fourier Analyser, Curve-Fitting, and Modal Tuning", *Report No. NASA-CR-161886, CAR-81-1*, Oct. 1981.

[190] Rades, M., "Identification Of Resonance Frequencies By A Singular Value Approach", *Buletinul Instit. Politehnic Bucaresti, Seria Transport. Aeronave*, Vol. 45, 1983, pp. 33-40.

[191] Ibáñez, P., Blakely, K. D., "Automatic Force Appropriation - A Review And Suggested Improvements", *Proceedings of the 2nd International Modal Analysis Conference (IMAC II)*, Orlando, Florida, U. S. A., 1984, pp. 903-907.

[192] Hunt, D. L., Peterson, E. L., Vold, H., Williams, R., "Optimal Selection Of Excitation Methods For Enhanced Modal Testing", *AIAA Paper*, No. 84-1068, presented at the AIAA Dynamics Conference, Palm Springs, California, May 1984, pp. 549-553.

[193] Allemang, R. J., Brown, D. L., Fladung, W., "Modal Parameter Estimation: A Unified Matrix Polynomial Approach", *Proceedings of the 12th International Modal Analysis Conference (IMAC XII)*, Vol. I, Honolulu, Hawaii, U. S. A., 1994, pp. 501-514.

[194] Zienkiewicz, "The Finite Element Method In Engineering Science", *MacGraw-Hill*, 1971.

[195] Ewins, D. J., "Whys And Wherefores Of Modal Testing", *Journal of the Society of Environmental Engineers*, Sept. 1979.

[196] Bishop, R. E. D., Johnson, D. C., "The Mechanics Of Vibration", *Cambridge University Press*, 1960.

[197] Timoshenko, S., Young, D. H., Weaver, W., "Vibration Problems In Engineering", *John Wiley & Sons*, 1974.

[198] Urgueira, A. P. V., "Dynamic Analysis Of Coupled Structures Using Experimental Data", *Ph.D. Thesis*, University of London, U.K., Oct. 1989.

[199] Duncan, W. J., "Mechanical Admittances And Their Applications To Oscillation Problems", *HMSO R and M2000*, 1947.

[200] Sykes, A. O., "Application Of Admittance And Impedance Concepts In The Synthesis Of Vibrating Systems", *ASME, Synthesis of Vibrating Systems*, 1971.

[201] Ewins, D. J., "Measurement And Application Of Mechanical Impedance Data", (3 Parts), *Journal of the Society of Environmental Engineers*, Vol. 14, No. 3, Vol. 15, No. 1, and Vol. 15, No. 2, 1976.

[202] Klosterman, A. L., "A Combined Experimental And Analytical Procedure For Improving Automotive System Dynamics", *SAE Paper*, No. 720093, 1972.

[203] Sainsbury, M. G., Ewins, D. J., "Vibration Analysis Of Damped Machinery Foundation Structure Using Dynamic Stiffness Coupling Technique", *ASME Paper*, No. 73-DET-136, 1973.

[204] Ewins, D. J.; Silva, J. M. M., Maleci, G., "Vibration Analysis Of A Helicopter Plus An Externally-Attached Structure", *Shock and Vibration Bulletin*, Vol. 50, No. 2, 1980.

[205] Heer, E., Lutes, L. D., "Application Of The Mechanical Receptance Coupling Principle To Spacecraft Systems", *Shock and Vibration Bulletin*, Vol. 38, No. 2, 1968.

[206] Ewins, D. J., Sainsbury, M. G., "Mobility Measurements For The Vibration Analysis Of Connected Structures", *Shock and Vibration Bulletin*, Vol. 42, No. 1, 1972.

[207] Ewins, D. J., Gleeson, P. T., "Experimental Determination Of Multi-Directional Mobility Data For Beams", *Journal of Sound and Vibration*, Vol. 45, No. 5, 1975.

[208] Lutes, L. D., Heer, E., "Receptance Coupling Of Structural Components Near A Component Resonance Frequency", *Jet Propulsion Laboratory Report*, Oct. 1968.

[209] Ewins, D. J., "Modal Test Requirements For Coupled Structue Analysis Using Experimentally-Derived Component Models", *Joint ASME/ASCE Applied Mechanics Conference*, Albuquerque, New Mexico, U.S.A., 1985.

[210] Gleeson, P. T., "Identification Of Spatial Models For The Vibration Analysis Of Lightly-Damped Structures", *Ph.D. Thesis*, Imperial College of Science and Technology, University of London, U.K., 1979.

[211] Imregun, M., Robb, D. A., Ewins, D. J., "Structural Modification And Coupling Dynamic Analysis Using FRF Data", *Proceedings of the 5th International Modal Analysis Conference (IMAC V)*, London, U. K., 1987, pp. 1136-1141.

[212] Przemieniecki, J. S., "Matrix Structural Analysis Of Substructures", *AIAA Journal*, Vol. 1, No. 1, 1963.

[213] Hurty, W. C., "Dynamic Analysis Of Structural Systems Using Component Modes", *AIAA Journal*, Vol. 3, No. 4, 1965.

[214] Craig, R. R., Bampton, M. C. C., "Coupling Of Substructures For Dynamic Analysis", *AIAA Journal*, Vol. 6, No. 7, 1968.

[215] Gladwell, G. M. L., "Branch-Mode Analysis Of Vibrating Systems", *Journal of Sound and Vibration*, Vol. 1, 1964.

[216] Goldman, R. L., "Vibration Analysis By Dynamic Partitioning", *AIAA Journal*, Vol. 7, No. 6, 1969.

[217] Hou, S., "Review Of Modal Synthesis Techniques And A New Approach", *Shock and Vibration Bulletin*, Vol. 40, No. 4, 1969.

[218] Craig, R. R., "Methods Of Component Mode Synthesis", *Shock and Vibration Digest*, Vol. 9, No. 11, 1977.

[219] Nelson, F. C., "A Review Of Substructure Analysis Of Vibrating Systems", *Shock and Vibration Digest*, Vol. 11, No. 11, 1979.

[220] Hurty, W. C., Collins, J. D., Hart, G. C., "Dynamic Analysis Of Large Structures By Modal Synthesis Techniques", *Computers and Structures*, Vol. 1, 1971.

[221] MacNeal, R. H., "A Hybrid Method Of Component Mode Synthesis", *Computers and Structures*, Vol. 1, 1971.

[222] Hintz, R. M., "Analytical Methods In Component Mode Synthesis", *AIAA Journal*, Vol. 13, No. 8, 1975.

[223] Rubin, S., "Improved Component-Mode Representation For Structural Dynamic Analysis", *AIAA Journal*, Vol. 13, No. 8, 1975.

[224] Craig, R. R., Chang, C. J., "On The Use Of Attachment Modes In Substructure Coupling For Dynamic Analysis", *Proceedings of the AIAA/ASME 18th SSDM Conference*, 1977.

[225] Martinez, D. R., Carne, T. G.,Miller, A. K., "Combined Experimental/Analytical Modeling Using Component Mode Synthesis", *SAND83-1889*, Sandia National Laboratories, Albuquerque, New Mexico, U.S.A., April 1984.

[226] Coppolino, R. N., "Employment Of Hybrid Experimental/Analytical Modeling In Component Mode Synthesis Of Aerospace Structures", *Joint ASME/ASCE Apllied Mechanics Conference*, Albuquerque, New Mexico, U.S.A., 1985.

[227] Urgueira, A. P. V., Lieven, N. N., Ewins, D. J., "A Generalised Reduction Method For Modal Testing", *Proceedings of the 8th International Modal Analysis Conference*, Kissimmee, Florida, U. S. A., 1990, pp. 22-27.

[228] Guyan, R. J., "Reduction Of Stiffness And Mass Matrices", *AIAA Journal*, Vol. 3, No. 2, 1965, pp. 280.

[229] Kuhar, E. J., Stahle, C. V., "Dynamic Transformation Method For Modal Synthesis", *AIAA Journal*, Vol. 12, No. 5, 1974.

[230] Larsson, P. O., "Methods Using Frequency-Response Functions For The Analysis Of Assembled Structures", *3th International Conference on Recent Adavances in Structural Dynamics*, University of Southampton, U.K., 1988.

[231] Urgueira, A. P. V.,"Using The S.V.D. For The Selection Of Independent Connection Coordinates In The Coupling Of Substructures", *Proceedings of the 9th International Modal Analysis Conference (IMAC IX)*, Florence, Italy, 1991, pp. 919-925.

[232] Urgueira, A. P. V., "Coupled Structure Analysis Using Incomplete Models", *Internal Report 87012*, Dynamics Section, Imperial College of Science and Technology, University of London, U.K., Nov. 1987.

[233] Sekimoto, S., "A Study On Truncation Error In Substructure Testing", *Proceedings of the 3rd International Modal Analysis Conference (IMAC III)*, Orlando, Florida, U. S. A., 1985, pp. 1220-1226.

[234] Gwinn, K. W., Lauffer, J. P., Miller, A. K., "Component Mode Synthesis Using Experimental Modes Enhanced By Mass Loading", *Proceedings of the 6th International Modal Analysis Conference (IMAC VI)*, Kissimmee, Florida, U. S. A., 1988, pp. 1088-1093.

[235] Jetmundsen, B., Bielawa, R. L., Flannelly, W. G., "Generalized Frequency Domain Substructure Synthesis", *Journal of the American Helicopter Society*, Jan. 1988.

[236] Urgueira, A. P. V., Ewins, D. J., "A Refined Modal Coupling Technique For Including Residual Effects Of Out-of-Range Modes", *Proceedings of the 7th International Modal Analysis Conference (IMAC VII)*, Las Vegas, Nevada, U.S.A., 1989, pp. 299-306.

[237] Lord Rayleigh, "Theory Of Sound", *Dover Publications*, 2nd edition, New York, 1945.

[238] Stetson, K. A., Palma, G. E., "Inverse Of First Order Perturbation Theory And Its Application To Structural Design", *AIAA Journal*, Vol. 14, April 1976, pp. 454-460.

[239] Sandstrom, R. E., Anderson, W. J., "Modal Perturbation Methods For Marine Structures", *Transactions of the Society of Naval Architects and Marine Engineers*, Vol. 90, 1982, pp. 41-54.

[240] Weissenburger, J. T., "Effects Of Local Modification On The Vibration Characteristics Of Linear Systems", *Journal of Applied Mechanics*, Vol. 35, 1968, pp. 327-332.

[241] Pomazal, R. J., Snyder, V. W., "Local Modifications Of Damped Linear Systems", *AIAA Journal*, Vol. 9, 1971, pp. 2216-2221.

[242] Hallquist, J. O., "An Efficient Method For Determining The Effects Of Mass Modifications In Damped Systems", *Journal of Sound and Vibration*, Vol. 44, No. 3, 1976, pp. 449-459.

[243] Ram, Y. M., Blech, J. J., "The Dynamic Behaviour Of A Vibratory System After Modification", *Journal of Sound and Vibration*, Vol. 150, No. 3, 1991, pp. 357-370.

[244] Wilkinson, J. H., "The Algebraic Eigenvalue Problem", *Oxford University Press*, 1965.

[245] Rosenbrock, H. H., "Sensitivity Of An Eigenvalue To Changes In The Matrix", *Electronics Letters*, Vol. 1, 1965, pp. 278-279.

[246] Rogers, L. C., "Derivatives Of Eigenvalues And Eigenvectors", *AIAA Journal*, Vol. 8, 1970, pp. 943-944.

[247] Vanhonacker, P., "Differential And Difference Sensitivities Of Natural Frequencies And Mode Shapes Of Mechanical Structures", *AIAA Journal*, Vol. 18, 1980, pp. 1511-1514.

[248] Wang, J., Heylen, W., Sas, P., "Accuracy Of Structural Modification Techniques", *Proceedings of the 5th International Modal Analysis Conference (IMAC V)*, London, U. K., 1987, pp. 65-71.

[249] To, W. M., Ewins, D. J., "Structural Modification Analysis Using Rayleigh Quotient Iteration", *International Journal of Mechanical Sciences*, Vol. 32, No. 3, 1990, pp. 169-179.

[250] Skingle, G. W., Ewins, D. J., "Sensitivity Analysis Using Resonance And Anti-Resonance Frequencies - A Guide To Structural Modification", *European Forum on Aeroelasticity and Structural Dynamics*, Aachen, Germany, 1988.

[251] Tusei, Y. G., Yee, Eric K. L., "A Method To Modify Dynamic Properties Of Undamped Mechanical Systems", *Transactions of the ASME, Journal of Dynamic Systems, Measurement, and Control*, Vol. 111, Sept. 1987, pp. 403-408.

[252] Yee, Eric K. L., Tusei, Y. G., "Method Of Shifting Natural Frequencies Of Damped Mechanical Systems", *AIAA Journal*, Vol. 29, No. 11, Nov. 1991, pp. 1973-1977.

[253] Li, T., He, J., "Optimisation Of Dynamic Characteristics Of A MDOF System By Mass And Stiffness Modification", *Proceedings of the 15th International Modal Analysis Conference (IMAC XV)*, Orlando, Florida, U.S.A, 1997, pp. 1270-1276.

[254] Li, Y., He, J., Lleonart, G., "Finite Element Implementation Of Structural Dynamic Modification", *Proceedings of International Mechanical Engineering Congress*, Perth, Australia, 1994, pp. 157-161.

[255] He, J., Li, Y., "Relocation Of Anti-Resonances Of A Vibratory System By Local Structural Changes", *International Journal of Analytical and Experimental Modal Analysis*, Vol. 10, No. 4, 1995.

[256] Bucher, I., Braun, S., "The Structural Modification Inverse Problem: An Exact Solution", *Mechanical Systems and Signal Processing*, Vol. 7, No. 3, 1993, pp. 217-238.

[257] Skingle, G. W., "Structural Dynamic Modification Using Experimental Data", *Ph.D. Thesis*, Department of Mechanical Engineering, Imperial College of Science, Technology and Medicine, London, U.K., 1989.

[258] Ewins, D. J., "Modal Testing As An Aid To Vibration Analysis", *23rd Conference on Mechanical Engineering*, May 1990.

[259] Caesar B. et al, "Procedures For Updating Dynamic Mathematical Models", *Final Report*, ESA Contract Report, Prepared by Dornier Systems GMBH, EMSB - No. 23/86, May 1985.

[260] Ibrahim, S. R., Saafin, A. A., "Correlation Of Analysis And Test In Modelling Of Structures Assessment And Review", *Proceedings of the 5th International Modal Analysis Conference (IMAC V)*, London, U.K., 1987, pp. 1651-1660.

444

[261] Friswell, M. I., Mottershead, J. E., "Finite Element Updating In Structural Dynamics", *Kluwer Academic Publishers*, 1995.

[262] Lim, T. W., "Actuator/Sensor Placement For Modal Parameter Identification Of Flexible Structures", *International Journal of Analytical and Experimental Modal Analysis*, Vol. 8, No. 3, 1993, pp. 1-14.

[263] Lanczos, C., "An Iteration Method For The Solution Of The Eigenvalue Problem Of Linear Differential And Integral Operators", *Journal of Research of the National Bureau of Standards*, Vol. 45, 1950, pp 255-282.

[264] Bathe, K. J., Wilson, E. L., "Solution Methods For Eigenvalue Problems In Structural Mechanics", *International Journal of Numerical Methods in Engineering*, Vol. 6, 1972, pp. 213-226.

[265] Paz, M., "Dynamic Condensation", *AIAA Journal*, Vol. 22, No. 5, pp. 724-727.

[266] O'Callahan, J., Avitable, P., Riemer, R., "System Equivalent Reduction Expansion Process", *Proceedings of the 7th International Modal Analysis Conference (IMAC VII)*, Las Vegas, Nevada, U. S. A., 1989, pp. 29-37.

[267] O'Callahan, J., "A Procedure For An Improved Reduction System (IRS)", *Proceedings of the 7th International Modal Analysis Conference (IMAC VII)*, Las Vegas, Nevada, U. S. A., 1989, pp. 17-21.

[268] Avitable, P., Pechinsky, F., O'Callahan, J., "Study Of Vector Correlation Using Various Techniques For Model Reduction", *Proceedings of the 10th International Modal Analysis Conference (IMAC X)*, San Diego, California, U. S. A., 1992, pp. 572-583.

[269] Gysin, H-P., "Comparison Of Expansion Methods For FE Model Localisation", *Proceedings of the 8th International Modal Analysis Conference (IMAC VIII)*, Kissimmee, Florida, U. S. A., 1990, pp. 195-204.

[270] Kidder, R. L., "Reduction Of Structural Frequency Equations", *AIAA Journal*, Vol. 11, No. 6, 1973.

[271] Waters, T. P., Lieven, N. A. J., "A Modified Surface Spline For Modal Expansion", *International Journal of Analytical and Experimental Modal Analysis*, Vol. 10, No. 3, 1995, pp. 167-177.

[272] Ng'andu, A. N., Fox, C. H. J., Williams, E. J., "On The Estimation Of Rotational Degrees Of Freedom Using Spline Functions", *Proceedings of the 13th International Modal Analysis Conference (IMAC XIII)*, Nashville, Tennessee, U. S. A., 1995, pp. 791-797.

[273] Harder, R. L., Desmarais, R. N., "Interpolation Using Surface Splines", *Journal of Aircraft*, Vol. 9, 1972, pp. 189-191.

[274] Williams, E. J., Green, J. S., "A Spatial Curve Fitting Technique For Estimating Rotational Degrees Of Freedom", *Proceedings of the 8th International Modal Analysis Conference (IMAC VIII)*, Kissimmee, Florida, U. S. A., 1990, pp. 376-381.

[275] Lipkins, J., Vandeurzen, U., "The Use Of Smoothing Techniques For Structural Modification Applications", *Proceedings of 12th International Seminar on Modal Analysis*, 1987, S1-3.

[276] Lieven, N. A. J., Waters, T. P., "Error Location Using Normalised Cross Orthogonality", *Proceedings of the 12th International Modal Analysis Conference (IMAC XII)*, Honolulu, Hawaii, U. S. A., 1994, pp. 761-764.

[277] Allemang, R. J., Brown, D. L., "A Correlation Coefficient For Modal Vector Analysis", *Proceedings of the 1st International Modal Analysis Conference (IMAC I)*, Orlando, Florida, U. S. A., 1982, pp. 110-116.

[278] Cawley, P., Adams, R.D., "The Location Of Defects In Structures From Measurements Of Natural Frequencies", *Journal of Strain Analysis*, Vol. 12, No. 2, 1979, pp. 49-57.

[279] Chen, H.,P., Bicamic, "Damage Identification In Statically Determinate Space Trusses Using A Single Arbitrary Mode", *Proceedings of Structural Dynamics Modelling Test, Analysis and Correlation*, 1996, pp. 357-364.

[280] Pandey, A., K., Biswas, M., "Damage Diagnosis Of Truss Structures By Estimation Of Flexibility Change", *International Journal of Analytical and Experimental Modal Analysis*, Vol. 10, No. 2, 1995, pp. 104-117.

[281] Lieven, N. A. J., Ewins, D. J., "Spatial Correlation Of Modeshapes: The Coordinate Modal Assurance Criterion (COMAC)", *Proceedings of the 6th International Modal Analysis Conference (IMAC VI)*, Kissimmee, Florida, U. S. A., 1988, pp. 690-695.

[282] Fissette, E., Stavrinidis, C., Ibrahim, S., "Error Location And Updating Of Analytical Dynamic Models Using A Force Balance Method", *Proceedings of the 6th International Modal Analysis Conference (IMAC VI)*, Kissimmee, Florida, U. S. A., 1988, pp. 1063-1070.

[283] Larsson, P-O., Sas, P., "Model Updating Based On Forced Vibration Testing Using Numerically Stable Formulations", *Proceedings of the 10th International Modal Analysis Conference (IMAC X)*, San Diego, California, U. S. A., 1992, pp. 966-974.

[284] Lammens, S., Heylen, W., Sas, P., "Model Updating Using Experimental Frequency Response Functions: Case Studies", *Proceedings of Structural Dynamics Modelling Test, Analysis and Correlation*, 1993, pp. 195-204.

[285] He, J., Ewins, D. J., "Analytical Stiffness Matrix Correction Using Measured Vibration Modes", *International Journal of Analytical and Experimental Modal Analysis*, Vol. 3, No. 1, 1986.

[286] Fox, R. L., Kapoor, M. P., "Rates Of Change Of Eigenvalues And Eigenvectors", *AIAA Journal*, Vol. 6, No. 12, 1968, pp. 2426-2429.

[287] Jung, H., Ewins, D. J., "Error Sensitivity Of The Inverse Eigensensitivity Method For Model Updating", *Proceedings of the 10th International Modal Analysis Conference (IMAC X)*, San Diego, California, U. S. A., 1992, pp. 992-998.

[288] Lin, R. M., Lim, M. K., Du, H., "Improved Inverse Eigensensitivity Method For Structural Analytical Model Updating", *Transactions of the ASME, Journal of Vibration and Acoustics*, Vol. 117, 1995, pp. 192-198.

[289] Lammens, S., Heylen, W., Sas, P., "The Selection Of Updating Frequencies And The Choice Of A Damping Approach For Model Updating Procedures Using Experimental FRFs", *Proceedings of the 12th International Modal Analysis Conference (IMAC XII)*, Honolulu, Hawaii, U. S. A., 1994, pp.. 1383-1389.

[290] Gladwell, G. M. L., Ahmadian, H., "Families Of Acceptable Element Matrices For Finite Element Updating", *Mechanical Systems and Signal Processing*, 1995, Vol. 9, No. 6, pp. 601-614.

[291] D'Ambrogio, W., Fregolent. A., Salvini, P., "Reducing Noise Amplification Effects In The Direct Updating Of Nonconservative FE Models", *Proceedings of the 12th International Modal Analysis Conference (IMAC XII)*, Honolulu, Hawaii, U. S. A., 1992, pp. 738-744.

[292] Visser, W. J., Skingle, G. W., Nash, M., Imregun, M., Ewins, D. J., "Correlation And Updating Of A 3D Space Frame Structure", *Proceedings of Structural Dynamics Modelling Test, Analysis and Correlation*, 1993, pp. 447-458.

[293] Ben-Haim, Y., Prells, U., "Selective Sensitivity In The Frequency Domain, Part I: Theory", *Mechanical Systems and Signal Processing*, Vol. 7, 1993, pp. 461-475.

446

[294] Prells, U., Ben-Haim, Y., "Selective Sensitivity In The Frequency Domain, Part II: Applications", *Mechanical Systems and Signal Processing*, Vol. 7, 1993, pp. 551-574.

[295] Lammens, S., "Frequency Response Based Validation Of Dynamic Structural Finite Element Models", *Ph.D. Thesis*, Katholieke Universiteit Leuven, Belgium, 1995.

[296] Natke, H. G., "On Regularisation Methods Applied To The Error Localisation Of Mathematical Models", *Proceedings of the 9th International Modal Analysis Conference (IMAC IX)*, Florence, Italy, 1991, pp. 70-73.

[297] D'Ambrogio, W., Fregolent, A., "Natural Frequency Error Versus Response Residual In Dynamic Model Updating", *Proceedings of Structural Dynamics Modelling Test, Analysis and Correlation*, 1996, pp. 197-208.

[298] Lin, R. M., Ewins, D. J., "Model Updating Using FRF Data", *Proceedings of the 15th International Seminar on Modal Analysis*, 1990.

[299] Fritzen, C., Kiefer, T., "Localization And Correction Of Errors In Finite Element Models Based On Experimental Data", *Proceedings of the 17th International Seminar on Modal Analysis*, 1992, pp. 1581-1596.

[300] Bretl, J., "Updating Dynamic Models Using Response Sensitivities", *Proceedings of the 17th International Seminar on Modal Analysis*, 1992, pp. 683-695.

[301] D'Ambrogio, W., Fregolent, A., Salvini, P., "Updatability Conditions Of Non-Conservative FE Models With Noise On Incomplete Input-Output Data", *Proceedings of Structural Dynamics Modelling Test, Analysis and Correlation*, 1993, pp. 29-38.

[302] Lammens, S., Heylen, W., Sas, P., "Model Updating Using Experimental Frequency Response Functions: Case Studies", *Proceedings of Structural Dynamics Modelling Test, Analysis and Correlation*, 1993, pp. 195-204.

[303] Baruch, M., "Optimisation Procedure To Correct Stiffness And Flexibility Matrices Using Vibration Tests", *AIAA Journal*, Vol. 16, No. 11, 1978, pp. 1208-1210.

[304] Berman, A., Nagy, E. J., "Improvement Of A Large Analytical Model Using Test Data", *AIAA Journal*, Vol. 21, No. 8, 1983, pp. 1168-1173.

[305] Caesar, B., "Update And Identification Of Dynamic Mathematical Models", *Proceedings of the 8th International Modal Analysis Conference (IMAC VIII)*, Kissimmee, Florida, U. S. A., 1990, pp. 394-401.

[306] To, W. M , Lin, R. M., Ewins, D. J., "A Criterion For The Localisation Of Structural Modification Sites", *Proceedings of the 8th International Modal Analysis Conference (IMAC VIII)*, Kissimmee, Florida, U. S. A., 1990, pp. 756-762.

[307] Soma, A., Gola, M., "Model Updating Through Receptance Sensitivity", *Proceedings of 17th International Seminar on Modal Analysis*, 1992, pp. 17-31.

[308] Steuer, R. E., "Multiple Criteria Optimization: Theory, Computation And Application", *John Wiley & Sons*, 1986.

[309] Davis, L. (ed.), "Genetic Algorithms And Simulated Annealing", *Pitman*, 1987.

[310] Kirkpatrick, S., Gelatt, C. D., Vecchi, M. P., "Optimization By Simulated Annealing", *Science*, Vol. 220, No. 4598, 1983, pp. 671-680.

[311] Van Laarhoven, P. J. M., Aarts, E. H. L., "Simulated Annealing: Theory And Applications", *Kluwer Academic Publishers*, 1987.

[312] Holland, J., "Adaptation In Natural And Artificial Systems", *University of Michigan Press*, 1975.

[313] Goldberg, D. E., "Genetic Algorithms In Search, Optimisation And Machine Learning", *Addison-Wesley*, 1989.

[314] Nelder, J. A., Mead, R., "A Simplex Method For Function Minimization", *The Computer Journal*, Vol. 7, 1965, pp. 308-313.

[315] Rajeev, S., Krishnamoorthy, C. S., "Discrete Optimization On Structures Using Genetic Algorithms", *Journal of Structural Engineering A.S.C.E.*, Vol. 118, No. 5, 1992, pp. 1233-1249.

[316] Larson, C. B., Zimmerman, D. C., "Structural Model Refinement Using A Genetic Algorithm Approach", *Proceedings of the 11ᵗʰ International Modal Analysis Conference (IMAC XI)*, Kissimmee, Florida, U. S. A., 1993, pp. 1095-1101.

[317] Friswell, M. I., Penny, J. E. T., Garvey, S. D., "A Combined Genetic And Eigensensitivity Algorithm For The Location Of Damage In Structures", *Proceedings of the International Conference on Identification in Engineering Systems*, 1996, pp. 357-367.

[318] Mares, C., Surace, C., "Finite Element Model Updating Using A Genetic Algorithm", *DTA/NAFEMS 2ⁿᵈ International Conference on Structural Dynamics Modelling*, 1996, pp. 41-52.

[319] Dunn, S. A., "The Use Of Genetic Algorithms And Stochastic Hill-Climbing In Dynamic Finite-Element Model Identification", *Computers and Structures*, (In Press).

[320] Levin, R. I., Lieven, N. A. J., "Dynamic Finite Element Model Updating Using Simulated Annealing And Genetic Algorithms", accepted for publication in *Mechanical Systems and Signal Processing*.

[321] Sabourin, M., Mitiche, A., "Optical Character Recognition By A Neural Network", *Neural Networks*, Vol. 5, No. 5, 1992, pp. 843-852.

[322] Connor, J. T., Martin, R. D., Atlas, L. E., "Recurrent Neural Networks And Robust Time-Series Prediction", *IEEE Transactions on Neural Networks*, Vol. 5, No. 2, 1994, pp. 240-254.

[323] Bishop, C. M., "Neural Networks For Pattern Recognition", *Clarendon Press Oxford*, 1995.

[324] Rumelhart, D. E., Hinton, G. E., Williams, R. J., "Learning Internal Representations By Error Propagation", *In Rumelhart, McClelland and the PDP Research Group (Eds.)*, Parallel Distributed Processing: Explorations in the Microstructure of Cognition, Volume 1: Foundations, Cambridge, MA: MIT Press, 1986, pp. 318-362.

[325] Chen, S., Billings, S. A., "Neural Networks For Nonlinear Dynamic System Modelling And Identification", *International Journal of Control*, Vol. 56, No. 2, 1992, pp. 319-346.

[326] Atalla, M. J., Inman, D. J., "Model Updating Using Neural Networks", *DTA/NAFEMS Second International Conference - Structural Dynamics Modelling*, 1996, pp. 13-23.

[327] Levin, R. I., Lieven N. A. J., "Dynamic Finite Element Model Updating Using Neural Networks", accepted for publication in *Journal of Sound and Vibration* .

[328] Sidhu, J., Ewins, D. J., "Correlation Of Finite Element And Modal Test Studies Of A Practical Structure", *Proceedings of the 2ⁿᵈ International Modal Analysis Conference (IMAC II)*, Orlando, Florida, U. S. A., 1984, pp. 185-192.

[329] Lieven. N. A. J., Waters, T. P., "The Application Of High Density Meaurements To Dynamic Finite Element Reconciliation", *Proceedings of the 13ᵗʰ International Modal Analysis Conference (IMAC XIII)*, Nashville, Tennessee, U. S. A., 1995, pp. 322-328.

[330] Ibrahim, S. R., D'Ambrogio, W., Salvini, P., Sestieri, A., "Direct Updating Of Nonconservative Finite Element Models Using Measured Input-Output", *Proceedings of the 10ᵗʰ International Modal Analysis Conference (IMAC X)*, San Diego, California, U. S. A., 1992, pp. 202-210.

448

[331] Minas, C., Inman, D. J., "Correcting Finite Element Models With Measured Modal Results Using Eigenstructure Assignment Methods", *Proceedings of the 6th International Modal Analysis Conference (IMAC VI)*, Kissimmee, Florida, U. S. A., 1987, pp. 679-685.

[332] Berger, H., Chaquin, J.P., Ohayon, R., "Finite Element Model Adjustment Using Experimental Data", *Proceedings of the 2nd International Modal Analysis Conference (IMAC II)*, Orlando, Florida, 1984, pp. 638-642.

[333] Berger, H., Barthe, L., Ohayon, R., "Recalage D'un Modèle Par Elèments Finis À Partir De Données Expérimentales Du Type Vibratoire. Concept De Localisation", *R. T. DRET* , No. 7/3313 RY 070 R., May 1987.

[334] Berger, H., Barthe, L., Ohayon, R., "Parametric Updating Of A Finite Element Model From Experimental Modal Characteristics", *Proceedings of the European Forum on Aeroelasticity and Structural Dynamics*, Aachen, April 1989.

[335] Ladevèze, P., "Recalage De Modélisations Des Structures Complexes", *Note technique* n°33.11.01.4., AEROSPATIALE , Les mureaux, 1983.

[336] Ladevèze, P., Reynier, M., "A Localization Method Of Stiffness Errors For The Adjustment Of F.E. Models", Special Issue, *12th ASME Mechanical Vibration and Noise Conference*, Montreal, Sept. 1989.

[337] Ladevèze, P, Reynier, M., Barthe, B., Berger, H., Ohayon, R., Quetin, F., "Méthodes De Recalage De Modèles De Structures En Dynamique: Approche Par Réactions Dynamiques, Approche Par La Notion D'erreur En Relation De Comportement", *4ème Colloque GRECO-GIS, Calcul de Structures*, Giens, France, 1990.

[338] Reynier, M., "Sur Le Contrôle Des Modélisations Par Éléments Finis: Recalage À Partir D'essais Dynamiques", *Thèse de Doctorat d'Université Paris VI*, Laboratoire de Mécanique et Technologie, Cachan, France, Dec. 1990.

[339] Maia, N. M. M., Reynier, M., Ladevèze, P., "Error Localisation For Updating Finite Element Models Using Frequency Response Functions", *Proceedings of the 12th International Modal Analysis Conference (IMAC XII)*, Honolulu, Hawaii, U. S. A., 1994, pp. 1299-1308.

[340] Ladevèze, P., Reynier, M., Maia, N. M. M., "Error On The Constitutive Relation In Dynamics: Theory And Application For Model Updating", *"Inverse Problems in Engineering Mechanics*, ed. H. D. Bui, M. Tanaka *et al*, Nov. 1994, pp. 251-256.

[341] "The Fundamentals Of Signal Analysis", *Application Note* 243, Hewlett Packard, 1985.

[342] Wiener, N., "Nonlinear Problems In Random Theory", *Wiley*, 1958.

[343] He, J., "Identification of Structural Dynamic Characteristics", *Ph.D. Thesis*, Mechanical Engineering Department, Imperial College of Science and Technology, University of London, U.K., 1987.

[344] Simon, M., Tomlinson, G. R., "Applications Of The Hilbert Transforms In The Modal Analysis Of Non-linear Systems", *Journal of Sound and Vibration*, Vol. 96, No. 3, 1984.

[345] He, J., Lin, R. M., Ewins, D. J., "Evaluation Of Some Nonlinear Modal Analysis Methods", *13th International Seminar on Modal Analysis*, Leuven, Belgium, 1988.

[346] Lin, R. M., Ewins, D. J., Lim, M. K., "Identification Of Nonlinearity From Analysis Of Complex Modes", *International Journal of Analytical and Experimental Modal Analysis*, Vol. 8, No. 3, 1993.

[347] Beards, C. F., "Damping In Structural Joints", *Sound and Vibration Digest*, Vol. 11, No. 9, 1979.

[348] Lin, R. M., "Identification Of Dynamic Characteristics Of Nonlinear Structures", *Ph.D. Thesis*, Mechanical Engineering Department, Imperial College of Science, Technology and Medicine, University of London, U.K., 1991.

[349] Narayanan, S., "Transistor Distortion Analysis Using Volterra Series Representation", *Bell System Technology Journal*, Vol. 46, 1967.

[350] Maurer, R. E., Narayanan, S., "Noise Loading Analysis Of A Third-order System With Memory", *IEEE Transactions on Communication Technologies*, COM-16, 1968.

[351] Narayanan, S., "Application Of Volterra Series To Intermodulation Distortion Analysis Of Transistor Feedback Amplifier", *IEEE Transactions on Circuit Theory*, CT-17, 1970.

[352] Chouychai, T., "Dynamic Behaviour Of Nonlinear Structures", *Ph.D. Thesis* (in French), Sain-Ouen, Paris, 1986.

[353] Gifford, S. J., Tomlinson, G. R., "A Functional Series Approach In The Identification Of Nonlinear Structures", *Proceedings of the 5th International Modal Analysis Conference (IMAC V)*, London, U. K., 1987, pp. 593-600.

[354] Vihn, T., Liu, H., "Extension Of Modal Analysis Of Nonlinear Systems", *Proceedings of the 7th International Modal Analysis Conference (IMAC VII)*, Las Vegas, Nevada, U. S. A., USA, 1989, pp. 1379-1385.

[355] Edward, B., Stephen, O. R., "The Output Properties Of Volterra Systems (Nonlinear Systems With Memory) Driven By Harmonic And Gaussian Input", *Proceedings of the IEEE*, Vol. 59, No. 12, 1971.

[356] Schetzen, M., "Measurement Of Wiener Kernels Of A Nonlinear System By Cross-Correlation", *International Journal of Control*, Vol. 2, 1965.

[357] Schetzen, M., "The Volterra And Wiener Theories Of Nonlinear Systems", *John Wiley & Sons*, 1980.

[358] Lin, R. M., "Nonlinear Coupling Analysis Based On A Describing Function Method", *Internal Report* No. 88030, Dynamics Section, Mechanical Engineering Department, Imperial College of Science Technology and Medicine, University of London, U.K., 1988.

[359] Lin, R. M., Ewins, D. J., "Location Of Localised Stiffness Nonlinearity Using Measured Modal Data", *Mechanical Systems and Signal Processing*, Vol. 9, No. 3, 1995.

[360] Rosenberg, R., "On Nonlinear Vibration Of Systems With Many Degrees Of Freedom", *Advanced Applied Mechanics*, 1966.

[361] Rand, R., "Nonlinear Normal Modes In Two-Degree-Of-Freedom Systems", *Journal of Applied Mechanics*, 1971.

[362] Maia, N. M. M., "Fundamentals of Singular Value Decomposition", *Proceedings of the 9th International Modal Analysis Conference (IMAC IX)*, Florence, Italy, Vol. II, April 1991, pp. 1515-1521.

[363] Golub, G. H., Reinch, C., "Singular Value Decomposition And Least Squares Solutions", *Numerische Mathematik*, Vol. 14, 1970, pp. 403-420.

[364] Chan, T. F., "An Improved Algorithm For Computing The Singular Value Decomposition", *ACM Transactions on Mathematical Software*, Vol. 8, No. 1, Mar. 1982, pp. 72-83.

[365] Chan, T. F., "Algorithm 581 - An Improved Algorithm For Computing The Singular Value Decomposition [F1]", *ACM Transactions on Mathematical Software*, Vol. 8, No. 1, Mar. 1982, pp. 84-88.

[366] Paige, C. C., Saunders, M. A., "Towards A Generalized Singular Value Decomposition", *SIAM Journal of Numerical Analysis*, Vol. 18, No. 3, June 1981, pp. 398-405.

[367] Forsythe, G. E., Malcolm, M. A., Moler, C. B., "Computer Methods For Mathematical Computations", *Prentice-Hall*, 1977.

[368] Businger, P. A., Golub, G. H., "Algorithm 358 - Singular Value Decomposition Of A Complex Matrix [F1, 4, 5]", *Communications of the ACM*, Vol. 12, No. 10, Oct. 1969, pp. 564-565.

[369] Golub, G. H., Kahan, W., "Calculating The Singular Values And Pseudo-Inverse Of A Matrix", *SIAM Journal of Numerical Analysis*, Ser. B, Vol. 2, No. 2, 1965.

[370] Golub, G. H., Van Loan, C. F., "Matrix Computations", *North Oxford Academic*, 1983.

[371] Otte, D., "Development And Evaluation Of Singular Value Analysis Methodologies For Studying Multivariate Noise And Vibration Problems", *Ph.D. Thesis*, Katholieke Universiteit Leuven, Belgium, May 1994.

List of Abbreviations

ACRF	Advanced Characteristic Response Function
A/D	Analogue-to-Digital
ADC	Analogue-to-Digital Converter
ARMA	Autoregressive Moving-Average
ARMAV	Autoregressive Moving-Average Vector
ARMAX	Autoregressive Moving-Average with exogenous variables
ASD	Auto-Spectral Density
CAD	Computer-Aided Design
CE	Complex Exponential
CEFD	Complex Exponential Frequency Domain
CGN	Constrained Global Nonlinear
CMIF	Complex Mode Indicator Function
CMP	Correlated Mode Pairs
COMAC	Coordinate Modal Assurance Criterion
COPF	Complex Orthogonal Polynomial Function
CRF	Characteristic Response Function
DFT	Discrete Fourier Transform
DLS	Double Least-Squares
DOF	Degree-Of-Freedom
DSMA	Direct Simultaneous Modal Approximation
DSPI	Direct System Parameter Identification
EMA	Experimental Modal Analysis
EMM	Error Matrix Method
ERA	Eigensystem Realisation Algorithm
ERA-FD	Eigensystem Realisation Algorithm in the Frequency Domain

FAME	Force Appropriation for Modal Evaluation
FDPM	Frequency Domain Prony Method
FE	Finite Element
FEM	Finite Element Method
FFT	Fast Fourier Transform
FRF	Frequency Response Function
GAs	Genetic Algorithms
GRFP	Global Rational Fraction Polynomial
GSH	Gaukroger-Skingle-Heron
IC	Integrated Circuit
IFT	Inverse Fourier Transform
IRF	Impulse Response Function
IRS	Improved Reduction System
ITD	Ibrahim Time Domain
ISSPA	Identification of Structural System Parameters
LSCE	Least-Squares Complex Exponential
LVDT	Linear Variable Differential Transformer
MAC	Modal Assurance Criterion
MCF	Modal Confidence Factor
MDM	Matrix Decomposition Method
MDOF	Multi-Degree-Of-Freedom
MEM	Maximum Entropy Method
MIMO	Multi-Input Multi-Output
MISO	Multi-Input Single-Output
MLP	Multi-Layer Perceptron
MPSS	Multi-Phased Stepped-Sine
MPR	Multi-Point Random
MSF	Modal Scale Factor
MvMIF	Multi-variate Mode Indicator Function
NCO	Normalised Cross Orthogonality
NMD	Normalised Modal Difference
OLS	Orthogonal Least-Squares
PRCE	Polyreference Complex Exponential
PRFD	Polyreference Frequency Domain
PSD	Power Spectral Density
RBF	Radial Basis Function
RDD	Random Decrement Technique

RFM   Response Function Method
RFOP   Rational Fraction Orthogonal Polynomial
RFP   Rational Fraction Polynomial
RMS, rms   Root-Mean-Square
RVDT   Rotary Variable Differential Transformer

SA   Simulated Annealing
SDOF   Single Degree-Of-Freedom
SEREP   System Equivalent Reduction Expansion Process
SFD   Simultaneous Frequency Domain
SIMO   Single-Input Multi-Output
SISO   Single-Input Single-Output
SSG   Signal Source Generator
SSTD   Single-Station Time Domain
STD   Sparse Time Domain
SVD   Singular Value Decomposition

TFA   Transfer Function Analyser
TFRC   Transfer Function Real Condensation

UMPA   Unified Matrix Polynomial Approach

Index

Printed and bound in the UK by
CPI Antony Rowe, Eastbourne

Printed and bound by CPI Group (UK) Ltd, Croydon, CR0 4YY

16/04/2025

14658822-0005